Methods in Enzymology

Volume 263

PLASMA LIPOPROTEINS

Part C

Quantitation

METHODS IN ENZYMOLOGY

EDITORS-IN-CHIEF

John N. Abelson   Melvin I. Simon

DIVISION OF BIOLOGY
CALIFORNIA INSTITUTE OF TECHNOLOGY
PASADENA, CALIFORNIA

FOUNDING EDITORS

Sidney P. Colowick and Nathan O. Kaplan

Methods in Enzymology

Volume 263

Plasma Lipoproteins

Part C
Quantitation

EDITED BY

William A. Bradley

DEPARTMENT OF MEDICINE
UNIVERSITY OF ALABAMA AT BIRMINGHAM
BIRMINGHAM, ALABAMA

Sandra H. Gianturco

DEPARTMENT OF MEDICINE
UNIVERSITY OF ALABAMA AT BIRMINGHAM
BIRMINGHAM, ALABAMA

Jere P. Segrest

DEPARTMENTS OF MEDICINE/ARU AND BIOCHEMISTRY
UNIVERSITY OF ALABAMA AT BIRMINGHAM
BIRMINGHAM, ALABAMA

ACADEMIC PRESS
San Diego New York Boston London Sydney Tokyo Toronto

This book is printed on acid-free paper. ♾

Academic Press, Inc.
A Division of Harcourt Brace & Company
525 B Street, Suite 1900, San Diego, California 92101-4495

United Kingdom Edition published by
Academic Press Limited
24-28 Oval Road, London NW1 7DX

International Standard Serial Number: 0076-6879

International Standard Book Number: 0-12-182164-1

PRINTED IN THE UNITED STATES OF AMERICA
95 96 97 98 99 00 EB 9 8 7 6 5 4 3 2 1

Table of Contents

Contributors to Volume 263 ix

Preface xiii

Volumes in Series xv

Section I. Characterization of Plasma Lipoproteins

1. Apolipoproteins: Pathophysiology and Clinical Implications — Wolfgang Patsch and Antonio M. Gotto, Jr. — 3

2. Significance of Apolipoproteins for Structure, Function, and Classification of Plasma Lipoproteins — Petar Alaupovic — 32

Section II. Apolipoproteins

A. Apolipoprotein B

3. Quantitation of Apolipoprotein B by Chemical Methods — Gloria Lena Vega and Scott M. Grundy — 63

4. Simultaneous Quantification of Apolipoproteins B-100, B-48, and E Separated by SDS–PAGE — Nathalie Bergeron, Leila Kotite, and Richard J. Havel — 82

5. Apolipoprotein B-48 — Fredrik Karpe, Anders Hamsten, Kristine Uffelman, and George Steiner — 95

6. Chromatographic Method for Isolation and Quantification of Apolipoproteins B-100 and B-48 — Janet D. Sparks and Charles E. Sparks — 104

7. Identification and Characterization of Truncated Forms of Apolipoprotein B in Hypobetalipoproteinemia — Stephen G. Young, Elaine S. Krul, Sally McCormick, Robert V. Farese, Jr., and MacRae F. Linton — 120

8. Apolipoprotein B-48: Problems Related to Quantification — Diana M. Lee — 146

9. Immunochemical Separation of Apolipoprotein B-48- and B-100-Containing Lipoproteins — Ross W. Milne — 166

B. Apolipoprotein E

10. Quantitation of Apolipoprotein E — Elaine S. Krul and Thomas G. Cole — 170

C. Apolipoprotein C

11. Quantification of Apolipoprotein C-II by Immunochemical and Chromatographic Methods — Philip W. Connelly, Camilla Vezina, and Graham F. Maguire — 188

12. Immunochemical Methods for Quantification of Human Apolipoprotein C-III — Moti L. Kashyap — 208

D. Apolipoprotein [a]

13. Electrophoretic Methods for Quantitation of Lipoprotein [a] — John W. Gaubatz, Pavan Mital, and Joel D. Morrisett — 218

14. Quantitation of Lipoprotein (a) after Lysine-Sepharose Chromatography and Density Gradient Centrifugation — Gunther M. Fless and Margaret L. Snyder — 238

E. Apolipoprotein A

15. Two-Dimensional Nondenaturing Electrophoresis of Lipoproteins: Applications to High-Density Lipoprotein Speciation — Christopher J. Fielding and Phoebe E. Fielding — 251

16. Heterogeneity of High-Density Lipoproteins and Apolipoprotein A-I as Related to Quantification of Apolipoprotein A-I — Steven T. Kunitake, Patricia O'Connor, and Josefina Naya-Vigne — 260

17. Chromatographic Methods for Quantitation of Apolipoprotein A-I — G. M. Anantharamaiah and David W. Garber — 267

18. Purification, Isoform Characterization, and Quantitation of Human Apolipoprotein A-IV — Richard B. Weinberg, Rachel A. Hopkins, and Jennifer B. Jones — 282

19. Immunochemical Methods for Quantification of Apolipoprotein A-IV — Noel H. Fidge — 297

F. Apolipoprotein J

20. Quantitation of Plasma Apolipoprotein J — Sarah H. Jenkins, William D. Stuart, L. A. Bottoms, and Judith A. K. Harmony — 309

Section III. Lipases and Cholesteryl Ester Transfer Protein

21. Human Lipoprotein Lipase: Production *in Vitro*, Purification, and Generation of Polyclonal Antibody — RICA POTENZ, JING-YI LO, EVA ZSIGMOND, LOUIS C. SMITH, AND LAWRENCE CHAN — 319

22. Immunochemical Quantitation of Lipoprotein Lipase — EVA ZSIGMOND, JING-YI LO, LOUIS C. SMITH, AND LAWRENCE CHAN — 327

23. Sandwich Immunoassay for Measurement of Human Hepatic Lipase — ANDRÉ BENSADOUN — 333

24. Quantification of Cholesteryl Ester Transfer Protein: Activity and Immunochemical Assay — KEVIN C. GLENN AND MICHELLE A. MELTON — 339

25. Determination of Apolipoprotein mRNA Levels by Ribonuclease Protection Assay — ALANA MITCHELL AND NOEL FIDGE — 351

AUTHOR INDEX 365

SUBJECT INDEX 395

Contributors to Volume 263

Article numbers are in parentheses following the names of contributors.
Affiliations listed are current.

PETAR ALAUPOVIC (2), *Lipid and Lipoprotein Laboratory, Oklahoma Medical Research Foundation, Oklahoma City, Oklahoma 73104*

G. M. ANANTHARAMAIAH (17), *Departments of Medicine and Biochemistry, Atherosclerosis Research Unit, University of Alabama at Birmingham, Birmingham, Alabama 35294*

ANDRÉ BENSADOUN (23), *Division of Nutritional Sciences, Cornell University, Ithaca, New York 14853*

NATHALIE BERGERON (4), *Department of Medicine, Cardiovascular Research Institute, University of California, San Francisco, San Francisco, California 94143*

L. A. BOTTOMS (20), *College of Medicine, University of Cincinnati, Cincinnati, Ohio 45267*

LAWRENCE CHAN (21, 22), *Departments of Cell Biology and Medicine, Baylor College of Medicine, Texas Medical Center, Houston, Texas 77030*

THOMAS G. COLE (10), *Department of Internal Medicine, Division of Atherosclerosis, Nutrition, and Lipid Research, Washington University School of Medicine, St. Louis, Missouri 63110*

PHILIP W. CONNELLY (11), *Departments of Medicine, Biochemistry, and Clinical Biochemistry, University of Toronto, and St. Michael's Hospital, Toronto, Ontario, Canada M5B 1A6*

ROBERT V. FARESE, JR. (7), *Gladstone Institute of Cardiovascular Disease, Department of Medicine, University of California, San Francisco, San Francisco, California 94140*

NOEL H. FIDGE (19, 25), *Lipoprotein–Atherosclerosis Unit, Baker Medical Research Institute, Melbourne, Victoria 3181, Australia*

CHRISTOPHER J. FIELDING (15), *Department of Medicine, Cardiovascular Research Institute, University of California, San Francisco, San Francisco, California 94143*

PHOEBE E. FIELDING (15), *Department of Physiology, Cardiovascular Research Institute, University of California, San Francisco, San Francisco, California 94143*

GUNTHER M. FLESS (14), *Department of Medicine, University of Chicago, Chicago, Illinois 60637*

DAVID W. GARBER (17), *Departments of Medicine and Biochemistry, Atherosclerosis Research Unit, University of Alabama at Birmingham, Birmingham, Alabama 35294*

JOHN W. GAUBATZ (13), *Department of Medicine, Baylor College of Medicine, Houston, Texas 77030*

KEVIN C. GLENN (24), *Cardiovascular Disease Research Department, Searle Research and Development, Monsanto Company, St. Louis, Missouri 63167*

ANTONIO M. GOTTO, JR. (1), *Department of Medicine, Baylor College of Medicine, Houston, Texas 77030*

SCOTT M. GRUNDY (3), *Internal Medicine and Biochemistry Departments, University of Texas Health Science Center, Dallas, Texas 75235*

ANDERS HAMSTEN (5), *Atherosclerosis Research Unit, King Gustav V Research Institute, Department of Medicine, Karolinska Hospital, S-171 76 Stockholm, Sweden*

JUDITH A. K. HARMONY (20), *Department of Pharmacology and Cell Biophysics, University of Cincinnati College of Medicine, Cincinnati, Ohio 45267*

RICHARD J. HAVEL (4), *Department of Medicine, Cardiovascular Research Institute, University of California, San Francisco, San Francisco, California 94143*

RACHEL A. HOPKINS (18), *Department of Internal Medicine, Section of Gastroenterology, Bowman Gray School of Medicine, Winston-Salem, North Carolina 27157*

SARAH H. JENKINS (20), *Department of Pathology and Laboratory Medicine, College of Medicine, University of Cincinnati, Cincinnati, Ohio 45267*

JENNIFER B. JONES (18), *Department of Internal Medicine, Section of Gastroenterology, Bowman Gray School of Medicine, Winston-Salem, North Carolina 27157*

FREDRIK KARPE (5), *Atherosclerosis Research Unit, King Gustav V Research Institute, Department of Medicine, Karolinska Hospital, S-171 76 Stockholm, Sweden*

MOTI L. KASHYAP (12), *Department of Veteran Affairs Medical Center, University of California, Irvine, California 90822*

LEILA KOTITE (4), *Department of Medicine, Cardiovascular Research Institute, University of California, San Francisco, San Francisco, California 94143*

ELAINE S. KRUL (7, 10), *Cardiovascular Disease Research Department, Searle Research and Development, Monsanto Company, St. Louis, Missouri 63167*

STEVEN T. KUNITAKE (16), *Cardiovascular Research Institute, University of California, San Francisco, San Francisco, California, 94143*

DIANA M. LEE (8), *Research Division for Women's Health Center, Obstetrics and Gynecology Department, University of Oklahoma Health Sciences Center, Oklahoma City, Oklahoma 73104*

MACRAE F. LINTON (7), *Gladstone Institute of Cardiovascular Disease, Department of Medicine, University of California, San Francisco, San Francisco, California 94140*

JING-YI LO (21, 22), *Life Technologies, Corporate Research and Cell Biology R & D, Gaithersburg, Maryland 20884*

GRAHAM F. MAGUIRE (11), *St. Michael's Hospital and Departments of Medicine Biochemistry and Clinical Biochemistry, University of Toronto, Toronto, Ontario, Canada M5B 1A6*

SALLY MCCORMICK (7), *Gladstone Institute of Cardiovascular Disease, Department of Medicine, University of California, San Francisco, San Francisco, California 94140*

MICHELE A. MELTON (24), *Cardiovascular Disease Research Department, Searle Research and Development, Monsanto Company, St. Louis, Missouri 63167*

ROSS W. MILNE (9), *Departments of Pathology and Biochemistry, University of Ottawa Heart Institute, Ottawa, Ontario, Canada K1Y 4E9*

PAVAN MITAL (13), *Department of Medicine, Baylor College of Medicine, Houston, Texas, 77030*

ALANA MITCHELL (25), *Lipoprotein-Atherosclerosis Unit, Baker Medical Research Institute, Prahran, Victoria 3181, Australia*

JOEL D. MORRISETT (13), *Departments of Medicine and Biochemistry, Baylor College of Medicine, Houston, Texas 77030*

JOSEFINA NAYA-VIGNE (16), *Cardiovascular Research Institute, University of California, San Francisco, San Francisco, California 94143*

PATRICIA O'CONNOR (16), *Cardiovascular Research Institute, University of California, San Francisco, San Francisco, California 94143*

WOLFGANG PATSCH (1), *Department of Laboratory Medicine, Landes Krankenanstalten, A-5020 Salzburg, Austria*

RICA POTENZ (21), *Departments of Cell Biology and Medicine, Baylor College of Medicine, Houston, Texas 77030*

LOUIS C. SMITH (21, 22), *Departments of Cell Biology and Medicine, Baylor College of Medicine, Houston, Texas 77030*

MARGARET L. SNYDER (14), *Department of Medicine, University of Chicago, Chicago, Illinois 60637*

CHARLES E. SPARKS (6), *Department of Pathology and Laboratory Medicine, University of Rochester, School of Medicine and Dentistry, Rochester, New York 14642*

JANET D. SPARKS (6), *Department of Pathology and Laboratory Medicine, University of Rochester, School of Medicine and Dentistry, Rochester, New York 14642*

GEORGE STEINER (5), *Department of Medicine and Physiology, World Health Center for Study of Atherosclerosis in Diabetes, Toronto General Hospital, University of Toronto, Toronto, Ontario, Canada M5G 2C4*

WILLIAM D. STUART (20), *Department of Pharmacology and Cell Biophysics, College of Medicine, University of Cincinnati, Cincinnati, Ohio 45267*

KRISTINE UFFELMAN (5), *Department of Medicine and Physiology, World Health Center for Study of Atherosclerosis in Diabetes, Toronto General Hospital, University of Toronto, Toronto, Ontario, Canada M5G 2C4*

GLORIA LENA VEGA (3), *Department of Clinical Nutrition, and Center for Human Nutrition, University of Texas Health Science Center, Dallas, Texas 75235*

CAMILLA VEZINA (11), *St. Michael's Hospital, Toronto, Ontario, Canada M5B 1A6*

RICHARD B. WEINBERG (18), *Department of Internal Medicine, Section of Gastroenterology, Bowman Gray School of Medicine, Winston-Salem, North Carolina 27157*

STEPHEN G. YOUNG (7), *Gladstone Institute of Cardiovascular Disease, Department of Medicine, University of California, San Francisco, San Francisco, California 94140*

EVA ZSIGMOND (21, 22), *Departments of Cell Biology and Medicine, Baylor College of Medicine, Houston, Texas 77030*

Preface

This is the third of three *Methods in Enzymology* volumes on plasma lipoproteins dedicated to the methodologies used to isolate, characterize, and study the metabolism and molecular and cell biology of these complex macromolecular assemblies (Parts A and B, Volumes 128 and 129, respectively, were published in 1986). This volume emphasizes the quantitation of the major apolipoproteins and plasma lipid metabolizing proteins by a variety of methods that should be useful not only to lipoprotein specialists, but also to the growing number of investigators who bring new molecular/genetic technologies to this field and must be able to correlate their data to levels of apolipoproteins. The quantitative methods covered represent valid techniques that can be used in any reasonably equipped research laboratory.

The quantitation of apolipoproteins remains a difficult task with many pitfalls. Thus, throughout this volume, experts directly address the specific problems associated with the measurements of the apolipoproteins within different lipoproteins.

We particularly want to thank all the contributors for their time, effort, and patience in the production of this volume. In addition, we gratefully acknowledge the expert advice and help of the staff of Academic Press.

William A. Bradley
Sandra H. Gianturco
Jere P. Segrest

METHODS IN ENZYMOLOGY

VOLUME I. Preparation and Assay of Enzymes
Edited by SIDNEY P. COLOWICK AND NATHAN O. KAPLAN

VOLUME II. Preparation and Assay of Enzymes
Edited by SIDNEY P. COLOWICK AND NATHAN O. KAPLAN

VOLUME III. Preparation and Assay of Substrates
Edited by SIDNEY P. COLOWICK AND NATHAN O. KAPLAN

VOLUME IV. Special Techniques for the Enzymologist
Edited by SIDNEY P. COLOWICK AND NATHAN O. KAPLAN

VOLUME V. Preparation and Assay of Enzymes
Edited by SIDNEY P. COLOWICK AND NATHAN O. KAPLAN

VOLUME VI. Preparation and Assay of Enzymes (*Continued*)
Preparation and Assay of Substrates
Special Techniques
Edited by SIDNEY P. COLOWICK AND NATHAN O. KAPLAN

VOLUME VII. Cumulative Subject Index
Edited by SIDNEY P. COLOWICK AND NATHAN O. KAPLAN

VOLUME VIII. Complex Carbohydrates
Edited by ELIZABETH F. NEUFELD AND VICTOR GINSBURG

VOLUME IX. Carbohydrate Metabolism
Edited by WILLIS A. WOOD

VOLUME X. Oxidation and Phosphorylation
Edited by RONALD W. ESTABROOK AND MAYNARD E. PULLMAN

VOLUME XI. Enzyme Structure
Edited by C. H. W. HIRS

VOLUME XII. Nucleic Acids (Parts A and B)
Edited by LAWRENCE GROSSMAN AND KIVIE MOLDAVE

VOLUME XIII. Citric Acid Cycle
Edited by J. M. LOWENSTEIN

VOLUME XIV. Lipids
Edited by J. M. LOWENSTEIN

VOLUME XV. Steroids and Terpenoids
Edited by RAYMOND B. CLAYTON

VOLUME XVI. Fast Reactions
Edited by KENNETH KUSTIN

VOLUME XVII. Metabolism of Amino Acids and Amines (Parts A and B)
Edited by HERBERT TABOR AND CELIA WHITE TABOR

VOLUME XVIII. Vitamins and Coenzymes (Parts A, B, and C)
Edited by DONALD B. MCCORMICK AND LEMUEL D. WRIGHT

VOLUME XIX. Proteolytic Enzymes
Edited by GERTRUDE E. PERLMANN AND LASZLO LORAND

VOLUME XX. Nucleic Acids and Protein Synthesis (Part C)
Edited by KIVIE MOLDAVE AND LAWRENCE GROSSMAN

VOLUME XXI. Nucleic Acids (Part D)
Edited by LAWRENCE GROSSMAN AND KIVIE MOLDAVE

VOLUME XXII. Enzyme Purification and Related Techniques
Edited by WILLIAM B. JAKOBY

VOLUME XXIII. Photosynthesis (Part A)
Edited by ANTHONY SAN PIETRO

VOLUME XXIV. Photosynthesis and Nitrogen Fixation (Part B)
Edited by ANTHONY SAN PIETRO

VOLUME XXV. Enzyme Structure (Part B)
Edited by C. H. W. HIRS AND SERGE N. TIMASHEFF

VOLUME XXVI. Enzyme Structure (Part C)
Edited by C. H. W. HIRS AND SERGE N. TIMASHEFF

VOLUME XXVII. Enzyme Structure (Part D)
Edited by C. H. W. HIRS AND SERGE N. TIMASHEFF

VOLUME XXVIII. Complex Carbohydrates (Part B)
Edited by VICTOR GINSBURG

VOLUME XXIX. Nucleic Acids and Protein Synthesis (Part E)
Edited by LAWRENCE GROSSMAN AND KIVIE MOLDAVE

VOLUME XXX. Nucleic Acids and Protein Synthesis (Part F)
Edited by KIVIE MOLDAVE AND LAWRENCE GROSSMAN

VOLUME XXXI. Biomembranes (Part A)
Edited by SIDNEY FLEISCHER AND LESTER PACKER

VOLUME XXXII. Biomembranes (Part B)
Edited by SIDNEY FLEISCHER AND LESTER PACKER

VOLUME XXXIII. Cumulative Subject Index Volumes I–XXX
Edited by MARTHA G. DENNIS AND EDWARD A. DENNIS

VOLUME XXXIV. Affinity Techniques (Enzyme Purification: Part B)
Edited by WILLIAM B. JAKOBY AND MEIR WILCHEK

VOLUME XXXV. Lipids (Part B)
Edited by JOHN M. LOWENSTEIN

VOLUME XXXVI. Hormone Action (Part A: Steroid Hormones)
Edited by BERT W. O'MALLEY AND JOEL G. HARDMAN

VOLUME XXXVII. Hormone Action (Part B: Peptide Hormones)
Edited by BERT W. O'MALLEY AND JOEL G. HARDMAN

VOLUME XXXVIII. Hormone Action (Part C: Cyclic Nucleotides)
Edited by JOEL G. HARDMAN AND BERT W. O'MALLEY

VOLUME XXXIX. Hormone Action (Part D: Isolated Cells, Tissues, and Organ Systems)
Edited by JOEL G. HARDMAN AND BERT W. O'MALLEY

VOLUME XL. Hormone Action (Part E: Nuclear Structure and Function)
Edited by BERT W. O'MALLEY AND JOEL G. HARDMAN

VOLUME XLI. Carbohydrate Metabolism (Part B)
Edited by W. A. WOOD

VOLUME XLII. Carbohydrate Metabolism (Part C)
Edited by W. A. WOOD

VOLUME XLIII. Antibiotics
Edited by JOHN H. HASH

VOLUME XLIV. Immobilized Enzymes
Edited by KLAUS MOSBACH

VOLUME XLV. Proteolytic Enzymes (Part B)
Edited by LASZLO LORAND

VOLUME XLVI. Affinity Labeling
Edited by WILLIAM B. JAKOBY AND MEIR WILCHEK

VOLUME XLVII. Enzyme Structure (Part E)
Edited by C. H. W. HIRS AND SERGE N. TIMASHEFF

VOLUME XLVIII. Enzyme Structure (Part F)
Edited by C. H. W. HIRS AND SERGE N. TIMASHEFF

VOLUME XLIX. Enzyme Structure (Part G)
Edited by C. H. W. HIRS AND SERGE N. TIMASHEFF

VOLUME L. Complex Carbohydrates (Part C)
Edited by VICTOR GINSBURG

VOLUME LI. Purine and Pyrimidine Nucleotide Metabolism
Edited by PATRICIA A. HOFFEE AND MARY ELLEN JONES

VOLUME LII. Biomembranes (Part C: Biological Oxidations)
Edited by SIDNEY FLEISCHER AND LESTER PACKER

VOLUME LIII. Biomembranes (Part D: Biological Oxidations)
Edited by SIDNEY FLEISCHER AND LESTER PACKER

VOLUME LIV. Biomembranes (Part E: Biological Oxidations)
Edited by SIDNEY FLEISCHER AND LESTER PACKER

VOLUME LV. Biomembranes (Part F: Bioenergetics)
Edited by SIDNEY FLEISCHER AND LESTER PACKER

VOLUME LVI. Biomembranes (Part G: Bioenergetics)
Edited by SIDNEY FLEISCHER AND LESTER PACKER

VOLUME LVII. Bioluminescence and Chemiluminescence
Edited by MARLENE A. DELUCA

VOLUME LVIII. Cell Culture
Edited by WILLIAM B. JAKOBY AND IRA PASTAN

VOLUME LIX. Nucleic Acids and Protein Synthesis (Part G)
Edited by KIVIE MOLDAVE AND LAWRENCE GROSSMAN

VOLUME LX. Nucleic Acids and Protein Synthesis (Part H)
Edited by KIVIE MOLDAVE AND LAWRENCE GROSSMAN

VOLUME 61. Enzyme Structure (Part H)
Edited by C. H. W. HIRS AND SERGE N. TIMASHEFF

VOLUME 62. Vitamins and Coenzymes (Part D)
Edited by DONALD B. MCCORMICK AND LEMUEL D. WRIGHT

VOLUME 63. Enzyme Kinetics and Mechanism (Part A: Initial Rate and Inhibitor Methods)
Edited by DANIEL L. PURICH

VOLUME 64. Enzyme Kinetics and Mechanism (Part B: Isotopic Probes and Complex Enzyme Systems)
Edited by DANIEL L. PURICH

VOLUME 65. Nucleic Acids (Part I)
Edited by LAWRENCE GROSSMAN AND KIVIE MOLDAVE

VOLUME 66. Vitamins and Coenzymes (Part E)
Edited by DONALD B. MCCORMICK AND LEMUEL D. WRIGHT

VOLUME 67. Vitamins and Coenzymes (Part F)
Edited by DONALD B. MCCORMICK AND LEMUEL D. WRIGHT

VOLUME 68. Recombinant DNA
Edited by RAY WU

VOLUME 69. Photosynthesis and Nitrogen Fixation (Part C)
Edited by ANTHONY SAN PIETRO

VOLUME 70. Immunochemical Techniques (Part A)
Edited by HELEN VAN VUNAKIS AND JOHN J. LANGONE

VOLUME 71. Lipids (Part C)
Edited by JOHN M. LOWENSTEIN

VOLUME 72. Lipids (Part D)
Edited by JOHN M. LOWENSTEIN

VOLUME 73. Immunochemical Techniques (Part B)
Edited by JOHN J. LANGONE AND HELEN VAN VUNAKIS

VOLUME 74. Immunochemical Techniques (Part C)
Edited by JOHN J. LANGONE AND HELEN VAN VUNAKIS

VOLUME 75. Cumulative Subject Index Volumes XXXI, XXXII, XXXIV–LX
Edited by EDWARD A. DENNIS AND MARTHA G. DENNIS

VOLUME 76. Hemoglobins
Edited by ERALDO ANTONINI, LUIGI ROSSI-BERNARDI, AND EMILIA CHIANCONE

VOLUME 77. Detoxication and Drug Metabolism
Edited by WILLIAM B. JAKOBY

VOLUME 78. Interferons (Part A)
Edited by SIDNEY PESTKA

VOLUME 79. Interferons (Part B)
Edited by SIDNEY PESTKA

VOLUME 80. Proteolytic Enzymes (Part C)
Edited by LASZLO LORAND

VOLUME 81. Biomembranes (Part H: Visual Pigments and Purple Membranes, I)
Edited by LESTER PACKER

VOLUME 82. Structural and Contractile Proteins (Part A: Extracellular Matrix)
Edited by LEON W. CUNNINGHAM AND DIXIE W. FREDERIKSEN

VOLUME 83. Complex Carbohydrates (Part D)
Edited by VICTOR GINSBURG

VOLUME 84. Immunochemical Techniques (Part D: Selected Immunoassays)
Edited by JOHN J. LANGONE AND HELEN VAN VUNAKIS

VOLUME 85. Structural and Contractile Proteins (Part B: The Contractile Apparatus and the Cytoskeleton)
Edited by DIXIE W. FREDERIKSEN AND LEON W. CUNNINGHAM

VOLUME 86. Prostaglandins and Arachidonate Metabolites
Edited by WILLIAM E. M. LANDS AND WILLIAM L. SMITH

VOLUME 87. Enzyme Kinetics and Mechanism (Part C: Intermediates, Stereochemistry, and Rate Studies)
Edited by DANIEL L. PURICH

VOLUME 88. Biomembranes (Part I: Visual Pigments and Purple Membranes, II)
Edited by LESTER PACKER

VOLUME 89. Carbohydrate Metabolism (Part D)
Edited by WILLIS A. WOOD

VOLUME 90. Carbohydrate Metabolism (Part E)
Edited by WILLIS A. WOOD

VOLUME 91. Enzyme Structure (Part I)
Edited by C. H. W. HIRS AND SERGE N. TIMASHEFF

VOLUME 92. Immunochemical Techniques (Part E: Monoclonal Antibodies and General Immunoassay Methods)
Edited by JOHN J. LANGONE AND HELEN VAN VUNAKIS

VOLUME 93. Immunochemical Techniques (Part F: Conventional Antibodies, Fc Receptors, and Cytotoxicity)
Edited by JOHN J. LANGONE AND HELEN VAN VUNAKIS

VOLUME 94. Polyamines
Edited by HERBERT TABOR AND CELIA WHITE TABOR

VOLUME 95. Cumulative Subject Index Volumes 61–74, 76–80
Edited by EDWARD A. DENNIS AND MARTHA G. DENNIS

VOLUME 96. Biomembranes [Part J: Membrane Biogenesis: Assembly and Targeting (General Methods; Eukaryotes)]
Edited by SIDNEY FLEISCHER AND BECCA FLEISCHER

VOLUME 97. Biomembranes [Part K: Membrane Biogenesis: Assembly and Targeting (Prokaryotes, Mitochondria, and Chloroplasts)]
Edited by SIDNEY FLEISCHER AND BECCA FLEISCHER

VOLUME 98. Biomembranes (Part L: Membrane Biogenesis: Processing and Recycling)
Edited by SIDNEY FLEISCHER AND BECCA FLEISCHER

VOLUME 99. Hormone Action (Part F: Protein Kinases)
Edited by JACKIE D. CORBIN AND JOEL G. HARDMAN

VOLUME 100. Recombinant DNA (Part B)
Edited by RAY WU, LAWRENCE GROSSMAN, AND KIVIE MOLDAVE

VOLUME 101. Recombinant DNA (Part C)
Edited by RAY WU, LAWRENCE GROSSMAN, AND KIVIE MOLDAVE

VOLUME 102. Hormone Action (Part G: Calmodulin and Calcium-Binding Proteins)
Edited by ANTHONY R. MEANS AND BERT W. O'MALLEY

VOLUME 103. Hormone Action (Part H: Neuroendocrine Peptides)
Edited by P. MICHAEL CONN

VOLUME 104. Enzyme Purification and Related Techniques (Part C)
Edited by WILLIAM B. JAKOBY

VOLUME 105. Oxygen Radicals in Biological Systems
Edited by LESTER PACKER

VOLUME 106. Posttranslational Modifications (Part A)
Edited by FINN WOLD AND KIVIE MOLDAVE

VOLUME 107. Posttranslational Modifications (Part B)
Edited by FINN WOLD AND KIVIE MOLDAVE

VOLUME 108. Immunochemical Techniques (Part G: Separation and Characterization of Lymphoid Cells)
Edited by GIOVANNI DI SABATO, JOHN J. LANGONE, AND HELEN VAN VUNAKIS

VOLUME 109. Hormone Action (Part I: Peptide Hormones)
Edited by LUTZ BIRNBAUMER AND BERT W. O'MALLEY

VOLUME 110. Steroids and Isoprenoids (Part A)
Edited by JOHN H. LAW AND HANS C. RILLING

VOLUME 111. Steroids and Isoprenoids (Part B)
Edited by JOHN H. LAW AND HANS C. RILLING

VOLUME 112. Drug and Enzyme Targeting (Part A)
Edited by KENNETH J. WIDDER AND RALPH GREEN

VOLUME 113. Glutamate, Glutamine, Glutathione, and Related Compounds
Edited by ALTON MEISTER

VOLUME 114. Diffraction Methods for Biological Macromolecules (Part A)
Edited by HAROLD W. WYCKOFF, C. H. W. HIRS, AND SERGE N. TIMASHEFF

VOLUME 115. Diffraction Methods for Biological Macromolecules (Part B)
Edited by HAROLD W. WYCKOFF, C. H. W. HIRS, AND SERGE N. TIMASHEFF

VOLUME 116. Immunochemical Techniques (Part H: Effectors and Mediators of Lymphoid Cell Functions)
Edited by GIOVANNI DI SABATO, JOHN J. LANGONE, AND HELEN VAN VUNAKIS

VOLUME 117. Enzyme Structure (Part J)
Edited by C. H. W. HIRS AND SERGE N. TIMASHEFF

VOLUME 118. Plant Molecular Biology
Edited by ARTHUR WEISSBACH AND HERBERT WEISSBACH

VOLUME 119. Interferons (Part C)
Edited by SIDNEY PESTKA

VOLUME 120. Cumulative Subject Index Volumes 81–94, 96–101

VOLUME 121. Immunochemical Techniques (Part I: Hybridoma Technology and Monoclonal Antibodies)
Edited by JOHN J. LANGONE AND HELEN VAN VUNAKIS

VOLUME 122. Vitamins and Coenzymes (Part G)
Edited by FRANK CHYTIL AND DONALD B. MCCORMICK

VOLUME 123. Vitamins and Coenzymes (Part H)
Edited by FRANK CHYTIL AND DONALD B. MCCORMICK

VOLUME 124. Hormone Action (Part J: Neuroendocrine Peptides)
Edited by P. MICHAEL CONN

VOLUME 125. Biomembranes (Part M: Transport in Bacteria, Mitochondria, and Chloroplasts: General Approaches and Transport Systems)
Edited by SIDNEY FLEISCHER AND BECCA FLEISCHER

Volume 126. Biomembranes (Part N: Transport in Bacteria, Mitochondria, and Chloroplasts: Protonmotive Force)
Edited by Sidney Fleischer and Becca Fleischer

Volume 127. Biomembranes (Part O: Protons and Water: Structure and Translocation)
Edited by Lester Packer

Volume 128. Plasma Lipoproteins (Part A: Preparation, Structure, and Molecular Biology)
Edited by Jere P. Segrest and John J. Albers

Volume 129. Plasma Lipoproteins (Part B: Characterization, Cell Biology, and Metabolism)
Edited by John J. Albers and Jere P. Segrest

Volume 130. Enzyme Structure (Part K)
Edited by C. H. W. Hirs and Serge N. Timasheff

Volume 131. Enzyme Structure (Part L)
Edited by C. H. W. Hirs and Serge N. Timasheff

Volume 132. Immunochemical Techniques (Part J: Phagocytosis and Cell-Mediated Cytotoxicity)
Edited by Giovanni Di Sabato and Johannes Everse

Volume 133. Bioluminescence and Chemiluminescence (Part B)
Edited by Marlene DeLuca and William D. McElroy

Volume 134. Structural and Contractile Proteins (Part C: The Contractile Apparatus and the Cytoskeleton)
Edited by Richard B. Vallee

Volume 135. Immobilized Enzymes and Cells (Part B)
Edited by Klaus Mosbach

Volume 136. Immobilized Enzymes and Cells (Part C)
Edited by Klaus Mosbach

Volume 137. Immobilized Enzymes and Cells (Part D)
Edited by Klaus Mosbach

Volume 138. Complex Carbohydrates (Part E)
Edited by Victor Ginsburg

Volume 139. Cellular Regulators (Part A: Calcium- and Calmodulin-Binding Proteins)
Edited by Anthony R. Means and P. Michael Conn

Volume 140. Cumulative Subject Index Volumes 102–119, 121–134

Volume 141. Cellular Regulators (Part B: Calcium and Lipids)
Edited by P. Michael Conn and Anthony R. Means

Volume 142. Metabolism of Aromatic Amino Acids and Amines
Edited by Seymour Kaufman

Volume 143. Sulfur and Sulfur Amino Acids
Edited by William B. Jakoby and Owen Griffith

Volume 144. Structural and Contractile Proteins (Part D: Extracellular Matrix)
Edited by Leon W. Cunningham

Volume 145. Structural and Contractile Proteins (Part E: Extracellular Matrix)
Edited by Leon W. Cunningham

Volume 146. Peptide Growth Factors (Part A)
Edited by David Barnes and David A. Sirbasku

Volume 147. Peptide Growth Factors (Part B)
Edited by David Barnes and David A. Sirbasku

Volume 148. Plant Cell Membranes
Edited by Lester Packer and Roland Douce

Volume 149. Drug and Enzyme Targeting (Part B)
Edited by Ralph Green and Kenneth J. Widder

Volume 150. Immunochemical Techniques (Part K: *In Vitro* Models of B and T Cell Functions and Lymphoid Cell Receptors)
Edited by Giovanni Di Sabato

Volume 151. Molecular Genetics of Mammalian Cells
Edited by Michael M. Gottesman

Volume 152. Guide to Molecular Cloning Techniques
Edited by Shelby L. Berger and Alan R. Kimmel

Volume 153. Recombinant DNA (Part D)
Edited by Ray Wu and Lawrence Grossman

Volume 154. Recombinant DNA (Part E)
Edited by Ray Wu and Lawrence Grossman

Volume 155. Recombinant DNA (Part F)
Edited by Ray Wu

Volume 156. Biomembranes (Part P: ATP-Driven Pumps and Related Transport: The Na,K-Pump)
Edited by Sidney Fleischer and Becca Fleischer

Volume 157. Biomembranes (Part Q: ATP-Driven Pumps and Related Transport: Calcium, Proton, and Potassium Pumps)
Edited by Sidney Fleischer and Becca Fleischer

Volume 158. Metalloproteins (Part A)
Edited by James F. Riordan and Bert L. Vallee

Volume 159. Initiation and Termination of Cyclic Nucleotide Action
Edited by Jackie D. Corbin and Roger A. Johnson

Volume 160. Biomass (Part A: Cellulose and Hemicellulose)
Edited by Willis A. Wood and Scott T. Kellogg

Volume 161. Biomass (Part B: Lignin, Pectin, and Chitin)
Edited by WILLIS A. WOOD AND SCOTT T. KELLOGG

Volume 162. Immunochemical Techniques (Part L: Chemotaxis and Inflammation)
Edited by GIOVANNI DI SABATO

Volume 163. Immunochemical Techniques (Part M: Chemotaxis and Inflammation)
Edited by GIOVANNI DI SABATO

Volume 164. Ribosomes
Edited by HARRY F. NOLLER, JR., AND KIVIE MOLDAVE

Volume 165. Microbial Toxins: Tools for Enzymology
Edited by SIDNEY HARSHMAN

Volume 166. Branched-Chain Amino Acids
Edited by ROBERT HARRIS AND JOHN R. SOKATCH

Volume 167. Cyanobacteria
Edited by LESTER PACKER AND ALEXANDER N. GLAZER

Volume 168. Hormone Action (Part K: Neuroendocrine Peptides)
Edited by P. MICHAEL CONN

Volume 169. Platelets: Receptors, Adhesion, Secretion (Part A)
Edited by JACEK HAWIGER

Volume 170. Nucleosomes
Edited by PAUL M. WASSARMAN AND ROGER D. KORNBERG

Volume 171. Biomembranes (Part R: Transport Theory: Cells and Model Membranes)
Edited by SIDNEY FLEISCHER AND BECCA FLEISCHER

Volume 172. Biomembranes (Part S: Transport: Membrane Isolation and Characterization)
Edited by SIDNEY FLEISCHER AND BECCA FLEISCHER

Volume 173. Biomembranes [Part T: Cellular and Subcellular Transport: Eukaryotic (Nonepithelial) Cells]
Edited by SIDNEY FLEISCHER AND BECCA FLEISCHER

Volume 174. Biomembranes [Part U: Cellular and Subcellular Transport: Eukaryotic (Nonepithelial) Cells]
Edited by SIDNEY FLEISCHER AND BECCA FLEISCHER

Volume 175. Cumulative Subject Index Volumes 135–139, 141–167

VOLUME 176. Nuclear Magnetic Resonance (Part A: Spectral Techniques and Dynamics)
Edited by NORMAN J. OPPENHEIMER AND THOMAS L. JAMES

Volume 177. Nuclear Magnetic Resonance (Part B: Structure and Mechanism)
Edited by NORMAN J. OPPENHEIMER AND THOMAS L. JAMES

Volume 178. Antibodies, Antigens, and Molecular Mimicry
Edited by JOHN J. LANGONE

Volume 179. Complex Carbohydrates (Part F)
Edited by VICTOR GINSBURG

Volume 180. RNA Processing (Part A: General Methods)
Edited by JAMES E. DAHLBERG AND JOHN N. ABELSON

Volume 181. RNA Processing (Part B: Specific Methods)
Edited by JAMES E. DAHLBERG AND JOHN N. ABELSON

Volume 182. Guide to Protein Purification
Edited by MURRAY P. DEUTSCHER

Volume 183. Molecular Evolution: Computer Analysis of Protein and Nucleic Acid Sequences
Edited by RUSSELL F. DOOLITTLE

Volume 184. Avidin–Biotin Technology
Edited by MEIR WILCHEK AND EDWARD A. BAYER

Volume 185. Gene Expression Technology
Edited by DAVID V. GOEDDEL

Volume 186. Oxygen Radicals in Biological Systems (Part B: Oxygen Radicals and Antioxidants)
Edited by LESTER PACKER AND ALEXANDER N. GLAZER

Volume 187. Arachidonate Related Lipid Mediators
Edited by ROBERT C. MURPHY AND FRANK A. FITZPATRICK

Volume 188. Hydrocarbons and Methylotrophy
Edited by MARY E. LIDSTROM

Volume 189. Retinoids (Part A: Molecular and Metabolic Aspects)
Edited by LESTER PACKER

Volume 190. Retinoids (Part B: Cell Differentiation and Clinical Applications)
Edited by LESTER PACKER

Volume 191. Biomembranes (Part V: Cellular and Subcellular Transport: Epithelial Cells)
Edited by SIDNEY FLEISCHER AND BECCA FLEISCHER

Volume 192. Biomembranes (Part W: Cellular and Subcellular Transport: Epithelial Cells)
Edited by SIDNEY FLEISCHER AND BECCA FLEISCHER

Volume 193. Mass Spectrometry
Edited by JAMES A. MCCLOSKEY

Volume 194. Guide to Yeast Genetics and Molecular Biology
Edited by CHRISTINE GUTHRIE AND GERALD R. FINK

Volume 195. Adenylyl Cyclase, G Proteins, and Guanylyl Cyclase
Edited by ROGER A. JOHNSON AND JACKIE D. CORBIN

Volume 196. Molecular Motors and the Cytoskeleton
Edited by Richard B. Vallee

Volume 197. Phospholipases
Edited by Edward A. Dennis

Volume 198. Peptide Growth Factors (Part C)
Edited by David Barnes, J. P. Mather, and Gordon H. Sato

Volume 199. Cumulative Subject Index Volumes 168–174, 176–194

Volume 200. Protein Phosphorylation (Part A: Protein Kinases: Assays, Purification, Antibodies, Functional Analysis, Cloning, and Expression)
Edited by Tony Hunter and Bartholomew M. Sefton

Volume 201. Protein Phosphorylation (Part B: Analysis of Protein Phosphorylation, Protein Kinase Inhibitors, and Protein Phosphatases)
Edited by Tony Hunter and Bartholomew M. Sefton

Volume 202. Molecular Design and Modeling: Concepts and Applications (Part A: Proteins, Peptides, and Enzymes)
Edited by John J. Langone

Volume 203. Molecular Design and Modeling: Concepts and Applications (Part B: Antibodies and Antigens, Nucleic Acids, Polysaccharides, and Drugs)
Edited by John J. Langone

Volume 204. Bacterial Genetic Systems
Edited by Jeffrey H. Miller

Volume 205. Metallobiochemistry (Part B: Metallothionein and Related Molecules)
Edited by James F. Riordan and Bert L. Vallee

Volume 206. Cytochrome P450
Edited by Michael R. Waterman and Eric F. Johnson

Volume 207. Ion Channels
Edited by Bernardo Rudy and Linda E. Iverson

Volume 208. Protein–DNA Interactions
Edited by Robert T. Sauer

Volume 209. Phospholipid Biosynthesis
Edited by Edward A. Dennis and Dennis E. Vance

Volume 210. Numerical Computer Methods
Edited by Ludwig Brand and Michael L. Johnson

Volume 211. DNA Structures (Part A: Synthesis and Physical Analysis of DNA)
Edited by David M. J. Lilley and James E. Dahlberg

Volume 212. DNA Structures (Part B: Chemical and Electrophoretic Analysis of DNA)
Edited by David M. J. Lilley and James E. Dahlberg

VOLUME 213. Carotenoids (Part A: Chemistry, Separation, Quantitation, and Antioxidation)
Edited by LESTER PACKER

VOLUME 214. Carotenoids (Part B: Metabolism, Genetics, and Biosynthesis)
Edited by LESTER PACKER

VOLUME 215. Platelets: Receptors, Adhesion, Secretion (Part B)
Edited by JACEK J. HAWIGER

VOLUME 216. Recombinant DNA (Part G)
Edited by RAY WU

VOLUME 217. Recombinant DNA (Part H)
Edited by RAY WU

VOLUME 218. Recombinant DNA (Part I)
Edited by RAY WU

VOLUME 219. Reconstitution of Intracellular Transport
Edited by JAMES E. ROTHMAN

VOLUME 220. Membrane Fusion Techniques (Part A)
Edited by NEJAT DÜZGÜNEŞ

VOLUME 221. Membrane Fusion Techniques (Part B)
Edited by NEJAT DÜZGÜNEŞ

VOLUME 222. Proteolytic Enzymes in Coagulation, Fibrinolysis, and Complement Activation (Part A: Mammalian Blood Coagulation Factors and Inhibitors)
Edited by LASZLO LORAND AND KENNETH G. MANN

VOLUME 223. Proteolytic Enzymes in Coagulation, Fibrinolysis, and Complement Activation (Part B: Complement Activation, Fibrinolysis, and Nonmammalian Blood Coagulation Factors)
Edited by LASZLO LORAND AND KENNETH G. MANN

VOLUME 224. Molecular Evolution: Producing the Biochemical Data
Edited by ELIZABETH ANNE ZIMMER, THOMAS J. WHITE, REBECCA L. CANN, AND ALLAN C. WILSON

VOLUME 225. Guide to Techniques in Mouse Development
Edited by PAUL M. WASSARMAN AND MELVIN L. DEPAMPHILIS

VOLUME 226. Metallobiochemistry (Part C: Spectroscopic and Physical Methods for Probing Metal Ion Environments in Metalloenzymes and Metalloproteins)
Edited by JAMES F. RIORDAN AND BERT L. VALLEE

VOLUME 227. Metallobiochemistry (Part D: Physical and Spectroscopic Methods for Probing Metal Ion Environments in Metalloproteins)
Edited by JAMES F. RIORDAN AND BERT L. VALLEE

VOLUME 228. Aqueous Two-Phase Systems
Edited by HARRY WALTER AND GÖTE JOHANSSON

VOLUME 229. Cumulative Subject Index Volumes 195–198, 200–227

VOLUME 230. Guide to Techniques in Glycobiology
Edited by WILLIAM J. LENNARZ AND GERALD W. HART

VOLUME 231. Hemoglobins (Part B: Biochemical and Analytical Methods)
Edited by JOHANNES EVERSE, KIM D. VANDEGRIFF, AND ROBERT M. WINSLOW

VOLUME 232. Hemoglobins (Part C: Biophysical Methods)
Edited by JOHANNES EVERSE, KIM D. VANDEGRIFF, AND ROBERT M. WINSLOW

VOLUME 233. Oxygen Radicals in Biological Systems (Part C)
Edited by LESTER PACKER

VOLUME 234. Oxygen Radicals in Biological Systems (Part D)
Edited by LESTER PACKER

VOLUME 235. Bacterial Pathogenesis (Part A: Identification and Regulation of Virulence Factors)
Edited by VIRGINIA L. CLARK AND PATRIK M. BAVOIL

VOLUME 236. Bacterial Pathogenesis (Part B: Integration of Pathogenic Bacteria with Host Cells)
Edited by VIRGINIA L. CLARK AND PATRIK M. BAVOIL

VOLUME 237. Heterotrimeric G Proteins
Edited by RAVI IYENGAR

VOLUME 238. Heterotrimeric G-Protein Effectors
Edited by RAVI IYENGAR

VOLUME 239. Nuclear Magnetic Resonance (Part C)
Edited by THOMAS L. JAMES AND NORMAN J. OPPENHEIMER

VOLUME 240. Numerical Computer Methods (Part B)
Edited by MICHAEL L. JOHNSON AND LUDWIG BRAND

VOLUME 241. Retroviral Proteases
Edited by LAWRENCE C. KUO AND JULES A. SHAFER

VOLUME 242. Neoglycoconjugates (Part A)
Edited by Y. C. LEE AND REIKO T. LEE

VOLUME 243. Inorganic Microbial Sulfur Metabolism
Edited by HARRY D. PECK, JR., AND JEAN LEGALL

VOLUME 244. Proteolytic Enzymes: Serine and Cysteine Peptidases
Edited by ALAN J. BARRETT

VOLUME 245. Extracellular Matrix Components
Edited by E. RUOSLAHTI AND E. ENGVALL

VOLUME 246. Biochemical Spectroscopy
Edited by KENNETH SAUER

VOLUME 247. Neoglycoconjugates (Part B: Biomedical Applications)
Edited by Y. C. LEE AND REIKO T. LEE

VOLUME 248. Proteolytic Enzymes: Aspartic and Metallo Peptidases
Edited by ALAN J. BARRETT

VOLUME 249. Enzyme Kinetics and Mechanism (Part D: Developments in Enzyme Dynamics)
Edited by DANIEL L. PURICH

VOLUME 250. Lipid Modifications of Proteins
Edited by PATRICK J. CASEY AND JANICE E. BUSS

VOLUME 251. Biothiols (Part A: Monothiols and Dithiols, Protein Thiols, and Thiyl Radicals)
Edited by LESTER PACKER

VOLUME 252. Biothiols (Part B: Glutathione and Thioredoxin; Thiols in Signal Transduction and Gene Regulation)
Edited by LESTER PACKER

VOLUME 253. Adhesion of Microbial Pathogens
Edited by RON J. DOYLE AND ITZHAK OFEK

VOLUME 254. Oncogene Techniques
Edited by PETER K. VOGT AND INDER M. VERMA

VOLUME 255. Small GTPases and Their Regulators (Part A: Ras Family)
Edited by W. E. BALCH, CHANNING J. DER, AND ALAN HALL

VOLUME 256. Small GTPases and Their Regulators (Part B: Rho Family)
Edited by W. E. BALCH, CHANNING J. DER, AND ALAN HALL

VOLUME 257. Small GTPases and Their Regulators (Part C: Proteins Involved in Transport)
Edited by W. E. BALCH, CHANNING J. DER, AND ALAN HALL

VOLUME 258. Redox-Active Amino Acids in Biology
Edited by JUDITH P. KLINMAN

VOLUME 259. Energetics of Biological Macromolecules
Edited by MICHAEL L. JOHNSON AND GARY K. ACKERS

VOLUME 260. Mitochondrial Biogenesis and Genetics (Part A)
Edited by GIUSEPPE M. ATTARDI AND ANNE CHOMYN

VOLUME 261. Nuclear Magnetic Resonance and Nucleic Acids
Edited by THOMAS L. JAMES

VOLUME 262. DNA Replication
Edited by JUDITH L. CAMPBELL

VOLUME 263. Plasma Lipoproteins (Part C: Quantitation)
Edited by WILLIAM A. BRADLEY, SANDRA H. GIANTURCO, AND JERE P. SEGREST

VOLUME 264. Mitochondrial Biogenesis and Genetics (Part B)
Edited by GIUSEPPE M. ATTARDI AND ANNE CHOMYN

VOLUME 265. Cumulative Subject Index Volumes 228, 230–262 (in preparation)

VOLUME 266. Computer Methods for Macromolecular Sequence Analysis (in preparation)
Edited by RUSSELL F. DOOLITTLE

VOLUME 267. Combinatorial Chemistry (in preparation)
Edited by JOHN N. ABELSON

VOLUME 268. Nitric Oxide (Part A) (in preparation)
Edited by LESTER PACKER

VOLUME 269. Nitric Oxide (Part B) (in preparation)
Edited by LESTER Packer

Section I

Characterization of Plasma Lipoproteins

[1] Apolipoproteins: Pathophysiology and Clinical Implications

By WOLFGANG PATSCH and ANTONIO M. GOTTO, JR.

Introduction

The term apolipoprotein was initially coined by John Oncley[1] to refer to the protein constituents of plasma lipoproteins. Plasma apolipoproteins are of interest principally because of their central role in plasma lipid transport and, hence, their role in atherogenesis. Over the last several decades, a great deal has been learned about apolipoprotein structure and function. The regulation of apolipoprotein gene expression and the molecular mechanisms of apolipoprotein function are beginning to be revealed. Given the rate of progress in this research, therapeutic applications that target the expression or metabolism of specific apolipoproteins can be expected in the near future.

Central to the functions of all apolipoproteins (apo) is the ability to bind phospholipids that reside in specialized regions termed amphiphathic helices.[2] The characteristic feature of the amphipathic helix is the spatial arrangement of hydrophobic and hydrophilic amino acids. The hydrophobic face of the helix is intercalated between the fatty acyl chains of phospholipids, whereas the hydrophilic face is located close to the polar head groups of phospholipids. Such an orientation allows the interaction of protein domains with lipoprotein-modifying enzymes and cellular receptors that control the catabolism of lipoproteins (Lp) and their removal from the circulation.

In 1977, multiple repeats of 22 amino acids, each consisting of a tandem array of two 11-mers, were recognized as structural elements making up the amphiphatic helices in apoA-I,[3,4] apoA-II, and apoC-III.[4,5] A 22-mer periodicity has since been found in other apolipoproteins as well.[6] These 22-mer regions tend to begin or end with a proline, which serves as an

[1] J. L. Oncley, *in* "Brain Lipids and Lipoproteins and Leukodystrophies" (J. Folch Pi and H. Bauer, eds.), p. 5. Elsevier, Amsterdam, 1963.

[2] J. P. Segrest, J. D. Morrisett, R. L. Jackson, and A. M. Gotto, *FEBS Lett.* **38,** 274 (1974).

[3] A. D. McLachlan, *Nature* (*London*) **267,** 465 (1977).

[4] W. M. Fitch, *Genetics* **86,** 623 (1977).

[5] W. C. Barker and M. O. Dayhoff, *Comp. Biochem. Physiol.* **576,** 309 (1977).

[6] M. S. Boguski, N. Elshourbagy, J. M. Taylor, and J. I. Gordon, *Proc. Natl. Acad. Sci. U.S.A.* **81,** 5021 (1984).

α-helix breaker. Apolipoproteins A-I, A-II, A-IV, C-I, C-II, C-III, and E exhibit a common sequence block that comprises the last 33 codons of exon 3 (exon 2 in the apoA-IV gene, which has lost the first intron).[7] In addition to sequence similarities in the coding regions, these apolipoprotein genes display similarity in genomic organization (i.e., intron–exon location). Hence, they probably arose from an ancestral gene, resembling apoC-I, by multiple partial or complete gene duplication, translocation, unequal crossover, and perhaps gene conversion.[7,8] As a result, the genes have acquired distinct sequences that confer functional specificity (Table I) and permit differential regulation of expression in response to metabolic stimuli.

The apoB gene differs from other apolipoprotein genes in genomic organization, in that it contains 29 exons, 2 of which are extremely long.[9,10] In addition to the amphipathic helix motifs located mainly in the carboxy-terminal half, apoB contains unique proline-rich sequences predicted to form amphipathic β sheets. The evolutionary relationship between apoB and the other apolipoproteins is therefore not clear. ApoD is not a member of the apolipoprotein gene family as it differs from all other apolipoproteins in genomic organization, primary structure, and tissue sites of expression.[11,12] Similar arguments can be used to qualify apo[a] and minor apolipoproteins such as apoH and apoJ as atypical apolipoproteins.

Apolipoprotein Gene Expression

The primary control point in the regulation of differential gene expression is the frequency of transcription initiation, which depends on the interaction of transcription factors with promoter or enhancer elements. Such interactions control apolipoprotein gene expression during development and differentiation and govern tissue-specific gene expression and changes in expression in response to hormonal and metabolic stimuli. Sequences dictating the tissue-specific expression of apolipoprotein genes have been identified,[13,14] positive and negative regulatory cis elements have

[7] C.-C. Luo, W.-H. Li, M. N. Moore, and L. Chan, *J. Mol. Biol.* **187,** 325 (1986).

[8] S. K. Karathanasis, *Proc. Natl. Acad. Sci. U.S.A.* **82,** 6374 (1985).

[9] B. D. Blackhart, E. M. Ludwig, V. R. Pierotti, L. Caiati, M. A. Onasch, S. C. Wallis, L. Pownell, R. Pease, T. J. Knott, M. L. Chu, R. W. Mahley, J. Scott, B. J. McCarthy, and B. Levy-Wilson, *J. Biol. Chem.* **261,** 15364 (1986).

[10] E. H. Ludwig, B. D. Blackhart, V. R. Pierotti, L. Caiati, C. Fortier, T. Knott, J. Scott, R. W. Mahley, B. Levy-Wilson, and B. J. McCarthy, *DNA* **6,** 363 (1987).

[11] D. T. Drayna, J. W. McLean, K. L. Wion, J. M. Trent, H. A. Drabkin, and R. M. Lawn, *DNA* **6,** 199 (1987).

[12] D. Drayna, C. Fielding, J. McLean, B. Baer, G. Castro, E. Chen, L. Comstock, W. Henzel, W. Kohr, L. Rhee, K. Wion, and R. Lawn, *J. Biol. Chem.* **261,** 16535 (1986).

[13] K. Sastry, U. Seedorf, and S. K. Karathanasis, *Mol. Cell. Biol.* **8,** 605 (1988).

[14] T. Leff, K. Reue, A. Melian, H. Culver, and J. L. Breslow, *J. Biol. Chem.* **264,** 16132 (1989).

TABLE I
PROPERTIES AND FUNCTIONS OF APOLIPOPROTEINS[a]

Name	Chromosomal localization	Tissue expression	Length of mature protein (amino acids)	Functions
ApoA-I	11	Liver, intestine	243	Structural, activator of LCAT, ligand for HDL binding, promotes cholesterol efflux
ApoA-II	1	Liver (intestine)	77	Structural, activator of hepatic lipase
ApoA-IV	11	Liver, intestine	377	Modulator of LPL activity, LCAT activator, function in TG transport (?)
ApoB-100	2	Liver	4536	Structural, secretion of VLDL, ligand for LDL receptor
ApoB-48	2	Intestine	2152	Structural, secretion of chylomicrons
ApoC-I	19	Liver (intestine)	57	Activator of LCAT, inhibits removal of TGRL by LDL receptor-related protein
ApoC-II	19	Liver (intestine)	79	Activator of LPL
ApoC-III	11	Liver (intestine)	79	Inhibitor of LPL, inhibits uptake of TGRL by liver
ApoD	3	Liver, intestine, spleen, pancreas, brain, adrenal gland, kidney	169	Binding of heme-related compounds (?), potential radical scavenger, reverse cholesterol transport (?)
ApoE	19	Liver, macrophage, brain, perhaps ubiquitous	299	Ligand for LDL receptor and LDL receptor-related protein, reverse cholesterol transport, immunoregulatory properties, regulator of cell growth

[a] HDL, High-density lipoprotein; LCAT, lecithin–cholesterol acyltransferase (phosphatidylcholine–sterol *O*-acyltransferase); LDL, low-density lipoprotein; LPL, lipoprotein lipase; TG, triglyceride; TGRL, triglyceride-rich lipoproteins; VLDL, very-low-density lipoprotein.

been mapped within the proximal and distal promoters of several human apolipoprotein genes,[15–17] and a number of trans factors have been charac-

[15] J. D. Smith, A. Melian, T. Leff, and J. L. Breslow, *J. Biol. Chem.* **263,** 8300 (1988).

[16] P. Papazafiri, K. Ogami, D. P. Ramji, A. Nicosia, P. Monaci, C. Cladaras, and V. I. Zannis, *J. Biol. Chem.* **266,** 5790 (1991).

[17] B. Paulweber, A. R. Brooks, B. P. Nagy, and B. Levy-Wilson, *J. Biol. Chem.* **266,** 21956 (1991).

terized.[18] Several members of the steroid receptor superfamily have been shown to converge with high affinity at sites in the proximal promoter of the apoA-I and apoC-III genes. Retinoic acid-responsive factor α activates transcription from basal apoA-I promoters,[19] whereas hepatocyte nuclear factor 4 stimulates transcription from the apoC-III promoter.[20] Apolipoprotein A-I regulatory protein 1 and ErbA-related protein 3/chicken ovalbumin promoter-transcription factor (Ear3/COUP-TF) suppress transcriptional activity from both promoters.[21] Thus, the intracellular balance of these factors may be important in initiating transcription of the apoA-I and C-III genes.[22]

Similar interactions seem to be involved in the expression of the apoB and apoA-II genes.[23] However, little is known about the hierarchy of cis and trans regulatory elements in governing the expression of apolipoprotein genes *in vivo.* The genes encoding apolipoproteins A-I, C-III, and A-IV are closely linked within a 20-kbp DNA segment.[8] Because of the genomic organization of these genes, transcriptional interference may play a role in the basal and metabolite-induced transcription rate. That transcription from one promoter may cause suppression of transcription from an adjacent promoter has been demonstrated with constructs composed of two globin genes,[24] and such suppression has been implicated in the mechanism of insertional oncogenesis by the avian leukosis virus.[25]

A promoter polymorphism that may be associated with phenotypic effects has been described for the apoA-I gene. At a position 76 bp upstream of the transcription start site, either a guanosine (G) or an adenosine (A) is present. In several European populations, the less common A allele was associated with increased high-density lipoprotein (HDL) cholesterol and/or apoA-I.[26–28] Transient expression studies in Hep 3B (human hepato-

[18] V. I. Zannis, D. Kardassis, P. Cardot, M. Hadzopoulou-Cladaras, E. E. Zanni, and C. Cladaras, *Curr. Opin. Lipidol.* **3,** 96 (1992).

[19] J. N. Rottman, R. L. Widom, B. Nadal-Ginard, V. Mahdavi, and S. K. Karathanasis, *Mol. Cell. Biol.* **11,** 3814 (1991).

[20] F. M. Sladek, W. Zhong, E. Lai, and J. E. Darnell, Jr., *Genes Dev.* **4,** 2352 (1990).

[21] J. A. A. Ladias and S. K. Karathanasis, *Science* **251,** 561 (1991).

[22] M. Mietus-Snyder, F. M. Sladek, G. S. Ginsburg, C. F. Kuo, J. A. A. Ladias, J. E. Darnell, Jr., and S. K. Karathanasis, *Mol. Cell. Biol.* **12,** 1708 (1992).

[23] J. A. A. Ladias, M. Hadzopoulou-Cladaras, D. Kardassis, P. Cardot, J. Cheng, V. Zannis, and C. Cladaras, *J. Biol. Chem.* **267,** 15849 (1992).

[24] N. J. Proudfoot, *Nature* (*London*) **322,** 562 (1986).

[25] G. S. Payne, S. A. Courtneidge, L. B. Drittenden, A. M. Fadley, J. M. Bishop, and H. E. Varmus, *Cell* (*Cambridge, Mass.*) **23,** 311 (1981).

[26] M. Jeenah, A. Kessling, N. Miller, and S. Humphries, *Mol. Biol. Med.* **7,** 233 (1990).

[27] G. Sigurdsson, V. Gudnason, G. Sigursson, and S. E. Humphries, *Arterioscler. Thromb.* **12,** 1017 (1992).

[28] F. Pagani, A. Sidoli, G. A. Giudici, L. Barenghi, C. Vergani, and F. E. Baralle, *J. Lipid Res.* **31,** 1371 (1990).

cellular carcinoma) cells showed that the G allele exhibited higher promoter efficiency in constructs spanning 330 bp 5′ to the transcription start site. When regions further upstream were included in the constructs, the expression levels of reporter genes were similar for both alleles.[29] Unlike the case in the European populations mentioned, no effect of this polymorphism on plasma levels of apoA-I and HDL cholesterol was observed in a United States population,[30] perhaps reflecting differences in genetic backgrounds. Nevertheless, apoA-I production rates were lower in subjects from the United States with the A allele, a result consistent with *in vitro* expression studies. The equalizing effect of distal promoter sequences on expression levels of the A and G alleles demonstrates the need for cautious interpretation of *in vitro* results for the *in vivo* situation.

Although several metabolic conditions have been identified which alter mRNA abundance levels of individual apolipoproteins, the contribution of changes in gene transcription to changes in gene expression *in vivo* has been investigated in only a few studies. Among these studies, the effect of thyroid hormone provides insight into the complexity of apoA-I gene expression. With a single dose of thyroid hormone, transcriptional activity of the apoA-I gene in liver as well as the abundance of its mRNA rapidly rise. With chronic administration, transcriptional activity from the apoA-I gene decreases to 50% of control, but abundance levels of nuclear and total cellular apoA-I mRNA remain elevated, implying hormonal effects on mRNA maturation.[31] Selection of splice sites is not altered, nor are the primary transcript or other mRNA precursors posttranscriptionally edited as a result of chronic hormone administration. Rather, an apoA-I mRNA precursor containing introns 1 and 3 but not 2 is protected from degradation in chronically hyperthyroid rats.[32,33] In this setting, enhanced mRNA processing may increase hepatic apoA-I gene expression sevenfold. Because of the inverse association between nuclear apoA-I RNA abundance and transcription rate in euthyroid and chronically hyperthyroid rats, autoregulation of apoA-I gene transcription has been postulated.[31] Indeed, apoA-I gene transcription in rats with chronic hyperthyroidism is suppressed by a transcription elongation block that can be competitively relieved by

[29] R. Tuteja, N. Tuteja, C. Melo, G. Casari, and F. E. Baralle, *FEBS Lett.* **304,** 98 (1992).

[30] J. D. Smith, E. A. Brinton, and J. L. Breslow, *J. Clin. Invest.* **89,** 1796 (1992).

[31] W. Strobl, N. L. Gorder, Y.-C. Lin-Lee, A. M. Gotto, Jr., and W. Patsch, *J. Clin. Invest.* **85,** 659 (1990).

[32] Y.-C. Lin-Lee, W. Strobl, S. Soyal, M. Radosavljevic, M. Song, A. M. Gotto, Jr., and W. Patsch, *J. Lipid Res.* **34,** 249 (1993).

[33] S. M. Soyal, C. Seelos, Y.-C. Lin-Lee, S. Sanders, A. M. Gotto, Jr., D. L. Hachey, and W. Patsch, *J. Biol. Chem.* **270,** 3996 (1995).

RNA fragments.[33a] Thus, apoA-I mRNA synthesis not only depends on the rate of transcription initiation, but is also controlled by the rate of transcript elongation.

The biosynthesis of apoB-48 depends on posttranscriptional events that occur within the nucleus. ApoB-48 mRNA is transcribed from the same gene as apoB-100 mRNA. However, the cytidine at nucleotide 6666 is changed to a uridine by a mechanism termed posttranscriptional editing. This substitution produces an in-frame stop codon which replaces the CAA codon for Gln-2153 in apoB-100.[34,35] Most likely, a sequence-specific cytidine deaminase or transaminase reaction is involved in apoB RNA editing.[36,37] Several models of apoB mRNA editing have been proposed that are based on sequence requirements defined with mutagenized transcripts.[38–40] Apolipoprotein B mRNA is the only mRNA species known to undergo editing by cellular extracts, but a similar mechanism has been postulated for the substitution of glutamine by arginine in glutamate receptors in the brain.[41] Because the amino acid change occurs in the putative channel-forming segment of the receptor, it may alter the properties of ion flow. Other mammalian RNAs may also be substrates for this editing activity, because tissues that do not express apoB exhibit editing activity.[42] A cDNA clone has been isolated from rat small intestine which encodes a protein essential for apoB mRNA editing.[43]

Control of apolipoprotein gene expression also occurs at the translational level,[44] and several posttranslational modifications have been described, including proteolytic processing, N- and O-glycosylation, acylation, and phosphorylation. Both apoA-I and apoA-II are synthesized as prepro-

[33a] Y.-C. Lin-Lee, S. M. Soyal, A. Surguchov, S. Sanders, W. Strobl, and W. Patsch, *J. Lipid Res.* **36,** 1586 (1995).

[34] L. M. Powell, S. C. Wallis, R. J. Pease, Y. H. Edwards, T. J. Knott, and J. Scott, *Cell (Cambridge, Mass.)* **50,** 831 (1987).

[35] S.-H. Chen, G. Habib, C.-Y. Yang, Z.-W. Gu, B. R. Lee, S.-A. Weng, S. R. Silberman, S.-J. Cai, J. Deslypere, M. Rosseneu, A. M. Gotto, Jr., and L. Chan, *Science* **238,** 363 (1989).

[36] K. Boström, Z. Garcia, K. S. Poksay, D. F. Johnson, A. J. Lusis, and T. L. Innerarity, *J. Biol. Chem.* **265,** 22446 (1990).

[37] P. E. Hodges, N. Navaratnam, J. C. Greeve, and J. Scott, *Nucleic Acids Res.* **19,** 1197 (1991).

[38] J. W. Backus and H. C. Smith, *Nucleic Acids Res.* **19,** 6781 (1991).

[39] P. Hodges and J. Scott, *Trends Biochem. Sci.* **17,** 77 (1992).

[40] L. Chan, *J. Biol. Chem.* **267,** 25621 (1992).

[41] B. Sommer, M. Kohler, R. Sprengel, and P. H. Seeburg, *Cell (Cambridge, Mass.)* **67,** 11 (1991).

[42] Teng B, M. Verp, J. Salomon, and N. O. Davidson, *J. Biol. Chem.* **266,** 20616 (1990).

[43] B. Teng, C. F. Burant, and N. O. Davidson, *Science* **260,** 1816 (1993).

[44] M. F. Go, G. Schonfeld, B. Pfleger, T. G. Cole, N. L. Sussman, and D. H. Alpers, *J. Clin. Invest.* **81,** 1615 (1988).

apolipoproteins. After cotranslational cleavage of the prepeptide, proapoA-I is rapidly secreted. The 6-amino acid-containing propeptide of apoA-I is rapidly cleaved by a metalloprotease.[45] Hence, mature apoA-I is the major apoA-I isoform in plasma, and additional minor isoforms are the result of deamidation.[46] In contrast, the 5-amino acid-containing propeptide of apoA-II seems to be removed intracellularly.[47] The physiological significance of the short propeptides is not understood, but they do not seem to affect intracellular transport, secretion, lipid association, or function of apoA-I. Other modifications such as acylation[48] and phosphorylation[49] may play roles in intracellular trafficking and lipoprotein assembly. O-Glycosylation is not required for secretion of apoE,[50] but it affects the ability of apoA-II to associate with HDL.[51]

Posttranslational events are thought to regulate hepatic secretion of apoB, the principal apolipoprotein of triglyceride (TG)-rich lipoproteins and low-density lipoproteins (LDL). Even though several cis and trans regulatory elements in the apoB gene have been characterized that enhance or suppress transcription in *in vitro* systems or in various cell cultures, the contribution of changes in apoB transcription to altered apoB gene expression is not known, and only a few *in vivo* studies demonstrate regulation of apoB expression at the pretranslational level.[52] In contrast, several hormonal and metabolic stimuli that alter hepatic apoB production at the translational or posttranslational level have been identified. Studies in primary rat hepatocyte cultures showed that apoB secretion is inhibited by insulin, even though cellular apoB concentrations are not reduced and hepatic TG synthesis is actually increased by the hormone.[53] In pulse–chase

[45] C. J. Edelstein, J. I. Gordon, K. R. Tocsas, H. F. Sims, A. W. Strauss, and A. M. Scanu, *J. Biol. Chem.* **258,** 11430 (1983).
[46] G. Ghiselli, M. F. Gohde, S. Tanenbaum, S. Krishnan, and A. M. Gotto, Jr., *J. Biol. Chem.* **260,** 15662 (1985).
[47] K. J. Lackner, S. B. Edge, R. E. Gregg, J. M. Hoeg, and H. B. Brewer, Jr., *J. Biol. Chem.* **260,** 703 (1985).
[48] J. M. Hoeg, M. S. Meng, R. Ronan, S. J. Demosky, T. Fairwell, and H. B. Brewer, Jr., *J. Lipid Res.* **29,** 1215 (1988).
[49] Z. H. Beg, J. A. Stonik, J. M. Hoeg, S. J. Demosky, Jr., T. Fairwell, and H. B. Brewer, Jr., *J. Biol. Chem.* **264,** 6913 (1989).
[50] E. E. Zanni, A. Kouvatsi, M. Hadzopoulou-Cladaras, M. Krieger, and V. I. Zannis, *J. Biol. Chem.* **264,** 9137 (1989).
[51] A. T. Remaley, A. W. Song, U. K. Schumacher, M. S. Meng, and H. G. Brewer, Jr., *J. Biol. Chem.* **286,** 6785 (1993).
[52] L. K. Hennessy, J. Osada, J. M. Ordovas, R. J. Nicolosi, A. F. Stucchi, M. E. Brousseau, and E. J. Schaefer, *J. Lipid Res.* **33,** 351 (1992).
[53] W. Patsch, S. Franz, and G. Schonfeld, *J. Clin. Invest.* **71,** 1161 (1983).

studies, the transport of apoB from the endoplasmic reticulum to the Golgi was inhibited by the hormone.[54]

Studies in several laboratories have now established that degradation of apoB in the endoplasmic reticulum or in a pre-Golgi compartment is a principal mechanism by which apoB production is controlled (reviewed in Refs. 55 and 56). Large domains of apoB may be exposed to the cytoplasm during translocation into the endoplasmic reticulum,[57] and the rate of translocation may be influenced by sequences mediating stopping and restarting of translocation.[58] Such pause transfer sequences may be necessary for proper folding, formation of disulfide bonds, and association with lipids. Acquisition of a critical amount of lipid may be required for apoB to become competent for transport to the Golgi rather than being degraded. The microsomal triglyceride transfer complex consisting of disulfide isomerase and a second protein, probably representing its functional unit, may play an essential role in lipid assembly of apoB-containing lipoproteins.[59,60]

The inhibitory effect of insulin on apoB secretion may be of clinical relevance. Inhibition of very-low-density lipoprotein (VLDL) secretion during meal absorption (when portal insulin levels are high) would facilitate catabolism of intestinal lipoproteins that share a common clearance pathway with hepatic TG-rich lipoproteins. Downregulation of hepatic insulin receptors *in vitro* diminishes the inhibitory effect of the hormone on hepatic apoB secretion.[61] Insulin resistance may therefore enhance the magnitude of postprandial lipemia, which has been shown to identify patients with angiographically verified CAD (coronary artery disease) with the same or greater accuracy than any other variable of lipid transport measured in the fasting state.[62,63]

[54] W. Patsch, Y.-C. Lin-Lee, A. M. Gotto, Jr., and J. R. Patsch, *Arteriosclerosis* (*Dallas*) **6,** 536a (1983).

[55] J. L. Dixon and H. N. Ginsberg, *J. Lipid Res.* **34,** 167 (1993).

[56] J. D. Sparks and C. E. Sparks, *Curr. Opin. Lipidol.* **4,** 177 (1993).

[57] S. L. Chuck, Z. Yao, B. D. Blackhart, B. J. McCarthy, and V. R. Lingappa, *Nature* (*London*) **346,** 382 (1990).

[58] S. L. Chuck and V. R. Lingappa, *Cell* (*Cambridge, Mass.*) **68,** 9 (1992).

[59] J. R. Wetterau, K. A. Combs, L. R. McLean, S. N. Spinner, and L. P. Aggerbeck, *Biochemistry* **30,** 9728 (1991).

[60] J. R. Wetterau, L. P. Aggerbeck, M.-E. Bouma, C. Eisenberg, A. Munck, M. Hermier, J. Schmitz, G. Gay, D. J. Rader, and R. E. Gregg, *Science* **258,** 999 (1992).

[61] W. Patsch, A. M. Gotto, Jr., and J. R. Patsch, *J. Biol. Chem.* **261,** 9603 (1986).

[62] P. H. E. Groot, W. A. J. J. van Stiphuit, X. H. Drauss, H. Jansen, A. van Tol, E. van Ramshorst, S. Chin-On, S. R. Cresswell, and L. Havekes, *Arterioscler. Thromb.* **11,** 653 (1991).

[63] J. R. Patsch, G. Miesenbock, T. Hopferwiese, V. Muehlberger, E. Knapp, J. K. Dunn, A. M. Gotto, Jr., and W. Patsch, *Arterioscler. Thromb.* **12,** 1336 (1992).

Apolipoprotein Functions in Plasma Lipid Transport

Apolipoprotein A-I

Apolipoprotein A-I is the main apolipoprotein of HDL, and its plasma concentration is inversely associated with the incidence of CAD.[64] The mechanisms whereby apoA-I might protect against atherosclerosis are not fully understood, but the role of the apolipoprotein in reverse cholesterol transport may be key. By binding to a cell surface protein, operationally termed an HDL receptor,[65] apoA-I promotes translocation of cholesterol from intracellular pools to the cell membranes,[66] facilitates transfer of cholesterol from cell membranes to nascent HDL, and traps cholesterol via lecithin–cholesterol acyltransferase (LCAT)-mediated esterification in the core of HDL particles.[67]

Apolipoproteins C-I, A-IV, and E have also been shown to activate LCAT *in vitro,*[68–70] albeit to a lesser extent than apoA-I. Both amino- and carboxy-terminal cyanogen bromide fragments of apoA-I activate LCAT.[71] According to studies with synthetic peptides[72] and mutant apoA-I forms,[70] the stimulatory effect of carboxy-terminal regions of apoA-I reflects their lipid binding ability. For full activation of LCAT, the region spanning amino acid residues 66–121[73] or 95–121[74] may be required.

Other mechanisms by which apoA-I may provide protection against atherosclerosis have been suggested. These include stimulation of endothe-

[64] J. J. Maciejko, D. R. Holmes, B. A. Kottke, A. R. Zinsmeister, D. M. Dinh, and S. J. T. Mao, *N. Engl. J. Med.* **309,** 385 (1983).

[65] G. L. McKnight, J. Reasoner, T. Gilbert, K. O. Sundquist, B. Hokland, P. A. McKernan, J. Champagne, C. J. Johnson, M. C. Bailey, R. Holly, P. J. O'Hara, and J. F. Oram, *J. Biol. Chem.* **267,** 12131 (1992).

[66] G. Nowicka, T. Bruening, A. Boettcher, G. Kahl, and G. Schmitz, *J. Lipid Res.* **31,** 1947 (1990).

[67] C. J. Fielding, V. G. Shore, and P. E. Fielding, *Biochem. Biophys. Res. Commun.* **46,** 1493 (1972).

[68] A. K. Soutar, C. W. Garner, H. N. Baker, J. T. Sparrow, R. L. Jackson, A. M. Gotto, Jr., and L. C. Smith, *Biochemistry* **14,** 3057 (1975).

[69] A. Steinmetz and G. Utermann, *J. Biol. Chem.* **260,** 2258 (1985).

[70] N. Zorich, A. Jonas, and H. J. Pownall, *J. Biol. Chem.* **260,** 8831 (1985).

[71] H. J. Pownall, A. Hu, A. M. Gotto, Jr., J. J. Albers, and J. T. Sparrow, *Proc. Natl. Acad. Sci. U.S.A.* **77,** 3154 (1980).

[72] A. Minnich, X. Collet, A. Roghan, C. Cladaras, R. L. Hamilton, C. J. Fielding, and V. I. Zannis, *J. Biol. Chem.* **267,** 16553 (1992).

[73] G. M. Anantharamaiah, Y. V. Venkatachalapathi, C. G. Brouillette, and J. P. Segrest, *Arteriosclerosis (Dallas)* **10,** 95 (1990).

[74] C. L. Banka, D. J. Bonnet, A. S. Black, R. S. Smith, and L. K. Curtiss, *J. Biol. Chem.* **266,** 23886 (1991).

lial cell proliferation,[75] inhibition of smooth muscle cell proliferation,[76] and interference with the uptake of oxidized LDL by macrophages.[77]

Some genetic defects affecting plasma levels of apoA-I are understood at the molecular level. Mutations on chromosome 11 that preclude the synthesis of apoA-I have been described. These include an inversion of 5.5 kbp containing part of the structural genes encoding apoA-I and apoC-III[78] and a gene complex deletion.[79] Premature CAD and very low HDL cholesterol levels are characteristic features in subjects homozygous for such mutations. Apolipoprotein A-I_{Milano}, which contains a cysteine instead of an arginine at position 173 of mature apoA-I, was the first of an increasing number of apoA-I structural variants to be described.[80,81] This amino acid substitution alters the lipid binding properties of apoA-I, allows for disulfide bonding with other cysteine-containing proteins such as apoA-II and apoE, and may have profound effects on the substrate properties of HDL for lipoprotein-modifying enzymes. Importantly, the susceptibility of CAD is not increased in patients with this variant, even though HDL cholesterol is reduced. Other variants are associated with reduced LCAT activation,[82] low HDL cholesterol,[83] and amyloidotic neuropathy,[84] but an association with premature CAD is not characteristic for structural variants of apoA-I.

In Tangier disease, the structural gene of apoA-I is normal,[85] but apoA-I plasma levels are very low due to increased catabolism.[86] Because of defective retroendocytosis, HDL may be degraded in lysosomes, resulting in

[75] J. P. Tauber, J. Cheng, and D. Gospodarowicz, *J. Clin. Invest.* **66,** 696 (1980).

[76] B. F. Burkey, N. Vlaxic, D. France, T. E. Hughes, M. Drelich, X. Ma, M. B. Stemerman, and J. R. Paterniti, *Circulation* **86**(Suppl. 1), I-472 (1992).

[77] S. Parthasarathy, J. Barnett, and L. G. Fong, *Biochim. Biophys. Acta* **1044,** 275 (1990).

[78] S. K. Karathanasis, E. Ferris, and I. A. Haddad, *Proc. Natl. Acad. Sci. U.S.A.* **84,** 7198 (1987).

[79] J. M. Ordovas, D. K. Cassidy, F. Civeira, C. L. Bisgaier, and E. J. Schaefer, *J. Biol. Chem.* **264,** 16339 (1989).

[80] G. Franceschini, C. R. Sirtori, A. Capurso, K. H. Weisgraber, and R. W. Mahley, *J. Clin. Invest.* **66,** 892 (1980).

[81] K. H. Weisgraber, T. P. Bersot, R. W. Mahley, G. Franceschini, and C. R. Sirtori, *J. Clin. Invest.* **66,** 901 (1980).

[82] G. Utermann, J. Haas, A. Steinmetz, R. Paetzold, S. C. Rall, and K. H. Weisgraber, *Eur. J. Biochem.* **144,** 326 (1984).

[83] A. van Eckardstein, H. Funke, A. Henke, K. Atland, A. Benninghoven, and G. Assmann, *J. Clin. Invest.* **84,** 1722 (1989).

[84] W. C. Nichols, F. E. Dwulet, J. Liepnieks, and M. D. Bensen, *Biochem. Biophys. Res. Commun.* **156,** 762 (1988).

[85] S. C. Makrides, N. Rutz-Opazo, M. Hayden, A. L. Nussbaum, J. L. Breslow, and V. I. Zannis, *Eur. J. Biochem.* **173,** 465 (1988).

[86] G. Assmann, G. Schmitz, and H. B. Brewer, Jr., *in* "The Metabolic Basis of Inherited Disease" (C. R. Scriver, A. L. Beaudet, W. S. Sly, and D. Valle, eds.), p. 1267. McGraw-Hill, New York, 1989.

defective removal of cellular cholesterol.[87] Symptomatic atherosclerotic disease has not been reported to occur before age 40,[88] but other biochemical abnormalities such as a low LDL cholesterol level and reduction in the number and reactivity of platelets may reduce the susceptibility for CAD in Tangier patients.

Other inborn errors of metabolism such as fish-eye disease (partial deficiency of LCAT)[89,90] or lipoprotein lipase deficiency are also associated with low plasma apoA-I levels, but they do not greatly enhance the risk of premature CAD. Hence, the inverse relation between plasma levels of apoA-I and risk for CAD is not absolute. This conclusion is also supported by observations in experimental animals. Overexpression of human apoA-I in transgenic mice decreases the severity of diet-induced atherosclerosis,[91] but targeted disruption of the apoA-I gene in mice is not associated with an increased formation of atherosclerotic lesions.[92] These contrasting findings could, at least in part, relate to structural and functional differences between human and mouse apoA-I.[93] However, a mutant chicken strain with 70–90% less plasma apoA-I does not exhibit an increased susceptibility to spontaneous or diet-induced atherosclerosis relative to control chickens.[94] Thus, apoA-I plasma levels per se cannot be equated with protection from CAD, and the inverse relationship observed in population studies between plasma apoA-I and incidence of CAD may, at least in part, reflect a metabolic state that confers protection from CAD.

Apolipoprotein A-II

In humans, apoA-II is the second main apolipoprotein of HDL. Mature human apoA-II exists as a dimer of two 77-amino acid chains linked by a disulfide bridge at position 6 in the sequence.[95] In several animal species, apoA-II circulates as a monomer because the residue at position 6 is not

[87] G. Schmitz, G. Assmann, H. Robenek, and B. Brennhausen, *Proc. Natl. Acad. Sci. U.S.A.* **82,** 6305 (1985).

[88] E. J. Schaefer, *Arteriosclerosis (Dallas)* **4,** 303 (1982).

[89] G. Assmann, A. von Eckardstein, and H. Funke, *Curr. Opin. Lipidol.* **2,** 110 (1991).

[90] L. A. Carlson, *Eur. J. Clin. Invest.* **12,** 41 (1982).

[91] E. M. Rubin, R. M. Krauss, E. A. Spangler, J. G. Verstuyft, and S. M. Clift, *Nature (London)* **353,** 265 (1991).

[92] N. Maeda, *Curr. Opin. Lipidol.* **4,** 90 (1993).

[93] E. M. Rubin, B. Y. Ishida, S. M. Clift, and R. M. Krauss, *Proc. Natl. Acad. Sci. U.S.A.* **88,** 434 (1991).

[94] F. Poernama, R. Subramanian, M. E. Cook, and A. D. Attie, *Arterioscler. Thromb.* **12,** 601 (1992).

[95] H. B. Brewer, Jr., S. E. Lux, R. Ronan, and K. M. John, *Proc. Natl. Acad. Sci. U.S.A.* **69,** 1304 (1972).

a cysteine.[96] Unlike the case in humans, plasma concentrations of apoA-II are very low in animal species expressing the monomer, most likely because of a shortened plasma residence time.

Human apoA-II is an activator of hepatic lipase activity. Although early studies were contradictory, reporting both stimulation[97] and nonspecific inhibition[98] of hepatic lipase activity, Mowri *et al.*[99] showed that lipolysis of TG and phospholipids in postprandial HDL_2 containing both apoA-I and apoA-II was higher than in postprandial HDL_2 containing apoA-I but devoid of apoA-II. Reconstitution studies of apoA-I-containing HDL_2 with apoA-II clearly established the stimulatory effect of apoA-II.[99] These studies explain the preferential conversion of apoA-II-containing HDL_2 to HDL_3 in the postprandial state.[100] However, apoA-II is not obligatory for the lipolysis of phospholipids and triglycerides in HDL. Whereas some studies in myocardial survivors suggest a protective role of apoA-II in CAD,[101] apoA-II deficiency due to a mutation at the splice junction of exon 3 has little influence on lipoprotein profiles or on the occurrence of CAD.[102]

A gene termed Ath-1 near the apoA-II locus on chromosome 1 determines HDL levels and atherosclerosis susceptibility in mouse strains.[103] However, the Ath-1 and apoA-II genes are distinct. In fact, in some mouse strains increased expression of apoA-II is associated with enhanced fatty streak development despite increased HDL cholesterol.[104] These results are consistent with our understanding of the relationship between HDL subpopulations containing apoA-II (LpA-I/A-II) but devoid of apoA-II (LpA-I) and CAD.[105] Hepatic lipase is thought to play a critical role in

[96] P. N. Herbert, H. G. Windmueller, T. P. Bersot, and R. S. Shulman, *J. Biol. Chem.* **249,** 5718 (1979).

[97] C. E. Jahn, J. C. Osborne, Jr., E. J. Schaefer, and H. B. Brewer, Jr., *Eur. J. Biochem.* **131,** 25 (1983).

[98] M. Shinomiya, N. Sasaki, R. L. Barnhart, K. Shirai, and R. L. Jackson, *Biochim. Biophys. Acta* **713,** 292 (1982).

[99] H.-O. Mowri, W. Patsch, S. C. Smith, A. M. Gotto, Jr., and J. R. Patsch, *J. Lipid Res.* **33,** 1269 (1992).

[100] J. R. Patsch, S. Prasad, A. M. Gotto, Jr., and G. Bengtsson-Olivecrona, *J. Clin. Invest.* **74,** 2017 (1984).

[101] G. Fager, O. Wiklund, and S. O. Olofsson, *Arteriosclerosis* (*Dallas*) **1,** 273 (1981).

[102] S. S. Deeb, K. Takata, R. Peng, G. Kajiyama, and J. J. Albers, *Am. J. Hum. Genet.* **46,** 822 (1990).

[103] B. Paigen, D. Mitchell, K. Reue, A. Morrow, A. J. Lusis, and R. C. LeBoeuf, *Proc. Natl. Acad. Sci. U.S.A.* **87,** 3763 (1987).

[104] M. Mehrabian, J.-H. Qiao, R. Hyman, D. Ruddle, C. Laughton, and A. J. Lusis, *Arterioscler. Thromb.* **13,** 1 (1993).

[105] P. Puchois, A. Kandoussi, P. Fievet, J. L. Fourrier, M. Bertrand, E. Koren, and J. C. Fruchart, *Atherosclerosis* **68,** 35 (1987).

remnant removal, perhaps by altering the expression of apoE on the surface of particles.[106] By competing with remnants for hepatic lipase, apoA-II-containing HDL may indirectly affect the clearance of these atherogenic lipoprotein particles from the circulation.

High-density lipoproteins are heterogeneous with respect to physical, chemical, and functional properties. The molar ratios of apoA-I to apoA-II differ among particles within the HDL_3 density spectrum. The ratios are too high in some HDL_2 fractions to allow the presence of one molecule of apoA-II on each particle,[107] and LpA-I/A-II and LpA-I can be isolated by immunochromatography.[108] There is evidence to suggest that plasma levels of LpA-I reflect the protective function of HDL better than plasma levels of LpA-I/A-II. The well-established male–female difference in plasma levels of HDL_2 was shown to result from increased concentrations of LpA-I in the HDL_2 fraction of females,[109] and the presence of CAD was inversely correlated with levels of LpA-I, but not LpA-I/A-II, in a study comparing patients with angiographically verified CAD with controls.[105] Transgenic mice overexpressing human apoA-I exhibit enhanced protection against diet-induced atherosclerosis when compared with mice overexpressing both apoA-I and apoA-II.[110]

Mechanisms have been suggested to explain the contrasting relationships of LpA-I and LpA-I/A-II with atherosclerosis. Particles containing only apoA-I may be the physiological acceptor of cellular cholesterol,[111] but this small-sized LpA-I represents only a minor fraction of plasma LpA-I and thus cannot directly account for the inverse association between plasma levels of LpA-I and CAD. *In vitro* studies showed that LpA-I enhances cholesterol efflux from mouse adipocytes, whereas LpA-I/A-II may actually inhibit the LpA-I-stimulated efflux of cholesterol.[112] However, in several other mammalian cells, including rabbit aortic smooth muscle cells[113] and bovine aortic endothelial cells,[114] LpA-I and LpA-I/A-II seem to function equally well in removing cholesterol. The preferential association of LpA-I

[106] D. L. Brasaemle, K. Cornely-Moss, and A. Bensadoun, *J. Lipid Res.* **34,** 455 (1993).

[107] W. Patsch, G. Schonfeld, A. M. Gotto, Jr., and J. R. Patsch, *J. Biol. Chem.* **255,** 3178 (1980).

[108] M. C. Cheung and J. J. Albers, *J. Biol. Chem.* **259,** 12201 (1984).

[109] R. W. James, A. Proudfoot, and D. Pometta, *Biochim. Biophys. Acta* **1002,** 292 (1989).

[110] J. R. Schultz, J. G. Verstuyft, E. L. Gong, A. V. Nichols, and E. M. Rubin, *Circulation* **86**(Suppl. 1), I-472 (1992).

[111] G. R. Castro and C. J. Fielding, *Biochemistry* **27,** 25 (1988).

[112] R. Barbaras, P. Puchois, J. C. Fruchart, and G. Ailhoud, *Biochem. Biophys. Res. Commun.* **142,** 63 (1987).

[113] W. J. Johnson, E. P. C. Kilsdonk, A. van Tol, M. C. Phillips, and G. H. Rothblat, *J. Lipid Res.* **32,** 1993 (1991).

[114] S. Oikawa, A. J. Mendez, J. F. Oram, and E. L. Bierman, *Biochim. Biophys. Acta* **1165,** 327 (1993).

with LCAT and CETP (cholesterol ester transfer protein) may nevertheless enhance its ability to promote cholesterol efflux *in vivo*.[115]

In normolipidemic male subjects, plasma levels of LpA-I/A-II are 2.5 times higher than levels of LpA-I, and the density distribution of LpA-I and LpA-I/A-II shows that a larger portion of LpA-I is found in HDL_2. With increasing levels of HDL_2, the proportion of LpA-I in HDL_2 increases. Plasma levels of LpA-I, but not LpA-I/A-II, are inversely correlated with the magnitude of postprandial lipemia, which may in part account for the inverse association of LpA-I with CAD.[115a] Thus, TG metabolism may play a role in establishing the type and quantity of HDL particles that are either indicative of protection from CAD or may have specific abilities to interact and/or return cholesterol from cells of the arterial wall.

Apolipoprotein A-IV

Apolipoprotein A-IV shares several functions with apoA-I. Like apoA-I, apoA-IV activates LCAT,[69] may promote cholesterol efflux from peripheral cells,[116] and may facilitate the uptake of HDL by the liver.[117] However, dietary perturbations that enhance TG transport upregulate the expression of the apoA-IV gene (but not the apoA-I gene) in several animal species. ApoA-IV mRNA abundance increases in rat enterocytes after a fat bolus,[118] and changes in intestinal apoA-IV mRNA abundance correlate with developmental stages characterized by rapid triglyceride metabolism.[119] In inbred strains of mice, hepatic apoA-IV mRNA abundance is regulated by dietary lipid content. Changes of apoA-IV mRNA abundance levels in response to an atherogenic diet vary by orders of magnitude.[120] In some mouse strains, a 12-bp deletion within a polymorphic tandem repeat region at the 3′ end of the apoA-IV gene seems to influence its mRNA metabolism, thereby altering apoA-IV gene expression.[121] A sucrose-rich diet that increases hepatic TG synthesis and secretion also in-

[115] T. Ohta, R. Nakamura, Y. Ikeda, M. Shinohara, A. Miyazaki, S. Horiuchi, and I. Matsuda, *Biochim. Biophys. Acta* **1165,** 119 (1993).

[115a] H.-O. Mowri, J. R. Patsch, A. Ritsch, B. Föper, S. Brown, and W. Patsch, *J. Lipid Res.* **35,** 291 (1994).

[116] A. Steinmetz, R. Barbaras, N. Ghalim, V. Clavey, J. C. Fruchart, and G. Ailhaud, *J. Biol. Chem.* **265,** 7859 (1990).

[117] E. Dvorin, N. L. Gorder, D. M. Benson, and A. M. Gotto, Jr., *J. Biol. Chem.* **261,** 15714 (1986).

[118] J. I. Gordon, D. P. Smith, D. H. Alpers, and A. W. Strauss, *Biochemistry* **21,** 5424 (1982).

[119] I. A. Haddad, J. M. Ordovas, T. Fitzpatrick, and S. Karathanasis, *J. Biol. Chem.* **261,** 13268 (1986).

[120] S. C. Williams, S. G. Grant, K. Reue, B. Carrasquillo, A. J. Lusis, and A. J. Kinniburgh, *J. Biol. Chem.* **264,** 19009 (1989).

[121] K. Reue and T. H. Leete, *J. Biol. Chem.* **266,** 12715 (1991).

creases apoA-IV gene expression at the transcriptional level.[122] In Zucker fatty rats, an animal model for genetic obesity and overproduction of VLDL, hepatic apoA-IV gene expression is increased due to enhanced apoA-IV mRNA maturation.[123] Even though the function of apoA-IV in TG metabolism is not fully understood, apoA-IV may enhance the catabolism of TG by facilitating the transfer of apoC-II to TG-rich lipoproteins.[124] Fujimoto *et al.* have made the interesting observation that intravenous infusion of apoA-IV suppresses food intake in rats. ApoA-IV may therefore act as a physiological signal for satiation.[125]

Apolipoprotein A-IV is a polymorphic protein. A number of allelic variations have been described that alter the isoelectric point and/or molecular weight of apoA-IV.[126] The apoA-IV-1 allele is the most common variant, with allelic frequencies ranging between 0.88 and 0.95. Apolipoprotein A-IV-2 is the second most common allele, with allelic frequencies as high as 0.12 in some populations, and results from a G to T substitution that changes glutamine at position 360 of the mature protein to a histidine. Physicochemical studies suggest that the amino acid substitution increases the α-helix content of the native protein, enhances its affinity for egg phospholipid vesicles, and increases its potency to activate LCAT.[127] The latter observation, however, is at variance with another report.[128] In addition to the apoA-IV-2 isoform, a number of other variants have been described that may affect the function of apoA-IV.[126] In some populations, the apoA-IV-2 isoform has been associated with higher HDL cholesterol and/or lower plasma triglyceride levels,[129,130] but these associations have not been found in other populations.[131,132] Hence, the impact of common

[122] M. Radosavljevic, Y.-C. Lin-Lee, S. M. Soyal, W. Strobl, C. Seelos, A. M. Gotto, Jr., and W. Patsch, *Atherosclerosis* **95,** 147 (1992).

[123] W. Strobl, B. Knerer, R. Gratzl, K. Arbeiter, Y.-C. Lin-Lee, and W. Patsch, *J. Clin. Invest.* **92,** 1766 (1993).

[124] I. J. Goldberg, C. A. Scheraldi, L. K. Yakoub, U. Saxena, and C. L. Bisgaier, *J. Biol. Chem.* **8,** 4266 (1990).

[125] K. Fujimoto, J. A. Cardelli, and P. Tso, *Am. J. Physiol.* **262,** G1002 (1992).

[126] P. Lohse and H. B. Brewer, Jr., *Curr. Opin. Lipidol.* **2,** 90 (1991).

[127] R. B. Weinberg, M. K. Jordan, and A. Steinmetz, *J. Biol. Chem.* **265,** 18372 (1990).

[128] H. Tenkanen, M. Lukka, M. Jauhiainen, J. Metso, M. Bauman, L. Peltonen, and C. Ehnholm, *Arterioscler. Thromb.* **11,** 851 (1991).

[129] H. J. Menzel, E. Boerwinkle, S. Schrangl-Will, and G. Utermann, *Hum. Genet.* **79,** 368 (1988).

[130] H. J. Menzel, G. Sigurdsson, E. Boerwinkle, S. Schrangl-Wills, H. Dieplinger, and G. Utermann, *Hum. Genet.* **84,** 344 (1990).

[131] H. Tenkanen, P. Koskinen, J. Metso, M. Baumann, M. Lukka, R. Kauppinen-Makelin, K. Kontula, M. R. Taskinen, M. Manttari, V. Manninen, and C. Ehnholm, *Biochim. Biophys. Acta* **1138,** 27 (1992).

[132] C. L. Hanis, T. C. Douglas, and D. Hewett-Emmet, *Hum. Genet.* **86,** 323 (1991).

genetic variation in the structural gene of apoA-IV on plasma lipid transport is not clear.

Apolipoprotein B

Apolipoprotein B occurs in two forms termed apoB-100 and apoB-48. In humans, apoB-48 is produced only by the intestine. Apolipoprotein B-100 originates from the liver even though the intestine may be competent for apoB-100 production.[133] ApoB-48 consists of the amino-terminal half of apoB-100, contains 2152 amino acid residues, and is devoid of the binding domain for the LDL receptor.[34,35] It is an essential component of chylomicrons and their remnants, but it can invariably be detected in plasma from fasting subjects.

Apolipoprotein B-100, which contains 4536 amino acid residues,[134–137] is the major apolipoprotein of VLDL, IDL (intermediate-density lipoprotein), and Lp[a] and the sole apolipoprotein of LDL. ApoB-100 contains two regions that show similarity to the LDL receptor binding domain of apoE, but only the region spanning amino acid residues 3359–3367 exhibits a high degree of conservation among animal species and maintains a strong net positive charge.[138] Furthermore, a synthetic peptide encompassing this region mediates the interaction of VLDL, rendered binding-incompetent, with the LDL receptor.[135] However, apoB sequences remote from these regions may modulate the expression of the LDL receptor binding domain in native lipoproteins.[139]

Among all apolipoproteins, apoB-100 has been most clearly associated with atherosclerosis. Several rare mutations in the apoB gene have been identified that influence plasma levels of apoB. Familial defective apoB-100 is a dominant disorder resulting from a mutation in the coding sequence of the apoB gene that changes the arginine at position 3500 to gluta-

[133] R. P. F. Dullaart, B. Speelberg, H. J. Schuurman, R. W. Milne, L. M. Havekes, Y. L. Marcel, H. J. Geuze, M. M. Hulshof, and D. W. Erkelens, *J. Clin. Invest.* **78,** 1397 (1986).

[134] J. Knott, R. J. Pease, L. M. Powell, S. C. Wallis, S. C. Rall, Jr., T. L. Innerarity, B. Blackhart, W. H. Taylor, Y. L. Marcel, R. Milne, D. Johnson, M. Fuller, A. J. Lusis, B. J. McCarthy, R. W. Mahley, B. Levy-Wilson, and J. Scott, *Nature (London)* **323,** 738 (1986).

[135] C. Y. Yang, S. H. Chen, S. H. Gianturco, W. A. Bradley, J. T. Sparrow, M. Tanimura, W. H. Li, D. A. Sparrow, H. DeLoof, M. Rosseneu, F. S. Lee, Z. W. Gu, A. M. Gotto, Jr., and L. Chan, *Nature (London)* **323,** 738 (1986).

[136] S. W. Law, S. M. Grant, K. Higuchi, A. Hospattankar, K. Lackner, N. Lee, and H. B. Brewer, Jr., *Proc. Natl. Acad. Sci. U.S.A.* **83,** 8142 (1986).

[137] C. Cladaras, M. Hadzoopoulou-Cladaras, R. T. Nolte, D. Atkinson, and V. I. Zannis, *EMBO J.* **5,** 3495 (1986).

[138] A. Law and J. Scott, *J. Lipid Res.* **31,** 1109 (1990).

[139] K. G. Parhofer, A. Daugherty, M. Kinoshita, and G. Schonfeld, *J. Lipid Res.* **31,** 2001 (1990).

mine.[140–142] Even though the mutation is not within the receptor binding domain per se, receptor binding activity is reduced to less than 5% of control. Hence, the substitution of glutamine for arginine may affect the competence for binding through conformational changes as suggested by ^{13}C nuclear magnetic resonance (NMR) studies.[143] In familial hypobetalipoproteinemia, more than 20 different mutations have been described in as many probands.[144] Affected individuals express different lengths of apoB. Apolipoprotein B species spanning between 31 and 46% of the amino-terminal portion of apoB can associate with HDL. With increasing length of the truncated apoB species produced, lipoprotein association of apoB normalizes. The frequency of apoB mutations causing hypobetalipoproteinemia is probably around 0.5% in general populations, and this condition may protect from CAD.

Studies using monoclonal antibodies for apoB to define allele-specific differences suggested that mutations at the apoB gene locus may significantly affect LDL cholesterol in as many as 10 to 20% of subjects in the population at large.[145] However, haplotype analysis in 23 informative families suggested that the apoB gene locus is not the major locus influencing plasma apoB levels.[146] Common disorders associated with elevated apoB level, such as familial combined hyperlipemia (FCH) or the lipid abnormality associated with small LDL, are not linked to the apoB gene locus.[147] In addition to the LDL receptor gene locus, loci harboring the genes for lipoprotein lipase, apoA-I/apoC-III/apoA-IV, and apoE (see below) have been shown to influence apoB levels in a large segment of the population.

Elevations of apoB-100 are characteristic of several syndromes and/or diseases including hyper-apoB-emia,[148] FCH,[149] and the atherogenic lipo-

[140] G. L. Vega and S. M. Grundy, *J. Clin. Invest.* **78,** 1410 (1986).

[141] T. L. Innerarity, K. H. Weisgraber, K. S. Arnold, R. W. Mahley, R. M. Krauss, G. L. Vega, and S. M. Grundy, *Proc. Natl. Acad. Sci. U.S.A.* **84,** 6919 (1987).

[142] L. F. Soria, E. H. Ludwig, H. R. G. Clarke, G. L. Vega, S. M. Grundy, and B. J. McCarthy, *Proc. Natl. Acad. Sci. U.S.A.* **86,** 587 (1989).

[143] S. Lund-Katz, T. L. Innerarity, K. S. Arnold, L. K. Curtiss, and M. C. Phillips, *J. Biol. Chem.* **266,** 2701 (1991).

[144] S. G. Young, *Circulation* **82,** 1574 (1990).

[145] D. Gavish, E. A. Brinton, and J. L. Breslow, *Science* **244,** 72 (1989).

[146] J. Coresh, T. H. Beaty, P. O. Kwiterovich, Jr., and S. E. Antonarakis, *Am. J. Hum. Genet.* **50,** 1038 (1992).

[147] M. A. Austin, E. Wijsman, S. Guo, R. M. Krauss, J. D. Brunzell, and S. Deeb, *Genet. Epidemiol.* **8,** 287 (1991).

[148] A. D. Sniderman, S. Shapiro, D. Marpole, B. Skinner, B. Teng, and P. O. Kwiterovich, *Proc. Natl. Acad. Sci. U.S.A.* **77,** 604 (1980).

[149] J. L. Goldstein, H. G. Schrott, W. R. Hazzard, E. L. Bierman, and A. G. Motulski, *J. Clin. Invest.* **52,** 1544 (1973).

protein phenotype.[150] Hyper-apoB-emia was common in patients who suffered from premature CAD.[148] Levels of apoB in these patients were as high as in subjects with heterozygous familial hypercholesterolemia (FH), but LDL cholesterol concentrations were much lower. Familial combined hyperlipemia was identified in familes of myocardial infarction survivors.[149] Although the original studies suggested a single gene defect with autosomal dominant inheritance, more recent studies indicate heterogeneity of FCH. Heterozygosity for LPL deficiency may be present in a subset of families,[151] and a mutation in or close to the gene cluster encoding apolipoproteins A-I, C-III, and A-IV has been implicated in other families with FCH.[152] Mild elevations of plasma TG, VLDL, IDL, and small size LDL and low levels of HDL, in particular HDL_2, constitute the atherogenic lipoprotein phenotype.[150]

Complex segregation analysis in normolipidemic and hyperlipidemic kindred suggested that a single locus designated ATHS (atherosclerosis susceptibility locus) controls the size of LDL.[153] This locus was mapped to the short arm of chromosome 19 at or near the LDL receptor locus and in close proximity to the insulin receptor locus.[154] Although the atherogenic lipoprotein phenotype may be common in subjects with insulin resistance, the insulin receptor gene locus did not appear to be causative because of recombinants with the ATHS locus. The researchers postulated that sequence variations in the LDL receptor gene underlie the expression of the atherogenic lipoprotein phenotype, a hypothesis that awaits rigorous testing. Small size LDL associated with reduced levels of HDL_2, increased IDL concentrations, and increased magnitude of postprandial lipemia distinguishes individuals heterozygous for a missense mutation in the LPL gene from unaffected family members.[155] Furthermore, it is well known that hypertriglyceridemia alters the structure of LDL by a mechanism that involves TG transfer to LDL in exchange for cholesteryl ester and

[150] M. A. Austin, M. C. King, K. M. Vranizan, and R. M. Krauss, *Circulation* **82,** 495 (1990).

[151] S. P. Babirak, P.-H. Iverius, W. Y. Fujimoto, and J. D. Brunzell, *Arteriosclerosis* (*Dallas*) **9,** 326 (1989).

[152] P. Wojciechowski, M. Farral, P. Cullen, T. M. E. Wilson, J. D. Bayliss, B. Farren, B. A. Griffin, M. H. Caslake, C. J. Packard, J. Shepherd, T. Thakker, and J. Scott, *Nature* (*London*) **349,** 161 (1991).

[153] M. A. Austin, M. C. King, K. M. Vranizan, B. Newman, and R. M. Krauss, *Am. J. Hum. Genet.* **43,** 838 (1988).

[154] P. M. Nishina, J. P. Johnson, J. K. Naggert, and R. M. Krauss, *Proc. Natl. Acad. Sci. U.S.A.* **89,** 708 (1992).

[155] G. Miesenbock, B. Holzl, B. Foger, E. Brandstatter, B. Paulweber, F. Sandhofer, and J. R. Patsch, *J. Clin. Invest.* **91,** 448 (1993).

subsequent lipolysis of core TG.[156] A similar mechanism has been implicated in lowering of HDL_2 cholesterol.[157] It is therefore conceivable that several genetic defects that increase TG in either the fasting or postprandial state may be associated with the atherogenic lipoprotein phenotype.

Endocytosis of LDL via the LDL receptor does not lead to foam cell formation because cholesterol internalized with LDL downregulates the expression of LDL receptors.[158] Hence, postsecretory modifications rendering lipoproteins competent for endocytosis via a pathway distinct from the LDL receptor pathway must be invoked to explain a role of LDL in foam cell development. Modification of LDL with malondialdehyde, a product of arachidonic acid metabolism, causes foam cell formation *in vitro.*[159] Oxidation of LDL, a biologically plausible modification, also leads to foam cell formation (reviewed by Steinberg *et al.*[160]). Incubation of LDL with endothelial cells, smooth muscle cells, and macrophages in culture induces oxidation of polyunsaturated fatty acids. Lipid peroxides formed may fragment fatty acyl chains and may attach covalently to apoB or fragments thereof, thereby rendering the modified particles competent for endocytosis by the scavenger receptor and perhaps other routes of endocytosis. Immunologic studies demonstrating oxidatively modified low-density lipoproteins in atherosclerotic lesions[161] and animal experiments showing a protective effect of the antioxidant drug probucol[162] provide indirect evidence for a causative role of LDL oxidation in atherogenesis. Other modifications of LDL such as glycation also alter the interaction of LDL with cellular receptors and may play a role in atherogenesis as well.[163,164]

Apolipoproteins C

Apolipoprotein C-I is the smallest apolipoprotein. Apolipoprotein C-I can activate LCAT,[68] may inhibit the binding of β-VLDL to the LDL

[156] R. J. Deckelbaum, E. Granot, Y. Oschry, L. Rose, and S. Eisenberg, *Arteriosclerosis (Dallas)* **4,** 225 (1984).

[157] J. R. Patsch, S. Prasad, A. M. Gotto, Jr., and W. Patsch, *J. Clin. Invest.* **80,** 341 (1987).

[158] M. S. Brown and J. L. Goldstein, *Science* **232,** 34 (1986).

[159] A. M. Fogelman, I. Schechter, J. Seager, M. Hokom, J. S. Child, and P. A. Edwards, *Proc. Natl. Acad. Sci. U.S.A.* **77,** 2214 (1980).

[160] D. Steinberg, S. Parthasarathy, T. E. Carew, J. C. Khoo, and J. L. Witztum, *N. Engl. J. Med.* **320,** 915 (1989).

[161] S. Yla-Herttuala, W. Palinski, M. E. Rosenfeld, S. Parthasarathy, T. E. Carew, S. Butler, J. L. Witztum, and D. Steinberg, *J. Clin. Invest.* **84,** 1086 (1989).

[162] T. Kita, Y. Nagano, M. Yokode, K. Ishii, N. Kume, A. Ooshima, H. Yoshida, and C. Kawai, *Proc. Natl. Acad. Sci. U.S.A.* **84,** 5928 (1987).

[163] U. P. Steinbrecher and J. L. Witztum, *Diabetes* **33,** 130 (1984).

[164] J. L. Witztum, U. P. Steinbrecher, Y. A. Kesaniemi, and M. Fisher, *Proc. Natl. Acad. Sci. U.S.A.* **81,** 3204 (1984).

receptor-related protein,[165] and may inhibit the phospholipase A_2-mediated hydrolysis of phospholipids.[166]

Apolipoprotein C-II is the cofactor of lipoprotein lipase.[167] Studies with proteolytic fragments and synthetic peptides indicate that the carboxy-terminal end of apoC-II (residues 56–79) is essential for activator function.[168] This finding is consistent with *in vivo* observations. In nonfunctional apoC-$II_{Toronto}$, residues 69 through 79 are replaced by a hexapeptide through frameshift mutation.[169,170] Furthermore, sequence comparison among species reveals a high degree of conservation of the carboxy-terminal block encoded by exon 4.[7] Several mutations in the apoC-II gene have been described which result in fasting chylomicronemia due to a deficiency or defective function of the protein.[171] Even though the amount of apoC-II transcribed from the normal allele is sufficient for full activation of lipoprotein lipase, the heterozygous state for certain apoC-II mutations may nevertheless cause hypertriglyceridemia in subjects harboring a mutation in a second gene that influences TG metabolism. Such a genetic interaction has been described in carriers of the apoC-$II_{Toronto}$ allele and the ε4 allele.[172]

Apolipoprotein C-III circulates in plasma in association with TG-rich lipoproteins and HDL. Apolipoprotein C-III inhibits the activity of lipoprotein lipase,[173] decreases apoE-mediated remnant removal,[174,175] and therefore reduces the catabolism of TG-rich lipoproteins. The catabolism of TG-rich lipoproteins is accelerated in patients lacking apoC-III due to an inversion of the apoA-I and apoC-III genes,[176] and overexpression of human apoC-III in transgenic mice is associated with hypertriglyceridemia.[177] In

[165] K. H. Weisgraber, R. W. Mahley, R. C. Kowal, J. Herz, J. L. Goldstein, and M. S. Brown, *J. Biol. Chem.* **265,** 22453 (1990).
[166] J. Poensgen, *Biochim. Biophys. Acta* **1042,** 188 (1990).
[167] J. C. LaRosa, R. I. Levy, P. Herbert, S. E. Lux, and D. S. Fredrickson, *Biochem. Biophys. Res. Commun.* **41,** 57 (1970).
[168] L. C. Smith, P. K. J. Kinnunen, R. L. Jackson, A. M. Gotto, Jr., and J. T. Sparrow, *Ann. N.Y. Acad. Sci.* **348,** 213 (1980).
[169] W. C. Breckenridge, J. A. Little, G. Steiner, A. Chow, and M. Poapst, *N. Engl. J. Med.* **298,** 1265 (1978).
[170] P. W. Connelly, G. F. Maguire, T. Hofmann, and J. A. Little, *Proc. Natl. Acad. Sci. U.S.A.* **84,** 270 (1987).
[171] S. Santamarina-Fojo, *Curr. Opin. Lipidol.* **3,** 186 (1992).
[172] R. A. Hegele, W. C. Breckenridge, D. W. Cox, G. F. Maguire, J. A. Little, and P. W. Connelly, *Arterioscler. Thromb.* **11,** 1303 (1991).
[173] C. S. Wang, W. J. McConnathy, H. U. Kloer, and P. Alaupovic, *J. Clin. Invest.* **75,** 384 (1985).
[174] F. Shelburne, J. Hanks, W. Meyers, and S. Quarfordt, *J. Clin. Invest.* **65,** 652 (1980).
[175] E. Windler and R. J. Havel, *J. Lipid Res.* **26,** 556 (1985).
[176] H. N. Ginsberg, N. A. Le, I. A. Goldberg, J. C. Gibson, A. Rubinstein, P. Wang-Iverson, R. Norum, and W. V. Brown, *J. Clin. Invest.* **78,** 1287 (1986).
[177] Y. Ito, N. Azrolan, A. O'Connell, A. Walsh, and J. L. Breslow, *Science* **249,** 790 (1990).

transgenic mice, increased levels of apoC-III displace apoE from VLDL, thereby reducing removal of particles from the circulation.[178] Conceivably, increased expression of apoC-III may play a role in human hypertriglyceridemias. A number of studies have shown an association between an *Sst*I polymorphic site in the 3′ untranslated region of the apoC-III gene and hypertriglyceridemia.[179,180] Five polymorphic sites in the promoter of the apoC-III gene have been identified, which were in strong linkage disequilibrium with one another and with the variant *Sst*I site.[181] Two of the three common promoter haplotypes exhibited strong associations with the presence or absence of hypertriglyceridemia.

Apolipoprotein D

Apolipoprotein D is a member of the lipocalin superfamily of proteins which contain an eight-stranded antiparallel β basket and possess binding sites for small hydrophobic ligands. Potential physiological ligands of apoD have been reviewed.[182] Molecular modeling studies suggest that apoD may have a greater affinity for heme-related compounds than for cholesterol and cholesteryl ester.[183] Apolipoprotein D comigrates with apoA-I, LCAT, and CETP in two-dimensional nondenaturing gels,[184] but the specific function of apoD in reverse cholesterol transport awaits further definition.

Apolipoprotein E

Apolipoprotein E serves as a ligand for the LDL receptor and mediates removal of chylomicrons from the circulation.[185] It may also facilitate the CETP-mediated acquisition of cholesteryl ester by VLDL.[186] Human apoE is polymorphic. The three common isoforms resulting from allelic variation differ in amino acid residues 112 and 158. Apolipoprotein E-4 has arginine

[178] K. Aalto-Setälä, E. A. Fisher, X. Chen, T. Chajek-Shaul, T. Hayek, R. Zechner, A. Walsh, R. Ramakrishnan, H. N. Ginsberg, and J. L. Breslow, *J. Clin. Invest.* **90,** 1889 (1992).

[179] A. Rees, C. C. Shoulders, J. Stocks, D. J. Galton, and F. E. Baralle, *Lancet* **1,** 444 (1983).

[180] C. C. Shoulders, P. J. Harry, L. Lagrost, S. E. White, N. F. Shaw, J. D. North, M. Gilligan, P. Gambert, and M. J. Ball, *Atherosclerosis* **87,** 239 (1991).

[181] M. Dammerman, L. A. Sandkuul, J. L. Halaas, W. Chung, and J. L. Breslow, *Proc. Natl. Acad. Sci. U.S.A.* **90,** 4562 (1993).

[182] R. W. Milne, E. Rassart, and Y. L. Marcel, *Curr. Opin. Lipidol.* **4,** 100 (1993).

[183] M. C. Peitsch and M. S. Boguski, *New Biol.* **2,** 197 (1990).

[184] O. L. Francone, A. Gurakar, and C. J. Fielding, *J. Biol. Chem.* **264,** 7066 (1989).

[185] R. W. Mahley and S. C. Rall, Jr., *in* "The Metabolic Basis of Inherited Disease" (C. R. Scriver, A. L. Beaudet, W. S. Sly, and D. Valle, eds.), p. 1195. McGraw-Hill, New York, 1989.

[186] M. Kinoshita, H. Arai, M. Fukasawa, T. Watanabe, K. Tsukamoto, Y. Hashimoto, K. Inoue, K. Kurokawa, and T. Teramoto, *J. Lipid Res.* **34,** 261 (1993).

and apoE-2 has cysteine in both positions, whereas apoE-3, the most common isoform, has cysteine and arginine at positions 112 and 158, respectively.[185] The apoE-2 isoform exhibits less than 5% of the receptor binding activity displayed by apoE-3. *In vitro,* the receptor binding affinity of apoE-4 is similar to that of apoE-3, but *in vivo* apoE-4-containing lipoproteins are cleared more rapidly than those containing apoE-3.[187] Homozygosity for apoE-2 is the most common predisposing factor for familial type III hyperlipoproteinemia.

Although the frequencies of apoE alleles differ among populations, the effects of the apoE polymorphism on plasma lipid levels are consistent.[188] The average effect of the $\varepsilon 2$ allele is to lower plasma and LDL cholesterol by 10 to 20 mg/dl, whereas the average effect of the $\varepsilon 4$ allele is to raise plasma cholesterol by 5 to 10 mg/dl. Reduced binding affinity of the apoE-2 isoform for hepatic receptors and subsequent upregulation of LDL receptors is a likely mechanism for the effect of the $\varepsilon 2$ allele on LDL cholesterol concentrations.[189] In addition, reduced formation of LDL may contribute to the lower LDL cholesterol in subjects carrying the $\varepsilon 2$ allele, as conversion of VLDL to LDL may be impeded by the apoE-2 isoform.[190]

Consistent with the defective binding of the apoE-2 isoform, several studies have shown reduced clearance of postprandial lipoproteins from the circulation. However, the $\varepsilon 2$ allele does not predispose to CAD, but may in fact confer moderate protection.[191,192] This association, commonly attributed to the effect of the $\varepsilon 2$ allele on LDL cholesterol, would qualify the importance of remnants in atherogenesis. In a recent study in 474 subjects, the $\varepsilon 2$ allele was associated with delayed clearance of postprandial retinyl palmitate but had no effect on postprandial TG concentrations.[193] Thus, impaired terminal catabolism of remnants does not seem to be atherogenic, and other factors such as enhanced VLDL production by the liver or impaired lipolysis of TG-rich lipoproteins, irrespective of their origin, must be implicated in the atherogenicity of remnant particles. This hypothesis is supported by studies showing that increased postprandial TG levels

[187] R. E. Gregg, L. A. Zech, E. J. Schaefer, D. Stark, D. Wilson, and H. B. Brewer, Jr., *J. Clin. Invest.* **78,** 815 (1986).

[188] D. M. Hallmann, E. Boerwinkle, N. Saha, C. Sandholzer, H. J. Menzel, A. Czazar, and G. Utermann, *Am. J. Hum. Genet.* **49,** 338 (1991).

[189] M. S. Weintraub, S. Eisenberg, and J. L. Breslow, *J. Clin. Invest.* **80,** 1571 (1987).

[190] C. Ehnholm, R. W. Mahley, D. A. Chappell, K. H. Weisgraber, E. Ludwig, and J. L. Witztum, *Proc. Natl. Acad. Sci. U.S.A.* **81,** 5566 (1984).

[191] A. M. Cummings and F. W. Robertson, *Clin. Genet.* **25,** 310 (1984).

[192] J. E. Hixson and PDAY Research Group, *Arterioscler. Thromb.* **11,** 1237 (1991).

[193] E. Boerwinkle, S. A. Brown, A. R. Sharrett, G. Heiss, and W. Patsch, *Am. J. Hum. Genet.* **54,** 341 (1994).

are predictive of CAD.[62,63] Furthermore, a defect in TG metabolism is required to precipitate type III hyperlipoproteinemia in ε-2 homozygotes. In subjects with pronounced postprandial lipemia, the amount of cholesteryl esters entering the remnant pathway is increased and conceivably their atherogenic potential is enhanced. In type III hyperlipoproteinemia, the atherogenic potential of the cholesteryl ester-rich remnants may be further increased because of their prolonged residence time (due to the receptor binding deficiency of the apoE-2 isoform).

Absence of apoE due to an acceptor splice site mutation of the apoE gene[194] or apoE deficiency due to synthesis of a truncated apoE species[195] is characterized by remnant accumulation in plasma. Triglyceride levels in these disorders are lower than in the common familial type III disease. This is consistent with the concept that a second gene defect which results in overproduction or impaired catabolism of TG-rich lipoproteins is necessary for precipitation of the more common type III hyperlipoproteinemia in E2/2 subjects. Heterozygosity for some rare apoE mutants may be associated with type III hyperlipoproteinemia as well.[185] In these patients, the mutant protein may exhibit a greater affinity for apoB-containing lipoproteins than the protein resulting from the normal allele.

The role of remnants in atherogenesis as well as the importance of apoE in remnant removal has been demonstrated in animals in which plasma concentrations of apoE have been altered. Infusion of apoE into Watanabe heritable hyperlipidemic (WHHL) rabbits reduced atherosclerotic lesions despite no or only minor effects on plasma lipids.[196] Conversely, targeted disruption in both copies of the apoE gene in mice results in hyperlipemia and severe atherosclerosis, which are further aggravated by a high-fat diet.[197,198]

According to studies by Weintraub *et al.,* the ε4 allele is associated with faster clearance of postprandial retinyl palmitate than the ε3 allele.[189] A more rapid transfer of apoE-4 to chylomicrons was thought to underlie the faster removal of alimentary lipoproteins from the circulation, which results in downregulation of hepatic LDL receptors and attendant elevations of LDL cholesterol.[189] Even though these studies provided mechanistic insight,

[194] C. Cladaras, M. Hadzopoulou-Cladaras, B. K. Felber, G. Pavlakis, and V. I. Zannis, *J. Biol. Chem.* **262,** 2310 (1987).

[195] P. Lohse, H. B. Brewer III, M. S. Meng, S. I. Skarlatos, J. C. LaRosa, and H. B. Brewer, Jr., *J. Lipid Res.* **33,** 1583 (1992).

[196] N. Yamada, I. Inoue, M. Kawamura, K. Harada, Y. Watanabe, H. Shimano, T. Gotoda, M. Shimada, K. Kohzaki, T. Tsukada, M. Shiomi, Y. Watanabe, and Y. Yazaki, *J. Clin. Invest.* **89,** 706 (1992).

[197] S. H. Zhang, R. L. Reddick, J. A. Piedrahita, and N. Maeda, *Science* **258,** 468 (1992).

[198] A. S. Plump, J. D. Smith, T. Hayek, K. Aalto-Setälä, A. Walsh, J. G. Verstuyft, E. M. Rubin, and J. L. Breslow, *Cell (Cambridge, Mass.)* **71,** 343 (1992).

they were probably not representative of the effect of the ε4 allele on postprandial lipid metabolism in the population at large. When compared to the E3/3 subjects, the E4/3 and E4/4 subjects tended to have lower mean plasma TG in the study of Weintraub *et al.* Information combined across multiple studies indicates that the ε4 allele increases plasma TG levels[199] and, compared to E-3/3, subjects carrying the ε4 allele exhibit a greater postprandial response of plasma TG and retinyl palmitate when their fasting TG is also higher.[200]

The mechanism whereby the ε4 allele affects fasting plasma TG is not known. However, increased association of apoE-4 with TG-rich lipoproteins may render these particles amenable to endocytosis at an earlier stage of the delipidation cascade, thereby increasing hepatic fatty acid concentrations, a factor known to drive hepatic VLDL secretion (Fig. 1). A common mechanism, namely, hepatic uptake of lipid-rich remnants, may therefore underlie the associations of the ε4 allele with hypercholesterolemia and hypertriglyceridemia. Depending on other genetic and/or environmental factors regulating VLDL production and catabolism, including its conversion to LDL, two different phenotypes may thus be expressed in subjects carrying the ε4 allele. Associations of the ε4 allele with type V hyperlipoproteinemia[201] and hypertriglyceridemia in obesity[202] support the concept that this allele is conducive to enhanced TG secretion by the liver. Hence, not only elevated LDL cholesterol, but also impaired metabolism of TG-rich lipoproteins may contribute to the association of the ε4 allele with CAD.[203]

Minor Apolipoproteins

Other proteins associated with circulating lipoproteins include apoH, apoJ, an esterase-like activity identified as paraoxonase,[204] and serum amyloid.[205] Apolipoprotein H (2-glycoprotein I) is a negative acute-phase protein that may exert multiple inhibitory effects in the coagulation pathway including the activation of Hageman factor.[206] Apolipoprotein H may play

199 J. Dallongeville, S. Lussier-Cacan, and J. Davignon, *J. Lipid Res.* **33,** 447 (1992).

200 A. J. Brown and D. C. K. Roberts, *Arterioscler. Thromb.* **11,** 1737 (1991).

201 G. Ghiselli, E. J. Schaefer, L. A. Zech, R. E. Gregg, and H. B. Brewer, Jr., *J. Clin. Invest.* **70,** 474 (1982).

202 F. Fumeron, D. Rigaud, M. C. Bertiere, S. Bardon, C. DeLy, and M. Apfelbaum, *Clin. Genet.* **34,** 1 (1988).

203 M. S. Nieminen, K. J. Mattila, K. Aalto Setälä, T. Kuusi, K. Kontula, R. Kauppinen-Mäkelin, C. Ehnhol, M. Jauhiainen, M. Valle, and M.-R. Taskinen, *Arterioscler. Thromb.* **12,** 58 (1992).

204 M. C. Blattern, R. W. James, S. Messmer, F. Barja, and D. Pometta, *Eur. J. Biochem.* **211,** 871 (1993).

205 N. Eriksen and E. P. Benditt, *Proc. Natl. Acad. Sci. U.S.A.* **77,** 6860 (1980).

206 M. L. Henry, B. Everson, and O. D. Ratnoff, *J. Lab. Clin. Med.* **111,** 519 (1988).

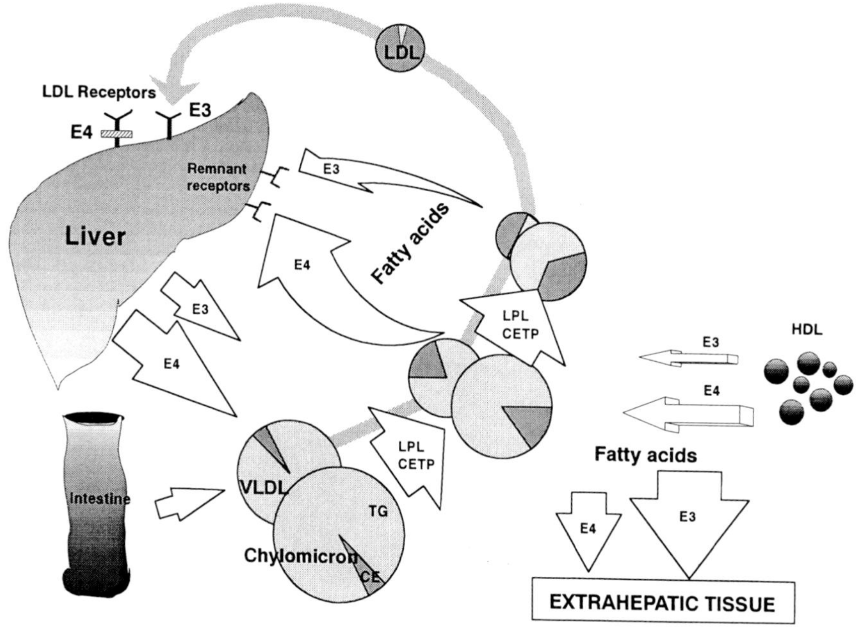

FIG. 1. Metabolic scenario that may explain why the ε4 allele increases both triglycerides and LDL cholesterol in populations at large. Owing to more rapid transfer of apoE-4 to TG-rich lipoproteins, these particles may become competent for removal during an earlier stage of delipidation. As a result, the proportion of fatty acids entering the liver and the production of VLDL are increased in subjects carrying the ε4 allele. Effective catabolism of VLDL can compensate for VLDL overproduction, and LDL cholesterol increases because LDL receptor expression may be downregulated or the proportion of VLDL converted to LDL may increase. When VLDL production is enhanced by additional factors or when catabolism is impaired, an increase in plasma triglycerides is the result.

a role in both the induction and interaction of lupus anticoagulant activity with phospholipid.[207] The urinary excretion of apoH may have clinical utility as an indicator of proximal renal tubule dysfunction.[208] Apolipoprotein J (clusterin, serum protein 40/40, NA1/NA2) inhibits complement-mediated cell lysis[209] and may therefore play a role in maintaining endothelial integ-

[207] H. P. McNeil, R. J. Simpson, C. N. Chesterman, and S. A. Krilis, *Proc. Natl. Acad. Sci. U.S.A.* **87,** 4120 (1990).

[208] M. Lapsley, P. A. Sanson, C. T. Marlow, F. V. Flynn, and A. G. Norden, *J. Clin. Pathol.* **44,** 812 (1991).

[209] B. F. Murphy, I. Kirszbaum, I. D. Walker, and A. J. F. D'Apice, *J. Clin. Invest.* **81,** 1858 (1988).

rity. Possible other functions of this protein as well as its structure and tissue expression have been reviewed.[210]

Apolipoprotein [a]

Apolipoprotein [a] belongs to a gene family that includes genes encoding clotting factors, structural proteins, and growth factors.[211] Domains shared by these proteins are protease-like domains, kringle units, calcium binding domains, and epidermal growth factor precursor domains. The majority of apo[a] is covalently linked to apoB-100 by a disulfide bridge to form the protein moiety of lipoprotein [a], which resembles LDL in its lipid composition.[212,213] Apolipoprotein [a] exists in polymorphs distinguished by molecular weights.[214,215] The molecular basis for the size variation of apo[a] is primarily due to multiple apo[a] alleles that differ in the number of kringle type 2 (plasminogen kringle type 4) repeats.[216–218] Minor variability in apo[a] size might be due to differences in glycosylation, as carbohydrates make up 25–40% of the apo[a] weight.[213,215]

A number of exogenous factors such as hormones and drugs can affect levels of Lp[a],[219,220] but genetic factors play the dominant role in determining plasma concentrations of Lp[a]. More than 90% of the variability of Lp[a] plasma levels in general populations is determined by size and sequence variation within the apo[a] gene.[221] An inverse relationship exists between plasma Lp[a] levels and apo[a] size.[214,215] Estimates of the variance of Lp[a] plasma levels explained by apo[a] size range from 19 to 70% among

[210] T. C. Jordan-Starck, D. P. Witte, B. J. Aronow, and J. A. K. Harmony, *Curr. Opin. Lipidol.* **3,** 75 (1992).

[211] J. McLean, J. Tomlinson, W. Kuang, D. Eaton, E. Chen, G. Fless, A. Scanu, and R. Lawn, *Nature (London)* **330,** 132 (1987).

[212] J. W. Gaubatz, C. Heideman, A. M. Gotto, Jr., J. D. Morrisett, and G. H. Dahlen, *J. Biol. Chem.* **258,** 4582 (1983).

[213] G. M. Fless, M. E. ZumMallen, and A. M. Scanu, *J. Biol. Chem.* **261,** 8712 (1986).

[214] G. Utermann, H. G. Kraft, H. J. Menzel, T. Hopferwieser, and C. Seitz, *J. Clin. Invest.* **80,** 458 (1987).

[215] J. W. Gaubatz, K. I. Ghanem, J. Guevara, Jr., M. L. Nava, W. Patsch, and J. D. Morrisett, *J. Lipid Res.* **31,** 603 (1990).

[216] J. E. Hixson, M. L. Britten, G. S. Manis, and D. L. Rainwater, *J. Biol. Chem.* **264,** 6013 (1989).

[217] D. Gavish, N. Azrolan, and J. L. Breslow, *J. Clin. Invest.* **84,** 2021 (1989).

[218] C. Lackner, E. Boerwinkle, C. C. Leffert, T. Rahmig, and H. H. Hobbs, *J. Clin. Invest.* **87,** 2077 (1991).

[219] A. Gurakar, J. M. Hoeg, G. M. Kostner, N. M. Papadopoulos, and H. B. Brewer, Jr., *Atherosclerosis* **57,** 293 (1985).

[220] A. A. Nabulsi, A. R. Folsom, A. White, W. Patsch, G. Heiss, K. K. Wu, and M. Szklo, *N. Engl. J. Med.* **328,** 1069 (1993).

[221] E. Boerwinkle, C. C. Leffert, J. Lin, C. Lackner, G. Chiesa, and H. H. Hobbs, *J. Clin. Invest.* **90,** 52 (1992).

different populations.[222] Thus, the sequences encoding the number of apo[a] kringle 2 units as well as additional sequence variation at the apo[a] locus are responsible for the control of plasma Lp[a] levels.

Studies by Utermann *et al.* suggested a multiplicative interaction between the LDL receptor gene locus and Lp[a] plasma levels.[223] Biochemical data on the role of the LDL receptor in removal of Lp[a] are at variance, however.[224–226] In addition, drugs known to reduce levels of LDL by increasing LDL receptor expression have no effect on Lp[a] plasma levels.[227] Furthermore, changes of plasma Lp[a] do not parallel the changes of LDL cholesterol that occur in response to diets known to alter the expression of LDL receptors.[228] Turnover studies have shown that plasma Lp[a] concentrations are determined by its production rather than its catabolism,[229] a finding that has been extended to subjects expressing the same apo[a] polymorph.[230] It is thus conceivable that the interaction between the LDL receptor gene locus and Lp[a] plasma levels reflects enhanced Lp[a] production driven by hepatic cholesterol biosynthesis. An interaction between the apoE gene locus and plasma Lp[a] levels has been reported as well.[231] The ε2 allele seems to reduce Lp[a] concentrations in plasma, whereas the ε4 allele has the opposite effect. This interaction is similar to that between the apoE polymorphism and LDL cholesterol levels and may thus be interpreted to reflect the effects of the apoE polymorphism on removal of Lp[a] by the LDL receptor. However, the ε4 allele may promote hepatic lipoprotein production (see above), whereas the opposite may apply for the ε2 allele.

Because the abundance of apo[a] mRNA in human and primate livers

[222] C. Sandholzer, D. M. Hallman, G. Sigursson, C. Lackner, A. Csaszar, E. Boerwinkle, and G. Utermann, *Hum. Genet.* **86,** 607 (1991).

[223] G. Utermann, F. Hoppichler, H. Dieplinger, M. Seed, G. Thompson, and E. Boerwinkle, *Proc. Natl. Acad. Sci. U.S.A.* **86,** 4171 (1989).

[224] K. Maartmann-Moe and K. Berg, *Clin. Genet.* **20,** 352 (1981).

[225] V. W. Armstrong, B. Harrach, H. Robenek, M. Helmhold, A. K. Walli, and D. Seidel, *J. Lipid Res.* **31,** 429 (1990).

[226] S. A. Hofmann, D. L. Eaton, M. S. Brown, W. J. McConathy, J. L. Goldstein, and R. L. Hammer, *J. Clin. Invest.* **85,** 1542 (1990).

[227] J. Thiery, V. W. Armstrong, J. Schleef, C. Creutzfeldt, W. Creutzfeldt, and D. Seidel, *Klin. Wochenschr.* **66,** 462 (1988).

[228] S. A. Brown, J. Morrisett, J. R. Patsch, R. Reeves, A. M. Gotto, Jr., and W. Patsch, *J. Lipid Res.* **32,** 1281 (1991).

[229] F. Krempler, G. M. Kostner, A. Roscher, F. Haslauer, K. Bolzano, and F. Sandhofer, *J. Clin. Invest.* **71,** 1431 (1983).

[230] D. J. Rader, W. Cain, L. A. Zech, D. Usher, and H. B. Brewer, Jr., *J. Clin. Invest.* **91,** 443 (1993).

[231] P. De Knijff, A. Kaptein, D. Boomsma, H. M. G. Princen, R. R. Frants, and L. M. Havekes, *Atherosclerosis* **90,** 169 (1991).

varies,[216,232] transcriptional events are likely to account for part of the variability in apo[a] gene expression. Indeed, the 5′ flanking regions of the apo[a] gene from two subjects with 30-fold differences in plasma Lp[a] levels exhibited 5-fold differences in ability to drive transcription of a reporter gene.[233] However, apo[a] size correlates with apo[a] mRNA size but does not correlate with mRNA abundance. Studies in baboons showed that Lp[a] may be undetectable in plasma despite the presence of apo[a] mRNA in liver.[234] Hence, translational or posttranslational events must significantly contribute to apo[a] production and plasma concentrations.

The association between elevated levels of Lp[a] and atherosclerotic disease, which has been established in numerous epidemiological studies, has been reviewed.[235] Transgenic mice expressing a human apo[a] gene develop fatty streak lesions only when fed a high-fat diet.[236] Interestingly, only 5% of the apo[a] expressed in transgenic mice was lipid associated. In humans, high-fat diets increased LDL cholesterol, thought to potentiate the atherogenicity of Lp[a].[237]

The mechanisms whereby Lp[a] accelerates atherosclerosis or precipitates clinically overt disease is incompletely understood. Both apo[a] and apoB are found in atherosclerotic lesions[238] and in vein grafts,[239] but the ratio of apo[a] to apoB in aortic tissue is higher than that in plasma.[240] The high abundance of apo[a] in arterial tissues may relate to its ability to interact with intima components such as proteoglycans, glycosaminoglycans, and fibronectin,[241,242] thereby initiating or promoting atherogenesis.

[232] D. P. Wade, B. L. Knight, K. Harders-Spengel, and A. K. Soutar, *Atherosclerosis* **91,** 63 (1991).

[233] D. P. Wade, J. G. Clarke, G. E. Lindahl, A. C. Liu, B. R. Zysow, K. Meer, K. Schwartz, and R. M. Lawn, *Proc. Natl. Acad. Sci. U.S.A.* **90,** 1369 (1993).

[234] A. L. White, D. L. Rainwater, J. E. Hixson, L. E. Estlack, and R. E. Lanford, *Chem. Phys. Lipids* **67168,** 123 (1994).

[235] J. D. Morrisett, J. R. Guyton, J. W. Gaubatz, and A. M. Gotto, Jr., *in* "Plasma Lipoproteins" (A. M. Gotto, ed.), p. 129. Elsevier, Amsterdam, 1987.

[236] R. M. Lawn, D. P. Wade, R. E. Hammer, G. Chiesa, J. G. Verstuyft, and E. M. Rubin, *Nature* (*London*) **360,** 670 (1992).

[237] V. M. Armstrong, P. Cremer, E. Eberle, A. Manke, F. Schulze, H. Wieland, H. Krezer, and D. Seidel, *Atherosclerosis* **62,** 249 (1986).

[238] R. L. Nachman, D. Gavish, N. Azrolan, and T. B. Clarkson, *Arterioscler. Thromb.* **11,** 32 (1991).

[239] G. L. Cushing, J. W. Gaubatz, M. L. Nava, B. J. Burdick, T. M. A. Bocan, J. R. Guyton, D. Weilbaecher, M. E. DeBakey, G. M. Lawrie, and J. D. Morrisett, *Arteriosclerosis* (*Dallas*) **9,** 593 (1989).

[240] E. B. Smith and S. Cochran, *Atherosclerosis* **84,** 173 (1990).

[241] M. Bihari-Varga, E. Gruber, M. Rotheneger, T. Zechner, and G. Kostner, *Arteriosclerosis* (*Dallas*) **8,** 851 (1988).

[242] C. Ehnholm, M. Jauhiainen, and J. Metso, *Eur. Heart J.* **11**(Suppl. E), 190 (1990).

Nevertheless, a causal relationship between the Lp[a] accumulation in the arterial wall and atherogenesis has not been established. Owing to the lower β-carotene content, Lp[a] may be more easily oxidized than LDL. Oxidized Lp[a] is prone to aggregation, is avidly taken up by monocyte–macrophages, and may therefore induce foam cell formation.[243,244]

An additional mechanism (or mechanisms) to explain the association of Lp[a] with prevalent CAD may relate to its thrombogenic potential. Because of structural homology with plasminogen,[211] apo[a] may compete with plasminogen in a number of ways,[245,246] which would all result in reduced fibrinolysis. An inverse association of apo[a] size with Lp[a] levels and with angiographically documented CAD has been described.[247] However, no difference in apo[a] polymorph size or apo[a] phenotypes was found between subjects with preclinical atherosclerosis and controls in the Atherosclerosis Risk in Communities (ARIC) study. Preclinical disease was defined as increased carotid intima media thickness in the absence of symptomatic coronary or cerebrovascular disease.[248] Thus, plasma Lp[a] concentration, irrespective of apo[a] size, may be implicated in atherogenesis (i.e., arterial intima media thickening), whereas small apo[a] polymorphs may increase the thrombogenic potential of Lp[a] that may be important in vasoocclusive disease.

Other Apolipoprotein Functions

In addition to playing a role in lipoprotein metabolism, several apolipoproteins may exhibit functions that are not or are only indirectly related to plasma lipid transport. Of particular interest are functions that may affect coagulation, since thrombosis is central to the pathogenesis of complications of atherosclerosis.[249] Abnormalities in TG transport are associated with factor VII activation and impaired fibrinolysis and may thus be conducive to thrombogenesis.[250,251] Impaired fibrinolysis may be due to TG-

[243] M. Naruszewicz, E. Selinger, and J. Davignon, *Metabolism* **41,** 1215 (1992).

[244] M. Naruszewicz, E. Selinger, R. Dufour, and J. Davignon, *Metabolism* **41,** 1225 (1992).

[245] L. A. Miles, G. M. Fless, E. G. Levin, A. M. Scanu, and E. F. Plow, *Nature* (*London*) **339,** 303 (1989).

[246] K. A. Hajjar, D. Gavish, J. L. Breslow, and R. L. Nachman, *Nature* (*London*) **339,** 301 (1989).

[247] M. Seed, F. Hoppichler, D. Reaveley, S. McCarthy, G. R. Thompson, E. Boerwinkle, and G. Utermann, *N. Engl. J. Med.* **322,** 1494 (1990).

[248] S. A. Brown, J. D. Morrisett, E. Boerwinkle, R. Hutchinson, and W. Patsch, *Arterioscler. Thromb.* **13,** 1558 (1993).

[249] V. Fuster, L. Badimon, J. J. Badimon, and J. H. Chesebro, *N. Engl. J. Med.* **326,** 242 (1992).

[250] G. J. Miller, J. C. Martin, K. A. Mitropoulos, B. E. Reeves, R. L. Thompson, T. W. Meade, J. A. Cooper, and J. K. Cruickshank, *Atherosclerosis* **86,** 163 (1991).

[251] K. A. Mitropoulos, G. J. Miller, B. E. Reeves, H. C. Wilkes, and J. K. Cruickshank, *Atherosclerosis* **76,** 203 (1989).

induced increases of plasminogen activator inhibitor-1.[252] An association of the apoA-IV gene locus with Lp[a] plasma levels has been demonstrated.[253] Subjects heterozygous for the apoA-IV-2 allele exhibited not only lower Lp[a] levels, but also higher plasma concentrations of the fibrin depredation product D-dimer when compared to subjects homozygous for the apoA-IV-1 allele. A study in Japanese subjects showed that apoA-I binds and stabilizes prostacyclin, which would be expected to reduce vasoconstriction and thrombus formation.[254] However, these results were not confirmed in another population.[255] Apolipoprotein A-II may inhibit the association of coagulation factor III with factor VIIa,[256] and the role of Lp[a] and apoH has been mentioned above.

Apolipoprotein E may facilitate repair of traumatized nerves and may exhibit immunoregulatory properties.[257] A reported association between the ε4 allele and late-onset familial Alzheimer disease is of potential importance.[258]

252 A. Hamsten, B. Siman, U. De Faire, and M. Blomback, *N. Engl. J. Med.* **313,** 1557 (1985).
253 A. von Eckardstein, J. Heinrich, H. Funke, H. Schulte, R. Schonfeld, E. Kohler, A. Steinmetz, and G. Assmann, *Arterioscler. Thromb.* **13,** 240 (1993).
254 Y. Yui, T. Aoyama, H. Morishita, M. Takahashi, Y. Takatsu, and C. Kawai, *J. Clin. Invest.* **82,** 803 (1988).
255 A. L. Tsai, M. J. Hsu, W. Patsch, and K. K. Wu, *Biochim. Biophys. Acta* **1115,** 131 (1991).
256 S. D. Carson, *J. Biol. Chem.* **262,** 718 (1987).
257 R. W. Mahley, *Science* **240,** 622 (1988).
258 W. J. Strittmatter, A. M. Saunders, D. Schmechel, M. Pericak-Vance, J. Enghild, G. S. Salvesen, and A. D. Roses, *Proc. Natl. Acad. Sci. U.S.A.* **90,** 1977 (1993).

[2] Significance of Apolipoproteins for Structure, Function, and Classification of Plasma Lipoproteins

By PETAR ALAUPOVIC

Introduction

Historically, the present view of the chemical nature and physiological function of plasma lipoproteins evolved through three developmental phases coinciding with and shaped by the availability of electrophoretic, ultracentrifugal, and immunologic methodologies. The purpose of this chapter is to describe these events briefly and to emphasize the growing recognition of apolipoproteins as protein constituents essential for the structural integrity, functional specificity, and classification of plasma lipoproteins.

Soluble plasma lipoproteins are recognized as a unique class of conjugated proteins. They form a polydisperse system of macromolecular complexes consisting of neutral lipids and phospholipids bound through noncovalent linkages to specific proteins called apolipoproteins. Exploration of the chemical nature of plasma lipoproteins characterized, among others, by the capacity to maintain water-insoluble lipids in a soluble form began almost a century ago with the pioneering studies of Schulz,[1] Nerking,[2] and Shimidzu[3] who observed that complete extraction of plasma lipids by diethyl ether could only be achieved after prior proteolytic degradation of plasma proteins. During subsequent years, several investigators provided further evidence that plasma lipids are associated with proteins by detecting both the neutral lipids and phospholipids in globulin and albumin fractions isolated from serum or plasma by fractional precipitation with concentrated salt solutions or dialysis against water.[4-9]

The notion that serum lipids are chemically bound to proteins was fully substantiated in the late 1920s by Macheboeuf, who first isolated from horse serum a lipoprotein of constant composition.[10,11] In these experiments, Macheboeuf selected as a starting material the albumin fraction that remained in solution after precipitation of globulin by half-saturation of serum with ammonium sulfate. Subsequent acidification of the soluble albumin fraction to pH 3.8 resulted in precipitation of a lipid–protein complex referred to as "acid-precipitable cenapse" with physical and chemical properties characteristic of a high-density lipoprotein. The lipid–protein complex, soluble at neutral or alkaline pH, contained 59.1% protein, 22.7% phospholipids, and 17.9% cholesterol esters. Remarkably, its chemical composition remained constant after nine consecutive dissolutions in distilled water and reprecipitations at pH 3.8. Subsequent studies showed that the lipoprotein exhibited a single sedimenting boundary on ultracentrifugation and migrated as a single band with α-globulin mobility in free electrophoresis.[12] Another important discovery resulting from these studies was the finding that the protein moiety of the lipoprotein differed from al-

[1] F. N. Schulz, *Pfluegers Arch. Ges. Physiol.* **65,** 299 (1897).
[2] J. Nerking, *Pfluegers Arch. Ges. Physiol.* **85,** 330 (1901).
[3] I. Shimidzu, *Biochem. Z.* **28,** 237 (1910).
[4] W. B. Hardy, *J. Physiol.* (*London*) **33,** 251 (1905).
[5] H. C. Haslam, *Biochem. J.* **7,** 492 (1913).
[6] H. Chick, *Biochem. J.* **8,** 404 (1914).
[7] K. Frankenthal, *Z. Immunitaesforsch.* **42,** 501 (1925).
[8] A. H. T. Theorell, *Biochem. Z.* **175,** 297 (1926).
[9] J. A. Gardner and H. Gainsborough, *Biochem. J.* **21,** 141 (1927).
[10] M. Macheboeuf, *Bull. Soc. Chim. Biol.* **11,** 268 (1929).
[11] M. Macheboeuf, *Bull. Soc. Chim. Biol.* **11,** 485 (1929).
[12] M. Macheboeuf and P. Rebeyrotte, *Discuss. Faraday Soc.* **6,** 62 (1949).

bumin and any other known globulin. These accomplishments established the constancy of composition and specificity of lipid-binding proteins as the major chemical criteria for defining lipid–protein interactions and provided the basis for further studies of their physical and chemical properties.

Electrophoretic Behavior and Classification of Plasma Lipoproteins

During the 1930s and 1940s, evidence gradually accumulated which indicated that all serum lipids are combined with specific proteins into two major lipoprotein classes differing in physical and chemical properties. Tiselius and co-workers demonstrated that the major part of serum lipids migrated in the electric field with α_1- and β-globulin mobilities.[13] This important finding confirming the existence of lipid–protein associations and indicating the occurrence of, at least, two specific lipid-binding proteins was fully substantiated by Cohn and associates, who first developed a reproducible, preparative procedure for the isolation of serum lipoproteins.[14] In the procedure, fractionation of human plasma in ethanol–water mixtures at low ionic strength and low temperatures resulted in the precipitation of two lipid–protein complexes differing in physical properties and lipid composition.[15] Owing to the characteristic electrophoretic mobilities, these two lipoproteins were named α_1- and β-lipoproteins. The α_1-lipoproteins were characterized by lower lipid and higher protein contents and by a lower molecular weight than β-lipoproteins. The physical properties and chemical composition of α_1-lipoproteins were similar to those of "acid-precipitable cenapses" isolated from horse serum by Macheboeuf and co-workers.[10,11] Both lipoproteins were considered to have undisclosed functions in the transport and exchange of lipids during their passage between the blood and tissues.

The recognition of α_1- and β-lipoproteins as two distinct lipoprotein classes provided the first classification of plasma lipoproteins on the basis of electrophoretic behavior as the main distinguishing criterion applicable to human as well as animal lipoproteins. However, the relatively wide boundaries and variations in electrophoretic mobilities displayed in free electrophoresis by α_1- and β-lipoproteins suggested a further compositional complexity of the two lipoprotein classes.[16] Application of zonal electrophoresis to analysis of serum lipoproteins revealed lipid-stained bands not only

[13] G. Blix, A. Tiselius, and H. Svensson, *J. Biol. Chem.* **137,** 485 (1941).

[14] E. J. Cohn, L. E. Strong, W. L. Hughes, Jr., D. J. Mulford, J. N. Ashworth, M. Melin, and H. L. Taylor, *J. Am. Chem. Soc.* **63,** 459 (1946).

[15] F. R. N. Gurd, J. L. Oncley, J. T. Edsall, and E. J. Cohn, *Discuss. Faraday Soc.* **6,** 70 (1949).

[16] G. Sandor, *in* "Serum Proteins in Health and Disease" (G. Sandor, ed.), p. 71. Williams & Wilkins, Baltimore, Maryland, 1966.

in the α_1- and β-globulin regions but also at the origin and in the α_2- (pre-β), prealbumin, and γ-positions substantiating the already suspected heterogeneity of electrophoretically separated lipoproteins.[17-20]

The major achievements of the electrophoretic era were the confirmation of the lipoprotein hypothesis and the discovery of two major classes of lipoproteins coinciding in electrophoretic mobilities with those of α_1- and β-globulins. However, the simple and elegant notion of two lipoprotein classes with distinct protein moieties and distinct chemical, physical, and, possibly, physiological properties was shattered in the ensuing developmental phase dominated by the introduction of ultracentrifugal methodology.

Ultracentrifugal Behavior and Classification of Plasma Lipoproteins

The solubility properties and electrophoretic behavior of plasma lipoproteins are similar to those of simple proteins. However, owing to the presence of lipids, lipoproteins have relatively low density in comparison with simple proteins. In the 1950s, this unique physical property of plasma lipoproteins had been used successfully for developing alternative methods for preparative isolation and characterization of lipoproteins. While working in Svedberg's laboratory, Pedersen discovered that the so-called x-protein, a density-dependent macromolecular component appearing in the analytical ultracentrifuge as a separate peak between those of albumin and globulin,[21] could be isolated by ultracentrifugal flotation in a medium of high solvent density as a lipid–protein complex of relatively low density and β-globulin electrophoretic mobility.[22] This finding marked the beginning of a new developmental phase in the analysis and preparative isolation of plasma lipoproteins.

In the late 1940s and early 1950s, Gofman and co-workers observed that plasma lipoproteins represent a wide spectrum of particle sizes and hydrated densities characterized by regions of maximal and minimal concentrations along a density gradient ranging from 0.92 to 1.21 g/ml.[23-26] By

[17] B. Swahn, *Scand. J. Clin. Invest.* **4,** 98 (1952).
[18] H. G. Kunkel and R. J. Slater, *J. Clin. Invest.* **31,** 677 (1952).
[19] W. G. Dangerfield and E. B. Smith, *J. Clin. Pathol.* **8,** 132 (1955).
[20] G. B. Phillips, *Proc. Soc. Exp. Biol. Med.* **100,** 19 (1959).
[21] A. S. McFarlane, *Biochem. J.* **29,** 660 (1935).
[22] K. O. Pedersen, *J. Phys. Colloid Chem.* **51,** 156 (1947).
[23] J. W. Gofman, F. T. Lindgren, and H. Elliott, *J. Biol. Chem.* **179,** 973 (1949).
[24] F. T. Lindgren, H. A. Elliot, and J. W. Gofman, *J. Phys. Colloid Chem.* **55,** 80 (1951).
[25] J. W. Gofman, O. DeLalla, F. Glazier, N. K. Freeman, F. T. Lindgren, A. V. Nichols, B. Strisower, and A. R. Tamplin, *Plasma* **2,** 413 (1954).
[26] O. F. DeLalla and J. W. Gofman, *Methods Biochem. Anal.* **1,** 459 (1954).

utilizing regions of minimal concentrations as demarcation boundaries, lipoproteins were separated by fractional ultracentrifugal flotation in salt solutions of successively increased densities into four major classes designated as chylomicrons ($d < 0.94$ g/ml), very-low-density lipoproteins (VLDL, $d = 0.94$–1.006 g/ml), low-density lipoproteins (LDL, $d = 1.006$–1.063 g/ml), and high-density lipoproteins (HDL, $d = 1.063$–1.21 g/ml).[24–27] Distribution studies showed a marked heterogeneity of each major lipoprotein density class with respect to the particle size and hydrated density.[28,29] The relatively high degree of this heterogeneity necessitated a further subdivision of VLDL,[28,30,31] LDL,[32–34] and HDL[25,27,33,35] into several arbitrarily selected density subclasses.

Application of analytical and preparative ultracentrifugation to studies of plasma lipoproteins has been of great practical and theoretical importance for further developments in the field. Preparative ultracentrifugation in its various modifications[28,35–39] became the most convenient procedure for the isolation of a variety of lipoprotein density classes, and hydrated density replaced electrophoretic mobility as the main criterion for characterizing and classifying plasma lipoproteins. Compositional studies showed that changing proportions of neutral lipids and phospholipids and changing lipid/protein ratios of individual lipoprotein particles were the main contributors to the density and size heterogeneity of lipoprotein density classes.[27,29,30,33] The results of immunochemical characterization of ultracentrifugally separated lipoprotein density classes confirmed the electrophoretically established occurrence of at least two distinct protein moieties, one of which was found to be a characteristic constituent of α_1-lipoproteins or HDL and the other of β-lipoproteins

[27] F. T. Lindgren and A. V. Nichols, *in* "Plasma Proteins" (F. W. Putman, ed.), Vol. 2, p. 1. Academic Press, New York, 1960.
[28] F. T. Lindgren, A. V. Nichols, F. T. Upham, and R. D. Wills, *J. Phys. Chem.* **66,** 2007 (1962).
[29] A. M. Ewing, N. K. Freeman, and F. T. Lindgren, *Adv. Lipid Res.* **3,** 25 (1965).
[30] A. Gustafson, P. Alaupovic, and R. H. Furman, *Biochemistry* **4,** 596 (1965).
[31] S. H. Quarford, A. Nathans, M. Dowdee, and H. L. Hilderman, *J. Lipid Res.* **13,** 435 (1972).
[32] J. L. Oncley, K. W. Walton, and D. G. Cornwell, *J. Am. Chem. Soc.* **79,** 4666 (1957).
[33] A. V. Nichols, *Adv. Biol. Med. Phys.* **11,** 109 (1967).
[34] D. M. Lee and P. Alaupovic, *Biochemistry* **9,** 2244 (1970).
[35] R. J. Havel, H. A. Eder, and J. H. Bragdon, *J. Clin. Invest.* **34,** 1345 (1955).
[36] F. T. Lindgren, L. C. Jensen, and F. T. Hatch, *in* "Blood Lipids and Lipoproteins: Quantification, Composition and Metabolism" (G. J. Nelson, ed.), p. 181. Wiley(Interscience), New York, 1972.
[37] H. G. Wilcox, D. C. Davis, and M. Heimberg, *J. Lipid Res.* **12,** 160 (1971).
[38] B. H. Chung, T. Wilkinson, J. C. Geer, and J. P. Segrest, *J. Lipid Res.* **21,** 284 (1980).
[39] D. M. Lee and D. Downs, *J. Lipid Res.* **23,** 14 (1982).

or LDL[40–43]; the former protein was referred to as α_1-protein and the latter as β-protein. Thus, in the late 1950s, the soluble plasma lipoproteins were viewed as macromolecular complexes consisting of neutral lipids, phospholipids, and at least two distinct lipid-binding proteins linked through noncovalent interactions to form a dynamic system of particle populations heterogeneous with respect to size, hydrated density, electric charge, and lipid–protein composition.

Despite structural and compositional heterogeneity, major lipoprotein density classes have been tacitly accepted as the basic chemical and metabolic entities of the lipid transport system. Some of the major reasons for the acceptance of this conceptual view included the emphasis on lipids as important chemical determinants of the macromolecular distributions of lipoproteins[28,33] and potential injurious agents in the development of atherosclerosis,[44–49] the availability of relatively simple ultracentrifugal procedures for the preparative isolation of lipoprotein density classes,[26,35] and the already developed methodology for the measurement of plasma lipids including cholesterol,[50] phospholipids,[51] and triglycerides.[52] This view of plasma lipoproteins was further strengthened by the disclosure of a metabolic relationship between major lipoprotein density classes and subclasses[25,33,53] and their clinical usefulness in characterizing hyperlipoproteinemias.[25,33,44–46,54]

The utilization of hydrated density as the main criterion for classifying lipoproteins expanded the number of discernible lipoprotein classes and emphasized lipid rather than protein constituents as the primary determinants of the newly recognized polydispersity of the plasma lipoprotein

[40] H. G. Kunkel, *Fed. Proc.* **9,** 193 (1950).
[41] L. Levine, D. L. Kauffman, and R. K. Brown, *J. Exp. Med.* **102,** 105 (1955).
[42] L. DeLalla, L. Levine, and R. K. Brown, *J. Exp. Med.* **106,** 261 (1957).
[43] A. Scanu, L. A. Lewis, and I. H. Page, *J. Exp. Med.* **108,** 185 (1958).
[44] J. W. Gofman, H. B. Jones, F. T. Lindgren, T. P. Lyon, H. A. Elliott, and B. Strisower, *Circulation* **2,** 161 (1950).
[45] H. B. Jones, J. W. Gofman, F. T. Lindgren, T. P. Lyon, D. M. Graham, B. Strisower, and A. V. Nichols, *Am. J. Med.* **11,** 358 (1951).
[46] E. M. Russ, H. A. Eder, and D. P. Barr, *Am. J. Med.* **11,** 468 (1951).
[47] D. P. Barr, E. M. Russ, and H. A. Eder, *Am. J. Med.* **11,** 480 (1951).
[48] L. A. Lewis and I. H. Page, *Circulation* **7,** 707 (1953).
[49] H. A. Eder, *Am. J. Med.* **23,** 269 (1957).
[50] W. M. Sperry and M. Webb, *J. Biol. Chem.* **187,** 97 (1950).
[51] C. H. Fiske and I. Subbarow, *J. Biol. Chem.* **66,** 375 (1925).
[52] E. Van Handel and D. B. Zilversmit, *J. Lab. Clin. Med.* **50,** 152 (1957).
[53] D. Gitlin, D. G. Cornwell, D. Nakasoto, J. L. Oncley, W. L. Hughes, Jr., and C. H. Janeway, *J. Clin. Invest.* **37,** 172 (1958).
[54] J. W. Gofman, L. Rubin, J. P. McGinley, and H. B. Jones, *Am. J. Med.* **17,** 514 (1954).

system. However, in the late 1950s, the paucity of data concerning the nature and composition of protein moieties and the recognition of their importance for structural stability of lipoproteins emphasized the need for better understanding of this somewhat neglected aspect of lipoprotein chemistry. The ensuing discovery of an unsuspected number of apolipoproteins added another dimension to the complexity of plasma lipoproteins and necessitated the introduction of immunologic techniques for unraveling their relationship to polydisperse lipoprotein density classes.

Isolation of Apolipoproteins and Their Structural and Functional Properties

Studies on Apolipoproteins A, B, and C

Studies on the immunologic properties[40–43] and N-terminal amino acids[55,56] of major lipoprotein density classes confirmed the electrophoretic evidence for the occurrence of two distinct apolipoproteins, one of which was found to be characteristic of HDL and the other of LDL. Furthermore, it was considered that these two apolipoproteins also constituted the protein moieties of chylomicrons and VLDL. However, these and other studies[57–60] also suggested that at least the lower density lipoproteins, including chylomicrons and VLDL, may contain additional protein constituents. Shore identified in VLDL three N-terminal amino acids, namely, glutamic acid, thought to be the N terminus of β-protein, and serine and threonine.[56] In a study on the protein moiety of human chylomicrons, Rodbell and Fredrickson described the "fingerprints" of α_1- and β-protein referred to as A and B and a third "fingerprint" of an unknown protein referred to as C.[61]

In the early 1960s, our approach to the study of protein heterogeneity of chylomicrons and VLDL was based on the assumption that chylomicron and VLDL subfractions might contain different apolipoproteins.[62] However, the N-terminal amino acid analyses of five chylomicron and VLDL subfractions showed that each subfraction was characterized by the same N-terminal amino acid pattern as the entire density range (d < 1.006

[55] J. Avigan, R. Redfield, and D. Steinberg, *Biochim. Biophys. Acta* **20,** 557 (1956).
[56] B. Shore, *Arch. Biochem.* **71,** 1 (1957).
[57] F. Aladjem, M. Lieberman, and J. W. Gofman, *J. Exp. Med.* **105,** 49 (1957).
[58] W. W. Briner, J. W. Riddle, and D. G. Cornwell, *J. Exp. Med.* **110,** 113 (1959).
[59] D. Gitlin, *Science* **117,** 591 (1953).
[60] M. Rodbell, *Science* **127,** 701 (1958).
[61] M. Rodbell and D. S. Fredrickson, *J. Biol. Chem.* **234,** 562 (1959).
[62] P. Alaupovic, *Prog. Biochem. Pharmacol.* **4,** 91 (1968).

g/ml) used as the starting material for the fractionation. Failure to detect density subfractions characterized by distinct apolipoprotein composition suggested that the identification and separation of apolipoproteins ought to be attempted with delipidized preparations of intact chylomicrons and VLDL. Because the available delipidization procedures based on the extraction of lipoproteins with polar solvents resulted in water-insoluble proteins or poor yields of lipid–protein residues, Gustafson developed a mild delipidization procedure utilizing *n*-heptane as a nonpolar solvent.[63] The resulting phospholipid–protein residues of chylomicrons and VLDL were extracted by aqueous buffers and separated by Pevikon zone electrophoresis and ultracentrifugation into three fractions which differed in physical, chemical, and immunologic characteristics.[64,65] Two of the phospholipid–protein residues were found to correspond to α_1-protein and β-protein, respectively, whereas the third phospholipid–protein residue contained a protein characterized by threonine and serine as N-terminal amino acids and a peptide pattern different from those of α_1-protein and β-protein. In accordance with the suggestion by Oncley that the protein moiety of HDL or α_1-protein be referred to as apolipoprotein A (apoA) and that of LDL or β-protein as apolipoprotein B (apoB), we proposed that the newly recognized protein of VLDL be called apolipoprotein C (apoC).

In the meantime, the biochemical characterization of two inherited hypolipoproteinemias, namely, abetalipoproteinemia[66] and Tangier disease,[67] as apolipoprotein deficiency diseases provided the most impressive evidence for the essential role of apolipoproteins in the formation of lipoprotein density classes. In abetalipoproteinemia, the lack of apoB results in the near absence of all apoB-containing lipoprotein density classes (chylomicrons, VLDL, and LDL), whereas in Tangier disease the severe deficiency of apoA results in an equally severe deficiency of HDL.

The detection of threonine and serine as the N-terminal amino acids of apoC suggested a protein consisting of either two nonidentical polypeptides or two different proteins.[68] However, Brown *et al.*[69,70] resolved this dilemma by demonstrating that apoC was composed of three nonidentical

[63] A. Gustafson, *J. Lipid Res.* **6,** 512 (1965).

[64] A. Gustafson, P. Alaupovic, and R. H. Furman, *Biochim. Biophys. Acta* **84,** 767 (1964).

[65] A. Gustafson, P. Alaupovic, and R. H. Furman, *Biochemistry* **5,** 632 (1966).

[66] H. B. Salt, O. H. Wolff, J. K. Lloyd, A. S. Forsbrooke, and A. H. Cameron, *Lancet* **2,** 325 (1960).

[67] D. S. Fredrickson, P. H. Altrocchi, L. V. Avioli, and D. S. Goodman, *Ann. Intern. Med.* **55,** 1016 (1961).

[68] W. J. McConathy and P. Alaupovic, *Circulation* **40** (Suppl. 3), 17 (1969).

[69] W. V. Brown, R. I. Levy, and D. S. Fredrickson, *J. Biol. Chem.* **244,** 5687 (1969).

[70] W. V. Brown, R. I. Levy, and D. S. Fredrickson, *J. Biol. Chem.* **245,** 6588 (1970).

polypeptides initially named according to their C-terminal amino acids as apoLP-Ser, apoLP-Glu, and apoLP-Ala. The apoLP-Ser and apoLP-Glu polypeptides were characterized by threonine and apoLP-Ala by serine as the N-terminal amino acid. Furthermore, apoLP-Ala was shown to exist in three isomorphic forms differing with respect to the number of terminal neuraminic acid residues.[70]

Although it was initially thought that α_1-protein or apoA represents a single protein, Shore and Shore established that apoA consists of two nonidentical polypeptides named according to the C-terminal amino acids apoLP-Thr and apoLP-Gln.[71,72] This important discovery also provided an explanation for the initial failure to recognize the apolipoprotein heterogeneity of HDL. The apoLP-Gln was found to have a blocked N terminus, which prevented its recognition on the basis on N-terminal amino acid analysis. The apoLP-Thr characterized by aspartic acid as the N-terminal amino acid was subsequently shown to contain glutamine as the C-terminal amino acid and was renamed apoLP-Gln.[73] In contrast to the protein heterogeneity of apoA, there was no evidence for polypeptide heterogeneity of apoB, characterized by glutamic acid as the single N-terminal amino acid.[55,56,65]

ABC Nomenclature

The discovery of apoA and apoC polypeptides necessitated the introduction and acceptance of a nomenclature that adequately and unambiguously denoted the already known apolipoproteins as well as those which might be discovered in the future. To this end, we have introduced the so-called ABC nomenclature in which apolipoproteins are designated by capital letters, their constitutive polypeptides by Roman numerals, and the polymorphic forms of either apolipoproteins or polypeptides by Arabic numbers.[74,75] We also suggested that any plasma protein considered as a potential apolipoprotein should conform to certain criteria before being recognized as an integral constituent of lipid transport and incorporated into this system of nomenclature.[75] Despite some theoretical objections, the proponents of the C-terminal amino acid nomenclature accepted the simplicity and practicality of the ABC nomenclature and urged its adoption

[71] B. Shore and V. G. Shore, *Biochemistry* **7,** 2773 (1968).
[72] B. Shore and V. G. Shore, *Biochemistry* **8,** 4510 (1969).
[73] G. Kostner and P. Alaupovic, *FEBS Lett.* **15,** 320 (1971).
[74] P. Alaupovic, *Protides Biol. Fluids Proc. Colloq.* **19,** 9 (1972).
[75] P. Alaupovic, *Prog. Lipid Res.* **30,** 105 (1991).

in 1973.[76] In designing this nomenclature we assumed that some apolipoproteins may consist of nonidentical polypeptides or subunits and, thus, introduced Roman numerals for their designation. For example, because of the frequently observed clustering of apoA and apoC polypeptides, they were considered to be the nonidentical polypeptides of apoA and apoC, respectively. Therefore, the two apoA polypeptides were called apoA-I (apoLP-GlnI) and apoA-II (apoLP-GlnII), whereas those of apoC were called apoC-I (apoLP-Ser), apoC-II (apoLP-Glu), and apoC-III (apoLP-Ala). The isomorphic forms of apoC-III were named apoC-III-0, apoC-III-1, and apoC-III-2.

Discovery of Minor Apolipoproteins

The continuing search for additional apolipoproteins led to the discovery of a minor apolipoprotein first detected in very-high-density lipoproteins (VHDL, $d > 1.21$ g/ml)[77] and LDL,[34] which was referred to as "thin-line" polypeptide. This apolipoprotein, occurring mainly in HDL,[78,79] was named apoD. Kostner described the same protein and called it apoA-III because it was frequently detected in association with apoA-I and apoA-II.[80] Although there were some discrepancies in the amino acid composition of apoD and apoA-III, Albers *et al.* confirmed the originally reported amino acid composition,[81] and this apolipoprotein remained recognized as apoD.

The next apolipoprotein isolated from triglyceride-rich lipoproteins was independently discovered by three groups of investigators.[82–84] Because of its relatively high content of arginine, this apolipoprotein was initially called the arginine-rich polypeptide. In accordance with the ABC nomenclature it is now called apoE. Olofsson *et al.* isolated from HDL a minor apolipoprotein characterized by a relatively low isoelectric point (pI = 3.7).[85] This acidic apolipoprotein, frequently found in association with apoA-I and apoA-II,[86] was designated apoF. Another minor apolipoprotein differing

[76] D. S. Fredrickson, *Circulation* **87** (Suppl. 3), 1 (1993).
[77] P. Alaupovic, S. S. Sanbar, R. H. Furman, M. L. Sullivan, and S. L. Walraven, *Biochemistry* **5,** 4044 (1966).
[78] W. J. McConathy and P. Alaupovic, *FEBS Lett.* **37,** 178 (1973).
[79] W. J. McConathy and P. Alaupovic, *Biochemistry* **15,** 515 (1976).
[80] G. Kostner, *Biochim. Biophys. Acta* **336,** 383 (1974).
[81] J. J. Albers, M. C. Cheung, S. L. Ewens, and J. H. Tollefson, *Atherosclerosis* **39,** 395 (1981).
[82] F. A. Shelburne and S. H. Quarfordt, *J. Biol. Chem.* **249,** 1428 (1974).
[83] G. Utermann, *Hoppe-Seylers Z. Physiol. Chem.* **356,** 1113 (1975).
[84] V. G. Shore and B. Shore, *Biochemistry* **12,** 502 (1973).
[85] S. O. Olofsson, W. J. McConathy, and P. Alaupovic, *Biochemistry* **17,** 1032 (1978).
[86] E. Koren, W. J. McConathy, and P. Alaupovic, *Biochemistry* **21,** 5347 (1982).

in chemical, electrophoretic, and immunologic properties from other apolipoproteins was identified in HDL and VHDL; this glucosamine-rich apolipoprotein was called apoG.[87]

A minor apolipoprotein, first identified in rat HDL,[88] was named apoA-IV presumably on the assumption that all apolipoproteins discovered in this density class ought to be considered as nonidentical polypeptides of apoA. Although this designation was incorrect according to the rules of ABC nomenclature,[74,75] it was retained as originally proposed.[88] In humans, a homologous apolipoprotein was first identified in triglyceride-rich lipoproteins of intestinal origin[89] and only later in HDL.[90] However, more than 90% of apoA-IV was found to occur in its lipid-free form.[89] In 1961, Schultze *et al.*[91] described a plasma glycoprotein termed β_2-glycoprotein I that was later proposed to play a role in VLDL metabolism.[92] Polz and Kostner demonstrated that, indeed, β_2-glycoprotein I was an integral protein constituent of human VLDL and demonstrated that its distribution among lipoprotein density classes was similar to that of apoA-IV[93]; the major part (65–75%) of β_2-glycoprotein I was present in the lipid-free form, whereas 7–9% occurred in chylomicrons and VLDL, 1–2% in LDL, and 16–18% in HDL. Because of its association with chylomicrons and other lipoproteins, β_2-glycoprotein I was thought to satisfy the criteria to be classified as an apolipoprotein and was named apoH.

The HDL of patients affected with severe trauma and a variety of other pathological conditions were found to contain the serum amyloid protein A (SAA).[94] Similar, if not identical, polypeptides were also detected in HDL of patients treated with antibiotics (the so-called threonine-poor polypeptides)[95] or parenteral glucose ("glucose-induced" or S-peptides).[96] Although the SAA proteins were found to occur in negligible concentrations in plasma lipoproteins of asymptomatic subjects, there were several arguments favoring their consideration as integral protein constituents of the lipid transport system[95] including their increased levels in pathological conditions, presence in major lipoprotein density classes, capacity to displace apoA-I on incubation with HDL, and formation of lipoprotein particles in

[87] M. Ayrault-Jarrier, J. F. Alix, and J. Polonovski, *Biochimie* **60,** 65 (1978).
[88] J. B. Swaney, H. Reese, and H. A. Eder, *Biochem. Biophys. Res. Commun.* **59,** 513 (1974).
[89] G. Utermann and U. Beisiegel, *Eur. J. Biochem.* **99,** 333 (1979).
[90] C. L. Bisgaier, O. P. Sachdev, L. Megna, and R. M. Glickman, *J. Lipid Res.* **26,** 11 (1985).
[91] H. E. Schultze, K. Heid, and H. Haupt, *Naturwissenschaften* **148,** 719 (1961).
[92] M. Burstein and P. Legmann, *Protides Biol. Fluids Proc. Colloq.* **25,** 407 (1978).
[93] E. Polz and G. M. Kostner, *FEBS Lett.* **102,** 183 (1979).
[94] G. Husby and J. B. Natvig, *J. Clin. Invest.* **53,** 1054 (1974).
[95] V. G. Shore, B. Shore, and S. B. Lewis, *Biochemistry* **17,** 2174 (1978).
[96] C. L. Malmendier, J. Christophe, and J. P. Aweryckx, *Clin. Chim. Acta* **99,** 167 (1979).

the presence or absence of apoA-I. On the basis of these arguments, the isomorphic group of SAA proteins is now referred to as apoI. The most recently discovered member of the growing apolipoprotein family was also found to be present in its lipoprotein form in HDL and VHDL.[97] This protein, named apoJ, is a glycoprotein consisting of two disulfide-linked subunits representing the posttranslationally cleaved amino acid carboxyl-terminal halves of a single-chain precursor protein; the two subunits are designated apoJα and apoJβ or, according to the ABC nomenclature, apoJ-I and apoJ-II. The apolipoprotein nature of apoJ was established by the isolation from HDL of its corresponding lipoprotein species by immunoaffinity chromatography on an immunosorber with antibodies to apoJ. APoJ is identical to the previously discovered human plasma complement lysis inhibitor or serum protein 40,40 (SP-40,40), a regulator of complement function.[98]

Lipoprotein(a) [Lp(a)] first described as a genetic variant of LDL[99] represents a unique family of lipoprotein particles formed through the interaction of LDL and a highly glycosylated, hydrophilic protein called apolipoprotein(a) or apo(a).[100,101] The protein moiety of Lp(a) consists of apoB covalently bound through a disulfide linkage to apo(a).[101] The structural, metabolic, and antigenic properties of Lp(a) are mainly due to apo(a), which was shown to possess a high degree of homology to plasminogen and, thus, to belong structurally to a superfamily of proteins including the proteases of fibrinolytic and coagulation systems.[101,102] Increased levels of Lp(a) are strongly associated with premature coronary artery disease.[100] It has been speculated that the pathophysiological potential of Lp(a) might be due both to its atherogenic component exemplified by apoB or LDL-like characteristics and the putative thrombogenic character represented by plasminogen-like apo(a).[101,102] Although structurally an intriguing protein,[101] apo(a) does not appear to form its own family of lipoprotein particles and has no specific function in lipid transport processes. For this reason, it is considered that apo(a) may not fulfill the criteria of an apolipoprotein at the present time.[101]

[97] H. V. DeSilva, W. D. Stuart, C. R. Duvic, J. R. Wetterau, M. J. Ray, D. G. Ferguson, H. W. Albers, W. R. Smith, and J. A. K. Harmony, *J. Biol. Chem.* **265,** 14292 (1990).

[98] B. F. Murphy, L. Kirszbaum, I. D. Walker, and A. J. F. D'Apice, *J. Clin. Invest.* **81,** 1858 (1988).

[99] K. Berg, *Acta Pathol. Microbiol. Scand.* **59,** 369 (1963).

[100] G. H. Dahlén, J. R. Guyton, M. Attar, J. A. Farmer, J. A. Kantz, and A. M. Gotto, Jr., *Circulation* **74,** 758 (1986).

[101] G. Utermann, *Science* **246,** 904 (1989).

[102] J. Loscalzo, *Arteriosclerosis* (*Dallas*) **10,** 672 (1990).

Polymorphisms of Apolipoproteins

A system for naming the polymorphic forms of plasma apolipoproteins has not been generally accepted, with the possible exception of a rigorous nomenclature proposed for phenotypes and genotypes of apoE polymorphs.[103] According to the ABC nomenclature, polymorphic forms are to be designated by Arabic numerals. For example, the three isomorphic forms of apoC-III were designated apoC-III-0, apoC-III-1, and apoC-III-2 on the basis of the number of terminal sialic acid residues. In another proposal,[104] assignment of isomorphic forms was based on separation by two-dimensional electrophoresis. The mature isomorphic form is designated by numeral zero as, for example, apoA-I_0; the other isomorphic forms are then coded by the number of negative or positive unit charges when compared to the mature form (apoA-I_{+1}, apoA-I_{+2}, apoA-I_{-1}, etc.).

As with other proteins (e.g., hemoglobin) it has become customary to identify apolipoprotein mutants with single amino acid changes by the name of the city or locality of its original discovery and the amino acid substitution as, for example, apoA-I_{Milano} ($Arg^{173} \rightarrow$ Cys). The truncated forms of apoB are named according to their apparent molecular weights or mobilities in sodium dodecyl sulfate–polyacrylamide gel electrophoresis relative to that of intact apoB designated as apoB-100. The apolipoprotein mutants and their effect on lipid transport processes have been expertly discussed in reviews.[105,106]

Structural Properties of Apolipoproteins

The anticipated role of apolipoproteins in maintaining the structural integrity and functional specificity of lipoprotein particles has been fully substantiated by extensive studies on their chemical and metabolic properties. Since the first reported amino acid sequence of apoA-II by Brewer *et al.*[107] in 1972, the primary structures of almost all other apolipoproteins (A-I, A-IV, B, C-I, C-II, C-III, D, E, I, and J) have been determined either by direct protein sequencing or by deduction from complementary DNAs.

[103] V. I. Zannis, J. L. Breslow, G. Utermann, R. W. Mahley, K. H. Weisgraber, R. J. Havel, J. L. Goldstein, M. S. Brown, G. Schonfeld, W. R. Hazzard, and C. Blum, *J. Lipid Res.* **23,** 911 (1982).

[104] D. L. Sprecher, L. Taam, and H. B. Brewer, Jr., *Clin. Chem.* **30,** 2084 (1984).

[105] G. Assman, A. von Eckardstein, and H. Funke, *in* "Structure and Function of Apolipoproteins" (M. Rosseneu, ed.), p. 85. CRC Pres, Boca Raton, Florida, 1992.

[106] P. Talmud, *in* "Structure and Function of Apolipoproteins" (M. Rosseneu, ed.), p. 123. CRC Press, Boca Raton, Florida, 1992.

[107] H. B. Brewer, S. E. Lux, R. Ronnan, and K. M. John, *Proc. Natl. Acad. Sci. U.S.A.* **69,** 1304 (1972).

Through the application of recombinant DNA techniques all of the corresponding apolipoprotein genes have been isolated, sequenced, and the chromosomal localization determined. This enormous accomplishment, including elucidation of the gene structure and amino acid sequence of apoB, one of the largest single-chain proteins (4536 amino acid residues), has been described in several reviews.[108–110]

These structural studies also revealed that apolipoproteins may be classified into three distinct groups on the basis of structural characteristics including the lipid-binding domains. The first group of seven soluble apolipoproteins (A-I, A-II, A-IV, C-I, C-II, C-III, and E) are members of an apolipoprotein multigene family, each of which is characterized by the occurrence of repeated amphipatic helical regions considered to be structural units essential for the lipid-binding properties.[109,111] The second group is represented by the insoluble apoB and various truncated forms including apoB-48 as the main intestinal form of this apolipoprotein, which is produced through a unique apoB–mRNA editing process whereby a glutamine codon corresponding to amino acid residue 2153 is converted to a stop codon.[112–114] In contrast to soluble apolipoproteins, the lipid-binding capability of apoB is considered to be due mainly, but not exclusively, to the occurrence of repeating hydrophobic proline-rich sequences with a tendency for β-sheet formation.[110,115] Apolipoproteins D, H, I, and J constitute the third group of apolipoproteins characterized by little or no structural homologies with either soluble apolipoproteins or apoB.

The apoD protein was found to display significant homology with the retinol-binding protein and other members of the α_2-microglobulin superfamily which transport small hydrophobic ligands.[116,117] Contrary to

[108] J. L. Breslow, *Physiol. Rev.* **68,** 85 (1988).

[109] W. H. Li, M. Tanimura, C. C. Luo, S. Datta, and L. Chan, *J. Lipid Res.* **29,** 245 (1988).

[110] M. Rosseneu, B. Vanloo, L. Lins, M. DePauw, J.-M. Ruysschaert, and R. Brasseur, *in* "Structure and Function of Apolipoproteins" (M. Rosseneu, ed.), p. 159. CRC Press, Boca Raton, Florida, 1992.

[111] J. P. Segrest, R. L. Jackson, J. D. Morrisett, and A. M. Gotto, Jr., *FEBS Lett.* **38,** 247 (1974).

[112] A. V. Hospattankar, K. Higuchi, S. W. Law, N. Meglin, and H. B. Brewer, Jr., *Biochem. Biophys. Res. Commun.* **148,** 279 (1987).

[113] S. H. Chen, G. Habib, C. J. Yang, Z. W. Gu, B. R. Lee, S. A. Wang, S. R. Silberman, S. J. Cai, J. P. Deslypere, M. Rosseneu, A. M. Gotto, Jr., W. H. Li, and L. Chan, *Science* **238,** 363 (1987).

[114] L. M. Powell, S. C. Wallis, R. J. Pease, I. H. Edwards, T. J. Knott, and J. Scott, *Cell (Cambridge, Mass.)* **50,** 831 (1987).

[115] C.-Y. Yang and H. J. Pownell, *in* "Structure and Function of Apolipoproteins" (M. Rosseneu, ed.), p. 63. CRC Press, Boca Raton, Florida, 1992.

[116] D. Drayna, J. W. McLean, K. L. Wion, J. M. Trent, H. A. Drabkin, and R. M. Lawn, *DNA* **6,** 199 (1987).

[117] M. C. Peitsch and M. S. Boguski, *New Biol.* **2,** 197 (1990).

the proposed role for apoD in cholesterol transport, the negative results of binding assays showed that cholesterol is not a preferred ligand for apoD.[117] The paucity of amphipathic helices in the primary structure of apoD suggested that its binding to lipoprotein surfaces may occur through hydrophobic interactions. It is quite possible that the recently identified apoD–apoA-II and apoD–apoB heterodimers may indirectly but significantly increase the lipoprotein-binding affinity of apoD.[118] Apolipoprotein H displays a significant structural homology with the superfamily of complement control proteins and several noncomplement mosaic proteins.[119] Its primary structure is characterized by five repeating units, each consisting of about 60 amino acids rich in proline and cysteine residues.[120] The secondary structure of apoH consists mainly of β-sheet and random coil configurations,[121] which are probably responsible for its binding propensity for neutral lipids.[122] Apolipoprotein I has an amphipathic helical segment of 26 amino acid residues which seems to be the main lipid binding domain of this "acute phase" apolipoprotein.[123] The lipid-binding capacity of apoJ resides most probably in three recognized amphipathic helical segments, one of which, consisting of 18 amino acid residues, occurs in apoJ-I and the other two, containing 17 and 18 amino acids, respectively, in apoJ-II.[124] The primary structures and lipid-binding domains of apoF and apoG remain to be determined in future studies.

Functional Properties of Apolipoproteins

Apolipoproteins play essential roles in the formation and stabilization of lipoprotein particles. However, in addition to their structural function, apolipoproteins also perform crucial roles in the metabolic conversions of lipoproteins including secretion, activation of lipolytic enzymes, retardation of premature removal, and recognition of binding or removal sites on

[118] F. Blanco-Vaca, D. P. Via, C.-I. Yang, J. B. Massey, and H. J. Pownall, *J. Lipid Res.* **33,** 1785 (1992).

[119] K. B. M. Reik and A. J. Day, *Immunol. Today* **10,** 177 (1989).

[120] J. Lozier, N. Takahashi, and F. W. Putnam, *Proc. Natl. Acad. Sci. U.S.A.* **81,** 3640 (1984).

[121] N. S. Lee, H. B. Brewer, Jr., and J. C. Osborne, Jr., *J. Biol. Chem.* **258,** 4765 (1983).

[122] E. Polz and G. M. Kostner, *Biochem. Biophys. Res. Commun.* **90,** 1305 (1979).

[123] C. L. Malmendier, J.-F. Lontie, and C. Delcroix, *Clin. Chim. Acta* **170,** 169 (1987).

[124] H. V. deSilva, J. A. K. Harmony, W. D. Stuart, C. M. Gil, and J. Robbins, *Biochemistry* **29,** 5380 (1990).

hepatic and extrahepatic cellular surfaces.[108,125–130] Thus, apoB-48 and apoB-100 seem to play essential roles in the secretion of lipoprotein particles.[131] Apolipoprotein C-II functions as the activator of lipoprotein lipase (LPL),[132] ApoH enhances the apoC-II-initiated activation of LPL,[133] and apoA-I is the principal activator of lecithin–cholesterol acyltransferase (LCAT; phosphatidylcholine–sterol *O*-acyltransferase) reaction.[134] Apolipoprotein C-III is an *in vitro* noncompetitive inhibitor of LPL[135] and interferes with or retards the hepatic uptake of intact or partially delipidized triglyceride-rich lipoproteins.[136,137] In the presence of apoA-II as an integral protein constituent, some triglyceride-rich particles become less efficient substrate for LPL.[138] Apolipoprotein A-II may also antagonize the potential role of apoA-I-containing lipoprotein particles in the removal of peripheral cholesterol.[139] Apolipoproteins B and E function as ligands for the LDL receptor and apoE as the ligand for the putative chylomicron remnant receptor.[128] It has been suggested that apoA-I may be the recognition signal for the HDL receptor.[128] The specific roles of other apolipoproteins in lipid transport processes have not been established.

Identification and characterization of a number of polymorphic forms and genetic variants of almost all known apolipoproteins and their contribution to the elucidation of structure–function relationships and roles of

[125] P. J. Dolphin, *Can. J. Biochem. Cell Biol.* **63,** 850 (1985).

[126] R. J. Havel and R. L. Hamilton, *Hepatology* (*Baltimore*) **8,** 1689 (1988).

[127] R. J. Havel, *Am. J. Physiol.* **253,** E1 (1987).

[128] U. Beisiegel, *in* "Structure and Function of Apolipoproteins" (M. Rosseneu, ed.), p. 269. CRC Press, Boca Raton, Florida, 1992.

[129] P. J. Dolphin, *in* "Structure and Function of Apolipoproteins" (M. Rosseneu, ed.), p. 295. CRC Press, Boca Raton, Florida, 1992.

[130] K.-A. Rye and P. J. Barter, *in* "Structure and Function of Apolipoproteins" (M. Rosseneu, ed.), p. 401. CRC Press, Boca Raton, Florida, 1992.

[131] H. B. Brewer, Jr., R. E. Gregg, J. M. Hoeg, and S. S. Fojo, *Clin. Chem.* **34,** B4 (1988).

[132] R. J. Havel, V. G. Shore, B. Shore, and D. M. Bier, *Circ. Res.* **27,** 596 (1970).

[133] Y. Nakaya, E. J. Schaefer, and H. B. Brewer, Jr., *Biochem. Biophys. Res. Commun.* **95,** 1168 (1980).

[134] C. J. Fielding, V. G. Shore, and P. E. Fielding, *Biochem. Biophys. Res. Commun.* **46,** 1493 (1972).

[135] C.-S. Wang, W. J. McConathy, H. U. Kloer, and P. Alaupovic, *J. Clin. Invest.* **75,** 384 (1985).

[136] E. Windler and R. J. Havel, *J. Lipid Res.* **26,** 556 (1985).

[137] H. N. Ginsberg, N.-A. Le, I. J. Goldberg, J. C. Gibson, A. Rubinstein, P. Wang-Iverson, R. Norum, and W. V. Brown, *J. Clin. Invest.* **78,** 1287 (1986).

[138] P. Alaupovic, C. Knight-Gibson, C.-S. Wang, D. Downs, E. Koren, H. B. Brewer, Jr., and R. E. Gregg, *J. Lipid Res.* **32,** 9 (1991).

[139] A. Barkia, P. Puchois, N. Ghalim, G. Torpier, R. Barbaras, G. Ailhaud, and J.-C. Fruchart, *Atherosclerosis* **87,** 135 (1991).

apolipoproteins in lipid transport processes have been expertly described in several reviews.[105,106,110]

Apolipoproteins as Markers for Identifying and Classifying Plasma Lipoproteins

The discovery of a surprisingly large number of apolipoproteins raised important questions regarding their concentrations, distribution throughout the lipoprotein density spectrum, and localization on individual lipoprotein particles. Answers to these questions were made possible mainly through the introduction and application of qualitative and quantitative immunologic techniques.[75] Quantitative determination of apolipoproteins showed that they were widely distributed throughout the entire lipoprotein density spectrum. Furthermore, these studies established that apolipoproteins occur in nonequimolar ratios, excluding the possibility that all lipoprotein particles within a defined density range have the same apolipoprotein composition.[75,140] In conjunction with immunochemical patterns of nonidentity or partial identity between apolipoproteins, these findings indicated that each major lipoprotein density class consists of several distinct lipoprotein species separable by affinity chromatography on concanavalin A,[84,141] heparin-Sepharose,[142,143] and immunosorbers with antibodies to specific apolipoproteins.[144] Thus, the identification of lipoprotein particles with similar density properties but different apolipoprotein composition demonstrated that major lipoprotein density classes are heterogeneous not only as to physical properties and lipid–protein composition but also as to the content and composition of discrete lipoprotein species. Kinetic studies have confirmed the marked heterogeneity of major lipoprotein density classes and suggested that the presence of discrete lipoprotein particles is the probable cause underlying this hetrogeneity.[143,145–147] In addition to their well-recognized structural and metabolic roles, apolipoproteins have been recognized as the most useful markers for identifying plasma lipoproteins and for studying their interactions in lipid transport processes.

[140] P. Alaupovic, W. J. McConathy, J. Fesmire, M. Tavella, and J. M. Bard, *Clin. Chem.* **34,** B13 (1988).

[141] W. J. McConathy and P. Alaupovic, *FEBS Lett.* **41,** 174 (1974).

[142] F. A. Shelburne and S. H. Quarford, *J. Clin. Invest.* **60,** 944 (1977).

[143] P. E. Fielding and C. J. Fielding, *J. Biol. Chem.* **261,** 5233 (1986).

[144] P. Alaupovic and E. Koren, *in* "Analyses of Fats, Oils and Lipoproteins" (E. G. Perkins, ed.), p. 599. American Oil Chemists' Society, Champaign, Illinois, 1991.

[145] N. Yamada, D. S. Shames, J. B. Strudemire, and R. J. Havel, *Proc. Natl. Acad. Sci. U.S.A.* **83,** 3479 (1986).

[146] J. Sheperd and C. J. Packard, *Am. Heart J.* **113,** 503 (1987).

[147] P. J. Nestel, *Am. Heart J.* **113,** 518 (1987).

The chemical and metabolic heterogeneity of lipoprotein density classes prompted the introduction of an alternative classification system of lipoproteins based on the chemical uniqueness of apolipoproteins as a differentiating criterion.[62,74] Accordingly, plasma lipoproteins are viewed as a system of discrete lipoprotein families or particles defined by the specific apolipoprotein composition.[75] Lipoprotein families that contain a single apolipoprotein are referred to as simple lipoprotein families, and those characterized by two or more apolipoproteins as complex lipoprotein families. All simple and complex lipoprotein families are polydisperse systems of particles differing in densities, sizes, and lipid/protein ratios, but characterized by the same qualitative apolipoprotein composition. Lipoprotein families are named according to the apolipoprotein constituents.

The apoA- (apoA-I plus apoA-II) and apoB-containing lipoproteins are the two major classes of lipoprotein families. Lipoproteins that contain apolipoproteins other than apoA or apoB constitute a third, minor group of lipoprotein families. The apoA-containing lipoproteins, found mainly within the HDL density segment, consist of phospholipid-rich lipoprotein A-I (LP-A-I), lipoprotein A-I:A-II (LP-A-I:A-II), and lipoprotein A-II (LP-A-II). Varying proportions of each of the three lipoprotein families consist of subfamilies characterized by the presence of minor apolipoproteins such as apoA-IV, apoC peptides, apoD, apoE, etc. The apoB-containing lipoprotein families, occurring mainly in VLDL, intermediate-density lipoprotein (IDL), and LDL, consist of cholesterol ester-rich lipoprotein B (LP-B) along with the triglyceride-rich lipoproteins B:C (LP-B:C), B:C:E (LP-B:C:E), B:E (LP-B:E), and A-II:B:C:D:E (LP-A-II:B:C:D:E or LP-A-II:B complex). Each of the apoC-containing lipoprotein families may also contain small amounts of subfamilies characterized by the presence of only one or two of the apoC peptides resulting most probably from partial lipolytic degradation of the parent lipoprotein families.

Separation and Characterization of Discrete Apolipoprotein A- and B-Containing Lipoprotein Families

Composition of Lipoprotein Families

The separation of apoA- and apoB-containing lipoprotein families may be achieved by nonimmunologic methods.[84,141–143] However, subfractionation of each of the two lipoprotein classes requires the application of specific immunologic procedures such as sequential immunoprecipitation or immunoaffinity chromatography.[75,144,148] The apoA- and apoB-containing

[148] J. C. Fruchart and G. Ailhaud, *Clin. Chem.* **38,** 793 (1992).

lipoprotein families have several characteristic features. Phospholipids are the major lipid constituents of all three apoA-containing lipoproteins[149]; LP-A-I:A-II particles have the lowest percentage content of phospholipids and the highest cholesterol ester/free cholesterol ratio, whereas LP-A-II particles have the highest percentage of phospholipids and the lowest cholesterol ester/free cholesterol ratio. In normolipidemic subjects, approximately 30–35% of LP-A-I and LP-A-I:A-II particles contain apoC peptides, apoD, apoE, and other minor apolipoproteins. However, LP-A-I:A-II particles contain 70–90% of the total HDL content of these minor apolipoproteins, suggesting that they may function as the major acceptor of these apolipoproteins.[150] On the other hand, LP-A-II particles seem to contain only apoD as the other apolipoprotein, possibly bound to apoA-II through a disulfide linkage.[118]

The LP-B particles are characterized by having cholesterol esters and LP-B:C, LP-B:C:E, and LP-A-II:B complex particles by having triglycerides as the main neutral lipid component. The LP-B:E particles have approximately same percentage of cholesterol esters and triglycerides. The complex triglyceride-rich apoB-containing lipoprotein particles have density properties similar to VLDL and IDL, whereas cholesterol ester-rich LP-B and LP-B:E occur mainly as IDL and LDL. However, the former lipoproteins may also be detected in LDL and the latter lipoproteins in VLDL. Their distribution depends on both the relative contents of triglycerides and cholesterol esters and the lipid/protein ratios, which are related to the metabolic state of individual particles. Although the percentage of triglycerides decreases, the percentages of cholesterol esters and protein increase with increasing density of lipoprotein particles.[95] The apolipoprotein composition of complex apoB-containing lipoproteins also changes in that the percentage of apoB increases whereas percentages of apoC and apoE decrease with increasing densities.[75]

Metabolic Properties of Lipoprotein Families

Although initially perceived as a chemical classification system,[62,74] the lipoprotein family concept gained further significance when studies showed that lipoprotein particles defined by their specific apolipoprotein composition also possess distinct metabolic properties. For example, the turnover

[149] E. D. Bekaert, P. Alaupovic, C. Knight-Gibson, and M. Ayrault-Jarrier, *Prog. Lipid Res.* **30,** 151 (1991).

[150] E. D. Bekaert, P. Alaupovic, C. Knight-Gibson, P. Blackett, and M. Ayrault-Jarrier, *Pediatr. Res.* **29,** 315 (1991).

rate of apoA-I in LP-A-I is faster than that of apoA-I in LP-A-I:A-II.[151] The LP-A-I particles, but not LP-A-I:A-II particles, seem to be responsible for the sterol efflux from cultured human fibroblasts[152] or mouse adipocytes,[139] and for esterifying and transferring cholesterol in plasma in association with LCAT and cholesterol transfer protein.[153,154] The LP-A-I particles bind with greater affinity to various cell membranes and have greater capacity to form complexes with apolipoproteins I and J than LP-A-I:A-II particles.[123,124] On the other hand, LP-A-I:A-II particles may function as the major donors or acceptors of apoC peptides and apoE in the formation and degradation of triglyceride-rich lipoproteins.[150] The LP-B:C and LP-B:C:E particles are more efficient substrates for LPL than LP-A-II:B complex despite similar triglyceride and apoC-III contents.[138] The LP-B:E particles bind to LDL receptors with greater affinity than LP-B.[155,156] The binding to LDL receptors on HeLa cells of LP-B:C and LP-B:C:E particles with E2/E2 phenotype was found to be negligible, suggesting that, in contrast to LP-B and LP-B:E, these lipoprotein families have little or no effect on regulation of 3-hydroxy-3-methylglutaryl-coenzyme A (HMG-CoA) reductase activity.[156]

The available information about formation of lipoprotein particles is still scarce. To gain initial information on the liver production of apoA- and apoB-containing lipoproteins, we have studied the chemical composition of these lipoproteins secreted into the medium of Hep G2 cells. The apoA-containing lipoproteins were found to consist of approximately equal amounts of LP-A-I and LP-A-I:A-II particles, both of which were characterized by a higher free cholesterol/cholesterol ester ratio than the corresponding lipoprotein particles in the plasma.[157] Cheung *et al.* have confirmed this finding and showed that both lipoprotein families consist of discoidal and spherical particles, with the majority of LP-A-I:A-II particles having relatively high (9.2–17.0 nm) and LP-A-I particles having relatively low Stokes diameters (<8.0 nm).[158] These results suggest that LP-A-I and LP-A-I:

[151] D. J. Rader, G. Castro, L. A. Zech, J. C. Fruchart, and H. B. Brewer, Jr., *J. Lipid Res.* **32,** 1849 (1991).

[152] C. J. Fielding and P. E. Fielding, *Proc. Natl. Acad. Sci. U.S.A.* **78,** 3911 (1981).

[153] M. C. Cheung, A. C. Wolf, K. D. Lum, J. H. Tollefson, and J. J. Albers, *J. Lipid Res.* **27,** 1135 (1986).

[154] O. L. Francone, A. Gurakar, and C. Fielding, *J. Biol. Chem.* **264,** 7066 (1989).

[155] E. Koren, P. Alaupovic, D. M. Lee, N. Dashti, H. U. Kloer, and G. Wen, *Biochemistry* **26,** 2734 (1987).

[156] G. Agnani, J.-M. Bard, L. Candelier, S. Delattre, J.-C. Fruchart, and V. Clavey, *Arterioscler. Thromb.* **11,** 1021 (1991).

[157] N. Dashti, E. Koren, and P. Alaupovic, *Biochem. Biophys. Res. Commun.* **163,** 574 (1989).

[158] M. Cheung, K. D. Lum, C. G. Brouillette, and C. L. Bisgaier, *J. Lipid Res.* **30,** 1429 (1989).

A-II particles are formed in the liver, but must undergo modifications in the plasma compartment to acquire structural and compositional properties characteristic of their plasma counterparts. The subfamilies of LP-A-I particles with apolipoproteins A-IV, apoC peptides, and apoH as additional protein constituents are probably formed in the intestine.[93,159]

The apoB-containing lipoproteins present in the Hep G2 culture medium were identified as LP-B and LP-B:E particles.[160] The content of apoC peptides was too small to be detected by immunoassays. The size distribution of both lipoprotein families was similar to that of plasma LDL. However, unlike the lipid composition of plasma LDL, both lipoprotein families were characterized by relatively high percentages of triglycerides (56–80%) and low percentages of cholesterol esters (15–22%).[160,161] These findings suggest that the nascent, triglyceride-rich LP-B and LP-B:E particles may serve as precursors for the extracellular formation of LP-B:C, LP-B:C:E, and LP-A-II:B complex particles, the three main lipoprotein vehicles for the transport of triglycerides in the plasma compartment. During lipolytic degradation of triglyceride-rich lipoproteins, the cholesterol-rich LP-B particles are formed as the finite remnants of this conversion process and the dissociated apoC peptides, apoE, and other minor apolipoproteins are transferred to LP-A-I:A-II and LP-A-I particles. However, the integration of these findings into a comprehensive view of lipid transport will require additional information about the chemical and metabolic properties of apolipoproteins and their lipoprotein families.

Apolipoprotein and Lipoprotein Family Profiles in Dyslipoproteinemias and Atherosclerosis

The function of lipoproteins is to transport exogenous and endogenous triglycerides and cholesterol from sites of absorption and formation to sites of storage and utilization. Under normal conditions, the input and output of triglyceride-rich lipoproteins are balanced with little or no change in plasma levels. However, an increased influx and/or a decreased efflux of triglyceride-rich lipoproteins from the plasma compartment have been identified as main pathophysiological mechanisms leading to hypertriglyceridemia.[162] The increased formation and/or decreased removal of cholesterol-

[159] A. M. Magun, T. A. Brasitus, and R. M. Glickman, *J. Clin. Invest.* **75,** 209 (1985).
[160] N. Dashti, P. Alaupovic, C. Knight-Gibson, and E. Koren, *Biochemistry* **26,** 4837 (1987).
[161] R. N. Thrift, T. M. Forte, B. E. Cahoon, and V. G. Shore, *J. Lipid Res.* **27,** 236 (1986).
[162] S. M. Grundy, *J. Lipid Res.* **25,** 1661 (1984).

rich lipoproteins results in hypercholesterolemia which may accompany hypertriglyceridemic states.

Deranged lipid transport processes are of great clinical significance, because they are one of the most important risk factors for the genesis and development of coronary artery disease.[25,45,47,163] Dyslipoproteinemias seem to be characterized by specific concentration profiles of apoA- and apoB-containing lipoprotein families.[75,140] In general, the apoA-containing lipoprotein families occur in higher concentrations in normolipidemic and hypercholesterolemic than hypertriglyceridemic subjects. The LP-B and LP-B:E particles are the main apoB-containing lipoproteins of normal subjects as well as hypercholesterolemic patients, whereas LP-B:C, LP-B:C:E, and LP-A-II:B complex particles are the characteristic apoB-containing lipoproteins of hypertriglyceridemic states. Apolipoprotein B is generally considered as a marker of atherogenic and apoA-I as a marker of nonatherogenic lipoproteins. Although as a group apoA-containing lipoproteins are thought to be nonatherogenic, studies have suggested that the "protective" capacity of LP-A-I may be greater than that of LP-A-I:A-II.[164,165] Because other studies have shown no significant difference in the nonatherogenic potential between LP-A-I and LP-A-I:A-II,[166,167] the clinical significance of LP-A-I as the protective factor of HDL remains to be resolved in future studies.

Equally challenging is the question of the relative atherogenicity of chemically and metabolically distinct apoB-containing lipoprotein particles.[75,168] Although all chemical characteristics of atherogenic lipoprotein particles have not yet been established, the circumstantial evidence suggests that size and specific structural arrangement of lipid and apolipoprotein constituents may be the main factors contributing to the atherogenicity of apoB-containing lipoproteins.[75,169] Current results suggest that the atherogenic potential of apoB-containing lipoproteins increases with decreasing size and density of lipoprotein particles, but that the optimal atherogenic

[163] R. B. Wallace and R. A. Anderson, *Epidemiol. Rev.* **9,** 95 (1987).

[164] P. Puchois, A. Kandoussi, P. Fievet, J. L. Fourrier, M. Bertrand, E. Koren, and J. C. Fruchart, *Atherosclerosis* **68,** 35 (1987).

[165] H. J. Parra, D. Arveiler, A. E. Evans, J. P. Cambou, P. Amouyel, A. Bingham, D. McMaster, P. Schaffer, P. Douste-Blazy, G. Luc, J. L. Richards, P. Ducimetière, J. C. Fruchart, and F. Cambien, *Arterioscler. Thromb.* **12,** 701 (1992).

[166] M. Coste-Burel, F. Mainard, L. Chivot, J. L. Auget, and Y. Madec, *Clin. Chem.* **36,** 1889 (1990).

[167] M. C. Cheung, B. G. Brown, A. C. Wolf, and J. J. Albers, *J. Lipid Res.* **32,** 383 (1991).

[168] P. Alaupovic and D. H. Blankenhorn, *Klin. Wochenschr.* **68** (Suppl. 22), 38 (1990).

[169] B. G. Nordestgaard, A. Tybjaerg-Hansen, and B. Lewis, *Arteroscler. Thromb.* **12,** 6 (1992).

size and composition reside in lipoproteins of intermediate density (1.006–1.040 g/ml).[45,169–171] This density range includes not only cholesterol-rich LP-B but also intact and partially delipidized triglyceride-rich LP-B:C, LP-B:C:E, and LP-A-II:B complex particles. The results of two clinical trials, the cholesterol-lowering atherosclerosis study (CLAS)[172] and the mevinolin atherosclerosis regression study (MARS),[173] have shown that increased levels of some triglyceride-rich lipoprotein particles are closely associated with the progression of atherosclerotic disease in moderately hypercholesterolemic subjects despite their cholesterol-rich LP-B particles having been significantly reduced by drug therapy with either niacin–colestipol in the former or lovastatin in the latter trial. Although these studies indicate that both cholesterol-rich and triglyceride-rich apoB-containing lipoprotein particles contribute to the development of coronary artery disease, the compositional and structural factors determining the atherogenic potential and, thus, the relative atherogenicity of apoB-containing lipoproteins remain to be determined.

Atmeh *et al.* first showed that lipid-lowering drugs may affect the levels of plasma lipoprotein families in a specific manner by establishing that probucol decreases and nicotinic acid increases the levels of LP-A-I particles with little or no effect on the levels of LP-A-I:A-II particles.[174] A similar effect on LP-A-I particles was also observed in hypercholesterolemic patients treated with cholestyramine[175] or simvastatin.[176] The combination of nicotinic acid and colestipol[177] as well as inhibitors of HMG-CoA reductase[175,176,178] lower the concentrations of LP-B and, possibly, LP-B:E particles, but have little or no effect on the levels of LP-B:C, LP-B:C:E, and LP-A-II:B complex particles. On the other hand, fibrates such as gemfi-

[170] T. A. Musliner and R. M. Krauss, *Clin. Chem.* **34,** B78 (1988).

[171] B. G. Nordestgaard, S. Stender, and K. Kjeldsen, *Scand. J. Clin. Lab. Invest.* **47,** (Suppl. 186), 7 (1987).

[172] D. H. Blankenhorn, P. Alaupovic, E. Wickham, H. P. Chin, and S. P. Azen, *Circulation* **81,** 470 (1990).

[173] H. N. Hodis, W. J. Mack, J. M. Pogoda, P. Alaupovic, L. C. Hemphill, and D. H. Blankenhorn, *J. Am. Coll. Cardiol.* **21,** 71A (1993).

[174] R. F. Atmeh, J. Shepard, and C. J. Packard, *Biochim. Biophys. Acta* **751,** 175 (1983).

[175] J. M. Bard, H. J. Parra, P. Douste-Blazy, and J. C. Fruchart, *Metabolism* **39,** 269 (1990).

[176] J. M. Bard, H. J. Parra, R. Camare, G. Luc, O. Ziegler, C. Dachet, E. Bruckert, P. Douste-Blazy, P. Drouin, B. Jacotot, J. L. DeGennes, U. Keller, and J. C. Fruchart, *Metabolism* **41,** 498 (1992).

[177] M. Tavella, P. Alaupovic, D. H. Blankenhorn, and H. P. Chin, *Arteriosclerosis* (*Dallas*) **7,** 494a (1987).

[178] P. Alaupovic, C. Knight-Gibson, W. J. Mack, D. M. Kramsch, H. N. Hodis, and D. H. Blankenhorn, *Circulation* **86-I,** 743 (1992).

brozil[179] or fenofibrate[180] significantly reduce the levels of LP-B:C and LP-B:C:E particles but have negligible effect on the levels of LP-B particles.

The recognition of apolipoproteins as the essential structural and functional constituents of discrete lipoprotein particles has created a need for developing specific, sensitive, and accurate assays for quantitative determination in whole plasma and in isolated lipoprotein fractions. Because of the structural complexity of plasma lipoproteins, quantitative determination of apolipoproteins has been based almost exclusively on highly specific immunologic procedures including radioimmunoassay, radial immunodiffusion, electroimmunoassay, enzyme immunoassays, immunonephelometry, and immunoturbidimetry.[181–184] The very fact that this volume is dedicated to the detailed description of immunoassays and their application to the measurement of individual apolipoproteins shows clearly the growing interest in determining and utilizing apolipoprotein profiles as a measure for the quantitative assessment of lipid transport processes. As the normal functioning of these processes depends on an optimal concentration of apolipoproteins or, more precisely, their corresponding lipoprotein families, any derangement of these processes will lead to changes in concentrations commensurate with the underlying genetic or metabolic defects on the one hand and environmental factors on the other. Consequently, various derangements of lipid transport should be characterized by specific concentration profiles of lipoprotein families or apolipoproteins as unique chemical markers.

Because of space constraints, detailed review and discussion of the available data on apolipoprotein levels in normal and dyslipoproteinemic subjects is not possible except for some generalizations regarding the past accomplishments and future directions. Analyses of plasma apolipoproteins in normolipidemic subjects have demonstrated sex- and age-related differences as well as fluctuations in apolipoprotein concentrations which affect

[179] C. Corder, M. Tavella, and P. Alaupovic, *Arteriosclerosis* (*Dallas*) **7,** 515a (1987).

[180] S. Lussier-Cacan, J. M. Bard, L. Boulet, A. C. Nestruck, A. M. Grothe, J. C. Fruchart, and J. Davignon, *Atherosclerosis* **78,** 167 (1989).

[181] Proceedings of the Workshop on Apolipoprotein Quantification (K. Lippel, ed.), U.S. Department of Health and Human Services, NIH Publication No. 83-1266, Bethesda, Maryland, 1983.

[182] P. S. Bachorik and P. O. Kwiterovich, Jr., *Clin. Chim. Acta* **178,** 1 (1988).

[183] J. J. Albers, J. D. Brunzell, and R. H. Knopp, *Clin. Lab. Med.* **9,** 137 (1989).

[184] C. Labeur, J. Shepherd, and M. Rosseneu, *Clin. Chem.* **36,** 691 (1990).

mainly apolipoproteins A-I, A-II, and B[140,182,183,185–187] with little or no effect on the levels of apoC peptides, apoD, and apoE.[140] Several studies have also indicated that genetic and environmental differences between geographic regions may have a substantial effect on apolipoprotein levels.[140,188,189]

We have suggested on the basis of studies from this and other laboratories that apolipoprotein profiles of dyslipoproteinemic states may be classified into three separate categories.[140] The first category is characterized by the absence of an immunochemically detectable plasma apolipoprotein. Apolipoprotein profiles in this category represent an absolute test for identifying a lipid transport disorder (e.g., abetalipoproteinemia, apoC-II deficiency, apoA-I/apoC-III deficiency). The second category is characterized by a partial deficiency of one or more apolipoproteins. Although apolipoprotein profiles in this category (e.g., hypobetalipoproteinemia, Tangier disease) cannot be considered to be absolute diagnostic criteria, they may be specific enough for use in combination with other lipid and apolipoprotein variables to differentiate between hypolipoproteinemic disorders. The third category is characterized by increased levels of one or several apolipoproteins encompassing all the primary and secondary hyperlipoproteinemias. In this category, determinations of apolipoproteins cannot be used as differential diagnostic tests, but offer useful clues about the chemical nature of abnormally increased lipoproteins and, in some cases, about the biochemical derangements which cause these abnormalities.

The measurement of apolipoproteins is not limited to diagnostic purposes, but can be utilized as a means for studying the formation, interaction, and degradation of various lipoprotein classes. The quantitative determination of apoC-III in HDL and VLDL plus LDL and its expression as apoC-III ratio (apoC-III-HDL/apoC-III-VLDL+LDL) has been shown to be a useful means of expressing the efficiency of processes responsible for the catabolism of triglyceride-rich lipoproteins.[190] Another example

[185] B. A. Kottke, P. P. Moll, V. M. Michels, and W. H. Weidman, *Mayo Clin. Proc.* **66,** 1198 (1991).

[186] I. Jungner, G. Walldius, I. Holme, W. Kolar, and E. Steiner, *Int. J. Clin. Lab. Res.* **21,** 247 (1992).

[187] S. A. Brown, R. Hutchinson, J. Morrisett, E. Boerwinkle, C. E. Davis, A. M. Gotto, Jr., and W. Patsch, *Arterioscler. Thromb.* **13,** 1139 (1993).

[188] R. P. Donahue, D. R. Jacobs, Jr., S. Sidney, L. E. Wagenknecht, J. J. Albers, and S. B. Hulley, *Arteriosclerosis* (*Dallas*) **9,** 656 (1989).

[189] H. Kesteloot, V. O. Oviasu, A. O. Obasohan, A. Olomu, C. Cobbaert, and W. Lissens, *Atherosclerosis* **78,** 33 (1989).

[190] P. Alaupovic, *Can. J. Biochem.* **59,** 565 (1981).

is provided by a study on the *in vitro* lipolytic degradation of triglyceride-rich lipoproteins, a process monitored by the measurement of apoC peptides and apoE released during the course of reaction.[191] Quantification of apolipoproteins has also been used in several studies on the formation and secretion of the corresponding lipoprotein particles in Hep G2 cells,[157,158,160,161,192,193] Caco-2 cells,[194,195] and rat liver cells.[196–198]

In the early 1950s, one immediate consequence of the recognized association between deranged lipid transport and development of coronary artery disease was the need for selecting a lipoprotein or a lipoprotein constituent that could serve as the best possible predictor of the presence and severity of atherosclerotic vascular disease. The dilemma which developed between selecting one of the low-density lipoprotein classes, as advocated by Gofman and co-workers, or serum cholesterol, the choice of the majority of investigators in the field, was resolved in the favor of cholesterol as the most useful predictor of coronary artery disease.[76,199] However, further studies have shown that the levels of cholesterol in two major lipoprotein density classes correlate better with the presence (LDL-cholesterol) or absence (HDL-cholesterol) of coronary artery disease than do the concentrations of total cholesterol in plasma.[163,200]

The next step in evaluating the association between lipoproteins and atherosclerosis was to explore the predictive capacity of apolipoproteins, which had been shown to be better markers than lipids for identifying lipoprotein particles.[75] Results of numerous clinical studies have suggested that the levels of apolipoproteins A-I and B may be better predictors of coronary artery disease than the levels of total or lipoprotein cholesterol.[140,163,182,183,201,202] Among minor apolipoproteins, the levels of apoC-

[191] P. Alaupovic, C. S. Wang, W. J. McConathy, D. Weiser, and D. Downs, *Arch. Biochem.* **244,** 226 (1986).

[192] N. B. Jaritt, *FASEB J.* **4,** 161 (1990).

[193] J. Borén, M. Wettesten, A. Sjöberg, T. Thorlin, G. Bondjers, O. Wiklund, and S. O. Olofsson, *J. Biol. Chem.* **265,** 10556 (1990).

[194] T. E. Hughes, J. M. Ordovas, and E. J. Schaefer, *J. Biol. Chem.* **263,** 3425 (1988).

[195] N. Dashti, E. A. Smith, and P. Alaupovic, *J. Lipid Res.* **31,** 113 (1990).

[196] L. Wong, *J. Cell. Physiol.* **141,** 441 (1989).

[197] R. L. Hamilton, A. Moorehouse, and R. J. Havel, *J. Lipid Res.* **32,** 529 (1991).

[198] A. Rusiñol, H. Verkade, and J. E. Vance, *J. Biol. Chem.* **268,** 3555 (1993).

[199] The Technical Group and Committee on Lipoproteins and Atherosclerosis, *Circulation* **14,** 691 (1956).

[200] Report of the National Cholesterol Education Program Expert Panel on Detection, Evaluation, and Treatment of High Blood Cholesterol in Adults, *Arch. Intern. Med.* **148,** 36 (1988).

[201] A. Hamsten, *Acta Med. Scand.* **223,** 389 (1988).

[202] S. S. Levinson and S. G. Wagner, *Arch. Pathol. Lab. Med.* **116,** 1350 (1992).

III in HDL or VLDL plus LDL were found to have the strongest power in predicting the progression of atherosclerotic disease in patients treated with LDL-lowering drugs.[172,173] It should be pointed out, however, that there is not a unanimous opinion supporting apolipoproteins as better predictors of coronary artery disease than lipids.[202–205] One of the major objections to the routine measurement of apolipoproteins in clinical practice is the lack of an internationally acceptable standardization program. However, the acceptance by the World Health Organization (WHO) Expert Committee on Biological Standardization of an apoA-I international reference material,[206] soon to be followed by a similar apoB reference material, represents an important milestone in efforts to improve the accuracy and precision of apolipoprotein assays and to achieve optimum uniformity of apolipoprotein values among laboratories.

The selection of a most appropriate lipoprotein predictor of coronary artery disease may not be resolved even after the introduction and availability of a universally acceptable standardization program for lipid and apolipoprotein assays if the choice rests only with single lipoprotein constituents. Individual lipids and apolipoproteins exist and function in plasma only as integral constituents of discrete lipoprotein families. Although, qualitatively, all lipoprotein families have the same lipid constituents and several lipoprotein families share the same apolipoproteins, they display different metabolic properties and different atherogenic (apoB-containing lipoproteins) or nonatherogenic (apoA-containing lipoproteins) capacities. These differing properties depend on the quantitative composition and conformation of the integral lipid and protein components and the size and shape of the particles. Thus, quantification of a particular apoB-containing lipoprotein family of specific size and density exhibiting the greatest relative atherogenic potential should be a better predictor of atherosclerotic vascular disease than individual lipids or apolipoproteins. Exclusion of several discrete lipoproteins which are characterized by lower or negligible atherogenic capacities will increase predictive power.

Similar arguments in favor of particular low-density lipoprotein subclasses rather than individual lipid constituents as better predictors of coronary artery disease had been advanced by Gofman and co-workers[44,45] in the early 1950s and by Krauss and associates.[170,207] The main drawback of using lipoprotein density classes as predictors of coronary artery disease

[203] A. D. Sniderman and J. Silberberg, *Arteriosclerosis (Dallas)* **10,** 665 (1990).

[204] G. L. Vega and S. M. Grundy, *Arteriosclerosis (Dallas)* **10,** 668 (1990).

[205] W. G. Thompson, *South. Med. J.* **86,** 1994 (1993).

[206] S. M. Marcovina, J. J. Albers, L. O. Henderson, and W. H. Hannon, *Clin. Chem.* **39,** 773 (1993).

[207] M. A. Austin, M. C. King, K. M. Vranizan, and R. M. Krauss, *Circulation* **82,** 495 (1990).

is the presence within each density class of several discrete apoB-containing lipoprotein families differing in atherogenic properties.[75,168] Therefore, the measurement of individual apoA- and apoB-containing lipoprotein families of defined size ranges may provide a more accurate and precise assessment of the magnitude of coronary artery disease in individual subjects. When considered in conjunction with the selective effect of hypolipidemic drugs on lipoprotein families, the measurement of lipoprotein family profiles may also be used as a rational approach to the selection of appropriate therapies. Therapy could then be targeted at decreasing undesirable and/or increasing desirable concentrations of individual lipoprotein families.

Basic,[75,148,150,151,156–158,191,208] epidemiological,[165,209] and clinical[140,148,164–168,178,210–217] studies utilizing the lipoprotein family concept have already resulted in novel information about lipid transport processes in health and disease. However, an integrated view of lipid transport based on this concept will require further improvements in the available immunologic procedures for the isolation of lipoprotein particles as well as new information on genetic and other regulatory mechanisms responsible for lipoprotein particle formation and degradation, functional roles, and atherogenic potentials. It is equally important to improve the present methodology[140,216–219] and, if necessary, to devise new, relatively simple procedures for quantifying apoA- and apoB-containing lipoprotein parti-

[208] M.-R. Taskinen, J. Kahri, V. Koivisto, J. Shepherd, and C. J. Packard, *Diabetologia* **35,** 347 (1992).

[209] P. Puchois, N. Ghalim, G. Zylberberg, P. Fievet, C. Demarquilly, and J. C. Fruchart, *Arch. Intern. Med.* **150,** 1638 (1990).

[210] T. Ohta, S. Hattori, S. Nishiyama, A. Higashi, and I. Matsuda, *Metabolism* **38,** 834 (1989).

[211] T. Ohta, R. Nakamura, S. Nishiyama, M. Kodama, and I. Matsuda, *Pediatr. Res.* **28,** 42 (1990).

[212] J. J. Genest, Jr., J. M. Bard, J.-C. Fruchart, J. M. Ordovas, P. F. W. Wilson, and E. J. Schaefer, *Atherosclerosis* **90,** 149 (1991).

[213] E. D. Bekaert, P. Alaupovic, C. S. Knight-Gibson, M. J. Laux, J. M. Pelachyk, and R. A. Norum, *J. Lipid Res.* **32,** 1587 (1991).

[214] P. Alaupovic, J.-M. Bard, M. Tavella, and D. Shafer, *Diabetes* **41** (Suppl. 2), 18 (1992).

[215] P.-O. Attman, C. Knight-Gibson, M. Tavella, O. Samuelsson, and P. Alaupovic, *Miner. Electrolyte Metab.* **18,** 199 (1992).

[216] E. Koren, V. W. Armstrong, G. Mueller, P. R. Wilson, P. Schuff-Werner, J. Thiery, T. Eisenhauer, P. Alaupovic, and D. Seidel, *Atherosclerosis* **95,** 157 (1992).

[217] E. D. Bekaert, P. Alaupovic, C. S. Knight-Gibson, G. Franceschini, and C. R. Sirtori, *J. Lipid Res.* **34,** 111 (1993).

[218] E. Koren, P. Puchois, P. Alaupovic, J. Fesmire, A. Kandoussi, and J.-C. Fruchart, *Clin. Chem.* **33,** 38 (1987).

[219] H. J. Parra, H. Mezdur, N. Ghalim, J. M. Bard, and J. C. Fruchart, *Clin. Chem.* **36,** 1431 (1990).

cles including development of a universally acceptable standardization program. This volume demonstrates the steady progress in developing adequate procedures for measuring apolipoproteins and will also serve as a stimulus for similar future developments in the analytical methodology of lipoprotein particles.

Section II

Apolipoproteins

A. Apolipoprotein B
Chapters 3 through 9

B. Apolipoprotein E
Chapter 10

C. Apolipoprotein C
Chapters 11 and 12

D. Apolipoprotein [a]
Chapters 13 and 14

E. Apolipoprotein A
Chapters 15 through 19

F. Apolipoprotein J
Chapter 20

[3] Quantitation of Apolipoprotein B by Chemical Methods

By GLORIA LENA VEGA and SCOTT M. GRUNDY

Introduction

Human apolipoprotein B-100 (apoB) is a large glycoprotein composed of 4536 amino acids, and the protein is highly polymorphic.[1] A similar molecule is present in lipoproteins from other mammalian species, and it is one of the largest proteins found in mammals. Human apoB contains 25 cysteine residues per mole. Twelve of these are present in the first 500 amino acids, and cysteine residues at positions 3734 and 4190 appear to be "free." The carbohydrate constituents of apoB are mannose, *N*-acetylglucosamine, galactose, and *N*-acetylneuraminic acid; together they constitute 6 to 10% of the total protein mass. Apolipoprotein B is also covalently bound to palmitate and stearate through thioester bonds located at cysteine residues. The chemical link between the protein and fatty acids may be needed to help anchor the apolipoprotein to the lipoprotein.

Little is known about the secondary and tertiary structure of apoB. A mixture of α-helical and β-sheet structure has been reported when the molecule is embedded on the surface of low-density lipoprotein (LDL).[2] Apolipoprotein B is extremely hydrophobic. The amino acid composition has been determined by a number of laboratories, and there is general agreement on the results. The frequency of residues in decreasing order of frequency is the following: Leu > Glu > Asp > Ser > Lys > Thr, Ala, Ile > Val > Gly > Pro > Tyr > Arg > His > Met > $\frac{1}{2}$-Cys. Once apoB is removed from the environment of the lipoprotein, it becomes completely insoluble in an aqueous solution. This property contributed in the long delay in sequencing of apoB compared to other apolipoproteins. The initial approaches to the study of the primary sequence of apoB was to hydrolyze it into fragments that could be characterized chemically. Moreover, significant advances in the elucidation of the primary structure of apoB were made only after the gene encoding the apolipoprotein was cloned and nucleotide sequences deciphered.

[1] C. Y. Yang, Z. W. Gu, S. A. Weng, T. W. Kim, S. H. Chen, H. J. Pownall, P. M. Sharp, S. W. Liu, A. M. Gotto, Jr., and L. Chan, *Arteriosclerosis* (*Dallas*) **9,** 96 (1989).

[2] J. P. Kane, *Annu. Rev. Phys.* **45,** 637 (1983).

The apoB gene is located in chromosome 2.[3] The coding region of the gene consists of 43 kilobases, arranged into 29 exons. The carboxy-terminal side of the protein is encoded by exons 26 through 29, and the amino-terminal portion of the protein is encoded by the remaining 25 exons. The domain of apoB that binds to the LDL receptor is coded by exon 26. In mammals the apoB gene is expressed exclusively in the liver and intestine. The final product of the intestine, however, is an apoB that is 48% the size of hepatic apoB and has been designated apoB-48. Intestinal apoB-48 consists of 2152 amino acids, and this apolipoprotein is the product of apoB mRNA that has been shortened (edited) by posttranscriptional introduction of a stop codon.

Apolipoprotein B is one of the main cholesterol-carrying apolipoproteins, and for this reason it has engendered great interest. The presence of apoB in lipoproteins appears to be essential for rendering them atherogenic. In contrast, lipoproteins devoid of apoB seemingly do not elicit an atherogenic response.

The lipoproteins that contain apoB are LDL, very-low-density lipoprotein (VLDL), and intermediate-density lipoprotein (IDL). There is only one molecule of apoB per particle of LDL, VLDL, and IDL, although each of these lipoproteins vary greatly in lipid-carrying capacity. Apolipoprotein B is secreted from the liver primarily in association with VLDL particles, which are triglyceride-rich lipoproteins. As VLDL particles circulate, the triglycerides undergo lipolysis through the action of lipoprotein lipase (LPL), and cholesterol esters are acquired from other lipoproteins through the assistance of cholesterol ester transfer protein (CETP). These processes lead to formation of VLDL remnants which include IDL. The VLDL and IDL particles contain other apolipoproteins as well; these are the apoC species (C-I, C-II, and C-III) and apoE. The VLDL remnants are further catabolized into LDL. This conversion is mediated in part by hepatic triglyceride lipase (HTGL), although all of the mechanisms are not fully understood.

During the conversion of VLDL to LDL, apoB appears to undergo conformational changes on the surface of lipoprotein particles. These changes have been inferred from changes in immunoreactivity of apoB to monoclonal anti-apoB antibodies. It seems that the conformation necessary for receptor-mediated uptake is attained at the step of LDL formation.

[3] J. Scott, R. J. Pease, L. Powell, S. C. Wallis, B. J. McCarthy, R. W. Mahley, B. Levy-Wilson, and T. J. Knott, *Biochem. Soc. Trans.* **15,** 195 (1987); E. H. Ludwig, B. D. Blackhart, V. R. Pierotti, L. Caiati, C. Fortier, T. Knott, J. Scott, R. W. Mahley, B. Levy-Wilson, and B. J. McCarthy, *DNA* **6,** 363 (1987); B. D. Blackhart, E. M. Ludwig, V. R. Pierotti, L. Caiati, M. A. Onasch, S. C. Wallis, L. Powell, R. Pease, T. J. Knott, M. L. Chu, *et al., J. Biol. Chem.* **262,** 15364 (1986).

The precursors of LDL seemingly do not make use of apoB in receptor-mediated clearance. The concept of conformational changes in apoB also help to explain some of the difficulties encountered in the immunoquantitation of apoB in plasma. Most antibodies to apoB are raised against LDL; these often react poorly with apoB in VLDL,[4] especially in the presence of a high content of triglycerides.[5] These limitations in immunoquantitation of apoB in part account for the need for a specific chemical method for measuring apoB levels.

The primary function of hepatic apoB is to transport triglyceride from the liver for utilization by peripheral tissues. Intestinal apoB-48 serves the same function for newly absorbed triglycerides. This action also prevents accumulation of triglyceride in the liver (or intestine). In many species, VLDL remnants are rapidly removed by the liver, and only small amounts are converted to LDL. In humans a greater fraction of remnants are converted to LDL. The function of LDL is less obvious than that of VLDL, but three functions have been suggested. First, LDL may deliver cholesterol to liver and peripheral tissues; second, LDL carries vitamin E that is required for normal function of the central nervous system; and third, the unesterified cholesterol of LDL may help to maintain the integrity of the membranes of certain cells. The functions of LDL are implied from the clinical features of the rare disorder abetalipoproteinemia. In this disorder, a progressive neurological dysfunction develops that appears to be reversible by parenteral administration of vitamin E. The red blood cells become distorted in the form of acanthocytes; they may be the result of a deficiency of unesterified cholesterol in plasma. However, there is no indication that the lack of delivery of cholesterol to peripheral tissues leads to significant malfunction, and thus this function of LDL may not be essential to health.

Circulating LDL particles are taken up primarily by LDL receptors located on the surface of liver cells.[6] A genetic deficiency of these receptors retards LDL clearance and leads to the disorder called familial hypercholesterolemia (FH). A related condition is familial defective apoB (FDB) in which a substitution of Arg for Gln at amino acid 3500 renders it unrecognizable by LDL receptors. The resulting hypercholesterolemia likewise is characterized by delayed clearance of LDL from circulation.[7,8] Other causes

[4] S. G. Young, R. S. Smith, D. M. Hogle, L. K. Curtis, and J. L. Witztum, *Clin. Chem.* **32,** 1484 (1986).

[5] A. J. Lutalo-Bosa, J. L. Adolphonson, and J. J. Albers, *J. Lipid Res.* **26,** 995 (1985).

[6] J. L. Goldstein and M. S. Brown, *in* "Metabolic Basis of Inherited Disease" (C. R. Scriver, A. L. Beaudet, W. S. Sly, and D. Valle, eds.), p. 1215. McGraw-Hill, New York, 1989.

[7] G. L. Vega and S. M. Grundy, *J. Clin. Invest.* **78,** 1410 (1986).

[8] T. L. Innerarity, K. H. Weissgraber, K. S. Arnold, R. W. Mahley, R. M. Krauss, G. L. Vega, and S. M. Grundy, *Proc. Natl. Acad. Sci. U.S.A.* **84,** 6919 (1987).

of hypercholesterolemia are overproduction of VLDL by the liver, which leads to increased conversion of VLDL to LDL, and overloading of LDL particles with cholesterol esters. The former is due mainly to obesity, whereas the causes of the latter are not known. In all forms of hypercholesterolemia except in the latter, plasma levels of LDL apoB are abnormally high; in the latter, the LDL apoB levels are normal, and only LDL cholesterol levels are raised.[9]

Another condition that has focused attention on apoB is "hyperapobetalipoproteinemia" (hyperapo B).[10] This disorder originally was defined as an abnormally high level of LDL apoB, but the definition has been extended to signify an increase in serum total apoB levels. Hyperapobetalipoproteinemia occurs most commonly in patients with elevated LDL, IDL, and VLDL concentrations, and it is characterized by an increase in apoB levels out of proportion to cholesterol levels. Reports that patients with hyperapo B have a higher risk for coronary heart disease (CHD) than would be detected from measurements of total cholesterol or LDL cholesterol has been an impetus for the development of methods for the accurate measurement of plasma levels of apoB.

In clinical practice it would be convenient to measure apoB levels by immunological techniques; several such methods have been introduced, and others are under development. However, the development of immunological methods for determining apoB has been accompanied by several problems, and these have not been entirely overcome.[11] For this reason, our laboratory, which has measured apoB in humans for investigation of apoB metabolism, has chosen to employ chemical methods to quantify apoB. Our methods seemingly provide accurate and reproducible quantitation even in the presence of high plasma triglycerides. These advantages of the chemical measurement compensate for the lack of convenience and slow speed of the assay, particularly for research purposes. This chapter outlines the specific steps in chemical measurement of apoB.

Quantitation of Total Protein in Plasma Lipoproteins

The Lowry–Folin assay[12] is commonly used for quantitation of the total protein in lipoproteins. The method measures the tyrosine and tryptophan contents of proteins, although there is some additional color derived from

[9] G. L. Vega, M. A. Denke, and S. M. Grundy, *Circulation* **84,** 118 (1991).

[10] A. Sniderman, S. Shapiro, D. Marpole, B. Skinner, B. Teng, and P. O. Kwiterovich, Jr., *Proc. Natl. Acad. Sci. U.S.A.* **77,** 604 (1980).

[11] J. J. Albers, S. M. Marcovina, and H. Kennedy, *Clin. Chem.* **38,** 658 (1992).

[12] O. H. Lowry, N. J. Rosebrough, A. L. Farr, and R. J. Randall, *J. Biol. Chem.* **193,** 265 (1951).

peptide bonds assayed after chelation of copper ions under alkaline conditions.[13–15] The assay requires (a) removal of lipids that increase background absorbance (turbidity), and (b) comparison of the absorbance of the samples to that of a reference protein, usually bovine serum albumin. Beer's law of spectrophotometry is applied for quantitation.

Delipidation of Lipoprotein Samples after Reaction with Phenol

The presence of lipids in lipoproteins requires modification of chemical methods for quantitation of total protein. Most investigators have employed diethyl ether,[16,17] or chloroform[18,19] to remove turbidity from the lipoprotein samples. Other investigators have used sodium dodecyl sulfate,[20–24] Triton X-100, Nonidet P-40, or Tween 80. However, a number of these detergents have been shown to interfere with the standard biuret and Folin methods for protein measurement in the presence of lipid.[25] The blanks are high, and precipitates form on addition of phenol. Sodium dodecyl sulfate has prevailed as the reagent of choice. This detergent has proved to be effective in solubilizing large amounts of lipids in a short time, and it does not precipitate on addition of phenol.[26] The ease of solubilization of proteolipids also circumvents a problem that occurs with other detergents, namely, an increase in reactivity of the sample mixture with phenol owing to oxidation of lipids in alkali.[27]

Selection of Standard for Colorimetric Assay

The preferred methods for determining the amount of protein in any sample are estimation of dry weight, nitrogen content, or amino acid content after hydrolysis of the peptide. These methods, however, are not practical

[13] H. E. Schultze and F. J. Heremans, *in* "Molecular Biology of Human Proteins," Vol. 1, Section I, Chap. 2, and Section II, Chap. 1. Elsevier, New York, 1966.
[14] D. Watson, *Adv. Clin. Chem.* **8,** 237 (1965).
[15] L. Kirk, *Adv. Protein Chem.* **3,** 139 (1947).
[16] A. Gustafson, P. Alaupovic, and R. H. Furman, *Biochemistry* **4,** 596 (1965).
[17] T. Sata, R. J. Havel, and A. L. Jones, *J. Lipid Res.* **13,** 757 (1972).
[18] J. P. Kane, T. Sata, R. L. Hamilton, and R. J. Havel, *J. Clin. Invest.* **56,** 1622 (1975).
[19] H. H. Hess and E. Lewis, *J. Neurochem.* **12,** 205 (1965).
[20] G. L. Vega and S. M. Grundy, *J. Lipid Res.* **25,** 580 (1984).
[21] M. B. Lees and S. Paxam, *Anal. Biochem.* **47,** 184 (1972).
[22] G. L. Peterson, *Anal. Biochem.* **83,** 346 (1977).
[23] H. H. Hess, M. B. Lees, and J. E. Derr, *Anal. Biochem.* **85,** 295 (1978).
[24] M. A. K. Markwell, S. M. Haas, L. L. Bieber, and N. E. Tolbert, *Anal. Biochem.* **87,** 206 (1978).
[25] I. H. Mather and C. B. Tamplin, *Anal. Biochem.* **93,** 139 (1979).
[26] H. H. Hess, M. B. Lees, and J. E. Derr, *Anal. Biochem.* **85,** 295 (1978).
[27] J. Eichberg and L. C. Mokrach, *Anal. Biochem.* **30,** 386 (1969).

for routine analysis. For this reason, colorimetric assays have been developed.

One problem with colorimetric assays has been the selection of an appropriate standard. An ideal standard would be a pure protein that is not hygroscopic, that is not a glycoprotein or proteolipid, and that is water-soluble. The standard should also follow Beer's law of spectrophotometry for a wide range of concentrations, and its reactivity with the Folin reagent should be similar to that of the samples to be tested. As early as 1968 three proteins were consider for this purpose: bovine serum albumin, human serum albumin, and ovalbumin.[28] Bovine serum albumin (BSA) has come to be preferred, mainly because it is relatively inexpensive to purify. Moreover, it contains less than 0.05% carbohydrate, 0.5% lipid, and less than 0.02% heavy metals; its ash content is less than 0.1%.[29] The properties of a primary protein standard subsequently released by the National Bureau of Standards (NBS) [currently the National Institute of Standards and Technology (NIST)] have been detailed.[30,31] This standard met most of the criteria noted above, except for the chromogenicity factor. The precise concentration of the NIST standard is determined by Kjeldahl method, absorbance at 280 nm, and gravimetry. The standard is packaged in ampules of 1 ml of known concentration, and the solvent is saline. This standard has been employed in our laboratory for estimating total apolipoprotein in lipoprotein samples ever since its commercial release (ca. 1980).

A "Check-Sample" program conducted by the American Society of Clinical Pathologists[32] to determine the accuracy and precision of protein quantitation among laboratories demonstrated considerable variation in values reported for check samples. Moreover, they showed that the variability was primarily due to variability in the standards used for protein quantitation. In 1990, the Division of Environmental Health Laboratory Sciences, Center for Environmental Health and Injury Control of the Centers for Disease Control (CDC), and the Committee on Apolipoproteins of the International Federation of Clinical Chemistry also conducted a survey of apolipoprotein quantitation in 93 laboratories.[33] For laboratories using the

[28] T. Peters, Jr., *Clin. Chem.* **14,** 1147 (1968).

[29] B. T. Doumas, *Clin. Chem.* **21,** 1159 (1975).

[30] B. T. Doumas, D. D. Bayse, R. J. Carter, T. Peters, Jr., and R. Schaffer, *Clin. Chem.* **27,** 1642 (1981).

[31] National Institute of Standards and Technology Certificate of Standard Reference Material 927a. W. P. Reed, Chief of Standard Reference Program, Gaithersburg, Maryland, 1993.

[32] T. B. Doumas and J. V. Straumfjord, Jr, Check-Sample, Albumin. Commission on Continuing Education, American Society of Clinical Pathologists, Chicago, Illinois, 1973.

[33] L. O. Henderson, M. K. Powell, S. J. Smith, W. H. Hannon, G. R. Cooper, and S. M. Marcovina, *Clin. Chem.* **36,** 1911 (1990).

TABLE I
SUBSTANCES KNOWN TO INTERFERE WITH LOWRY–FOLIN ASSAY

Increase blank absorbance	Decrease color development
Hydrazine	Glycerol
Dithiothreitol	Triton X-100 (≥0.1%, v/v)
2-Mercaptoethanol	Ethylene glycol
Glutathione	Glycine (≥0.5%, w/v)
EDTA (≥0.5 m*M*)	Citrate (≥0.1 m*M*)
Penicillin	Ammonium sulfate
HEPES	
Tricine	
Acetylacetone	
Glucosamine	
Fructose	
Mannose (≥1 m*M*)	
Glucose (≥1 m*M*)	
Sucrose	
Triton X-100	

Lowry procedure, a coefficient of variation of 23% (among-laboratory and within-laboratory) was noted for LDL protein measurements. Moreover, the coefficient of variation averaged about 50% for LDL quantitation when methods other than the Lowry assay were used. Clearly this is an unacceptable variation, and the causes need to be addressed.

Substances That Interfere with Lowry–Folin Assay

A number of substances other than lipids have been reported to interfere with the Lowry–Folin assay (Table I). Some workers recognized this problem and attempted to use reagent blanks to solve the problem. However, it is not always possible to use an appropriate reagent blank because the concentration of the interfering substance may vary among samples, and sometimes its concentration may be too high to make any meaningful correction. Thus, Bensadoun and Weinstein[34] developed a method of quantitative precipitation of proteins for solutions containing protein masses in the range of 1 to 50 μg. These workers showed a quantitative recovery of protein after its precipitation from solutions containing a combination of sodium deoxycholate and trichloroacetic acid. The protein masses can then be estimated by the Lowry–Folin method. The procedure for precipitation

[34] A. Bensadoun and D. Weinstein, *Anal. Biochem.* **70,** 241 (1976).

TABLE II
PROCEDURE FOR REMOVING SUBSTANCES THAT INTERFERE WITH LOWRY–FOLIN ASSAY[a]

Stock solutions
2% Sodium deoxycholate
1 *N* Sodium hydroxide
5% Cupric sulfate pentahydrate
2% Sodium carbonate
24% Trichloroacetic acid
2% Potassium/sodium tartrate
Folin–Ciocalteu phenol reagent (2*N*)
1. Adjust sample volume to 3 ml with deionized water
2. Add 25 μl sodium deoxycholate solution
3. Mix vigorously and let stand for 15 min
4. Add 1 ml trichloroacetic acid solution
5. Mix and spin at 100,000 *g*-min in swinging-bucket rotor
6. Remove and discard supernatant
7. Redissolve pellet in 1.5 ml Lowry reagent[34]
8. Mix vigorously for 10 sec
9. Add 150 μl Folin–Ciocalteu reagent (1 *N*); mix vigorously
10. Allow color development for 45 min
11. Read absorbancies at 660 nm

[a] From Bensadoun and Weinstein.[35]

is summarized in Table II. Subsequently, Peterson confirmed and extended the method of Bensadoun and Weinstein, and related methods.[35–37]

Markwell Modification of Lowry Procedure

A commonly used modification of the Lowry–Folin method is that of Markwell *et al.*[38] This method has been compared with the original procedure of Lowry *et al.,* and the results correlate well.[20] The technique is detailed in Table III. The main advantage of the Markwell modification is that organic solvents are not required to delipidate lipoproteins after the colorimetric reaction; still, the method is sensitive and linear. Even so, this method is not impervious to interfering substances.

[35] G. L. Peterson, *Anal. Biochem.* **83,** 346 (1977).
[36] M. B. Lees and S. Paxman, *Anal. Biochem.* **47,** 184 (1972).
[37] G. R. Schacterle and R. L. Pollack, *Anal. Biochem.* **51,** 654 (1973).
[38] M. A. K. Markwell, S. H. Haas, N. E. Tolbert, and L. L. Bieber, this series, Vol. 72, p. 296.

TABLE III
MODIFICATION OF LOWRY–FOLIN PROTEIN ASSAY[a]

Stock reagents
- Solution 1: 2.0% Sodium carbonate, 0.4% (w/v) sodium hydroxide, 0.16% (w/v) sodium tartrate, 1.0% (w/w) sodium dodecyl sulfate
- Solution 2: 4.0% (w/v) $CuSO_4 \cdot 5H_2O$
- Solution 3: 100 parts of solution 1 mixed with 1 part of solution 2 on day of assay
  - Folin–Ciocalteu diluted to 1 *N* (v/v) with deionized water
  - 7% (w/v) solution of bovine serum albumin (BSA–NIST)

Preparation of BSA–NIST standard curve
1. Prepare 5% (w/v) BSA stock solution using 150 m*M* NaCl as diluant
2. Dilute stock solution serially to obtain standard solutions of 2.5% (50 μg/20 μl), 1.25% (25 μg/20 μl), 0.625% (12.5 μg/μl), and 0.312% (6.25 μg/μl)
3. Use 20 μl of each standard in duplicate for assay

Assay procedure
1. Adjust sample to 1 ml with 150 m*M* sodium chloride solution; aliquot of sample should contain 10–50 μg protein
2. Add 3 ml solution 3
3. Mix vigorously and incubate at room temperature for 10 min
4. Add 300 μl dilute Folin–Ciocalteu reagent
5. Mix tube contents vigorously as solution is added
6. Incubate tubes at room temperature for 45 min
7. Read absorbancies of samples, standards, and blanks at 660 nm; final reaction color is stable for at least 40 min

[a] The procedure detailed is for a manual assay using SMI pipettes. A different procedure has been developed for automation.[38]

Characteristics of Assay

The standard curve consists of serial dilutions of BSA (NIST) made in 150 m*M* NaCl; the protein mass ranges from 5 to 50 μg. Five different concentrations are used, and determinations are made in duplicate. Reagent blanks are included routinely. The standard curve is linear, with a correlation coefficient greater than or equal to 0.998. The fitted line generally has an intercept less than 0.001. A second-degree polynomial also can be fitted with a correlation coefficient of 1.000, and with an intercept less than 0.001. The coefficient of variation for absorbancies of the standards within assay and between assay is less than, or equal to 1.5%.

The sensitivity of the assay has been determined using LDL. The method seems suitable for measurement of apolipoprotein within the range of 5 to 50 μg (See Fig. 1 showing range of 0 to 30 μg). A quality control that consists of BSA standards of low protein concentration (0.5 mg/ml) and high protein concentration (1.5 mg/ml) is included in every assay. The within-day and between-day coefficients of variation are less than 3%.

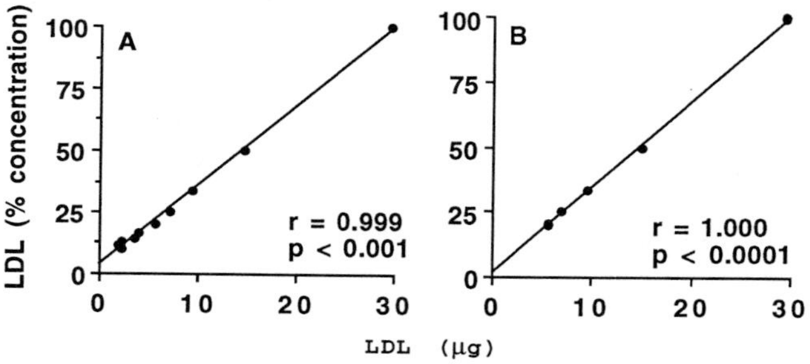

FIG. 1. Sensitivity of the modified Lowry–Folin assay[38] using LDL. An LDL protein solution containing 1.47 ± 0.01 mg/ml apolipoprotein was diluted serially with 150 m*M* NaCl. (A) Expected mass of LDL protein as a function of measured protein after serial dilution. A correlation coefficient of 0.999 for agreement between the expected and measured values was obtained over a dilution range of LDL between 0 and 90%. (B) Correlation of 1.00 for agreement between expected and measured values. A desirable agreement between expected and measured values is obtained from 5 to 30 μg.

Measurement of Very-Low-Density and Intermediate-Density Lipoprotein Apolipoprotein B

Indirect Method

The indirect method for measurement of apoB in VLDL requires determination of total protein as well as soluble non-apoB apolipoproteins after selective precipitation of apoB with organic solvents. The mass of apoB is estimated as the difference between total lipoprotein and soluble apolipoproteins. Kane *et al.*[39,40] introduced 1,1,3,3-tetramethylurea (TMU) for the rapid and selective precipitation of apoB in VLDL. At a concentration of 4.2 *M*, TMU selectively precipitates apoB, whereas the other apolipoproteins remain in solution in the supernatant. Total VLDL protein and soluble apolipoproteins are measured by the Lowry–Folin method. Kane *et al.*[40] determined the absolute mass of apoB gravimetrically and compared it to the mass determined by the Lowry procedure. A chromogenicity factor of 1.0 for apoB compared to BSA was found after correction for the carbohydrate content of 6.4%. For the TMU-soluble fraction, the chromogenicity factor was 1.16 compared to BSA after correction for a carbohydrate content of 5.3%. Unfortunately, there are disadvantages in using the TMU reagent: it has a short shelf-life, it causes a high blank reading in the Lowry assay, and the selective precipitation procedure requires an elevated

[39] J. P. Kane, *Anal. Biochem.* **53,** 350 (1973).

[40] J. P. Kane, T. Sata, R. L. Hamilton, and R. J. Havel, *J. Clin. Invest.* **56,** 1622 (1975).

temperature. Moreover, commercial production of the reagent was discontinued in the late 1970s.

Separately, Holmquist and Carlson[41] demonstrated that a 50% solution of 2-propanol was as effective as 4.2 *M* TMU in precipitating apoB and solubilizing apoC's and apoE in VLDL. Holmquist *et al.*[42] further compared the use of 2-propanol and TMU in the solubilization and quantitation of VLDL apolipoproteins. They obtained a correlation coefficient of 0.995 comparing the two methods for measurement of soluble (non-apoB) apolipoproteins. Furthermore, the extraction procedure did not require a long incubation time, and it could be carried out at room temperature. These workers, however, did not address the issue of differences in chromogenicities between albumin and the soluble apolipoproteins for quantitation with the Lowry method.

Direct Method

The direct method is based on the resolubilization of precipitated apoB and its direct measurement. Helenius and Simons[43] first attempted to solubilize apoB for estimation of molecular size. They showed that sodium dodecyl sulfate displaced all lipid from LDL, and a soluble protein (apo-LDL) could be isolated by gel-filtration chromatography.[44] Subsequently, they demonstrated that sodium deoxycholate and Nonidet P-40 delipidated LDL (1% residual lipids) and that these detergents effectively solubilized apoB when used at critical micellar concentrations. In addition, as shown by Bensadoun and Weinstein,[34] and confirmed by Peterson,[35] sodium deoxycholate does not interfere with the Lowry–Folin assay. In our laboratory, we have developed a method for direct quantitation of VLDL apoB by combining the principles derived from the work of Kane *et al.*,[39,40] Holmquist *et al.*,[41,42] and Helenius and Simons.[43,44] The method is detailed in Table IV. Briefly, the procedure involves (1) precipitation of apoB with 50% 2-propanol, (2) delipidation of precipitate, and (3) solubilization of precipitate with sodium deoxycholate. The precipitate contains only apoB, as shown by electrophoresis using 2 to 16% acrylamide gradient gels in the presence of 2% sodium dodecyl sulfate.

Comparison of the plasma levels of (VLDL+LDL)-apoB determined by the direct and indirect methods is shown in Fig. 2 for seven multiple measurements on seven normotriglyceridemic adults. Each had daily mea-

[41] L. Holmquist and K. Carlson, *Biochim. Biophys. Acta* **493,** 400 (1977).
[42] L. Holmquist, K. Carlson, and L. A. Carlson, *Anal. Biochem.* **88,** 457 (1978).
[43] A. Helenius and K. Simons, *Biochemistry* **10,** 2542 (1971).
[44] K. Simons and A. Helenius, *Fed. Eur. Biochem. Soc. Lett.* **7,** 59 (1970).

TABLE IV
SELECTIVE PRECIPITATION OF APOLIPOPROTEIN B-100 AND SOLUBILIZATION

Stock reagents
- 2.0% Sodium deoxycholate in 1 *N* NaOH
- 2-Propanol (A.C. reagent grade)

Procedure
1. Aliquot volume of lipoprotein (VLDL+IDL) containing ≥150 μg protein
2. Add equal volume of 150 m*M* NaCl, and vortex
3. Add equal volume of 2-propanol to mixture obtained in step 2
4. Spin at 3000 rpm for 30 min at room temperature
5. Remove supernatant; rinse precipitate with 50% 2-propanol, and repeat step 4
6. Add 2 ml of 2-propanol to precipitate, vortex
7. Repeat steps 4 and 5
8. Solubilize precipitate in 500 μl of 2.0% sodium deoxycholate in 1 *N* NaOH
9. Aliquot 100 μl in duplicate for Lowry–Folin assay
10. Proceed with protein assay as described in Table III.

surements of plasma lipids, lipoprotein cholesterol, apolipoprotein B in (VLDL+IDL), and LDL. With the direct method, the coefficient of physiological variation for VLDL plus IDL apoB averaged 11.3 ± 0.06%, whereas it was 9.92 ± 6.35% for the indirect method. These two values are not significantly different ($p = 0.54$, estimated by paired t-test). Nonetheless, the direct method has advantages over the indirect method; for example, a soluble form of apoB is available for additional analysis such as specific activity measurements for tracer kinetic studies, and variability in the chromogenicities of 2-propanol-soluble fractions due to changes in concentrations of apoE or apoC's need not be taken into account. The latter is

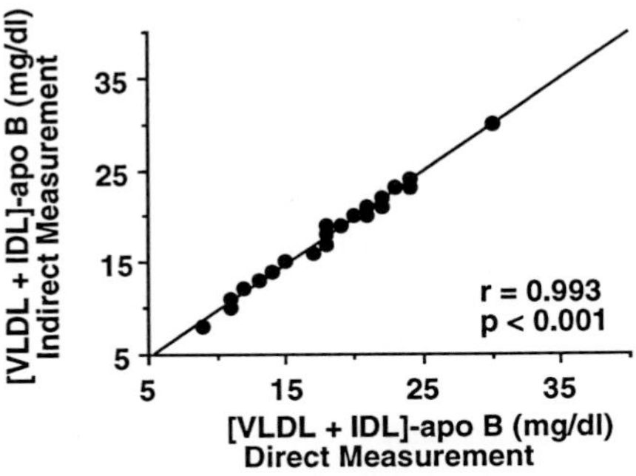

FIG. 2. Comparison of concentrations of (VLDL+IDL)-apoB determined by the direct method and indirect method in normotriglyceridemia.

a matter of concern, especially in samples obtained from patients with hypertriglyceridemia.

Measurement of Low-Density Lipoprotein Apolipoprotein

Determination of LDL protein by the Markwell modification of the Lowry–Folin method appears to be straightforward, but a number of factors must be considered. Foremost is whether measurement of total apolipoprotein in LDL is equivalent to measurement of apoB. Some workers have shown that apoB comprises more than 95% of the total LDL apolipoprotein when LDL is isolated in the density range of 1.019 to 1.063 g/ml.[45] Others, however, have identified apolipoproteins C-III, C-I, and E, and they have suggested that these apolipoproteins are significant constituents of LDL.[46] Moreover, traces of albumin are present in LDL preparations that are freshly isolated and not washed. However, albumin is considered a contaminant, rather than a constituent of the lipoprotein.

Redistribution of apolipoproteins is known to result from ultracentrifugation in a salt gradient, and the presence of apolipoproteins in LDL does not necessarily mean that they are physiologically attached. Artificial fragments of lipoproteins containing other apolipoproteins may float in the LDL density range. Besides, Gibson *et al.*[47] showed that plasma of normotriglyceridemic subjects contains discrete lipoprotein particles that harbor apolipoprotein E and that some of these particles overlap with LDL in buoyant density. Thus, LDL isolated by ultracentrifugation probably contains trace amounts of apolipoproteins other than apoB.

The chromogenicity factors for apoE, apoC-II, and apoC-III have not been determined with certainty for the Lowry procedure. However, these apolipoproteins differ markedly from apoB and from one another in their affinity for dyes such as Coomassie blue. Their presence could slightly distort measurement of LDL apoprotein content as estimated by the Lowry–Folin procedure. Nonetheless, for practical purposes, determination of LDL apolipoprotein appears to correspond closely to LDL apoB.

For clinical purposes and some clinical research protocols, "LDL" is defined by lipoproteins of density range 1.006 to 1.063 g/ml. This represents LDL obtained by "beta-quantification." However, this fraction contains LDL plus IDL, and IDL has apoE. Thus, only about 90% of LDL plus IDL protein consists of apoB.

[45] P. N. Herbert, T. M. Forte, R. S. Schulman, M. J. LaPiana, E. L. Gong, R. I. Levy, D. S. Fredrickson, and A. V. Nichols, *Prep. Biochem.* **5,** 93 (1975).

[46] D. M. Lee and P. Alaupovic, *Biochim. Biophys. Acta* **879,** 126 (1986).

[47] J. C. Gibson, A. Rubinstein, P. R. Bukberg, and V. W. Brown, *J. Lipid Res.* **24,** 886 (1986).

Estimation of Total Plasma Apolipoprotein B by Chemical Method in Normotriglyceridemia

Levels of plasma apolipoprotein B can be calculated as the sum of VLDL plus IDL apoB and LDL apoB. This requires the following steps: (a) ultracentrifugal isolation of VLDL plus IDL ($d < 1.019$ g/ml) and the plasma infranatant ($d > 1.019$ g/ml); (b) determination of total plasma cholesterol and cholesterol concentration in the corresponding plasma fractions ($d < 1.091$ g/ml, and $d > 1.019$ g/ml); (c) measurement of HDL cholesterol in plasma for samples with triglycerides less than 300 mg/dl or in the plasma infranatant ($d > 1.0063$ g/ml) for plasma samples with triglycerides greater than or equal to 300 mg/dl; (d) estimation of LDL cholesterol as the difference between total plasma cholesterol minus the sum of (VLDL+IDL) cholesterol and HDL cholesterol; (e) measurement of apoB in (VLDL+IDL) by the direct method, and (f) measurement of the ratio of cholesterol to apolipoprotein in LDL ($d = 1.020–1.065$ g/ml). Finally, the latter ratio is multiplied by the estimated LDL cholesterol (step d) to obtain LDL apolipoprotein B. Plasma total apoB is the sum of VLDL plus IDL apoB and LDL apoB.

Estimation of Total Plasma Apolipoprotein B by Non-High-Density Lipoprotein Cholesterol in Normotriglyceridemia

For clinical purposes, a close approximation of plasma total apoB can be made from cholesterol concentrations in LDL plus IDL plus VLDL fractions (non-HDL cholesterol). We have shown that this estimation is possible because a high correlation ($r = 0.951$, $p < 0.001$, for $n = 72$ subjects) was obtained between plasma concentration of apoB and non-HDL cholesterol for plasma samples of triglyceride levels below 300 mg/dl.[48] The regression equation was as follows: plasma apoB (mg/dl) = 4.62 + 0.62[non-HDL cholesterol (mg/dl)]. Furthermore, a reasonable estimate of total apoB concentrations in plasma of normotriglyceridemic subjects can be obtained by multiplying 0.65 by the non-HDL cholesterol concentrations. This factor also has proved useful for comparison of expected levels of plasma apoB and measured levels by immunoassays or by chemical methods. Values that deviate from the expected levels usually require further evaluation.

[48] G. L. Vega and S. M. Grundy, *Arteriosclerosis (Dallas)* **10,** 668 (1990).

Estimation of Total Plasma Apolipoprotein B in Hypertriglyceridemia

Measurement of total apoB in patients with hypertriglyceridemia may be of special clinical importance. Sniderman and co-workers[49] have reported that hypertriglyceridemic patients with elevated apoB levels commonly have premature CHD, whereas those with low apoB levels are much less prone to CHD. This observation suggests than an accurate measurement of total apoB concentration in hypertriglyceridemic patients would be both desirable and valuable. Unfortunately, a standardized method for measurement of serum apoB concentrations especially in hypertriglyceridemic patients is not available for clinical diagnosis of hyperapobetalipoproteinemia. For example, Lulato-Bosa *et al.*[5] demonstrated that some immunochemical methods overestimate levels of immunoreactive apoB in sera of hypertriglyceridemic subjects. Sniderman *et al.,* however, have not observed such a problem.[50] Thus, it is clear that considerable work remains before immunoassays can be used reliably for quantitation of apolipoprotein B in the presence of high levels of triglyceride.

In our laboratory, we use the direct chemical measurement of apoB as described in this chapter for measurement of concentrations in plasma of hypertriglyceridemic subjects. This method also has been employed in the study of lipoprotein kinetics in hypertriglyceridemic patients for the evaluation of underlying metabolic defects, and for effects of competitive inhibitors of 3-hydroxy-3-methylglutaryl-coenzyme A (HMG-CoA) reductase and of fibric acids on total apoB levels.[51] Measurement of VLDL plus IDL apoB in samples with triglyceride levels ranging from 300 to 600 mg/dl requires some modifications. First, one must ensure complete precipitation of apoB in VLDL plus IDL, and, second, it is necessary to precipitate apoB in LDL using 2-propanol.

Use of Chemical Method for Standardizing Immunoassays

The primary standard used by many laboratories for immunoquantitation of plasma apolipoprotein B is LDL apolipoprotein. The quantity of apolipoprotein in LDL is measured by the Markwell modification, with a serial dilution curve prepared as the standard for the immunoassay. The quality of the immunoassay therefore is dependent on the accuracy of

[49] A. Sniderman, C. Wolfson, B. Teng, F. A. Franklin, P. S. Bachorick, and P. O. Kwiterovich, Jr., *Ann. Intern. Med.* **97,** 833 (1982).

[50] A. Sniderman, H. Vu, and K. Cianflone, *Atherosclerosis* **89,** 109 (1991).

[51] G. L. Vega and S. M. Grundy, *JAMA, J. Am. Med. Assoc.* **264,** 2759 (1990).

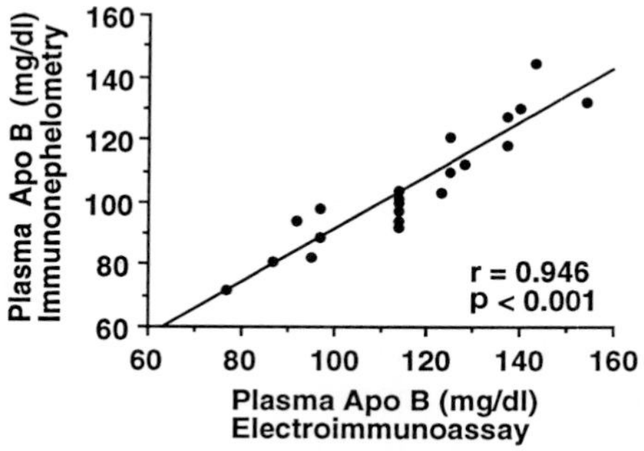

FIG. 3. Comparison of immunoquantitation of apoB in plasma of normotriglyceridemic subjects. The primary standard was an LDL preparation, washed twice at its native density. Apolipoprotein concentration was measured by the Markwell modification of the Lowry–Folin method. Bovine serum albumin (NIST) was employed as a standard in the Lowry assay.

chemical analysis of LDL apoB. Variability in chemical analysis of the standard accounts for differences among laboratories in immunoquantitation of apoB. For example, in one study of immunoquantitation,[33] the total coefficient of variation including within-laboratory and among-laboratory errors was found to be 23% for LDL. By standardizing the chemical procedure and using a bovine serum albumin solution calibrated at the NIST, a reasonable agreement can be obtained between different types of immunoassays. For example, measurement of apoB has been compared in our laboratory using the electroimmunoassay of Laurell and immunonephelometry. A correlation of 0.946 ($p < 0.001$) was obtained for total apoB in a series of samples from normotriglyceridemic patients using the two methods (Fig. 3).

The International Federation of Clinical Chemistry Committee on Apolipoproteins has developed a standardization approach for immunoquantitation of plasma apolipoproteins B and A-I.[52] An international reference standard was selected for apolipoprotein A-I. The absolute mass of the standard was determined by amino acid analysis. This material was used as a primary standard to calibrate secondary standards by laboratories participating in the standardization program. The interlaboratory coefficient of variation obtained for measurement of the same 50 samples by each of the participating laboratories was 3.6%. Thus, standardizing the primary standard proved to be effective in achieving accuracy among laboratories using different methods of immunoquantitation of apoA-I. A simi-

[52] S. M. Marcovina, J. J. Albers, L. O. Henderson, and W. H. Hannon, *Clin. Chem.* **39,** 773 (1993).

lar standardization remains to be developed for immunoquantitation of apolipoprotein B. The development of an appropriate standard that is stable and widely available appears to be the main obstacle.

Quantitation of Apolipoprotein B-48

Apolipoprotein B-48 is a constituent of chylomicrons, and it is also present among VLDL remnants of patients with hypertriglyceridemia and normolipidemia. Therefore, some attempts have been made in a number of laboratories to quantify apoB-48. This has proved to be difficult. Apolipoprotein B-48 coprecipitates with apolipoprotein B-100 during precipitation of apoB with 2-propanol. Antibodies raised against apoB-100 also cross-react with apoB-48. Therefore, quantitation of apoB-48 by a direct chemical assay or immunoassay is not possible at this time. Other methods have been introduced for this purpose, including high-performance liquid chromatography (HPLC)[53] and sodium dodecyl sulfate–polyacrylamide gel electrophoresis.[54]

Hidaka *et al.*[53] used two columns, each with a diameter of 7.9 mm and length of 25 cm. The columns were packed with hydrophilic silica gel (pore size of 300 Å). The elution buffer consisted of 0.2 mol/liter NaCl, 10 m*M* sodium phosphate buffer (pH 7.0), and 0.1% (w/v) sodium dodecyl sulfate. The elution temperature was 40°, and the flow rate was 0.4 ml/min. Elution profiles of VLDL apoB-100 and apoB-48 were monitored at 280 nm. The coefficient of variation for the assay was 2.5%. Using this method, Hidaka *et al.*[53] concluded that (1) apoB-48 and apoB-100 can be separated effectively and reproducibly by HPLC; (2) the assay was linear for each species of apoB, between 0 and 20 μg of total apoB; and (3) the percent apoB-48 in VLDL of 14 healthy individuals was 17.1 $\pm$ 5.6% of total apoB. However, these workers did not determine the mass of each apolipoprotein by amino acid analysis after separation of the two species of apoB. Nonetheless, HPLC appears to be a promising approach for the separation and measurement of apoB-48 and apoB-100.

Another approach for semiquantitation of the two species of apoB is densitometric scanning of Coomassie blue-stained proteins after separation by electrophoresis in a polyacrylamide gel matrix in the presence of sodium dodecyl sulfate. Poapst *et al.*[54] used this method and determined the chromogenicity factors for the two species of apoB. Of interest, these workers demonstrated that the chromogenicity of apoB-100 is greater than the chromogenicity of apoB-48. Also, they showed that the mass ratios of apoB-100 to apoB-48 in an undiluted sample is given by the product of the ratios

[53] H. Hidaka, H. Kojima, Y. Nakajima, *et al., Clin. Chem. Acta* **189,** 287 (1990).
[54] M. Poapst, K. Uffelman, and G. Steiner, *Atherosclerosis* **65,** 75 (1987).

of the slopes of lines that define staining of apoB-100 and apoB-48 as a function of apoB mass. The latter is determined by amino acid analysis for each species of apoB. The following equation was developed:

$$M_{100}/M_{48} = (c_{48}/c_{100})(d_{100}/d_{48})$$

where M is mass of either apoB-100 or apoB-48, c is the slope of the lines relating staining to measured aminoacyl mass of the apolipoprotein, and d is the slope of the lines relating staining to the sample dilution. A factor of 0.572 ± 0.028 was estimated for the ratio of c_{48}/c_{100}.

Determination of Specific Activities of Apolipoprotein B for Tracer Kinetic Studies

Specific activity measurements of apoB are needed for tracer kinetic studies. These measurements require specific, reproducible, and sensitive measurement of apoB mass in VLDL, IDL, and LDL. A number of methods have been developed, some of which are based on the procedures detailed above. Most methods involve selective precipitation of apolipoprotein B with organic solvents, and a number of solvents for solubilization of the apoB have been used. These methods[55–60] are summarized in Table V.

Other Protein Quantitation Methods

Protein quantitation, including measurements of apolipoprotein, has been approached using dye-binding methods or coupling of fluorescent probes. An example of each is detailed below.

Bradford Procedure

The principle of this method is the dye-binding affinity of proteins.[61] The dye used is Coomassie Brilliant Blue G-250, which binds to aromatic residues and to arginine residues. Coomassie Brilliant Blue has two different colors, red and blue. The red form is converted to blue on binding of the dye

[55] P. Bohlen, S. Stein, W. Dairman, and S. Underfriend, *Arch. Biochem. Biophys.* **155,** 213 (1973).

[56] L. L. Kilgore, J. L. Rogers, B. W. Patterson, N. H. Miller, and W. R. Fisher, **145,** 113 (1985).

[57] N. A. Le, J. S. Melish, B. C. Roach, H. N. Ginsberg, and W. V. Brown, *J. Lipid Res.* **19,** 578 (1978).

[58] G. Egusa, D. W. Brady, S. M. Grundy, and B. V. Howard, *J. Lipid Res.* **24,** 1261 (1983).

[59] R. L. Klein and D. B. Zilversmit, *J. Lipid Res.* **25,** 1380 (1984).

[60] G. L. Vega and S. M. Grundy, *Kidney Int.* **33,** 1160 (1988).

[61] M. M. Bradford, *Anal. Biochem.* **72,** 248 (1976).

TABLE V
METHODS EMPLOYED FOR MASS MEASUREMENTS OF APOLIPOPROTEIN B DURING DETERMINATION OF SPECIFIC ACTIVITIES IN TRACER KINETIC STUDIES[a]

Precipitating reagent	Solvent used for solubilizing apoB	Quantitative assay	Chromogenicity factor
TMU	Lowry reagent	Lowry–Folin	1[b]
IPA	1 *N* NaOH	Lowry–Folin	1[b]
BUT–IPE	0.5 *N* NaOH	Fluorescence	1[b]
SDS–PAGE	Electroelution into buffer	Fluorescence	1.5[c]
IPA	2% NaDCA	Markwell method	1[b]

[a] TMU, Tetramethylurea; IPA, 2-propanol; BUT–IPE, *n*-butanol/isopropyl ether; SDS–PAGE, polyacrylamide gel electrophoresis in sodium dodecyl sulfate; NaDCA, sodium deoxycholate. From Refs. 57–60.
[b] Chromogenicity factor compared to BSA.
[c] Chromogenicity factor compared to BSA using the Lowry–Folin method is 0.78 ± 0.03 (W. Fisher and V. N. Schumaker, this series, Vol. 128, p. 247).

to the protein.[62] The absorbance of the dye–protein complex is measured at 595 nm, and Beer's law may be employed for quantitation. The method appears to be very rapid and sensitive. However, proteins differ in affinity for the dye, as shown for equal masses of bovine serum albumin, β-gammaglobulin, egg albumin, and lysozyme.[63,64] Thus, application of Beer's law for quantitation requires the use of "chromogenicity factors." These have not been derived for the apolipoproteins relative to the standards used in the Bradford protein assay.

Fluorescamine Method

The fluorescamine method assays free amino groups in proteins using fluorescamine {4-phenylspiro[furan-2(3*H*),1′-phthalan]-3,3′-dione}.[55] The manual method has a sensitivity of 500 ng, and the semiautomatic assay has a sensitivity of 10 ng. The method involves mixing small amounts of a protein solution with fluorescamine (30 mg/100 ml dioxane) in the presence of a phosphate buffer at pH 8–9. The excitation wavelength is 390 nm, and the emission wavelength is 475 nm. Fluorescamine reacts with amines present in reagents and solvents; therefore, reagent blanks are needed. The method seemingly has a coefficient of variation of 5%; the assay is linear between 0.5 and 50 μg. However, quantitation of protein mass depends on

[62] P. Reisner, P. Nermes, and C. Bucholtz, *Anal. Biochem.* **64,** 509 (1975).
[63] S. M. Gotham, P. J. Fryer, and W. R. Paterson, *Anal. Biochem.* **173,** 353 (1988).
[64] C. P. Stowell, T. B. Kuhlenschmidt, and C. A. Hoppe, *Anal. Biochem.* **85,** 572 (1978).

the availability of free amines. Thus, equal masses of two proteins may yield different estimations of concentration by the fluorescamine method. Therefore, the method requires determination of correction factors. Nonetheless, this approach is very useful when measurements of proteins are needed at the level of nanogram quantities.

Conclusions

Quantitation of apolipoprotein B-100 is possible with chemical methods as detailed in this chapter. The method is accurate but somewhat laborious. It has not been automated for use as a diagnostic tool. The method depends on four factors: (1) use of a well-standardized bovine serum albumin solution as a primary standard for the Markwell modification of the Lowry procedure (the NIST solution is highly recommended for use in chemical quantitation of proteins); (2) selective precipitation of apoB by 2-propanol at a final concentration of 50%, (3) standardization of a chromogenicity factor for apoB-100 and apoB-48, and (4) complete resolubilization of apoB species in 2% sodium deoxycholate and 1 *N* NaOH solution. The chemical method is quite valuable in tracer kinetic studies of apoB where there is a need for determination of specific activities of the apolipoprotein.

Acknowledgments

This project was supported by GCRC Grant MO1-RR00633, Veterans Affairs Grants HL-29252 and MO-IRR00633 (NIH/DHS/DHHS), The Southwestern Medical Foundation, and the Moss Heart Foundation, Dallas, Texas, and an unrestricted Nutrition grant from Brystol-Myers Squibb.

[4] Simultaneous Quantification of Apolipoproteins B-100, B-48, and E Separated by SDS–PAGE

By NATHALIE BERGERON, LEILA KOTITE, and RICHARD J. HAVEL

Introduction

The two species of apolipoprotein (apo) B in humans, apoB-100 and apoB-48, define the hepatogenous and intestinal contributions to the secretion of triglyceride-rich lipoproteins (TRL) into the blood. Renewed interest in quantifying these structural and nonexchangeable apolipoproteins of TRL stems from accumulating evidence that remnants of VLDL and chylomicrons may be atherogenic.[1,2]

A number of methods have been proposed to quantify apoB-100 and apoB-48. Several procedures based on separation by sodium dodecyl sulfate–polyacrylamide gel electrophoresis (SDS–PAGE) have given inconsistent results. With electrophoresis in 3.3% polyacrylamide tube gels followed by Coomassie blue staining, Poapst *et al.*[3] reported a lower chromogenicity for apoB-48 than for apoB-100, as well as a nonlinear response between intensity of dye uptake and apolipoprotein mass. This method has been used to quantify apolipoproteins in human TRL[4] and has yielded apoB-100 and apoB-48 levels that are considerably higher than those reported by others using electrophoresis in slab gels.[5–7] Zilversmit and Shea,[8] using slab gels, found equal chromogenicities for rat apoB-48 and rat and human apoB-100 and a linear relationship between dye uptake and apolipoprotein mass. These observations are consistent with our findings[5,7] and those of others who have carried out electrophoresis of apoB[6] and other proteins[9] in slab gels, and they support the conclusion that apoB data obtained by tube gel electrophoresis, in which only a rim of dye penetrates into the gel,[9] should be viewed with caution.

A high-performance liquid chromatographic (HPLC) procedure[10] and an immunoblotting procedure[11] have also been proposed to quantify apoB-100 and apoB-48. Although HPLC on a silica gel column provides a good separation of apoB-100 and apoB-48, quantification of apolipoproteins with this method yields a postabsorptive apoB-48 concentration in TRL (13% of total apoB mass) that is severalfold higher than that measured by SDS slab gel electrophoresis.[5–7] This could reflect greater difficulty in solubilizing apoB before the sample is applied to the column. The immunoblotting procedure, in which an antiserum to the C terminus of apoB-48 is used,

[1] N. Phillips, D. Waters, and R. J. Havel, *Circulation* **88,** 2762 (1993).

[2] F. Karpe, G. Steiner, K. Uffelman, T. Olivecrona, and A. Hamsten, *Atherosclerosis* **106,** 83 (1994).

[3] M. Poapst, K. Uffelman, and G. Steiner, *Atherosclerosis* **65,** 75 (1987).

[4] F. Karpe, G. Steiner, T. Olivecrona, L. A. Carlson, and A. Hamsten, *J. Clin. Invest.* **91,** 748 (1993).

[5] B. O. Schneeman, L. Kotite, K. Todd, and R. J. Havel, *Proc. Natl. Acad. Sci. U.S.A.* **90,** 2069 (1993).

[6] F. Karpe and A. Hamsten, *J. Lipid Res.* **35,** 1311 (1994).

[7] L. Kotite, N. Bergeron, and R. J. Havel, *J. Lipid Res.* **36,** 890 (1995).

[8] D. B. Zilversmit and T. M. Shea, *J. Lipid Res.* **30,** 1639 (1989).

[9] R. Kahn and R. W. Rubin, *Anal. Biochem.* **67,** 347 (1975).

[10] H. Hidaka, H. Kojima, Y. Nakajima, T. Aoki, T. Nakamura, T. Kawabata, T. Nakano, Y. Harano, and Y. Shigeta, *Clin. Chim. Acta* **189,** 287 (1990).

[11] A. S. Peel, A. Zampelas, C. M. Williams, and B. J. Gould, *Clin. Sci.* **85,** 521 (1993).

holds promise for quantification of apoB-48, but it is currently only semi-quantitative and requires that other approaches be used to measure apoB-100.

A major advantage to the SDS–polyacrylamide slab gel electrophoretic method that we have developed is the ability to quantify apoB-100 and apoB-48 together with apoE in various lipoprotein fractions in humans and other species.

Preparation and Mass Measurements of Apolipoprotein Standards

Apolipoprotein B-100

Several methods can be used to prepare human apoB-100, the most common being gel-filtration chromatography and ultracentrifugation of low-density lipoproteins (LDL). The advantage of gel-filtration chromatography is the ability to separate apoB-100 from all other apolipoproteins present in either LDL or very-low-density lipoproteins (VLDL), so that standards can be prepared with the pure protein. Alternatively, ultracentrifugal isolation of "narrow-cut" LDL provides a readily available source of apoB-100 that can be used directly as an apoB-100 standard, because less than 3% of LDL protein is soluble in tetramethylurea (TMU).[12] We have shown that the chromogenicity of human LDL-apoB-100 isolated by gel filtration and that of human LDL-apoB-100 separated by ultracentrifugation are virtually the same,[7] so that either source can be used to prepare an apoB-100 standard.

Both approaches require the initial separation of narrow-cut LDL by sequential ultracentrifugation as follows. Blood from a normolipidemic donor is drawn into tubes containing 0.05% (w/v) EDTA, together with 0.03% benzamidine to prevent the scission of apoB-100 by kallikrein.[13] Plasma, separated by centrifugation (720 *g*) at 4°, is subjected to sequential ultracentrifugation[14] to isolate LDL ($1.025 < \sigma < 1.055$ g/ml). The LDL is recentrifuged for 18–20 hr at $\sigma = 1.055$ g/ml and dialyzed overnight at 4° against 0.15 *M* NaCl, 0.05% EDTA, 0.02% NaN_3, pH 7.4. Such a preparation of LDL normally contains less than 3% TMU-soluble protein[12] and can be used directly as an apoB-100 standard. Alternatively, LDL-apoB-100 can be isolated by gel-filtration chromatography.

Although LDL are the predominant and thus most convenient source

[12] J. P. Kane, *Anal. Biochem.* **53,** 350 (1973).

[13] A. D. Cardin, K. R. Witt, J. Chao, H. S. Margolius, V. H. Donaldson, and R. L. Jackson, *J. Biol. Chem.* **259,** 8522 (1984).

[14] R. J. Havel, H. Eder, and J. Bragdon, *J. Clin. Invest.* **34,** 1345 (1955).

of apoB-100 in humans, this is not the case for many other species so that apoB-100 can also be isolated from VLDL ($\sigma = 1.006$ g/ml), LDL, or total lipoproteins of $\sigma < 1.055$ g/ml by the same gel-filtration procedure that is used for human LDL, described as follows. Dialyzed, narrow-cut LDL (~10 mg total protein in 10–15 ml 0.15 *M* NaCl), isolated by ultracentrifugation, is delipidated overnight at $-20°$ with 20 volumes of ice-cold ethanol–diethyl ether (3:1, v/v), and the apolipoproteins are pelleted by centrifugation for 20 min at $-10°$, 720 *g*. The organic phase is discarded, and the pellet is extracted twice with cold anhydrous diethyl ether.[15] The moist protein pellet is then solubilized in a minimal volume (1 to 2 ml) of phosphate-buffered saline (PBS; 10 m*M* sodium phosphate, 0.15 *M* NaCl, pH 7.2) containing 10% glycerol, 2% SDS, and 0.01% EDTA, and a gentle stream of nitrogen is applied to the sample to remove excess ether. The sample is then applied to a 1.2 × 90 cm column packed with Ultrogel AcA 22 (Pharmacia, Piscataway, NJ) and is eluted at room temperature with PBS containing 0.1% SDS and 0.01% EDTA, at a flow rate of 4 ml/hr. The optical density of eluent fractions of 1 ml is monitored at 280 nm, and fractions from the peak of the apoB-100 elution profile are pooled.

In the final step, the mass of apoB-100 is estimated. Amino acid analysis is the method of choice for accurate mass measurements, but it is tedious and requires sophisticated equipment that is not routinely available in most laboratories. Furthermore, the mass of apoB-100 estimated by amino acid analysis agrees very well with that estimated by the Lowry procedure, with bovine serum albumin (BSA) as standard.[6] Our mass measurements are based on the Lowry procedure, with crystalline BSA as standard. We have found that BSA, kept in a vacuum desiccator over NaOH for at least 1 week, contains approximately 3% water of hydration. For narrow-cut LDL, apoB-100 is calculated as the difference between total LDL protein and TMU-soluble LDL protein,[12] measured by a modification[16] of the method of Lowry et al.,[17] where 3 ml chloroform is added to the incubated sample plus Lowry reagent mixture just before reading the optical density. This modification eliminates the turbidity created by the presence of lipoprotein lipid. For LDL- or VLDL-apoB-100 isolated by gel filtration in the presence of 0.1% SDS, the apolipoprotein mass is measured by the Lowry procedure.[17] We have found that the presence of up to 0.4% SDS in the sample has no effect on the absorbance of the BSA standard at 750 nm. Having estimated the mass of apoB-100, the purity of apoB-100 is then assessed by subjecting approximately 20 μg apoB-100 to electrophoresis in a 4%

[15] W. V. Brown, R. I. Levy, and D. S. Frederickson, *J. Biol. Chem.* **244,** 5687 (1969).

[16] T. Sata, R. J. Havel, and A. L. Jones, *J. Lipid Res.* **13,** 757 (1972).

[17] O. H. Lowry, N. J. Rosebrough, A. L. Farr, and R. J. Randall, *J. Biol. Chem.* **193,** 265 (1951).

polyacrylamide slab gel to verify that no bands other than apoB-100 are visible with Coomassie Brilliant Blue R-250 staining.

Apolipoprotein B-48

On the basis of work of Zilversmit and Shea,[8] who found equal chromogenicities for human and rat apoB-100, we have compared the chromogenicities of human apoB-100 to that of human and rat apoB-48.[7] The chromogenicity of human apoB-48 isolated by gel-filtration chromatography of VLDL obtained from a patient with familial dysbetalipoproteinemia was indistinguishable from that of apoB-48 from rat lymph chylomicrons. Furthermore, the chromogenicities of the two preparations of apoB-48 were the same as that of LDL-apoB-100 isolated either by gel filtration or ultracentrifugation. Consequently, and because ultracentrifugally isolated LDL is an abundant and readily available source of apoB-100, human LDL can be used to prepare a standard curve for quantification of both apoB-100 and apoB-48 in lipoproteins from either humans or rats. However, for quantification of these apolipoproteins in other species, the respective chromogenicities of apoB-100 and apoB-48 in that species should be determined.

Apolipoprotein E

To obtain sufficient amounts of human apoE for construction of standard curves, apoE is prepared from VLDL obtained from a patient with a hypertriglyceridemic disorder such as familial dysbetalipoproteinemia. For details on the isolation of apoE, the reader is referred to earlier procedures described by Havel *et al.*[18] and Rall *et al.*[19] The purity of apoE should also be assessed by SDS–PAGE and isoelectric focusing.

Preparation of Apolipoproteins and Lipoproteins for Electrophoresis

Preparation of Apolipoprotein and Lipoprotein Standards

For SDS slab gel electrophoresis, human apoB-100 and apoE isolated by gel-filtration chromatography are first diluted in a sample buffer containing 3% (w/v) SDS, 10% glycerol, 1.5% dithiothreitol (DTT), 1% mercaptoacetic acid, and 0.02% bromphenol blue in 0.12 *M* Tris, pH 6.8. For

[18] R. J. Havel, L. Kotite, J.-L. Vigne, J. P. Kane, P. Tun, N. Phillips, and G. C. Chen, *J. Clin. Invest.* **66,** 1351 (1980).

[19] S. C. Rall, K. H. Weisgraber, and R. W. Mahley, *in* "Isolation and Characterization of Apolipoprotein E" (J. P. Segrest and J. J. Albers, eds.), p. 273. Academic Press, Orlando, Florida, 1986.

LDL-apoB-100 isolated by ultracentrifugation, the LDL (~1 mg/ml) is first delipidated with 20 volumes of ethanol–ether as described above, the apolipoprotein pellet is solubilized with sample buffer, and a stream of nitrogen is applied to the sample to remove traces of ether. All samples are then heated in a boiling water bath for 3 min and cooled at room temperature. These apoB-100 and apoE preparations should be diluted so that portions of 0.2–30.0 μg apoB-100 or 1.0–8.0 μg apoE can be applied to SDS gels for construction of standard curves.

Preparation of Lipoprotein Samples

Because apolipoproteins are mostly insoluble in organic solvents, delipidation of lipoproteins with ethanol–ether is commonly used to precipitate and concentrate apolipoproteins. Using 15-ml siliconized glass centrifuge tubes, delipidation can be accomplished by gently injecting a given volume of a lipoprotein fraction (usually containing 150–500 μg total protein) into 10 volumes ice-cold ethanol ether, stored overnight at −20°. We have found that delipidating lipoproteins with 10 volumes of organic solvent, instead of 20 volumes,[15] is as efficient in removing lipids and precipitating apolipoproteins and allows lipoproteins to be delipidated in smaller centrifuge tubes. This modification facilitates protein pelleting and reduces the likelihood of apolipoprotein losses. After pelleting apolipoproteins by a 20-min centrifugation at −10°, 720 *g*, the organic phase is removed by aspiration, and the pellet is washed once with cold anhydrous diethyl ether and recentrifuged under the same conditions. Finally, the moist protein pellet is solubilized in 150–500 μl sample buffer (so that the final protein concentration is no more than 100 μg/100 μl), excess ether is removed, and the centrifuge tube is placed in a boiling water bath for 3 min.

This delipidation procedure can be used for all lipoprotein fractions separated by ultracentrifugation but requires that at least 75 μg total protein be delipidated in order to precipitate apolipoproteins without appreciable losses. We have found that considerable losses occur when smaller amounts of total protein are delipidated, presumably because more protein adheres to the sides of the delipidation tube or is lost when the organic phase is removed.[7] This problem is most often encountered with normal human VLDL collected in the fasting (postabsorptive) state, or with rodent lipoproteins (particularly VLDL and IDL) that typically contain very little protein. Apolipoprotein loss by delipidation of such lipoproteins can be minimized by adding 60 μg apotransferrin (Sigma, St. Louis, MO; human 98% pure) before delipidation to lipoproteins containing less than 75 μg total protein.[7] Apotransferrin (80 kDa) is a suitable carrier protein because it does not comigrate with apolipoproteins B-100, B-48, or E on 3–10% linear poly-

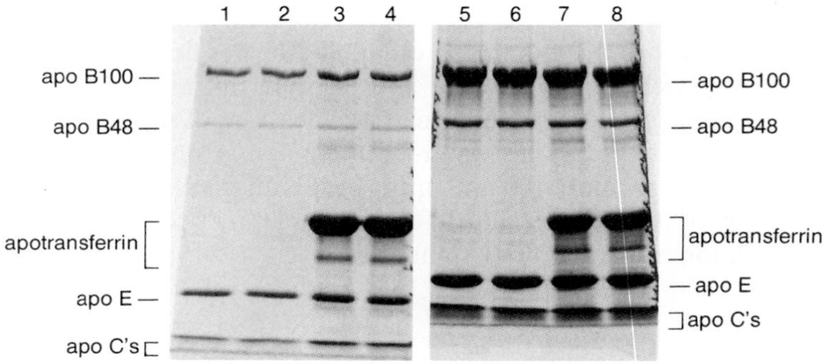

FIG. 1. Acrylamide SDS–PAGE slab gel (3–10% acrylamide) stained with Coomassie Brilliant Blue R-250, showing apoproteins of TRL delipidated without apotransferrin (lanes 1, 2, 5, and 6) and with 60 μg apotransferrin (lanes 3, 4, 7, and 8). For lanes 1–4, 24 μg TRL protein was delipidated and 6 μg was applied to slab gels. For lanes 5–8, 100 μg TRL protein was delipidated and 25 μg was applied to slab gels.

acrylamide SDS slab gels. A stock solution of apotransferrin (2 mg/ml in 0.15 *M* NaCl) can be kept at 4° for up to 2 weeks. As illustrated in Fig. 1, delipidating 24 μg TRL in the presence of 60 μg apotransferrin (lanes 3 and 4) greatly improved apolipoprotein recovery as compared with delipidation in the absence of apotransferrin (lanes 1 and 2). In contrast, when 100 μg TRL was delipidated, the recovery of apolipoproteins on slab gels was unaffected by addition of apotransferrin (compare Fig. 1, lanes 7 and 8, with lanes 5 and 6).

Electrophoresis of Apolipoprotein Standards and Lipoprotein Samples

Gel Casting

For optimal resolution of apoB-100, apoB-48, and apoE, apolipoproteins are separated by electrophoresis with the Laemmli system[20] in slab minigels prepared with a 3–10% linear gradient of polyacrylamide. The gel recipes are as follows: the 3% acrylamide resolving gel contains acrylamide (3%, w/v), bisacrylamide (0.08%), SDS (0.1%), *N,N,N′,N′*-tetramethylethylenediamine (TEMED; 0.05%), glycerol (0.03 g/ml), and freshly prepared ammonium persulfate (0.06%) in Tris (0.376 *M*), pH 8.9; the 10% acrylamide resolving gel contains acrylamide (10%), bisacrylamide (0.28%), SDS (0.1%), TEMED (0.05%), glycerol (0.10 g/ml), and ammonium persulfate

[20] U. K. Laemmli, *Nature* (*London*) **227,** 680 (1970).

(0.06%) in Tris (0.376 *M*), pH 8.9; and the 3% acrylamide stacking gel contains acrylamide (3%), bisacrylamide (0.08%), SDS (0.1%), TEMED (0.15%), and ammonium persulfate (0.1%) in Tris (0.127 *M*), pH 6.8. Using a two-chamber gradient gel mixer (Buchler, Fort Lee, NJ), two resolving gels are cast simultaneously and are immediately covered with a solution containing 0.1% SDS in 0.375 *M* Tris, pH 8.9, until gels have polymerized (20–30 min). The solution is then removed, and the stacking gel is poured just before electrophoresis. The electrophoresis apparatus is connected to a power supply (LKB Model 2197; Pharmacia-LKB, Piscataway, NJ) and run in the presence of a tank buffer containing 0.19 *M* glycine and 0.1% SDS in 0.25 *M* Tris, pH 8.9.

Electrophoresis

We have carried out gel electrophoresis with a Mini-PROTEAN II vertical slab gel apparatus (Bio-Rad Laboratories, Hercules, CA) in which two minigels are run simultaneously at 50 V/gel for the first 30 min and 75 V/gel for the remaining 60–90 min. Each minigel (8.5 × 5.5 × 0.15 cm) can accommodate 10 sample wells (1.2 × 0.5 cm), each able to contain 50–75 μl of sample. A major advantage of the minigel apparatus over the usual gel apparatus is the use of smaller amounts of reagents and, most of all, the ability to run gels in less than 2 hr. However, there are instances where a vertical gel apparatus for larger gels (e.g., 14 × 12.5 × 0.15 cm) may be needed. This vertical gel system is run at 15 mA/gel for the first 90 min and at 25 mA/gel for the remaining 4 hr. Although electrophoresis is much slower than with the minigel apparatus, the usual gel apparatus can accommodate 15 sample wells (2.7 × 0.6 cm), each able to contain 100–150 μl sample, which can be useful for electrophoresis of samples with a low protein concentration.

For electrophoresis of human apolipoprotein standards, at least six amounts of each apolipoprotein are applied in duplicate to slab gels, to cover a range of 0.2 to 30 μg for the apoB-100 standard, and 1.0 to 8–10 μg for the apoE standard. The apoB-100 must cover a wide protein range because it will be used to quantify both apoB-48 and apoB-100, whose masses in TRL typically differ by 30-fold. The standard curve for apoE need not cover such a wide protein range because the mass of apoE in normal TRL is typically 5-fold lower than that of total apoB.[5,7]

For electrophoresis of unknown lipoprotein fractions, 50-μl portions containing 30–75 μg total protein are applied in duplicate to SDS slab gels. To minimize between-run variation in dye uptake of apolipoproteins, a reference lipoprotein is included with each electrophoretic run. This reference lipoprotein can be apo-LDL, delipidated as described above and solubilized in 2% SDS–PBS. The total protein[17] and TMU-soluble protein[12] of apo-LDL are determined, and 30-μg portions (containing ~29 μg apoB-100) are stored in 1.5-ml microcentrifuge tubes at −70°. Before each run, a vial of reference apo-LDL is thawed at room temperature, diluted to a final volume of 250 μl with sample buffer, and placed in a boiling water bath for 3 min, and 50-μl portions (~5.8 μg apoB-100) are applied in duplicate to slab gels and run simultaneously with unknown samples. On the basis of the chromogenicity of apoB-100, the average mass of apoB-100 in the apoB-100 band is calculated. The ratio of the applied mass (~5.8 μg) to the calculated apoB-100 mass on the gel is then used to correct for between-run variation in dye uptake of apoB-100, apoB-48, and apoE in unknown lipoproteins. Reference apo-LDL can be substituted for reference VLDL because we have shown that the between-run dye uptake of the three proteins was concordant, and the between-run coefficients of variation in dye uptake of apoB-100 (10.3), apoB-48 (10.2), and apoE (10.1) from 320 applications of reference VLDL to slab gels were identical.[7]

After electrophoresis, minigels are stained for 18–22 hr in trays containing 100 ml (200 ml for larger gels) of 0.25% Coomassie Brilliant Blue R-250 (Sigma) in methanol–water–acetic acid (5 : 5 : 1, v/v/v) and destained for 7–8 hr with at least four changes of 150 ml (250 ml for larger gels) of methanol–water–acetic acid (5 : 5 : 1, v/v/v). Staining and destaining are done at room temperature and under constant agitation on an orbital shaker. After destaining, gels are placed between two sheets of cellophane gel wrap and dried at room temperature in a fume hood, overnight.

Measurement of Apolipoprotein Dye Uptake

Scanning

We have used a laser densitometer from Molecular Dynamics (Personal Densitometer, Sunnyvale, CA), equipped with an Image Quant Software package (version 3.2). When using this instrument, dried gels are placed on a glass plate, secured with cellophane tape, and scanned with the densitometer connected to a computer and equipped with software that automatically integrates the "volume" of stained apolipoproteins bands. In contrast to "area" integration, where a narrow portion of each gel lane is scanned, capturing only a fraction of the apolipoprotein band, with volume integration the entire area of the apolipoprotein band is encompassed in a rectangular object, regardless of its dimensions or shape. This feature facilitates accurate quantification of widely varying masses of apolipoproteins. Volume integration of stained bands in standards and unknown lipoproteins is accomplished in several steps. Each band is first enclosed in a rectangle.

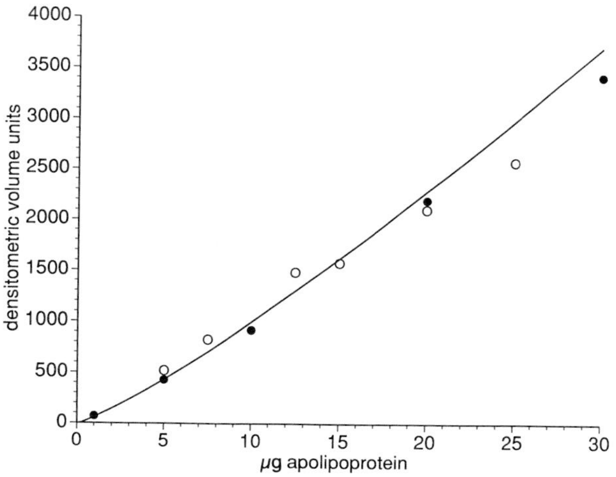

FIG. 2. Chromogenicities of human VLDL-apoB-100 (○) isolated by gel filtration and human LDL-apoB-100 (●) isolated by ultracentrifugation. Combined response of VLDL- and LDL-apoB-100: $y = 62.82x^{1.20}$ ($r = 0.998$).

Then, to correct for uneven destaining of acrylamide gels, "background" rectangles are drawn immediately above each band, and the density of each rectangle is integrated. Finally, the average density per pixel unit (0.077 mm^2) in the background rectangle is subtracted from each pixel unit within the stained band, and the resulting sum of all pixel units in the band is calculated and expressed in densitometric volume units. When lipoprotein samples contain very large amounts of apoB-100 relative to apoB-48, the lower edge of the apoB-100 band is closer to that of apoB-48 so that the background immediately above apoB-48 may be high. In such cases, background rectangles are drawn both above and below the apoB-48 band, and the average density of the two rectangles is used as background.

Chromogenicities of Apolipoproteins

Standard curves for apoB-100 and apoE are constructed by relating the intensity of dye uptake of each band to its known mass. The line of best fit is then calculated by the method of least squares, which in our case follows a power function $y = ax^b$ for all apolipoprotein standards.

As illustrated in Fig. 2, the chromogenicity of LDL-apoB-100 isolated by ultracentrifugation is similar to that of VLDL-apoB-100 isolated by gel filtration. We have also shown that the chromogenicities of LDL-apoB-100 isolated by either ultracentrifugation or gel filtration are similar.[7] These observations indicate that any of the above sources can be used as an apoB-

100 standard. Furthermore, the chromogenicities of apoB-100 in ultracentrifugally isolated LDL and apoB-48 from rat lymph chylomicrons are virtually identical (Fig. 3A). We also know from Kotite *et al.*[7] that the chromogenicities of rat and human apoB-48 are identical, which is expected given the close similarity of the derived amino acid sequences of rat and human apoB-48.[21,22] Taken together the above findings confirm that apoB-100 isolated by ultracentrifugation of LDL, or by gel filtration of LDL or VLDL, can be used as an apoB-100 standard to quantify both apoB-100 and apoB-48 from either human or rat lipoproteins. The chromogenicity of human apoE is approximately 2-fold higher than that of apoB-100 and apoB-48 (Fig. 3B), reflecting the greater number of positive charges on the protein.[23] Therefore, a separate standard curve is necessary to quantify apoE. Having established the chromogenicities of the apolipoprotein standards, apoB-100, apoB-48, and apoE are quantified in unknown lipoproteins by converting densitometric volume units of stained bands to micrograms protein on the basis of the standard curves for apoB-100 and apoE.

It should be emphasized that the relation between dye uptake and apolipoprotein mass may differ depending on the choice of densitometer and the method of integration (area versus volume). Yet, on the basis of previous observations[5] and those of Zilversmit and Shea[8] who used a one-dimensional light densitometer, this relationship should be linear over a wide range of 0.5 to 30 μg applied apolipoprotein mass. The deviation from linearity reported by Karpe and Hamsten[6] when more than 2.0 μg apoB-100 is applied to slab gels is inconsistent with these observations.

Scope of Method

The described procedure permits simultaneous quantification of apoB-100, apoB-48, and apoE from a single application of an apolipoprotein sample to SDS slab gels. By adding apotransferrin (60 μg) to lipoproteins with a low protein content (<75 μg), small amounts of lipoprotein can be delipidated without appreciable losses, and as little as 0.2 μg apoB-48 to as much as 30 μg apoB-100 applied to slab gels can be quantified.

Because the chromogenicities of human and rat apoB-100 and apoB-48 are the same, this method can be applied to the quantification of apoB-100 and apoB-48 in all lipoprotein fractions of both species and presumably other rodents. We have used our method to quantify apolipoproteins in

[21] B. Teng, D. D. Black, and N. O. Davidson, *Biochem. Biophys. Res. Commun.* **173,** 74 (1990).
[22] J. Greeve, I. Altkemper, J.-H. Dieterich, H. Greten, and E. Windler, *J. Lipid Res.* **34,** 1367 (1993).
[23] M. Tal, A. Silberstein, and E. Nusser, *J. Biol. Chem.* **260,** 9976 (1980).

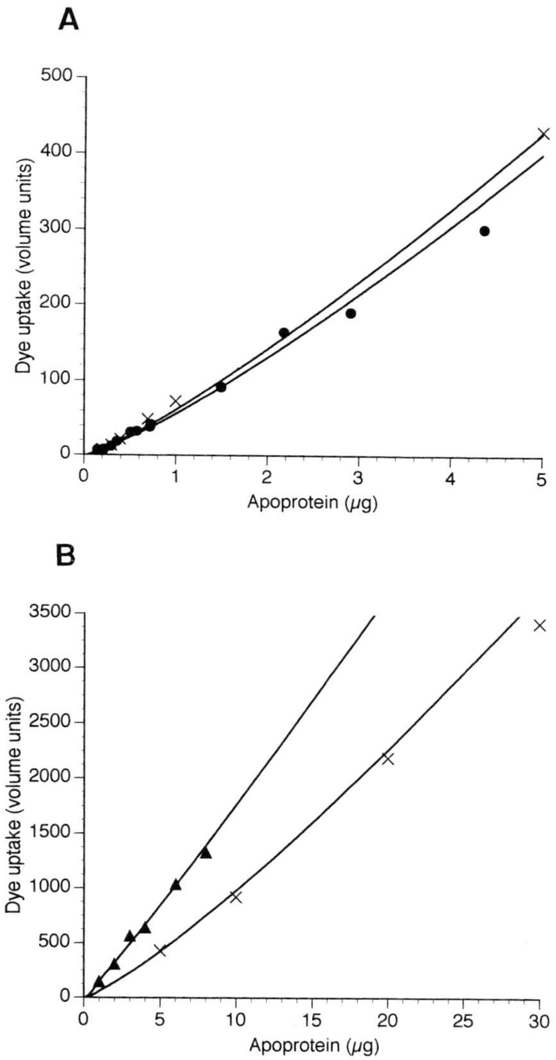

FIG. 3. Chromogenicities of (×) apoB-100 (0–5 μg) and (●) apoB-48 (0–4.4 μg) (A) and (×) apoB-100 (0–30 μg) and (▲) apoE (0–8 μg) (B). LDL apoB-100: $y = 61.28x^{1.20}$ ($r = 0.998$); rat chylomicron apoB-48: $y = 55.96x^{1.22}$ ($r = 0.989$); human apoE: $y = 150.64x^{1.06}$ ($r = 0.992$). [Reprinted from L. Kotite, N. Bergeron, and R. J. Havel, *J. Lipid Res.* **36,** 890 (1995), with permission.]

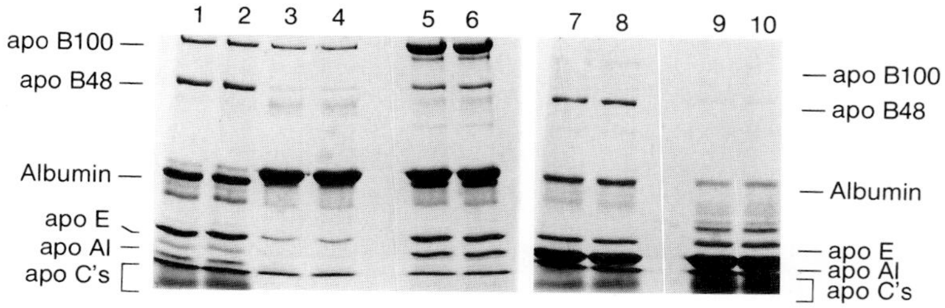

FIG. 4. Acrylamide SDS–PAGE slab gel (3–10% acrylamide) stained with Coomassie Brilliant Blue R-250, showing apoproteins of mouse lipoprotein fractions isolated by sequential ultracentrifugation (fractions were not recentrifuged). Lanes 1 and 2 contain VLDL ($\sigma <$ 1.006 g/ml); lanes 3 and 4, IDL (1.006 g/ml $< \sigma <$ 1.019 g/ml); lanes 5 and 6, LDL (1.019 g/ml $< \sigma <$ 1.055 g/ml); lanes 7 and 8, HDL_2 (1.055 g/ml $< \sigma <$ 1.085 g/ml); lanes 9 and 10, HDL_3 (1.085 g/ml $< \sigma <$ 1.21 g/ml). The heavy band above albumin in lanes 3–6 is apotransferrin. The amount of protein applied to the gel ranged from 20 to 50 μg.

various mouse lipoproteins (Fig. 4). As illustrated, apoB-48 is predominent in VLDL (Fig. 4, lanes 1 and 2), but is also present in intermediate-density lipoproteins (IDL; lanes 3 and 4), LDL (lanes 5 and 6), HDL_2 (lanes 7 and 8), and, in trace amounts, in HDL_3 (lanes 9 and 10). Apolipoprotein B-100 predominates in IDL and LDL, whereas apoA-1 is predominant in HDL_2 and HDL_3. Our method permits quantification of apoB-100, apoB-48, and apoE in mouse lipoproteins; the potential for quantifying other apolipoproteins in these gels is evident.

The described method is suitable for routine use in studies on the intestinal and hepatic contributions to TRL in the postabsorptive and postprandial state. The postabsorptive levels, expressed as means $\pm$ standard deviation (mg/dl), of apoB-100 (4.83 $\pm$ 2.19), apoB-48 (0.25 $\pm$ 0.12), and apoE (0.88 $\pm$ 0.47) in 16 healthy young men maintained on a controlled diet containing 38% of calories from fat[7] are consistent with earlier findings[5] and others.[6] The slab gel electrophoresis method for quantification of apoB-100 and apoB-48 compares favorably[7] with the 2-propanol precipitation method[24] and the TMU method[25] for quantification of total apoB mass in TRL. Obvious advantages to the described method include sensitivity and the ability to quantify apoB-48 and apoB-100, as well as apoE, simultaneously. Given these key features, this method provides a rapid, sensitive, and reliable means to quantify apoB-100, apoB-48 and apoE in a variety of metabolic conditions and in various lipoprotein fractions of humans and other species.

[24] G. Egusa, D. W. Brady, S. M. Grundy, and B. V. Howard, *J. Lipid Res.* **24,** 1261 (1983).
[25] J. P. Kane, T. Sata, R. L. Hamilton, and R. J. Havel, *J. Clin. Invest.* **56,** 1622 (1975).

[5] Apolipoprotein B-48

By FREDRIK KARPE, ANDERS HAMSTEN, KRISTINE UFFELMAN, and GEORGE STEINER

Introduction

Normally, apolipoprotein (apo) B exists in two molecular forms, apoB-100 and apoB-48. There is one single gene encoding both apoB-100 and apoB-48. The intestinal production of apoB-48 is the result of tissue-specific apoB messenger RNA (mRNA) editing.[1,2] In humans apoB-100 is synthesized in the liver and secreted into plasma as the major protein component of very-low-density lipoproteins (VLDL), whereas apoB-48 is secreted from the enterocytes of the small intestine in chylomicrons after absorption of dietary lipids. Of note, the plasma concentrations of both apoB-48 and apoB-100 in triglyceride-rich lipoproteins increase after fat intake in humans.[3–5]

Because there is only one apoB molecule per lipoprotein particle, apoB serves as a marker of lipoprotein particle number, and, in contrast to other plasma apolipoproteins and lipids, apoB does not exchange between different lipoprotein particles. For several reasons determination of apoB-48 concentrations in plasma is difficult. First, the amino acid sequence of apoB-48 is identical to the N-terminal 48% of apoB-100. Second, apoB-48 is present at very low concentrations in plasma compared with apoB-100. Third, the expression of apoB epitopes may vary depending on the lipid or apolipoprotein content of the lipoprotein particle. These factors render it difficult to use immunochemical techniques to quantify apoB-48. However, a specific polyclonal antibody for detection of apoB-48 has been described.[6] The production of this antibody was achieved by immunizing New Zealand White rabbits with a synthetic oligopeptide of identical se-

[1] L. M. Powell, S. C. Wallis, R. J. Pease, Y. H. Edwards, T. J. Knott, and J. Scott, *Cell (Cambridge, Mass.)* **50,** 831 (1987).
[2] S. H. Chen, G. Habib, C. Y. Yang, Z. W. Gu, B. R. Lee, S. A. Weng, S. R. Silberman, S. J. Cai, J. P. Deslypere, M. Rosseneu, A. M. Gotto, Jr., and L. Chan, *Science* **238,** 363 (1987).
[3] J. S. Cohn, J. R. McNamara, S. D. Cohn, J. M. Ordovas, and E. J. Schaefer, *J. Lipid Res.* **29,** 925 (1988).
[4] F. Karpe, G. Steiner, T. Olivecrona, L. A. Carlson, and A. Hamsten, *J. Clin. Invest.* **91,** 748 (1993).
[5] B. O. Schneeman, L. Kotite, K. M. Todd, and R. J. Havel, *Proc. Natl. Acad. Sci. U.S.A.* **90,** 2069 (1993).
[6] A. S. Peel, A. Zampelas, C. M. Williams, and B. J. Gould, *Clin. Sci.* **85,** 521 (1993).

METHODS IN ENZYMOLOGY, VOL. 263

quence as the C-terminal portion of apoB-48. This holds prospects for the future in terms of simple, specific quantification of apoB-48-containing lipoproteins. However, so far no wider application of this antibody has been published.

Attempts to measure the apoB-48 content of triglyceride-rich lipoproteins have therefore focused on analytical sodium dodecyl sulfate–polyacrylamide gel electrophoresis (SDS–PAGE).[4,7–9] A major issue in this approach to the quantification of apoB-48 is the relative chromogenicity of apoB-48 compared with apoB-100. Obviously, the peptide chain is partly identical between the two proteins, and the delipidated protein should therefore theoretically interact with dye in an identical and reproducible manner. Zilversmit and Shea[8] as well as Karpe and Hamsten[9] found that there was no difference in chromogenicity between apoB-48 and apoB-100 in comparisons where chyle and not plasma was used as source of chylomicrons and chylomicron remnants. However, a chromogenicity difference was found by Poapst *et al.*[7] The main difference between the methods for determining apoB-48 and apoB-100 is the use of slab gels in the first two procedures[8,9] and rod gels in the third.[7] The rod gels required application of a 10-fold more material than did the slab gels. This may be the basis for the variable chromogenicity observed when using rod gels.

Because both slab and rod gels provide accurate measurements of apoB-48 and apoB-100, two methods for determining the plasma concentration of apoB-48 and apoB-100 in isolated fractions of triglyceride-rich lipoproteins with use of analytical SDS–PAGE, either on slab or rod gels, are described.

Isolation of Triglyceride-Rich Lipoproteins for Subsequent Determination of Apolipoproteins B-48 and B-100

The following density gradient ultracentrifugation procedure for isolating subfractions of triglyceride-rich lipoproteins is suitable for SDS–PAGE on both slab and rod gels and is essentially as previously described.[4,10] Plasma is recovered by low-speed centrifugation (1750 *g*, 20 min, 1°). To minimize proteolytic degradation of apoB, 1.0 μl/ml plasma of phenylmethylsulfonyl fluoride (PMSF, Sigma, St. Louis, MO), 10 m*M* dissolved in 2-propanol, and 5 μl/ml plasma of aprotinin (Trasylol, Bayer, Leverkusen, Germany), 1400 μg/liter, are added. Subsequently, 140.4 mg solid NaCl is

[7] M. Poapst, K. Uffelman, and G. Steiner, *Atherosclerosis* **65,** 75 (1987).

[8] D. B. Zilversmit and T. M. Shea, *J. Lipid Res.* **30,** 1639 (1989).

[9] F. Karpe and A. Hamsten, *J. Lipid Res.* **35,** 1311 (1994).

[10] T. G. Redgrave and L. A. Carlson, *J. Lipid Res.* **20,** 217 (1979).

added per 1.0 ml plasma to increase the density to 1.10 kg/liter. Normally, a total volume of 4.0 ml of the *d* 1.10 kg/liter plasma is put in the bottom of a 13.4-ml polyallomer ultracentrifuge tube (Ultra-Clear, Beckman Instruments, Palo Alto, CA). Alternatively, 3.0 ml plasma can be mixed with 1.5 ml *d* 1.42 kg/liter NaBr, from which 4.0 ml is transferred to the ultracentrifuge tube. For the rod gel method, two such tubes are required to obtain enough material from each sample. For the slab gel method, 1.0 ml plasma is sufficient. In the latter case, a 1.0-ml portion of 1.10 kg/liter plasma can be mixed with 3.0 ml of 1.10 kg/liter NaCl in the tube. A density gradient consisting of 3.0 ml each of 1.065, 1.020, and 1.006 kg/liter NaCl solutions is then sequentially layered on top of the plasma.

Ultracentrifugation is performed in a SW40 Ti swinging bucket rotor (Beckman) at 40,000 rpm and 15° (Beckman L8-55 ultracentrifuge). Consecutive runs calculated to float Svedberg flotation rate (Sf) $>$ 400 (32 min), Sf 60–400 (3 hr 28 min), and Sf 20–60 (14–16 hr) particles are made. After each centrifugation, the top 0.5 ml of the gradient containing the respective lipoprotein subclasses is aspirated, and 0.5 ml of density 1.006 kg/liter salt solution is used to refill the tube before the next run. The Sf 12–20 fraction is recovered after the last ultracentrifugal run by slicing the tube 29 mm from the top after the Sf 20–60 lipoproteins have been aspirated. All salt solutions should be adjusted to pH 7.4 and contain 0.02% (w/v) NaN_3 and 0.01% Na_2EDTA. This method yields lipoprotein preparations almost completely devoid of plasma albumin.

Analytical SDS–PAGE Using Slab Gel Technique

Preparation of Samples

Isolated samples of apoB-containing lipoproteins are delipidated in a methanol/diethyl ether solvent system. Lipoprotein fractions with an NaCl concentration higher than approximately 0.5 *M* (d = 1.019 kg/liter) should be dialyzed prior to delipidation. A volume of 50–750 μl of the sample is injected into 4.0 ml methanol in a 10-ml round-bottom blass tube. Ideally, the sample should contain 1–25 μg apoB after delipidation. A glass syringe with Teflon gaskets is used; the needle tip can be slightly deformed in order to spread the sample in the methanol to ensure efficient delipidation. A volume of 4.0 ml ice-cold diethyl ether is then added to the methanol, and the delipidation mixture is centrifuged for 30 min at 4000 *g* at 1°. After gentle removal of solvent with a water suction device, another 4.0 ml diethyl ether is immediately added, and the sample is vortexed, pressure equalized, and centrifuged for 20 min under the same conditions, after which the diethyl ether is removed. Typically, a whitish haze is seen at the bottom of the tube. The protein material is dissolved in 50–500 μl of 0.15 *M* sodium phosphate, 12.5% (v/v) glycerol, 2% (w/v) sodium dodecyl sulfate (SDS; Bio-Rad, Hercules, CA), 5% (v/v) mercaptoethanol, 0.001% (w/v) bromphenol blue, pH 6.8, at room temperature during 30 min. The dissolved protein mixture is subsequently denatured at 80° for 10 min. After denaturation, the tubes are immediately centrifuged in order to retain the condensed water on the inner wall of the tube. Samples can then be frozen at −20° for later analytical SDS–PAGE.

Gel Casting

Polyacrylamide gels with a short stacking gel consisting of 3% acrylamide followed by a 3 to 20% acrylamide gradient can be cast using a two-chamber gradient mixer (GM-1, Pharmacia, Sollentuna, Sweden). The advantage of using gradient gels is that the contents of apoE and apoC's are also visualized after staining. First, a gel casting cassette (Hoefer Scientific, San Francisco, CA) prepared for 10 gels is filled from the bottom with 20 ml of a 3% acrylamide solution [acrylamide at 29.25 g/liter, bisacrylamide at 0.75 g/liter (Bio-Rad), Tris at 0.375 *M*, SDS at 0.1%, *N,N,N′,N′*-tetramethylethylenediamine (TEMED) at 1.0 μl/ml, and ammonium persulfate at 0.4 g/liter, prepared on ice]. Then, 45 ml of the 3% acrylamide solution and 45 ml of a 20% acrylamide solution (acrylamide at 195 g/liter, bisacrylamide at 5 g/liter, Tris at 0.375 *M*, SDS at 0.1%, TEMED at 0.6 μl/ml, and ammonium persulfate at 0.1 g/liter, prepared on ice) are filled into the two chambers of the gradient mixer. The acrylamide gradient is formed during about 5–10 min by filling the gel casting cassette from the bottom by hydrostatic pressure. The acrylamide solutions are prepared from filtered stock solutions stored at 4°, except for the ammonium persulfate, which is prepared on the day of gel casting. The polymerization time should be adjusted to 40 min by varying the volume of added ammonium persulfate. Teflon well-formers (Pharmacia) serrated with 15 prickles (3.0 mm width) are used. The gels are left at room temperature for at least 2 hr and then stored under moist conditions at 4° for no longer than 2 weeks.

Preparation of Reference Apolipoprotein B-100 for Slab Gel Method

Apolipoprotein B-100 derived from low-density lipoprotein (LDL) is used as a reference protein and for standard curve dilutions. The apoB-100 preparation is derived from LDL (1.030 < *d* < 1.040 kg/liter) that is isolated from fasting plasma samples by the density gradient ultracentrifugation procedure described above. The LDL subfraction is recovered from a 1.0-ml portion located 5.5 to 6.5 ml below the top of the gradient after 16 hr of ultracentrifugation. Fast and careful desalting can be achieved by passing the LDL through a PD-10 column (Pharmacia). The total protein content is then determined[11] with addition of SDS (final concentration 1%) to the reagent mixture to reduce turbidity. The mean of several determinations should be taken as the final protein value.

A glass syringe with Teflon gaskets (Unimetrics, Anaheim, CA) is calibrated to the volume calculated to contain 125.0 μg LDL protein. This volume of the LDL preparation (generally $\approx$100 μl) is immediately injected into 4.0 ml of methanol contained in a 10-ml glass tube with subsequent addition of 4.0 ml ice-cold diethyl ether in order to delipidate the protein. The tube is then centrifuged at 4000 *g* for 30 min at 1°. The methanol–diethyl ether is gently removed by water suction, and another 4.0 ml of ice-cold diethyl ether is added. After a second centrifugation under the same conditions for 20 min, the diethyl ether is removed. The protein pellet is subsequently dissolved at room temperature during 30 min in 500.0 μl of 0.15 *M* sodium phosphate, 12.5% glycerol, 2% SDS, 5% mercaptoethanol, 0.001% bromphenol blue, pH 6.8. The dissolved protein mixture is denatured at 80° for 10 min.

The properties of the LDL protein preparation are then tested on analytical SDS–PAGE (gel formula previously described) before it can be accepted as an apoB-100 standard. The following criteria may be used. First, to ensure that delipidation has been successful, the amount of protein that is larger than the sharply focused apoB-100 band or does not enter the gel is not allowed to exceed 1% of the apoB-100 band using the laser scan densitometer. Second, signs of degradation of apoB-100 resulting in several sharp bands smaller than apoB-100 with an added protein density exceeding 1% of the apoB-100 band are not accepted. A band localized in the size range of apoB-48 is normally not seen, whereas a thin band with an R_f corresponding to apoE could occasionally be encountered on extremely overloaded gels. However, the absorbance of this band should never exceed 0.5% of the apoB-100 band. If these criteria are fulfilled, more than 97.5% of the total LDL protein is confined to a single band on analytical SDS–PAGE, which by comparison with a commercially available high molecular weight standard protein mixture has an apparent molecular mass of about 500 kD, the molecular weight of apoB-100. Furthermore, a linear regression of the optical density of the apoB-100 band on the known amount of protein applied on the analytical SDS–PAGE gel, the latter ranging from 0.10 to 2.00 μg, is required to have an *r* value exceeding 0.95 and pass through the origin to be accepted.

Batches of the delipidated and denatured apoB-100 protein standard can be frozen and stored for 3 months at −80° and thawed immediately before use. Freezing and thawing of the denatured apoB-100 standard do

[11] O. H. Lowry, N. J. Rosebrough, A. L. Farr, and R. J. Randall, *J. Biol. Chem.* **193,** 265 (1951).

not change the slope between protein mass and dye uptake. Furthermore, delipidation and denaturation of LDL derived from different subjects result in apoB-100 standards with identical slopes.

Electrophoresis

Electrophoresis is performed using a vertical Hoefer Mighty Small II electrophoresis apparatus connected to an EPS 400/500 (Pharmacia, Stockholm, Sweden) power supply. The upper and lower electrophoresis buffers contain 25 m*M* Tris, 192 m*M* glycine, 0.2% SDS adjusted to pH 8.5. Preelectrophoresis is performed during 20 min at 60 V. Normally, both 2 and 20 μl of each sample are applied to obtain at least one protein band within the range of apoB-100 standard curve. The outer wells are not used. Six amounts (0.10–2.00 μg) of the apoB-100 standard are applied on most gels. When two gels are run simultaneously and thereafter fixed, stained, and destained in the same bowl, a standard curve is applied to only one of these gels. Electrophoresis is first run at 60 V for 20 min and then at 100 V for 2 hr. Gels are fixed in 12% (w/v) trichloroacetic acid for at least 30 min and stained in 0.2% (w/v) Coomassie blue G-250 (Serva, Heidelberg, Germany), 40% methanol, 10% (v/v) acetic acid for at least 4 hr in a glass petri dish. Destaining is made in 12% (v/v) methanol, 7% (v/v) acetic acid with four changes of destainer during 24 hr.

Gel Scanning

Gels are scanned using a laser scanner (Ultroscan XL, Pharmacia-LKB) connected to a personal computer equipped with software providing automatic integration of areas under the scanning curves (Gelscan XL, Pharmacia). The laser beam is adjusted for each lane to cover the whole band, which in most cases results in a laser track with a width of 5.6 mm and a length of about 15 mm. Correction for background intensity is made by scanning an empty lane. Boundaries for beginning and end of integration are set for each lane. Bands with an obvious optical density greater than the highest standard point are not evaluated. In most instances, the 2-μl application is used for determining the content of apoB-100 and the 20-μl application for determining apoB-48 in the sample. A standard curve is fitted by linear regression using the CricketGraph software (Apple, Cupertino, CA).

The chromogenicity of apoB-48 compared with apoB-100 and of apoB-100 derived from LDL, and Sf 12–20, Sf 20–60, and Sf 60–400 lipoproteins, respectively, has previously been evaluated using the present method.[9] No difference was recorded between the two proteins, nor between apoB-100 in lipoproteins of different size. With this procedure, the coefficients of variation for determination of plasma concentrations of apoB-48 and apoB-100 in isolated fractions of triglyceride-rich lipoproteins are low.[9]

Calculation of Apolipoprotein B-100 and B-48 Concentrations

Because the chromogenicity of apoB-100 and apoB-48 is equal using the slab gel method, the plasma concentrations of the two apoB species contained in triglyceride-rich lipoproteins are calculated directly with use of the equation derived from the linear regression between optical density and known apoB-100 content of the standard preparation.

Analytical SDS–PAGE Using Rod Gel Technique

Analytical SDS–PAGE using the rod gel technique requires two steps. First, the total amount of apoB-containing lipoproteins in the isolated lipoprotein fraction is determined immunologically. Second, the relative amount of apoB-48 and apoB-100 in the sample is determined by analytical SDS–PAGE using rod gels. To quantify the total amount of apoB in triglyceride-rich lipoproteins, we recommend the use of lipase treatment of the samples in order to expose antigenic determinants and to allow entry into an agarose gel when using rocket immunoelectrophoresis.[12]

The relative amounts of apoB-48 and apoB-100 in the isolated lipoprotein sample are quantified using an assay[7] that is based on the relative chromogenicities of these apolipoproteins after separation by electrophoresis in SDS–polyacrylamide gels as described by Weber and Osborne[13] and modified by Kane *et al.*[14] Aliquots of isolated lipoprotein fractions containing 200–500 μg total protein are first delipidated using the methanol–chloroform–diethyl ether system described by Herbert *et al.*,[15] and the protein residue is washed with diethyl ether. The delipidated precipitate is dissolved overnight in sample buffer (sodium phosphate at 50 m*M*, SDS at 1%, and mercaptoethanol at 1%, pH 7.2) and applied to 3.3% SDS–PAGE rod gels. Four or five dilutions of the delipidated lipoprotein–sample buffer preparation are run for each sample. A total apoB mass of approximately 20–60 μg is applied to each gel. Gels are stained with Coomassie Brilliant Blue R-250 (Eastman-Kodak Co., Rochester, NY) for 24 hr and then destained for 3 days in 10% acetic acid before scanning. Scanning is performed by an LKB Ultroscan laser densitometer (Pharmacia, Stockholm, Sweden)

[12] M. F. Reardon, M. E. Poapst, K. D. Uffelman, and G. Steiner, *Clin. Chem.* **27,** 892 (1981).
[13] K. Weber and M. Osborne, *J. Biol. Chem.* **244,** 4406 (1969).
[14] J. P. Kane, D. A. Hardman, and H. E. Paulus, *Proc. Natl. Acad. Sci. U.S.A.* **77,** 2465 (1980).
[15] P. N. Herbert, R. S. Shulman, R. I. Levy, and D. S. Fredrickson, *J. Biol. Chem.* **248,** 4941 (1973).

linked to a personal computer (Apple), and the area under each peak is integrated automatically by the instrument using the Gelscan software package (Pharmacia). Graphs separately relating the intensity of staining (area under the scanning curve) of the apoB-48 and of the apoB-100 to the lipoprotein sample dilution are constructed. The relative amounts of the two molecular weight forms of apoB in the sample can then be calculated. The absolute masses of apoB-48 and apoB-100 are derived from their relative quantities and the total apolipoprotein B mass determined immunologically.[12]

Calculation of Relative Amount of Apolipoproteins B-48 and B-100 Using SDS–PAGE with Rod Gels

Because several dilutions of the sample are run on separate gels, the dilution factor (f) of the sample can be plotted on the abscissa against the dye uptake by the respective apoB-100 and apoB-48 bands (A is the staining intensity expressed as the densitometrically scanned area). The relationships between the two for apoB-48 is expressed by Eq. (1):

$$A = d_{48}f + b_{48} \tag{1}$$

The slope of this "dilution" plot is d. Extension of the curve to the ordinate would give a y intercept designated b. If A is related to the actual mass of apoB-48 (determined, e.g., by amino acid analysis), then this relationship may be described by Eq. (2):

$$A = c_{48}m_{48} + b_{48} \tag{2}$$

In this case m_{48} is the actual mass of apoB-48 and c_{48} is the slope of the line. The y intercept is again given by b. It should be noted that b_{48} is the same for both dilution and mass plots. The actual mass of apoB-48 in any sample dilution (m_{48}) is equal to the mass of apoB-48 in the undiluted sample (M_{48}) times the dilution factor (f). In other words,

$$m_{48} = M_{48}f \tag{3}$$

Hence, by introducing Eq. (3) into Eq. (2),

$$A = c_{48}M_{48}f + b_{48}$$

Comparing this with Eq. (1), it can be seen that

$$d_{48} = c_{48}M_{48}$$

The corresponding equation for apoB-100 is

$$d_{100} = c_{100}M_{100}$$

From these last two equations it is evident that the ratio of the actual mass of apoB-48 to apoB-100 in an undiluted sample would be

$$\frac{M_{48}}{M_{100}} = \left(\frac{c_{100}}{c_{48}}\right)\left(\frac{d_{48}}{d_{100}}\right)$$

which is the equation used to calculate the relative concentration of apoB-48 and apoB-100 in a given sample.

Comparison and Validation of Methods

The present methods provide fairly simple, sensitive, and reliable means of determining the content of apoB-48 and apoB-100 in Sf > 12 lipoproteins by analytical SDS–PAGE with subsequent conventional protein staining and gel scanning. Using the slab gel technique, there is no difference in chromogenicity between apoB-48 and apoB-100.[7,8] However, it should be noted that chyle provided the source of the apoB-48 in the chromogenicity analysis. Ideally, apoB-48 should have been prepared from plasma. Furthermore, because apoB-100 derived from VLDL of different particle size has been shown to have different chromogenicity in a previous study,[7] apoB-48 from chylomicrons and chylomicron remnants of varying particle size should preferably be compared with apoB-100. This restriction notwithstanding, it is highly unlikely that there should be a difference in chromogenicity for apoB-48 derived from different sources using slab gels, since there was no difference for apoB-100 isolated from lipoproteins of varying particle size using the same technique.[9]

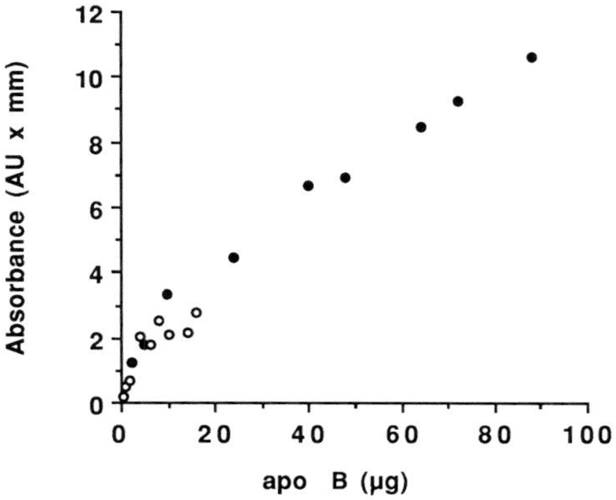

FIG. 1. Effect of massive overloading of isolated apoB-100 (filled circles) and apoB-48 (open circles) on densitometric absorbance determined by laser scanning. [From F. Karpe and A. Hamsten, *J. Lipid Res.* **35,** 1311 (1994).]

In contrast, there is a difference in chromogenicity between apoB-100 derived from different subclasses of triglyceride-rich lipoproteins using the rod gel technique.[7] This problem is circumvented by running the same sample several times at various dilutions, after which the amount of apoB-48 relative to apoB-100 can be calculated as described.

It appears that the difference in chromogenicity is a consequence of overloading of the rod gels, which results in a restricted dye–protein interaction.[9] This also pertains to slab gels. The effect of overloading a slab gel with a resulting nonlinear relationship between apoB protein mass and dye uptake is shown in Fig. 1. To eliminate an analytical error, the protein mass applied to the gel should therefore not exceed certain limits using the slab gel technique. It can be calculated that the maximum protein mass per unit surface area of resolution medium (polyacrylamide surface area in the bottom of the well) is around 0.5 μg/mm^2 for apoB-100. Apolipoprotein B-48 seems to be somewhat more sensitive to overloading, as the corresponding value is 0.2 μg/mm^2 for this protein.[9] These precautions might be critical when determining apoB-48 and apoB-100 masses by Coomassie blue staining of an analytical SDS–PAGE slab gel.

Acknowledgments

The studies forming the basis of the present chapter were supported by the Swedish Medical Research Council (8691), the Swedish Heart–Lung Foundation, the Knut and Alice Wallenberg Foundation, the Foundation for Old Servants, the King Gustaf V 80th birthday fund, Professor Nanna Svartz's Foundation, the Nordic Insulin Fund, the Swedish Margarine Industry Fund for Research on Nutrition, and the Heart and Stroke Foundation of Ontario.

[6] Chromatographic Method for Isolation and Quantification of Apolipoproteins B-100 and B-48

By Janet D. Sparks and Charles E. Sparks

Introduction

Apolipoprotein B (apoB) is a critical structural component of very-low-density lipoprotein (VLDL) and low-density lipoprotein (LDL). Apolipoprotein B is extremely large, and the full-length protein has a molecular mass in excess of 500 kDa. The molecule contains both amphipathic and α-helical structures characteristic of other apolipoproteins as well as a large proportion of β-strand structure which allows apoB to interact favorably with neutral lipid. There are two metabolically relevant forms of apoB,[1] one of higher molecular weight (apoB_H in rodents and apoB-100 in humans) and one of lower molecular weight (apoB_L in rodents and apoB-48 in humans). Apolipoprotein B_H corresponds to the entire coding region of the apoB gene and is present on hepatic lipoproteins in humans[2,3] and rats.[1,4] Apolipoprotein B_L is present on intestinal lipoproteins.[4,5] In humans, apoB_L has been designated apoB-48 because it is approximately 48% of the size of the higher molecular weight form (apoB-100) using sodium dodecyl sulfate (SDS) gel electrophoresis.[5] The two primary apoB variants are produced from a single-copy gene by a novel mRNA editing mechanism which introduces a translational "stop codon" within the apoB mRNA.[6-8] There is on average one apoB molecule per lipoprotein particle,[9] allowing quantitation of apoB to serve as a basis for estimation of the number of apoB–lipoprotein particles present in a sample.

Separation and quantitative analysis of apoB variants have been accomplished using a number of methods, the most popular being SDS–polyacrylamide gel electrophoresis (SDS–PAGE). Assay of radiolabeled apoB variants following SDS–PAGE involves cutting bands out, solubilization of the gel, and radioassay with appropriate quench and recovery factors. Alternatively, gels may be fixed and enhanced (if necessary), then fluorographs or autoradiographs of SDS gels made and scanned for quantitation. Separation and quantitative analysis of apoB variants by gel-filtration column chromatography in the presence of SDS (SDS columns) have certain unique advantages over SDS–PAGE methods. First, there is complete recovery of radioactivity applied to columns,[10] which is not the case with SDS–PAGE. In some gel systems, apoB aggregates can remain trapped in sample wells, leading to losses of radioactivity. Second, a larger volume of

[1] C. E. Sparks and J. B. Marsh, *J. Lipid Res.* **22,** 519 (1981).

[2] J. M. Rash, G. H. Rothblat, and C. E. Sparks, *Biochim. Biophys. Acta* **666,** 294 (1981).

[3] S. B. Edge, J. M. Hoeg, P. D. Schneider, and H. B. Brewer, Jr., *Metab. Clin. Exp.* **34,** 726 (1985).

[4] C. E. Sparks, O. Hnatiuk, and J. B. Marsh, *Can. J. Biochem.* **59,** 693 (1981).

[5] J. P. Kane, D. A. Hardman, and H. E. Paulus, *Proc. Natl. Acad. Sci. U.S.A.* **77,** 2465 (1980).

[6] S.-H. Chen, G. Habib, C.-Y. Yank, Z.-W. Gu, B. R. Lee, S.-A. Weng, S. R. Silberman, S.-J. Cai, J. P. Deslypere, M. Rosseneu, A. M. Gotto, Jr., W.-H. Li, and L. Chan, *Science* **238,** 363 (1987).

[7] L. M. Powell, S. C. Wallis, R. J. Pease, Y. H. Edwards, T. J. Knott, and J. Scott, *Cell (Cambridge, Mass.)* **50,** 831 (1987).

[8] A. V. Hospattankar, K. Higuchi, S. W. Law, N. Meglin, and H. B. Brewer, Jr., *Biochem. Biophys. Res. Commun.* **148,** 279 (1987).

[9] J. Elovson, J. E. Chatterton, G. T. Bell, V. N. Schumaker, M. A. Reuben, D. L. Puppione, J. R. Reeve, Jr., and N. L. Young, *J. Lipid Res.* **29,** 1461 (1988).

[10] C. E. Sparks and J. B. Marsh, *J. Lipid Res.* **22,** 514 (1981).

sample can be applied to columns compared with gels (3.5 ml versus 100 μl). Third, radioactivity in each column fraction can be directly quantitated by either γ or β counting, depending on the label. Fourth, a larger mass of apoB may be prepared for future use in procedures such as immunizations.

The SDS columns can also be used for the preparation of apoB standards for other techniques such as immunoblotting and for quantitative SDS–PAGE. An additional, important application of SDS columns is for the preparation of apoB for immunizations for polyclonal or monoclonal antibody production. Using SDS columns, each apoB variant can be purified from delipidated lipoprotein fractions and processed for use as an immunogen. This chapter outlines modifications of the SDS-column procedure originally described in 1981 by Sparks and Marsh[10] and includes details for the following specific applications: (1) separation of apoB variants on columns of cross-linked agarose; (2) preparation of apoB standards; (3) analysis of column fractions by SDS–PAGE; and (4) preparation of apoB as immunogen for antibody production.

Separation of Apolipoprotein B Variants on Columns of Cross-Linked Agarose

Column Setup

Materials

1.5 × 170 cm low-pressure glass chromatography columns (Bio-Rad Laboratories, Hercules, CA)
Buffer reservoir, 1 liter (Bio-Rad)
Transmission tubing, Tygon flexible plastic tubing, i.d. 0.050 (Norton Performance Plastics, Akron, OH)
19 × 7/8 inch Butterfly, 12 inch tubing infusion sets (Abbott Laboratories, North Chicago, IL)
Peristaltic pump (Tris 3-channel pump; Isco, Inc., Lincoln, NE) with 1/8 inch (i.d.) Silastic pump tubing and connectors

Procedure

1. Transmission tubing connects the flask containing the column buffer (A, Fig. 1) to the pump (B, Fig. 1) using connectors. Transmission tubing from the pump is then connected to the 19 × 7/8 inch Butterfly tubing (with the needle cut off). The Luer-lock of the Butterfly tubing completes the connection to the cap, which fits both the column reservoir and the column itself (C, Fig. 1).

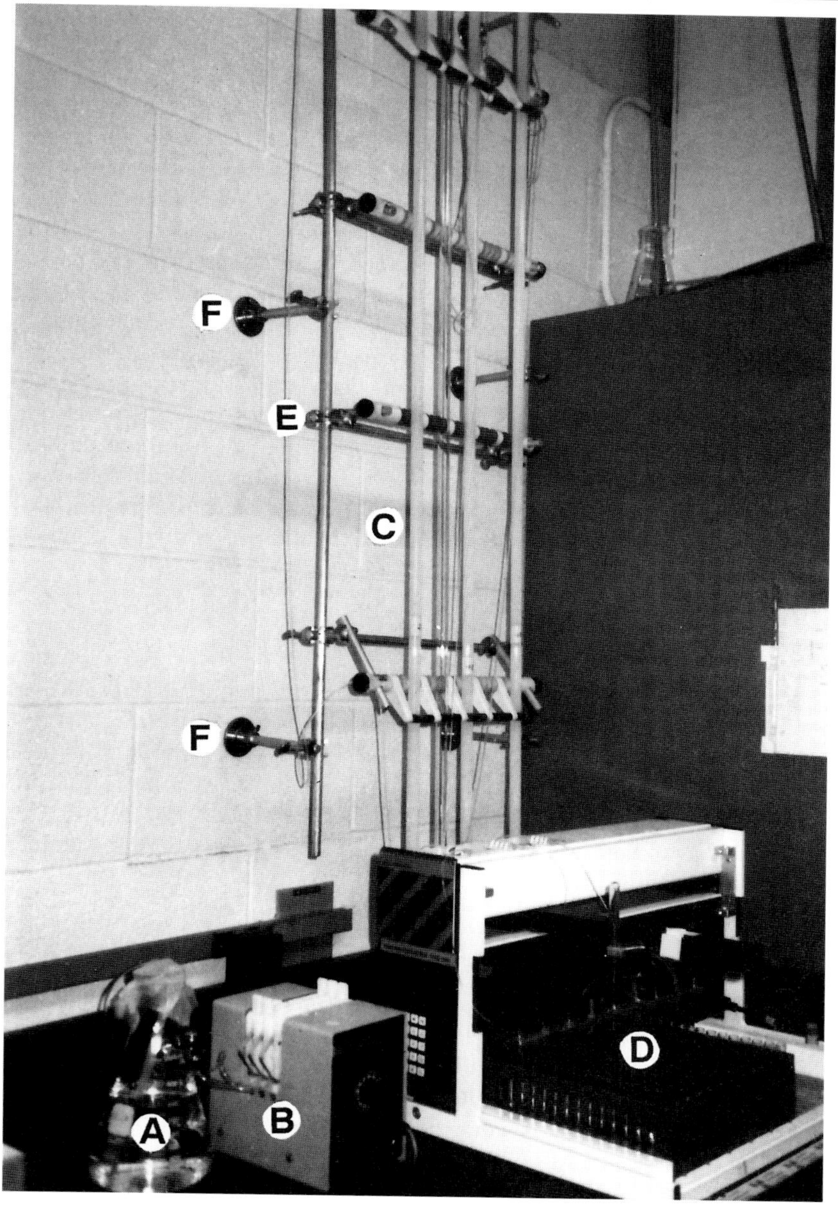

FIG. 1. Laboratory column setup, showing column buffer (A), peristaltic pump (B), 1.5 × 170 cm columns (C), and fraction collector (D). To conserve laboratory bench space, columns may be used in mounts (E) attached to the wall using brackets (F).

2. Another 19 × 7/8 inch Butterfly tubing (with the needle cut off) attaches via a Luer-lock to the bottom of the column, which is then connected by transmission tubing to a fraction collector (D, Fig. 1). In this configuration, with all connections being air-tight, the flow rate of the column is controlled by the pump as there is little operating pressure. Caution must be taken to prevent air leaks, which will produce an operating pressure equivalent to the height of the column and which can cause the column to run dry rapidly.
3. To conserve laboratory bench space, columns may be set up on mounts (E, Fig. 1) attached to the wall using brackets (F, Fig. 1). When columns are in use, the pump and fraction collector can be conveniently placed on the bench top. When not in use, columns can be stoppered and stored in position. The pump and fraction collector can be removed, leaving columns safely out of the way and bench space open. Columns can be stored indefinitely at room temperature in column buffer containing 1% (w/v) SDS and 2.5 m*M* EDTA. When restarting a column, they should be back-flowed for several hours and then run overnight in column buffer before applying samples.

Pouring Columns

Reagents

Stock EDTA: 0.25 *M* ethylenediaminetetraacetic acid (EDTA), pH 7.4; store at 4° after filtration sterilization (0.2 μm)

Column buffer: 0.019 *M* KH_2PO_4, 0.81 *M* Na_2PO_4, 1% (w/v) SDS, 2.5 m*M* EDTA (10 ml stock EDTA solution per liter), pH 7.4; room temperature

Beaded agarose: Sepharose CL-2B, CL-4B, CL-6B (Sigma Chemical Co., St. Louis, MO)

Procedure

1. Columns are poured and run at room temperature. Approximately 350 ml beaded agarose is washed three times in 500 ml column buffer.
2. The column is fitted with the 1-liter column reservoir and is tightly sealed by wrapping Parafilm around the connection between the reservoir and column. The column is then filled with buffer and allowed to stand for several hours to allow bubbles to escape.
3. The agarose is resuspended in a total volume of 500 ml (70% w/v settled gel) and poured into the 1-liter reservoir. The column is then flow-packed through the reservoir at a rate of about 5 ml/hr.
4. After the column has achieved the appropriate height (which takes about 24 hr for Sepharose CL-4B and CL-6B and 48 hr for Sepharose

CL-2B), the reservoir is removed. Columns are flowed overnight at about 5 ml/hr to equilibrate fully before use.

Preparation of Samples for SDS Columns

Materials and Reagents

Glass centrifuge tubes (40 ml) with Teflon-lined caps

Cold diethyl ether containing 0.001% (w/v) butylated hydroxytoluene as a preservative; stored at −20° in an explosion-proof freezer

Cold chloroform and methanol: stored at 4° in an explosion-proof refrigerator

Sample buffer: buffer used is similar in composition to the sample buffer described by Laemmli[11] and contains 62.5 m*M* Tris-HCl buffer, pH 6.8, 2% (w/v) SDS, 5% (v/v) 2-mercaptoethanol, 20% (v/v) glycerol with added 10 m*M* dithiothreitol

Procedure 1: Isolated Lipoproteins

1. A 1.0-ml aliquot of lipoprotein dialyzed against 0.15 *M* NaCl/2.5 m*M* EDTA to remove salt is added to 10 ml cold methanol and rapidly mixed. (The sample appears turbid at this point.) To the mixture is added 10 ml cold chloroform (4°), and after mixing a clear solution is formed.
2. To the chloroform–methanol solution is added 20 ml cold (−20°) diethyl ether. After mixing thoroughly the sample is placed at −20° overnight to allow precipitation of apolipoproteins. (Samples are stable at this point and can be stored at −20°.)
3. To complete delipidation, samples are removed from the freezer and are spun in a swinging-bucket rotor for 15 min at 1500 *g* in a refrigerated centrifuge precooled to 4°. (Caution should be taken if a non-explosion-proof centrifuge is used to ensure that no tubes break during centrifugation.)
4. In a solvent hood, the chloroform–methanol–ether is carefully removed by aspiration without disturbing the pellet. (Apolipoproteins appear as a thin film on the bottom of the glass test tube.) To remove residual lipids and chloroform–methanol, several aliquots (2–3 ml) of cold (−20°) diethyl ether are used to rinse down the sides of the test tube gently without disturbing the pellet. Should the pellet be disturbed, samples must be repelletted by centrifugation. Each wash is carefully removed by aspiration.

[11] U. K. Laemmli, *Nature (London)* **227,** 680 (1970).

5. The bottom of the test tube is warmed by holding in one's hand to evaporate residual diethyl ether. (When the apolipoprotein turns white, the sample is ready to be dissolved in sample buffer. If the pellet has a wet or gray appearance, additional diethyl ether washes are necessary.)
6. To the pellet is added sample buffer (up to 3 ml), and the pellet is allowed to dissolve completely at room temperature. This may occur immediately or may take several hours. (If chloroform–methanol has not been completely removed, the sample will not dissolve readily.) The sample is heated to 95°–100° for 10 min to denature proteins completely, and then the sample (up to 3.5 ml) is applied to the top of the column.

Procedure 2: Immunoprecipitated Lipoprotein Apolipoprotein B

1. Labeled apoB present in immunoprecipitates can be separated by SDS columns and readily quantitated by radioassay of column fractions using either β-scintillation counting or by γ counting, depending on radionuclide label. Radioassay in combination with protein assay of column fractions containing pure apoB variants allows for direct calculation of the specific radioactivity of apoB. Labeled apoB from immunoprecipitates is directly eluted into 1.0 ml sample buffer followed by heating to 95°–100° for 10 min.
2. Solid-phase immunoadsorbents such as protein A cells are removed by centrifugation when required. This elution process is repeated twice more, and supernatants containing labeled apoB are pooled. After a final spin to remove any particulate material, the sample can be applied to the column.

Procedure 3: Plasma or Serum Containing Radiolabeled Lipoprotein

1. Plasma or serum containing radiolabeled lipoproteins are prepared for column analysis as described in detail elsewhere.[1,10] Plasma or serum can be applied to columns directly after dilution with an equal volume of 10% (w/v) SDS, 10% (v/v) glycerol, 10% (v/v) 2-mercaptoethanol in 0.1 *M* Tris-HCl, pH 7.4.
2. Alternatively, plasma or serum containing labeled lipoproteins can be delipidated first followed by dissolving the air-dried pellet (up to 0.4 ml plasma or serum) in 1.0 ml sample buffer and incubating at room temperature for 18 hr. Samples are then heated to 95°–100° for 10 min with mixing, cooled, and applied to columns. Labeled or unlabeled neutral lipid will appear close to the void volume of the column without sample delipidation.

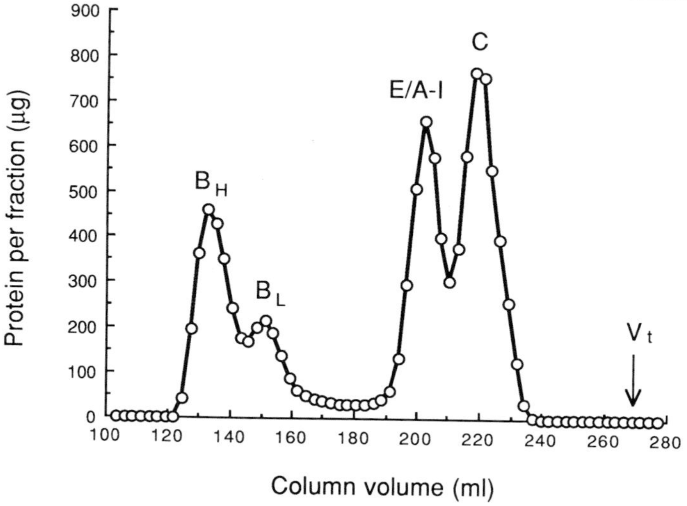

FIG. 2. SDS gel-filtration column chromatography of rat $d < 1.063$ g/ml lipoproteins on Sepharose CL-4B columns. Protein content was determined for each column fraction, and protein per fraction (μg) was plotted against column volume (ml). V_t was 274 ml, V_0 was 99.4 ml, and the flow rate was 5.3 ml/hr.

Preparation of Apolipoprotein B Standards

1. Lipoproteins are prepared from fresh rat sera (100–200 ml). Blood is collected and allowed to clot at 4°. After removal of cells, EDTA is added to serum at a final concentration of 2.5 to 5 m*M*. Lipoproteins are isolated by ultracentrifugation after adjustment of the salt density to 1.063 g/ml with solid sodium bromide using a 60Ti rotor (Beckman Instruments, Inc.) for 20 hr at 50,000 rpm, 14°. Lipoproteins are removed and then dialyzed against 4 liters of 0.15 *M* NaCl containing 2.5 m*M* EDTA using a 50,000 molecular weight cutoff membrane (Spectra/Por, Spectrum Medical Instruments, Inc., Houston, TX) at 4° using three changes every 4–6 hr.
2. After dialysis, lipoproteins are delipidated and apolipoproteins, dissolved in sample buffer, are applied to the Sepharose CL-4B column in a maximum total volume of 3.5 ml. After elution of the column, fractions can be analyzed for protein content[12] and protein per fraction can be plotted against column volume as seen in Fig. 2. To identify fractions containing pure apoB_H (apoB-100) and apoB_L

[12] M. K. Markwell, S. M. Haas, L. L. Bieber, and N. E. Tolbert, *Anal. Biochem.* **87,** 206 (1978).

(apoB-48), column fractions can be analyzed by SDS–PAGE methods as described below.

3. For standards, apoB variants from column fractions are separately pooled and assayed for protein content using the Markwell modification of the Lowry procedure.[12] After protein is analyzed, pools are diluted 1 : 1 with 2× sample buffer first and then diluted further with 1× sample buffer to adjust protein concentrations. Apolipoprotein B standards are divided into aliquots in small polypropylene tubes and stored frozen at $-20°$ or $-60°$. Rat $apoB_H$ and $apoB_L$ standards prepared in this manner give reproducible bands on SDS–PAGE for at least 4 years.

Analysis of Column Fractions by SDS–PAGE

1. Column fractions are analyzed after separation by SDS–PAGE using several modifications of the method of Laemmli.[11] We use the SE 600 gel apparatus, 1.5 mm × 14 cm wide × 16 cm long slab gels (Hoefer Scientific Instruments, San Francisco, CA), with the following modifications. The 2× treatment buffer for preparation of samples is adjusted to contain 20 m*M* dithiothreitol. The slab gel is composed of a gradient of acrylamide and Acrylaide instead of acrylamide and bisacrylamide. The gel monomer solution used to prepare gradients is a solution of 1% (w/v) Acrylaide cross-linker (FMC BioProducts, Rockland, ME) and 32% (w/v) acrylamide (33% total) as described previously.[13]
2. A piece of hydrated porous cellophane (SE 542, Hoefer) is placed on one side of one of the glass plates used to form the gel sandwich. Before use, the porous cellophane is soaked in 0.1% (w/v) SDS overnight and then extensively rinsed in distilled water. The wet cellophane is cut to fit the width of the gel sandwich and smoothed out onto the glass plate. Air bubbles are removed by wiping the surface with Kimwipes EX-L (Kimberly-Clark Corp., Roswell, CA). Excess cellophane on the top and bottom of the glass plate is wrapped around the other side of the glass plate (~2–3 cm), and air bubbles are also removed. Gel spacers are then placed on top of the cellophane-coated side, and the gel sandwich is assembled as usual. A gradient gel of 3.5 to 26% (w/v) Acrylaide–acrylamide is then poured into the gel sandwich mold. The stacking gel is 3.4% (w/v) total acrylamide consisting of a 50 : 50 mixture of diluted Acrylaide–acrylamide and bisacrylamide–acrylamide [Protogel, 30% (w/v)

[13] J. D. Sparks and C. E. Sparks, *J. Biol. Chem.* **265,** 8854 (1990).

acrylamide/0.8% (w/v) bisacrylamide (National Diagnostics, Manville, NJ)].

3. Column fractions are mixed with an equivalent volume of the modified 2× Laemmli treatment buffer and heated to 95°–100° for 10 min. Samples are applied and electrophoresed as usual. After electrophoresis, the gel sandwich is disassembled, and the gel, backed by porous cellophane, is placed directly in the electrophoretic transfer unit. (The cellophane backing of the gradient gel allows for easy transport and helps prevent distortion of the gel during placement in the transfer apparatus.)
4. Column-fractionated and SDS–PAGE-separated proteins are then transferred to polyvinylidene difluoride (PVDF) membranes (0.2 μm) using an electrophoretic transfer cell with plate electrodes (Bio-Rad Corporation, Richmond, CA) in a buffer of 25 m*M* Tris, pH 8.3, 192 m*M* glycine[14] containing 0.1% (w/v) SDS. (Because the electrophoresis tank buffer is identical in composition to the transfer buffer, preequilibration of the gel can be minimized, and we have found it to be unnecessary). An initial power range of 75 V, 0.5 A, is used for transfer for 3 hr with cooling to 0°–4°. (Typically, the current will rise to between 0.6 and 0.75 A during a 3-hr run). In control experiments, 3 hr of transfer removes greater than 95% of apoB from gels.
5. After transfer, PVDF membranes are stained using a colloidal gold total protein stain (Bio-Rad Corporation). As seen in Fig. 3, there is complete separation of apoB variants in the fractions presented. In samples prepared in contact with other detergents (Triton X-100), there may altered mobility on columns if there is incomplete removal of detergent. After staining (usually 10–20 min) the transfer membrane is washed in deionized water several times and allowed to air dry, becoming a permanent record of the column separation.

Preparation of Apolipoprotein B as Immunogen for Antibody Production

1. After column fractions containing the apoB variant of interest have been identified, appropriate fractions are pooled and most of the SDS is removed using either a dialysis procedure[13] or spin columns.
2. For the dialysis method, pooled column fractions are dialyzed against one to three changes of 4 liters of distilled and deionized water containing 25 g Amberlite MB-3 (Sigma) for 3–6 hr at room tempera-

[14] H. Towbin, T. Staehelin, and J. Gordon, *Proc. Natl. Acad. Sci. U.S.A.* **76,** 4350 (1979).

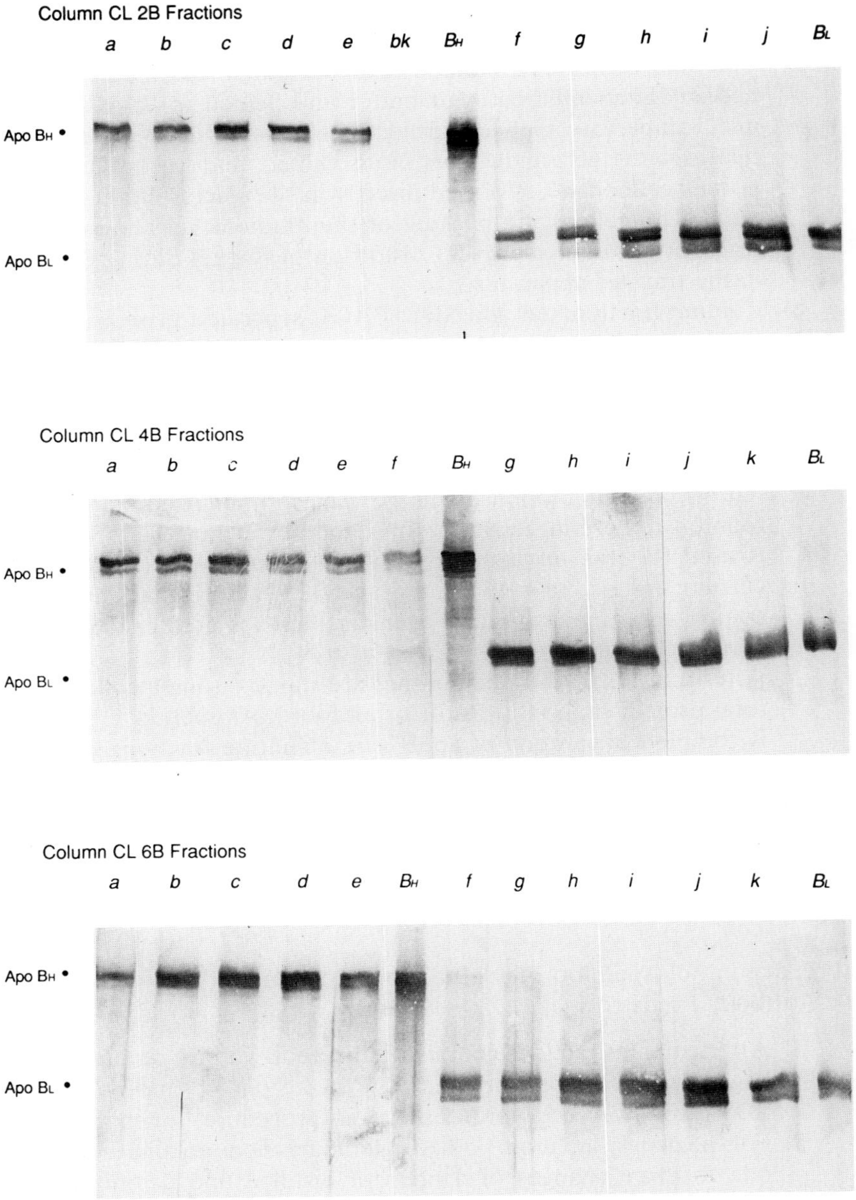

FIG. 3. Analysis by SDS–polyacrylamide gel electrophoresis of column fractions of rat apoB variants derived from delipidated $d < 1.063$ g/ml lipoproteins. For fractions from Sepharose CL-2B columns V_e values were as follows: a, 174.7 ml; b, 176.9 ml; c, 179.2 ml;

ture using a 50,000 molecular weight cutoff dialysis membrane. The resin is replaced after depletion when necessary. Dialysis is continued overnight at 4° against two changes of 4 liters of 0.01% (w/v) ammonium formate containing 0.01% (v/v) Triton X-100 without added resin. After dialysis, the apoB solution is removed and the inside membrane is carefully rinsed with filter-sterilized distilled and deionized water; the rinses are pooled with dialyzed apoB. The solution is then frozen in liquid nitrogen.

3. For the spin column method, pooled fractions containing apoB (maximum of 15 ml) are placed in C-100 Centriprep columns (Amicon Division, W. R. Grace & Co., Danvers, MA), which are spun in a swinging-bucket rotor at 500 *g* for 60 min at room temperature. After removal of the first filtrate, the retentate is made up to 15 ml using 0.01% (w/v) ammonium formate and the columns are respun. The retentate is again made up with 15 ml of 0.01% (w/v) ammonium formate and the column is respun. The retentate is then made up with 15 ml of 0.01% (w/v) ammonium formate containing 0.01% (v/v) Triton X-100 and respun, and this process is repeated one additional time. After the final spin, the retentate containing concentrated apoB plus several distilled and deionized water washes of the column membrane are pooled. The solution is then frozen in liquid nitrogen. Recovery of apoB using the spin column method is similar to that using the dialysis procedure. Recovery of apoB is better using C-30 Centriprep columns; however, more residual SDS remains in the sample.
4. Frozen apoB from either dialysis or spin column procedures is then lyophilized and reconstituted with 1.0 ml sterile, pyrogen-free saline. An emulsion of apoB is prepared using emulsion needles after addition of 1.0 ml Freund's adjuvant (either complete for primary immunizations or incomplete for secondary immunizations). A 2.0-ml emulsion prepared in this manner is used to immunize a single rabbit. It

d, 181.4 ml; e, 183.5 ml; f, 192.5 ml; g, 194.7 ml; h, 196.9 ml; i, 199.2 ml; j, 201.4 ml. Lane bk contains a column fraction before the void volume. For fractions from Sepharose CL-4B columns V_e values were as follows: a, 124.2 ml; b, 126.5 ml; c, 128.8 ml; d, 131.1 ml; e, 133.5 ml; f, 135.8 ml; g, 147.4 ml; h, 149.7 ml; i, 152.0 ml; j, 154.3 ml; k, 156.7 ml. For fractions from Sepharose CL-6B columns V_e values were as follows: a, 102.7 ml; b, 104.9 ml; c, 107.2 ml; d, 109.4 ml; e, 111.7. ml; f, 120.5 ml; g, 122.8 ml; h, 125.0 ml; i, 133.5 ml; j, 135.8 ml; k, 138.1 ml. Apolipoprotein B_H and B_L standards were column-purified and stored in sample buffer at −20° for over 1 year before electrophoresis. V_t values for Sepharose CL-2B, CL-4B, and CL-6B columns were 272.5, 265.7, and 271.5 ml, respectively. Flow rates for the three columns were 4.4, 4.6, and 4.4 ml/hr, respectively.

is convenient for apoB immunizations to store aliquots of apoB protein frozen in liquid nitrogen at −70° that are equivalent to a single immunization injection. For rabbit anti-rat apoB antibody production, we have used between 0.5 and 2 mg rat apoB per rabbit for primary immunizations and between 50 and 250 μg rat apoB per rabbit for secondary immunizations. Primary immunizations are given as multiple intramuscular injections, whereas booster injections can be given both intramuscularly and subcutaneously. Generally, booster injections are begun 6–8 weeks after primary immunizations and can be given every 2–3 months if necessary to maintain a high titer of antibody. Blood is collected 2 weeks after each booster injection.

Remarks and Comments

The SDS gel-filtration column chromatography method described here is a relatively simple method and has distinct advantages over other methods for the isolation and quantitation of apoB variants in lipoprotein fractions. Advantages include complete recovery of radiolabeled apoB,[10] ability to use larger volumes than those used in SDS–PAGE, and ease of radioassay of each apoB variant after separation. Using 1.5 × 170 cm columns, all of the chromatography media tested (2, 4, and 6% agarose) are able to separate human and rat apoB variants. Using Sepharose CL-6B columns the SDS micelle of apoB-100 is partially excluded, whereas using Sepharose CL-4B and CL-2B, apoB-100 permeates the gel completely (Fig. 4). Unlike human lipoproteins, column fractions containing rat $apoB_H$ also contain a significant amount of the smaller $apoB_H$ fragment (also termed P-II), which is believed to be derived from proteolytic cleavage of the carboxyl terminal of $apoB_H$[15] that occurs at the time of secretion by the liver.[15]

For the specific applications outlined in this chapter, Sepharose CL-4B was chosen as the most useful chromatographic medium for two reasons. First, $apoB_H$ (apoB-100) enters the CL-4B gel more completely than Sepharose CL-6B, which helps eliminate contaminants from $apoB_H$ fractions that may coelute in the first few column fractions. Second, only one-third of the total column elution volume is required to obtain both $apoB_H$ and $apoB_L$ compared with Sepharose CL-2B columns, which require nearly one-half to two-thirds of the total column elution volume.

In addition to the use of SDS columns for quantitative analysis of labeled apoB variants, there are two major applications that we describe here:

[15] M. A. Reuben, K. L. Svenson, M. H. Doolittle, D. F. Johnson, A. J. Lusis, and J. Elovson, *J. Lipid Res.* **29,** 1337 (1988).

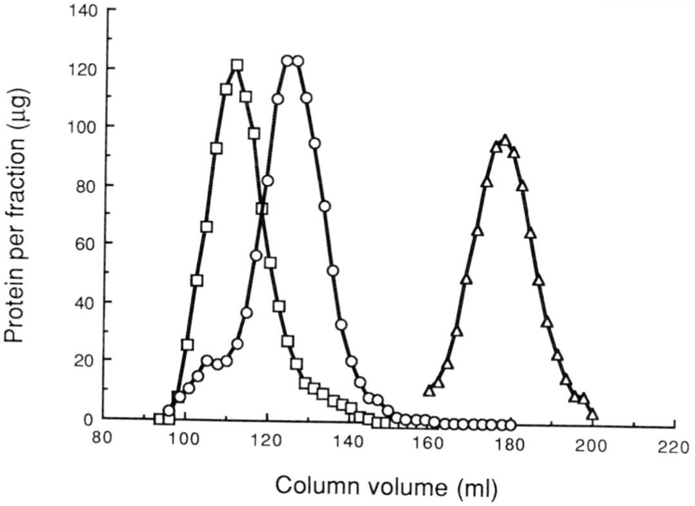

FIG. 4. Comparison of separations of 1 mg human LDL ($1.019 < d < 1.050$ g/ml) using SDS gel-filtration column chromatography on columns of Sepharose CL-2B (triangles), CL-4B (circles), and CL-6B (squares). Protein was determined in each column fraction, and protein per fraction (μg) was plotted against column volume (ml).

one, to prepare apoB standards and the other, to isolate apoB to use for immunizations for antibody production. In either application, column fractions are easily monitored by using SDS–PAGE, electrophoretic transfer onto PVDF membranes, and finally colloidal gold staining, which allows quick evaluation of apoB even in dilute column fractions. After pooling appropriate fractions, a more accurate measurement of protein can be made using a standard method such as the modification of the Lowry procedure suggested here. For standards, column fractions are diluted to achieve appropriate protein concentrations and stored frozen at $-20°$ or $-60°$. For antibody production, apoB from pooled column fractions is concentrated, most of the SDS is removed, and the solution is lyophilized and reconstituted for immunizations. Using column-isolated apoB, polyclonal antibodies can be prepared specifically against apoB_L, the amino-terminal half of apoB. Antisera produced by immunization with apoB_H contain, in addition to antibodies against amino-terminal determinants, antibodies against carboxyl-terminal determinants.

Gel-filtration column chromatography in the presence of SDS can be used to determine approximate molecular sizes of apoB under conditions of SDS solubilization using the relationship that the calculated K_{av} is inversely proportional to the logarithm of the molecular mass. The approximate molecular mass of rat apoB variants based on Sepharose CL-6B column mobility has been reported to be 500 kDa for $apoB_H$ and 250 kDa for $apoB_L$.[10] These values are similar to those reported for human apoB-100 of 549 kDa and apoB-48 of 264 kDa based on SDS–polyacrylamide gel electrophoresis[5] and are consistent with the direct calculation of 512 and 250 kDa based on corresponding sequences of amino acids encoded in apoB mRNA. From comparison of calculated K_{av} values of human and rat apoB variants using various percentage agarose columns (Table I), the sizes of the higher and lower molecular weight variants are similar.

When electrophoretic mobilities of column-fractionated apoB variants

TABLE I

K_{av} FOR APOLIPOPROTEIN B VARIANTS IN CROSS-LINKED AGAROSE[a]

ApoB variant	Cross-linked agarose concentration		
	2%	4%	6%
Human apoB-100[b]	0.292	0.175	0.056
Rat $apoB_H$	0.290	0.198 ± 0.017[c]	0.054 ± 0.007[d]
Human apoB-48[e]	0.456	0.341	0.162
Rat $apoB_L$	0.438	0.331 ± 0.017[c]	0.146 ± 0.005[d]

[a] On 1.5 × 170 cm columns K_{av} is the fraction of the stationary gel volume which is available for diffusion of a given protein species. K_{av} was calculated by the formula $K_{av} = (V_e - V_o)/(V_t - V_o)$ using the volume of elution (V_e) of Blue Dextran (molecular weight approximately 2×10^6) as the void volume (V_o) and the peak of 2-mercaptoethanol as the total column volume (V_t). Column flow rates were 4–5 ml/hr, and fractions were collected every 30 min. The absence of suitable standards to estimate the void volume in cross-linked agarose columns (especially 2 and 4%) makes calculation of K_{av} an approximation intended only for comparative purposes.

[b] Average of two column runs of apoB-100 derived from human LDL (1.019 g/ml < d < 1.050 g/ml) and from human Svedberg flotation rate (S_f) > 100 lipoproteins.

[c] Mean K_{av} ± S.D. of six column runs of $d < 1.063$ g/ml rat lipoproteins on three separately packed 4% cross-linked agarose columns. The mean V_e ± S.D. for $apoB_H$ was 134 ± 6 ml, and that for $apoB_L$ was 156 ± 7 ml.

[d] Average K_{av} of $apoB_H$ and $apoB_L$ on Sepharose CL-6B columns from Sparks and Marsh.[10]

[e] Single column run of $S_f > 100$ human lipoproteins. Approximately 1.5 mg protein was applied to each column, and the peak of apoB-48 was determined by examination of colloidal gold-stained, PVDF transfers of proteins in column fractions resolved by SDS–PAGE.

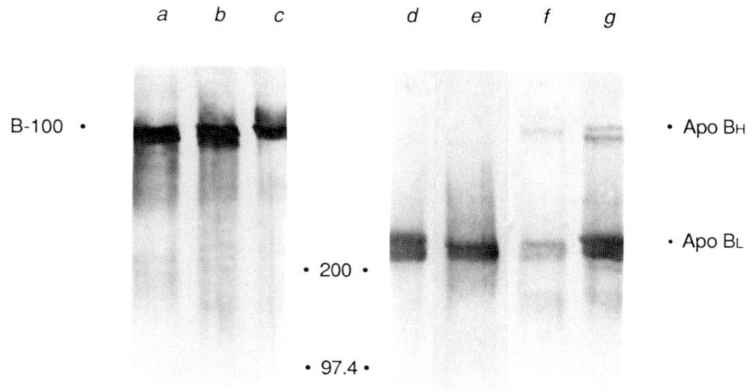

FIG. 5. Comparison of mobilities of SDS column-isolated rat and human apoB variants using SDS–PAGE, gel transfer, and colloidal gold staining techniques. Lane a, apoB-100 derived from human LDL; lane b, $apoB_H$ derived from rat lipoproteins; lane c, apoB-100 derived from human $S_f > 100$ lipoproteins; lane d, $apoB_L$ derived from rat lipoproteins; lane e, apoB-48 derived from human $S_f > 100$ lipoproteins; lane f, mixed apoB-100/apoB-48 fraction derived from human $S_f > 100$ lipoproteins; and lane g, mixed $apoB_H/apoB_L$ fraction from rat lipoproteins. The migration of molecular mass marker proteins, myosin heavy chain (200 kDa) and phosphorylase b (97.4 kDa) are indicated.

of rat and human lipoproteins are compared using SDS–PAGE (Fig. 5), human apoB-100 comigrates with rat $apoB_H$, as has been demonstrated by a number of investigators. Comparison of electrophoretic mobility in SDS gels of the lower molecular weight forms of apoB derived from human and rat lipoproteins suggests that possibly rat $apoB_L$ may be slightly larger than human apoB-48, and this is consistent with the slightly smaller K_{av} for rat $apoB_L$ on various chromatographic media (Table I). The reason for this small difference is not known, but the presence of a doublet in the $apoB_L$ band on gels may contribute to differences in K_{av} values. Whether the doublet is real or artifactual and whether there are slight differences in molecular weights of rat $apoB_L$ and human apoB-48 will be answered only when complete analysis of the amino acid sequence of rat apoB is achieved. In our hands, doublets of apoB are a consistent finding using high-resolution SDS–PAGE separations, which could also be explained by known differences in posttranslational modifications of apoB by glycosylation, fatty acylation, or phosphorylation.

Acknowledgments

We gratefully acknowledge Mary Bolognino and Joanne Cianci for excellent technical assistance. We thank Stephanie D. Kafonek, Johns Hopkins University, School of Medicine, Lipid Research Arteriosclerosis Unit, for providing human $S_f > 100$ lipoproteins. This work was supported by Grants HL 29837 from the National Institutes of Health, National Heart, Lung and Blood Institute (C. E. S.), and the Council for Tobacco Research—U.S.A. (J. D. S.).

[7] Identification and Characterization of Truncated Forms of Apolipoprotein B in Hypobetalipoproteinemia

By STEPHEN G. YOUNG, ELAINE S. KRUL, SALLY MCCORMICK, ROBERT V. FARESE, JR., and MACRAE F. LINTON

Introduction

In 1986, Young *et al.*[1,2] reported the presence of a truncated apolipoprotein (apo) B in the plasma of members of a kindred with hypobetalipoproteinemia. Subsequent studies revealed that the truncated apoB was caused by a frameshift mutation in the apoB gene.[3] Since then it has become apparent that a variety of apoB gene mutations that interfere with the translation of a full-length apoB-100 molecule can cause familial hypobetalipoproteinemia. Many of these mutations result in the production of a truncated species of apoB that can be detected within the plasma lipoproteins. A comprehensive review of familial hypobetalipoproteinemia that includes a summary of the historical aspects of the syndrome, clinical descriptions of the heterozygous and homozygous forms of the disorder, a list of apoB gene mutations causing hypobetalipoproteinemia, and a summary of the properties of lipoproteins containing truncated apoB species has been published.[4] Here, we summarize techniques that are useful in identification and characterization of truncated apoB species.

It has become clear that mutations causing truncated apoB species are not particularly rare, and it is possible that apoB mutations causing hypobetalipoproteinemia might occur in as many as 1 in 1000 healthy human subjects.[5,6] The percentage of all of the "hypobeta" apoB gene mutations

[1] S. G. Young, S. J. Bertics, L. K. Curtiss, B. W. Dubois, and J. L. Witztum, *J. Clin. Invest.* **79,** 1842 (1987).
[2] S. G. Young, S. J. Bertics, L. K. Curtiss, and J. L. Witztum, *J. Clin. Invest.* **79,** 1831 (1987).
[3] S. G. Young, S. T. Northey, and B. J. McCarthy, *Science* **241,** 591 (1988).
[4] M. F. Linton, R. V. Farese, Jr., and S. G. Young, *J. Lipid Res.* **34,** 521 (1993).
[5] P. M. Laskarzewski, P. Khoury, J. A. Morrison, K. Kelly, M. J. Mellies, and C. J. Glueck, *Metabolism* **31,** 558 (1982).
[6] G. E. Anderson, W. Trojaborg, and H. C. Lou, *Acta Paediatr. Scand.* **68,** 155 (1979).

that is associated with a truncated apoB species in the plasma, however, is unknown. Whether a "hypobeta" mutation is associated with a truncated apoB in the plasma appears to depend on the size of the truncated apoB that the mutation creates.[4] The apoB gene contains 29 exons and codes for a protein of 4536 amino acids.[7] Nonsense and frameshift mutations occurring toward the 3′ end of the gene (i.e., in exons 26–29, which code for apoB-100 amino acids 1306–4536) appear to be invariably associated with hypobetalipoproteinemia and a truncated apoB in the plasma lipoproteins.[4] Whether the truncated apoB is found primarily in the high-density lipoproteins (HDL) or in the very-low-density lipoproteins (VLDL) depends on the length of the truncated apoB; an inverse relationship exists between the buoyant density of the lipoprotein and the length of the truncated apoB.[4] So far, four mutations occurring 5′ to exon 26 have been reported.[8–10] These 5′ mutations, which are predicted to yield truncated apoB proteins less than 1300 amino acids in length, were not reported to be associated with detectable levels of a truncated apoB protein in the plasma. In those reports, however, the description did not make clear whether the search for a truncated apoB in the plasma was thorough. Several investigators have expressed truncated human apoB species in cultured cells by transfecting the cells with apoB cDNAs of various lengths.[11–14] These studies have revealed that apoB species as short as about 800 amino acids are secreted from cells in substantial amounts. Therefore, it is possible and perhaps likely that some nonsense mutations within the 5′ portion of the apoB gene ultimately may be found that are associated with detectable levels of truncated apoB species that are shorter than 1300 amino acids in length.

[7] B. D. Blackhart, E. H. Ludwig, V. R. Pierotti, L. Caiati, M. A. Onasch, S. C. Wallis, L. Powell, R. Pease, T. J. Knott, M.-L. Chu, R. W. Mahley, J. Scott, B. J. McCarthy, and B. Levy-Wilson, *J. Biol. Chem.* **261,** 15364 (1986).

[8] D. R. Collins, T. J. Knott, R. J. Pease, L. M. Powell, S. C. Wallis, S. Robertson, C. R. Pullinger, R. W. Milne, Y. L. Marcel, S. E. Humphries, P. J. Talmud, J. K. Lloyd, N. E. Miller, D. Muller, and J. Scott, *Nucleic Acids Res.* **16,** 8361 (1988).

[9] L.-S. Huang, H. Kayden, R. J. Sokol, and J. L. Breslow, *J. Lipid Res.* **32,** 1341 (1991).

[10] L. S. Huang, M. E. Ripps, S. H. Korman, R. J. Deckelbaum, and J. L. Breslow, *J. Biol. Chem.* **264,** 11394 (1989).

[11] Z. Yao, B. D. Blackhart, M. F. Linton, S. M. Taylor, S. G. Young, and B. J. McCarthy, *J. Biol. Chem.* **266,** 3300 (1991).

[12] H. Herscovitz, M. Hadzopoulou-Cladaras, M. T. Walsh, C. Cladaras, V. I. Zannis, and D. M. Small, *Proc. Natl. Acad. Sci. U.S.A.* **88,** 7313 (1991).

[13] D. L. Graham, T. J. Knott, T. C. Jones, R. J. Pease, C. R. Pullinger, and J. Scott, *Biochemistry* **30,** 5616 (1991).

[14] D. J. Spring, L. W. Chen-Liu, J. E. Chatterton, J. Elovson, and V. N. Schumaker, *J. Biol. Chem.* **267,** 14839 (1992).

Designation of Truncated Apolipoprotein B Species According to Centile System

Truncated apoB species have been named according to the centile system of nomenclature that was originally suggested by Kane and coworkers.[15] According to this system, a truncated apoB species that is 37% as large as the full-length apoB-100 is designated apoB-37. Initially the sizes of truncated apoB's were estimated according to the migration of the truncated proteins on sodium dodecyl sulfate (SDS)–polyacrylamide gels.[2] However, because the polymerase chain reaction technique[16] has made identifying the responsible apoB gene mutation relatively easy, apoB species now can be assigned more precise percentiles. For example, if a mutation is predicted to result in a truncated protein of 2050 amino acids, it would be designated apoB-45.2 (2050/4536 $\times$ 100).

Screening and Identifying Truncated Apolipoprotein B Species in Subjects with Hypolipidemia

We have had no difficulty identifying hypobetalipoproteinemia heterozygotes who have truncated apoB species in their plasma. To identify heterozygotes, we have screened healthy adults over the age of 30 who, on repeated measurement, have had total plasma cholesterol levels less than approximately 115 mg/dl and LDL cholesterol levels less than around 50 mg/dl.[4] These stringent cholesterol cutoff values undoubtedly have led us to overlook some heterozygotes, as our studies of hypobetalipoproteinemia kindreds have identified some heterozygotes with total plasma cholesterol levels greater than 190 mg/dl.[17] Adults with total plasma cholesterol levels of 30 to 70 mg/dl and LDL cholesterol levels of 1 to 10 mg/dl are likely to be hypobetalipoproteinemia homozygotes or compound heterozygotes. If a suspected homozygote is asymptomatic and has normal or near-normal plasma triglyceride levels, then one can be reasonably confident that one or more truncated forms of apoB will be found in the plasma.[4] Because the heterozygous form of hypobetalipoproteinemia is far more common than the homozygous, most of the truncated forms of apoB that have been studied have been identified in heterozygotes.

A screening strategy for apoB truncations in hypocholesterolemic subjects must consider the fact that the concentration of truncated apoB's in the plasma of heterozygotes is invariably quite low and is almost always

[15] J. P. Kane, D. A. Hardman, and H. E. Paulus, *Proc. Natl. Acad. Sci. U.S.A.* **77,** 2465 (1980).

[16] R. K. Saiki, D. H. Gelfand, S. Stoffel, S. J. Scharf, R. Higuchi, G. T. Horn, K. B. Mullis, and H. A. Erlich, *Science* **239,** 487 (1988).

[17] S. G. Young, S. T. Hubl, D. A. Chappell, R. S. Smith, F. Claiborne, S. M. Snyder, and J. F. Terdiman, *N. Engl. J. Med.* **320,** 1604 (1989).

less than 10% of the amount of apoB-100 that is produced by the normal allele.[4] The screening strategy also must take into account the propensity of apoB-100 to undergo proteolytic and oxidative breakdown. A cocktail of proteolytic inhibitors, such as the one used by Cardin *et al.*,[18] should be added to the blood immediately after the phlebotomy, particularly if lipoproteins are to be isolated. These inhibitors (and final concentrations) are as follows: aprotinin (100 kallikrein inhibitor units/ml), Polybrene (25 μg/ml), lima bean trypsin inhibitor (20 μg/ml), soybean trypsin inhibitor (20 μg/ml), benzamidine (2.0 m*M*), D-phenylalanyl-L-prolyl-L-arginine chloromethyl ketone (PPACK, 1.0 μM), and glutathione (0.02%). Probably the most important proteolytic inhibitor is aprotinin, and we occasionally have used it as the sole proteolytic inhibitor. We also recommend adding the cocktail to the lipoprotein fractions after they are isolated by ultracentrifugation. All screening studies should be performed rapidly after the phlebotomy because apoB degradation frequently proceeds within 4 to 5 days despite the addition of inhibitors. The propensity of apoB to undergo fragmentation does not mean, however, that the screening of older plasma samples is hopeless, as the degree of apoB breakdown appears to vary substantially from sample to sample. Occasionally, we have observed plasma samples, particularly from hypolipidemic subjects, that are free of apoB breakdown products even after prolonged storage.

To identify truncated apoB proteins in potential "hypobeta" heterozygotes, our strategy at the Gladstone Institute has been to isolate the VLDL fraction ($d < 1.006$ g/ml) and the HDL fraction ($d = 1.063$–1.21 g/ml) by ultracentrifugation and then to analyze the delipidated proteins from these fractions on stained SDS–polyacrylamide gels. Performing a second ultracentrifugation of the lipoprotein samples often is worth the effort, as it effectively removes the vast majority of irrelevant plasma proteins that frequently contaminate "single spin" lipoprotein preparations. We analyze the VLDL fraction because all truncations that are longer than apoB-37 will appear in substantial amounts in the VLDL,[2,19] and because apoB degradation products are rare in the VLDL fraction (Fig. 1, lane 1). Screening the HDL fraction is useful because apoB truncations shorter than apoB-37 but longer than apoB-31 are found almost exclusively in this fraction.[19–21]

[18] A. D. Cardin, K. R. Witt, J. Chao, H. S. Margolius, V. H. Donaldson, and R. L. Jackson, *J. Biol. Chem.* **259,** 8522 (1984).

[19] S. G. Young, S. T. Hubl, R. S. Smith, S. M. Snyder, and J. F. Terdiman, *J. Clin. Invest.* **85,** 933 (1990).

[20] S. P. A. McCormick, A. P. Fellowes, T. A. Walmsley, and P. M. George, *Biochim. Biophys. Acta* **1138,** 290 (1992).

[21] S. G. Young, C. R. Pullinger, B. R. Zysow, H. Hofmann-Radvani, M. F. Linton, R. V. Farese, Jr., J. F. Terdiman, S. M. Snyder, S. M. Grundy, G. L. Vega, M. J. Malloy, and J. P. Kane, *J. Lipid Res.* **34,** 501 (1993).

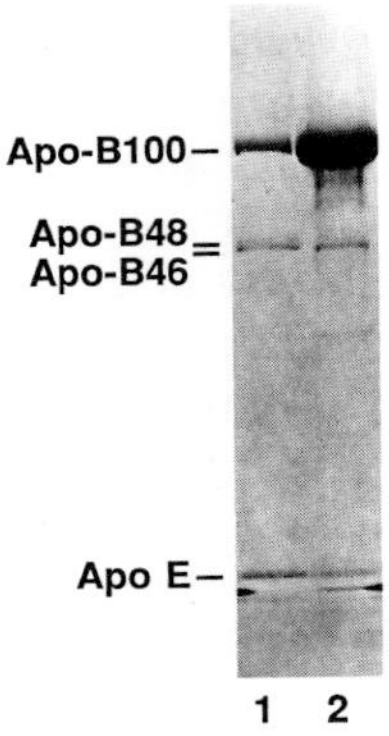

FIG. 1. Coomassie blue-stained 3–12% SDS–polyacrylamide gel demonstrating the presence of a truncated species of apoB, apoB-46, in the VLDL fraction (lane 1) and the LDL fraction (lane 2) of an apoB-46 heterozygote. Note the large amount of apoB-46 in VLDL and the small amount of apoB-46 in LDL, relative to apoB-100. [Reproduced with permission from R. V. Farese, Jr., M. F. Linton, and S. G. Young, *J. Int. Med.* **231,** 643 (1992).]

We have found that analyzing the low-density lipoprotein (LDL) fraction of potential heterozygotes is not particularly useful because it contains a large amount of apoB-100, and frequently apoB-100 breakdown products, and may not necessarily contain a truncated apoB even when the truncated protein is present in relatively high concentrations in the VLDL.[22] Figure 1, lane 1, illustrates the large amount of apoB-46 relative to apoB-100 in the VLDL of an apoB-46 heterozygote. Figure 1, lane 2, illustrates the low amount of apoB-46 relative to apoB-100 in the LDL of the same individual.

After isolating the lipoprotein fractions by ultracentrifugation, we dialyze the samples for several hours against a 1 m*M* EDTA buffer and then delipidate the samples with chloroform–methanol or ethanol–ether.[15] We then apply several different amounts of the apo-VLDL (1–5 μg) and the apo-HDL (2–20 μg) onto a 3–12% gradient SDS–polyacrylamide gel slab (1.5 mm thick) and electrophorese the samples until the bromphenol blue in the loading buffer reaches the bottom of the gel. We prefer to stain the gels with silver because it allows one to judge accurately the amount of apoB-100 degradation products within the sample, and it allows one to evaluate subjectively the appearance of protein bands; truncated apoB bands are typically sharp and have the same dark-brown color as the apoB-48 and apoB-100 bands. High-quality silver staining of polyacrylamide gels can be achieved using the Bio-Rad (Richmond, CA) silver staining kit,

[22] R. V. Farese, Jr., A. Garg, V. R. Pierotti, G. L. Vega, and S. G. Young, *J. Lipid Res.* **33,** 569 (1992).

provided that extremely pure water is used and that one rinses the yellow oxidizer from the gel completely (so that the gel is clear) before proceeding to the next step in the protocol. In general, truncated apoB bands, when present, are not subtle and are easy to identify. In the VLDL fraction, truncated apoB bands are typically 2- to 3-fold more intense than the apoB-48 band but 2- to 3-fold less intense than the apoB-100 band. In the HDL fraction, short apoB truncations such as apoB-31 or apoB-32 are present in very small amounts relative to the other HDL apolipoproteins such as apoA-I. Nevertheless, truncated apoB proteins within the HDL fraction are easy to identify because they migrate to a much higher molecular weight than all of the other HDL proteins.

Although examining lipoprotein fractions by silver-stained SDS–polyacrylamide gels is probably the best technique for screening plasma samples for apoB truncations, this procedure is labor-intensive and limited by one potential drawback: very short apoB truncations (shorter than apoB-30) probably would go undetected because they would be found only in the $d > 1.21$ g/ml fraction.[11] Because of these considerations, investigators at Washington University (St. Louis, MO) have adopted a Western blot screening procedure. In this method, 1 μl fresh plasma or the apoB immunoprecipitate from 5 μl plasma is electrophoresed on a 3–6% SDS–polyacrylamide gel slab, and the separated proteins are then transferred to a sheet of Immobilon P membrane (Millipore Corp., Bedford, MA). To ensure complete transfer, we recommend electrophoresis at 200 V for 12–24 hr when using a standard 25 m*M* Tris, 0.2 *M* glycine, 20% methanol transfer buffer. The high-voltage transfer requires the use of a refrigerated circulator and a cooling coil to prevent the transfer buffer from overheating, but it generally results in a nearly complete transfer of the apoB proteins, including the full-length apoB-100. Western blots then can be performed using either an apoB-specific polyclonal antibody or a monoclonal antibody that has an epitope within the amino-terminal 600 amino acids of apoB-100, such as 1D1, MB24, or C1.4.[23,24]

The chief advantage of the Western blotting method is its ability to identify very short apoB species, such as apoB-18 or apoB-23, that would not be present in any lipoprotein fraction. The chief drawback of Western blotting is the uncertainty regarding whether this technique is sufficiently sensitive to detect low concentrations of truncated apoB species among the relatively large amount of LDL-associated apoB-100 in the plasma.

[23] R. J. Pease, R. W. Milne, W. K. Jessup, A. Law, P. Provost, J.-C. Fruchart, R. T. Dean, Y. L. Marcel, and J. Scott, *J. Biol. Chem.* **265,** 553 (1990).

[24] E. S. Krul, Y. Kleinman, M. Kinoshita, B. Pfleger, K. Oida, A. Law, J. Scott, R. Pease, and G. Schonfeld, *J. Lipid Res.* **29,** 937 (1988).

Because most of the apoB in the plasma of heterozygotes is in the LDL fraction, results from this technique largely reflect the LDL. As noted earlier, most truncated apoB's are present in the VLDL, and at least some are completely absent from the LDL fraction. We are skeptical that apoB-83 or apoB-86, both of which were easily detectable in the VLDL by silver staining of SDS–polyacrylamide gels, would be detectable in the plasma of heterozygotes by Western blotting, as these truncated proteins are present in the plasma in only very small amounts relative to apoB-100.[4,22] Another disadvantage of Western blot screening is that most of the apoB-100 breakdown products are in the LDL fraction, which is, in effect, what is primarily being screened by this technique. For this reason, it is best to use fresh plasma or freshly frozen plasma samples that have been stored at $-70°$ for Western blotting.

Occasionally it is difficult to determine whether a small protein band on an SDS–polyacrylamide gel or a Western blot is an apoB-100 breakdown product or a bona fide truncated apoB. This problem arises most commonly when LDL fractions are examined by SDS–polyacrylamide gel electrophoresis or when whole plasma samples are screened by Western blotting. In these instances, perhaps the most reliable way to test for the presence of an apoB truncation is to obtain fresh plasma samples from the subject and the subject's parents and siblings. The presence of the same size protein band in the plasma of a hypolipidemic relative and the absence of this band in the plasma of a normocholesterolemic relative is good evidence for a bona fide truncated apoB. Another method to discriminate between an authentic truncation and a breakdown product is to assess the distribution of the protein band within the various lipoprotein fractions; this density distribution pattern can then be compared with density distribution information from previously described truncated apoB's (e.g., an apoB-75 should be present in the VLDL and LDL,[25] whereas an apoB-31 should be present only in the HDL and the $d > 1.21$ g/ml fraction[19]).

A final method that we have used when we have been uncertain about a band on an SDS–polyacrylamide gel is to electrophorese LDL or HDL fractions on nondenaturing gradient polyacrylamide gels (see procedure in the section Characterization of Lipoproteins Containing Truncated Apolipoprotein B Species, below). This technique, which separates lipopotein particles by size, allows one to assess whether a lipoprotein sample with a possible apoB truncation contains a population of abnormally small lipoprotein particles. On nondenaturing gels, LDL particles that contain a truncated apoB migrate to a significantly lower molecular weight than apoB-100-

[25] E. S. Krul, K. G. Parhofer, P. H. R. Barrett, R. D. Wagner, and G. Schonfeld, *J. Lipid Res.* **33,** 1037 (1992).

containing LDL particles; in contrast, LDL samples containing abundant apoB-100 proleolytic breakdown products will generally migrate to the same position as fresh LDL samples.

Estimating Size of Truncated Apolipoprotein B

The migration of a truncated apoB species on a 3–12% gradient SDS–polyacrylamide gel can be used to estimate accurately the size of a truncated apoB. We generally compare the migration of the truncated apoB with that of apoB-48 (2152 amino acids), apoB-100 (4536 amino acids), and the four main proteolytic fragments[18] present in thrombin-treated LDL (T1, 3249 amino acids; T2, 1287 amino acids; T3, 1952 amino acids, and T4, 1297 amino acids). It is useful but not essential to include some human lipoproteins containing truncated apoB's of known length. Silver staining of gels is preferable for these studies, as accurate and reproducible measurements can be obtained using small samples of proteins.

We assess the migration of each apoB by measuring the distance from the interface of the stacking and separating gels to the top of the truncated apoB band. The distance of migration for each apoB and thrombolytic fragment (in millimeters) is plotted against the length of each protein (in amino acids) on standard graph paper. A "line" drawn through the points derived from the protein standards is curvilinear, but we have found this crude standard curve sufficient to estimate the length of an "unknown" truncated apoB within at least 100 amino acids, and usually within 50. Alternatively, one can plot the log of the molecular weight versus migration distance to obtain a fairly straight line. Once the approximate size of the truncated apoB has been determined, Western blots using selected apoB-specific murine monoclonal antibodies or apoB-specific peptide antibodies may provide, in some cases, more precise estimates of the length of the truncated apoB.[4,26] A number of monoclonal antibodies with precisely localized epitopes[23,24] and one panel of apoB-specific peptide antibodies[27] have been described.

Identifying Mutation Responsible for Truncated Apolipoprotein B

Once the size of an apoB truncation has been estimated, this information can be used to search for the causative mutation in the appropriate region of the apoB gene. To date, all of the apoB mutations causing truncated

[26] M. F. Linton, V. Pierotti, and S. G. Young, *Proc. Natl. Acad. Sci. U.S.A.* **89,** 11431 (1992).

[27] T. L. Innerarity, S. G. Young, K. S. Poksay, R. W. Mahley, R. S. Smith, R. W. Milne, Y. L. Marcel, and K. H. Weisgraber, *J. Clin. Invest.* **80,** 1794 (1987).

apoB species have been either short deletions causing frameshifts or single nucleotide transitions or transversions causing nonsense mutations. Even though most of the mutations have been point mutations, we initially examine the genomic DNA by Southern blotting to test for the possibility of a large deletion.

If Southern blotting does not reveal an abnormality, we proceed to enzymatic amplification of the appropriate segment of the apoB gene using a thermostable DNA polymerase and the leukocyte genomic DNA of an affected subject.[16] We prefer *pfu* DNA polymerase (Stratagene, La Jolla, CA) over *Taq* polymerase (Perkin-Elmer Cetus, Emeryville, CA) because of its lower nucleotide misincorporation rate. Because most mutations are found in exon 26, we generally use oligonucleotide primers (24–27 nucleotides in length) to amplify a 1-kb segment coding for approximately 330 amino acids. To facilitate subcloning of the amplified DNA, we usually incorporate an *Eco*RI, *Pst*I, *Hin*dIII, or *Bam*HI site into the 5′ portion of each oligonucleotide primer; this usually can be accomplished by incorporating either one or two nucleotide mismatches into the oligonucleotides. We have found that highly specific amplification of any segment of the apoB gene from genomic DNA can be obtained with relatively minor adjustments in the amplification conditions.

After the identity of the amplified DNA fragment has been confirmed by restriction enzyme digestion, it can be cleaved with the appropriate restriction endonucleases and cloned into M13 for sequencing. To identify a point mutation in the amplified DNA of a heterozygote, we prefer cloning into M13 for sequencing over direct double-stranded sequencing of the amplified DNA fragment. Although we have had successes with direct double-stranded sequencing of amplified DNA,[17] we have not yet used a DNA sequencing protocol for amplified fragments that is sufficiently reproducible to justify the time investment. Cloning the amplified DNA fragment into M13 can be accomplished within several days using commercially available M13 cloning kits. Sequencing ten M13 clones is sufficient to find the responsible mutation in a heterozygote. To date, we always have confirmed mutations by sequencing both DNA strands, although this precaution is probably unnecessary if the DNA sequencing gels are of high quality and if the mutation has been identified in multiple M13 clones. If the mutation changes a restriction site in the apoB gene, then the existence of the mutation should be confirmed by the enzymatic digestion of the amplified DNA.[17,22,28]

[28] W. A. Groenewegen, E. S. Krul, and G. Schonfeld, *J. Lipid Res.* **34,** 971 (1993).

Quantifying Amount of Truncated Apolipoprotein B in Plasma

Because the concentration of the truncated apoB in the plasma of heterozygotes is invariably very low relative to the concentration of apoB-100 produced by the normal allele (we estimate that the amount of the truncated protein is almost always less than 10% of the amount of the apoB-100, and sometimes less than 2%), precise quantification of the amount of the truncated protein in the plasma is difficult. One approach is to measure the total amount of apoB (all apoB species, including the truncated apoB) in plasma by a radioimmunoassay (RIA)[29] using a monoclonal antibody specific to the amino terminus of apoB, and then to measure the amount of apoB-100 with an RIA employing a monoclonal antibody specific to the carboxyl terminus of apoB. The difference between the two RIA measurements is, in theory, a quantification of the amount of the truncated protein in the plasma. Although we have used this method,[1] it has been unsatisfying because of the inherent inaccuracies involved with two different radioimmunoassays, each having an error of approximately 10%. Furthermore, this method is dependent on the assumption that the amino-terminal monoclonal antibody will bind to the apoB-100 and to the truncated apoB with identical affinities. This is definitely not the case with certain antibodies, such as MB3 (with an epitope near apoB-100 amino acid 1000[23]), which binds to apoB-31 (1425 amino acids) and to apoB-32 (1449 amino acids) much more avidly than to apoB-100.

Another method to measure the amount of a truncated apoB in the plasma is to determine the amount of apoB in the plasma using a standard RIA and then to assess the relative amounts of the truncated apoB and apoB-100 in the plasma by Western blotting. For this technique, 1 μl plasma is electrophoresed on an SDS–polyacrylamide gel and then electrophoretically transferred to a sheet of nitrocellulose membrane. A Western blot is then performed using an ^{125}I-labeled monoclonal antibody that has an epitope within the amino-terminal portion of apoB-100. The number of ^{125}I counts per minute (cpm) in the truncated apoB band and the apoB-100 band can be assessed with a gel scanner or by removing the bands from the membrane and counting them in a γ counter.[22] The ratio of the cpm in the truncated band to the cpm in the apoB-100 band can be multiplied by the total apoB concentration in the plasma (measured by RIA) to estimate the concentration the truncated apoB. The principal problem with this technique is that the electrophoretic transfer efficiencies of truncated

[29] S. G. Young, S. J. Bertics, L. K. Curtiss, D. C. Casal, and J. L. Witztum, *Proc. Natl. Acad. Sci. U.S.A.* **83,** 1101 (1986).

apoB's and apoB-100 can differ, even on different lanes within the same gel. Also, monoclonal antibodies may not bind to truncated apoB's and to apoB-100 with identical affinities.

Characterization of Lipoproteins Containing Truncated Apolipoprotein B Species

Nondenaturing Gel Electrophoresis

Nondenaturing gradient polyacrylamide gel electrophoresis, which separates lipoproteins on the basis of particle diameter, is a useful method to demonstrate the size of lipoproteins containing truncated apoB's. This method has been employed to demonstrate that truncated apoB's exist on lipoprotein particles that are independent of those containing apoB-100. A detailed description of the method for sizing lipoproteins by nondenaturing gradient polyacrylamide gels has been described.[30] We have used commercially available 4–30% or 2–16% polyacrylamide gels from Pharmacia Fine Chemicals (Uppsala, Sweden) or 2.5–16% gels from Isolab, Inc. (Akron, OH). Techniques for pouring gradient polyacrylamide gels have been described.[31] To prepare the lipoproteins for electrophoresis, LDL or HDL samples (~8 μl containing ~1–4 μg protein) are added to 2 μl electrophoresis tank buffer containing 40% (w/v) sucrose and 0.05% bromphenol blue. [The electrophoresis tank buffer is Tris base (90 m*M*), boric acid (80 m*M*), sodium azide (3 m*M*), and EDTA (3 m*M*), pH 8.35.] We have found that dialyzing the lipoprotein samples to remove salts prior to electrophoresis is unnecessary. High molecular weight size standards, available from Pharmacia should be included on each gel (1 or 2 μl for a silver-stained gel). Electrophoresis of the lipoprotein samples (10 μl) is carried out at 10° for 24 hr at 125 V (constant voltage).

Because the amount of smaller particles containing the truncated apoB is very low relative to the amount of apoB-100-containing particles in the LDL fraction or apoA-I-containing particles in the HDL fraction, silver staining is required for detection. We have not been successful in visualizing apoB-37- or apoB-46-containing HDL particles on Coomassie blue-stained gels, even when 10–15 μg HDL protein has been applied to the gel. To confirm the identity of particles containing truncated apoB species and to study their apolipoprotein content, one can electrophoretically transfer the separated lipoproteins from the polyacrylamide gel to a sheet of nitrocellu-

[30] A. V. Nichols, R. M. Krauss, and T. A. Musliner, this series, Vol. 128, p. 417.

[31] D. L. Rainwater, D. W. Andres, A. L. Ford, W. F. Lowe, P. J. Blanche, and R. M. Krauss, *J. Lipid Res.* **33,** 1876 (1992).

lose membrane for Western blotting with apolipoprotein-specific antibodies. We have found that a 12.5 m*M* Tris, 0.1 *M* glycine buffer without methanol substantially improves transfer efficiency. We transfer at 100 to 200 V for 24 to 36 hr using a refrigerated circulator and a cooling coil to prevent overheating of the transfer buffer. Figure 2 is a Western blot demonstrating that apoB-67-containing LDL particles are much smaller than apoB-100-containing LDL particles, with virtually no overlap in size. The sizes of apoB-31-, apoB-37-, and apoB-46-containing particles in the HDL fractions of affected heterozygotes are shown in Fig. 3. The size of these particles overlaps minimally with apoA-I-containing HDL particles, but not with apoB-100-containing lipoproteins.

Gel Filtration

Lipoproteins containing truncated apoB proteins can be isolated and characterized by gel-filtration techniques. At the Gladstone Institute (San

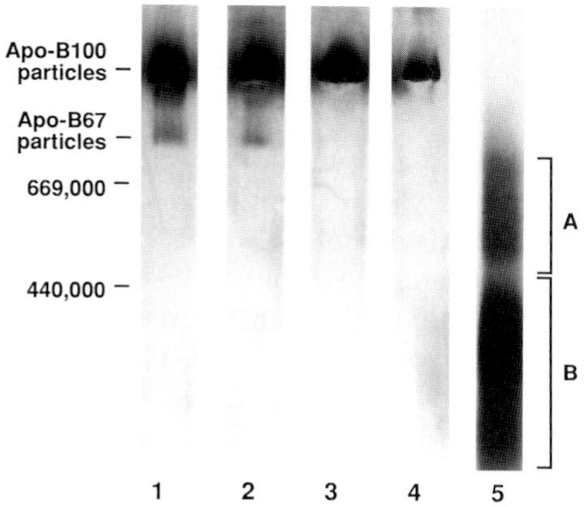

FIG. 2. Silver-stained nondenaturing 2–16% gradient polyacrylamide gel demonstrating the size of apoB-67-containing LDL particles. Lanes 1 and 2, LDL from two different apoB-67 heterozygotes; lane 3, LDL of a normolipidemic control subject; lane 4, LDL of an unaffected family member; lane 5, HDL fraction of an apoB-37 heterozygote. For each lane, 5–10 μg protein was loaded. On Western blots, the apoB-100-containing particles reacted with apoB-specific antibodies to epitopes in amino-terminal and carboxyl-terminal regions (data not shown). The apoB-67-containing particles reacted only with antibodies to epitopes located between amino acids 1 and 3080, and not with the carboxyl-terminal-directed antibodies (data not shown). For lane 5, migration of apoB-37-containing particles within the HDL fraction is indicated by bracket A; migration of apoA-I-containing particles in the HDL fraction is indicated by bracket B. Migration of size standards, namely, thyroglobulin (669,000) and ferritin (440,000), is indicated. (Reproduced from the *Journal of Clinical Investigation,* 1991, **87,** 1748–1754, by copyright permission of the American Society for Clinical Investigation.)

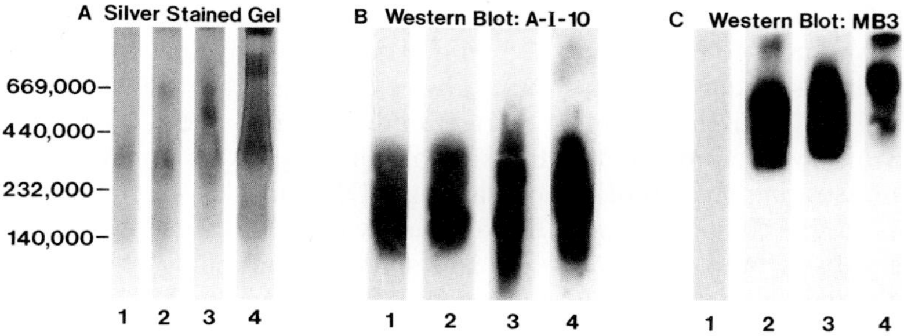

FIG. 3. Nondenaturing 4–30% polyacrylamide gradient gel demonstrating the size of apoB-31-, apoB-37-, and apoB-46-containing lipoproteins in the HDL fraction. Lane 1 shows HDL from a normolipidemic control subject; lane 2, HDL from an apoB-31 heterozygote; lane 3, HDL from an apoB-37 heterozygote; lane 4, HDL from an apoB-46 heterozygote. (A) Gel stained with silver; (B) Western blot using an apoA-I-specific monoclonal antibody, A-I-10 [L. K. Curtiss and T. S. Edgington, *J. Biol. Chem.* **260,** 2982 (1985)]; (C) Western blot using an apoB-specific monoclonal antibody, MB3 [Ref. 23 and L. K. Curtiss and T. S. Edgington, *J. Biol. Chem.* **257,** 15213 (1982)]. Migration of the size standards, namely, thyroglobulin (669,000), ferritin (440,000), catalase (232,000), and lactate dehydrogenase (140,000), is indicated. (Reproduced from the *Journal of Clinical Investigation,* 1990, **85,** 933–942, by copyright permission of the American Society for Clinical Investigation.)

Francisco, CA), we have used molecular sieve chromatography on an agarose column to separate the apoB-31- and apoB-100-containing lipoproteins from the plasma of an apoB-31 heterozygote.[19] At Washington University, fast performance liquid chromatography (FPLC) using Superose 6 columns (Pharmacia LKB, Uppsala, Sweden) has been used to characterize lipoproteins containing truncated apoB's. The FPLC method separates lipoproteins by size and can be used to separate VLDL, LDL, and HDL particles in the plasma within 3 hr.[25,32,33] The FPLC column fractions can be analyzed for lipids or apolipoproteins to generate distribution profiles. Such FPLC studies on the plasma samples of hypobetalipoproteinemia heterozygotes have revealed that lipoproteins containing truncated apoB's and those containing apoB-100 significantly differ in size. Lipoprotein separation using the FPLC method correlates reasonably well with the lipoprotein size separations observed with nondenaturing gradient polyacrylamide gel electrophoresis.

For FPLC, plasma samples are chromatographed at room temperature on two 25-ml Superose 6 HR 10/30 columns connected in series to an FPLC

[32] E. S. Krul, M. Kinoshita, P. Talmud, S. E. Humphries, S. Turner, A. C. Goldberg, K. Cook, E. Boerwinkle, and G. Schonfeld, *Arteriosclerosis (Dallas)* **9,** 856 (1989).

[33] T. Cole, R. Kitchens, A. Daugherty, and G. Schonfeld, *FPLC BioCommuniqué* **4,** 4 (1988).

P-500 pump through an LCC-500 liquid chromatography controller.[33] The columns are equilibrated with and stored in 1.0 m*M* EDTA, 0.02% sodium azide in 0.15 *M* NaCl, pH 8.0, prior to application of the plasma samples. All FPLC solutions are sterilized by filtration through 0.22-μm filters and degassed prior to use. At the Lipid Research Center at Washington University, the very same columns have been used for plasma lipoprotein separations for 9 consecutive years. Periodically, the outlet filters on either the first or second columns have to be replaced. The column pressures tend to increase with time; however, we have found that the column running pressures can be restored to no more than 1 MPa by passing buffer through inverted columns.

Immediately prior to injecting the plasma samples onto the FPLC columns, 2 ml plasma (obtained from a fasted subject) is centrifuged at 12,000 *g* for 10 min at room temperature. Following this short centrifugation, a small white pellicle consisting of triglyceride-rich lipoproteins occasionally can be observed floating on the surface of a plasma sample. This material should be carefully remixed with the supernatant fluid without disturbing any sediment at the bottom of the tube. The plasma samples are applied to the FPLC columns using a 1.5-ml sample loop and eluted in the descending direction at the rate of 0.5 ml/min. (With columns that have been used several hundred times, the flow rate should be reduced to approximately 0.35 ml/min to avoid excessive operating pressures.) After injection of the plasma sample, the first 12 ml eluent is diverted to waste and the subsequent eluent is collected in 60 fractions of 0.5 ml. Each FPLC run is continued until a total volume of 60 ml has passed through the column in order to assure the complete elution of all plasma peptides and salts. The elution of proteins from the column is monitored by measuring the absorbance at 280 nm.

Fractions from the column are routinely analyzed for cholesterol and triglycerides using enzymatic assay systems (Wako Pure Chemicals, Richmond, VA). Phospholipids can be measured enzymatically (Wako Pure Chemicals) or according to standard techniques.[34] To identify the apoB proteins, aliquots (35 μl) are removed from each fraction and analyzed by 3–6% gradient SDS–polyacrylamide gel electrophoresis. Following electrophoresis of the samples, the separated proteins are transferred to a sheet of Immobilon P membrane for immunoblotting. The immunoblots initially are incubated with an apoB-specific monoclonal antibody to the amino terminus, such as antibody C1.4,[24] and subsequently incubated with an ^{125}I-labeled goat anti-rabbit second antibody.[32] Autoradiographs of the Western

[34] G. R. Bartlett, *J. Biol. Chem.* **234,** 466 (1959).

blots can be scanned using a laser densitometer (SigmaScan by Jandel Scientific, San Rafael, CA) to determine the areas under the peaks corresponding to either apoB-100 or the truncated apoB species in each gel lane (each lane representing a single column fraction). The total areas in all lanes are summed for both apoB-100 and the truncated apoB species. Then the results for each lane (or column fraction) can be expressed as a percent of the total densitometric area and can be used to generate a distribution profile for each apoB species (Fig. 4).

Density Gradient Ultracentrifugation of Plasma

When characterizing lipoproteins containing truncated apoB's, density gradient ultracentrifugation offers some advantages when compared with gel filtration or conventional sequential density ultracentrifugation. First, density gradient ultracentrifugation is a preparative method, unlike gel

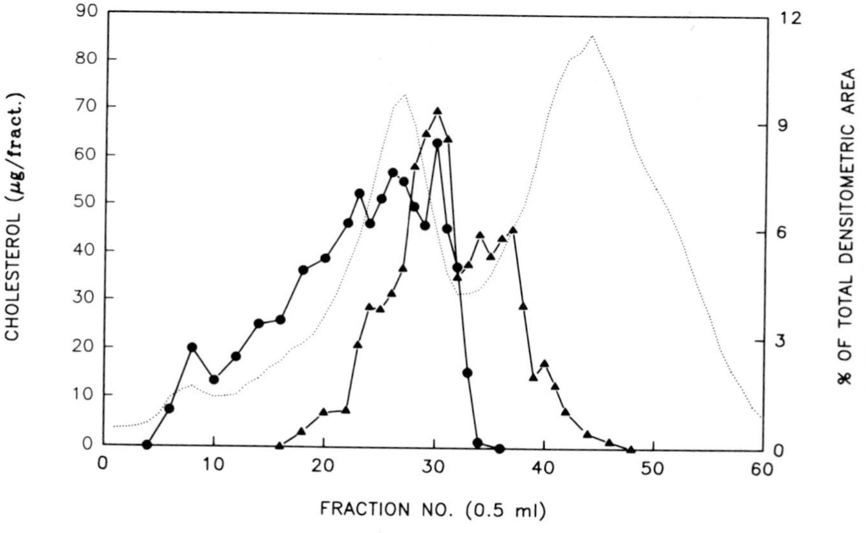

FIG. 4. Separation of apoB-100 from apoB-31 by FPLC on Superose 6 columns. Plasma was obtained from an apoB-31 heterozygote; 1.5 ml was applied onto two Superose 6 HR 10/30 columns as described in the text. Column fractions were analyzed for total cholesterol (dotted line), apoB-100 (circles), and apoB-31 (triangles). The VLDL, LDL, and HDL from normal subjects normally elute in fractions 5–9, 25–27, and 44–47, respectively. In this apoB-31 heterozygote, a significant proportion of the apoB-100 eluted in the VLDL–LDL size range, with the apoB-100 peak occurring within fraction 26. Most of the apoB-31 eluted between the LDL- and HDL-cholesterol peaks. Two peaks of apoB-31 are distinguishable, a larger peak at fraction 30 and a smaller, broader peak between fractions 34 and 47.

filtration on FPLC Superose 6 columns, which is primarily an analytical method. [FPLC can be preparative, but an initial ultracentrifugation step is required to concentrate the lipoproteins in a small enough volume (<1.5 ml) to permit resolution on the Superose 6 columns.] Second, density gradient ultracentrifugation requires only a single 24-hr ultracentrifugation step and does not require exposure of the apoB within the plasma to high salt concentrations, making it a gentler procedure for isolating lipoproteins than sequential ultracentrifugation.

The density gradient ultracentrifugation method is performed as described elsewhere[35] with some modifications. The inhibitors D-Phe-L-Pro-L-Arg chloromethyl ketone (PPACK) and aprotinin are added to the plasma, as outlined above, to prevent cleavage of apoB by kallikrein or thrombin. The plasma is adjusted to a density of 1.040 g/ml with solid potassium bromide (KBr). The density gradient is prepared in 40-ml QuickSeal tubes (Beckman Instruments, Fullerton, CA) as follows: 6 ml solution A (0.195 *M* NaCl, 1 m*M* Na_2EDTA, d = 1.006 g/ml) is introduced into the tube using a syringe equipped with a 3.5-inch-long 18-gauge needle from which the end bevel has been removed. The initial 6 ml solution A is carefully underlayered with 10 ml solution A adjusted to d = 1.020 g/ml with KBr. Next, 14 ml plasma (adjusted to d = 1.040 g/ml with KBr) is underlayered beneath the 1.020 g/ml solution. Finally, 10 ml solution A adjusted to d = 1.210 g/ml with KBr is underlayered beneath the plasma layer. The tube is centrifuged in a Beckman 50.2 Ti rotor for 24 hr at 45,000 rpm (12°). At the end of the ultracentrifugation, the rotor is allowed to coast to a stop. To remove the various density fractions, two winged infusion sets (Terumo, Elkton, MD) equipped with 3/4-inch, 23-gauge needles are used to pierce the top and bottom of the QuickSeal tube. Once inserted, the needles are not further adjusted. The gradient is removed from the top of the centrifuge tube by pumping solution A (adjusted to d = 1.30 g/ml with NaBr) into the bottom of the tube with a peristaltic pump at a rate of 1 ml/min. Fifty fractions of approximately 1 ml are then collected using a fraction collector. Each fraction is analyzed for lipids, apoB-100, and the truncated apoB as described for FPLC (Fig. 5). Consecutive pairs of tubes are pooled, and the density is measured using a DMA 35 densitometer (PAAR, Graz, Austria). We have found that the density gradient ultracentrifugation procedure yields highly reproducible lipid and apoB profiles.

As illustrated in Figs. 4 and 5, the lipid and apoB profiles derived from FPLC and density gradient ultracentrifugation separation of lipoproteins can be quite different. For example, by FPLC, the apoB-31-containing particles elute between normal LDL and HDL size (Fig. 4), in agreement

[35] D. M. Lee and D. Downs, *J. Lipid Res.* **23,** 14 (1982).

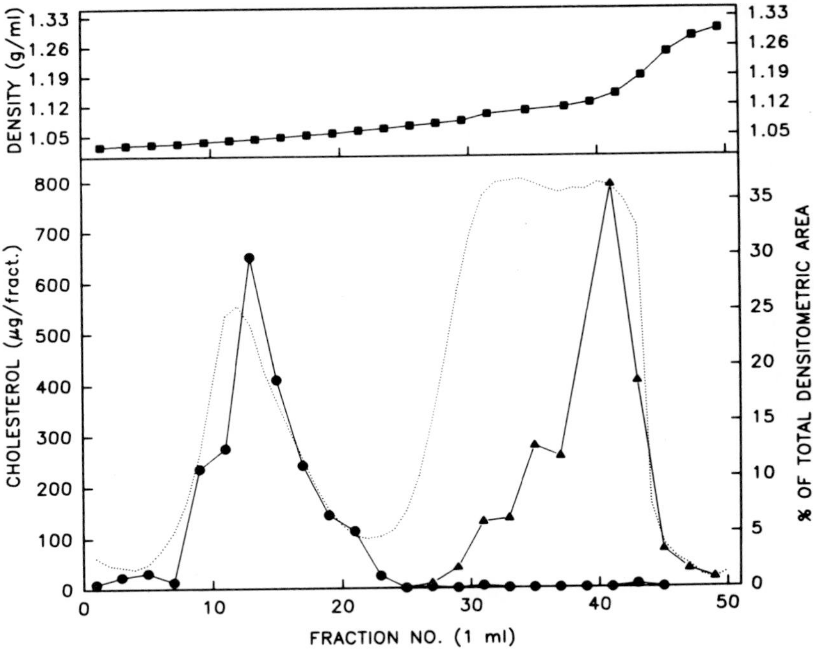

FIG. 5. Separation of apoB-100 and apoB-31 from the plasma of an apoB-31 heterozygote by density gradient ultracentrifugation. Density fractions were analyzed for total cholesterol (dotted line), apoB-100 (circles), and apoB-31 (triangles). The density profile of the gradient is shown at the top of the graph. The two major peaks of cholesterol, at fractions 10–14 and fractions 28–43, correspond to the peaks of LDL and HDL, respectively. The VLDL floats at the top of the tube in fractions 1–2. Density gradient ultracentrifugation results in the complete separation of apoB-100 and apoB-31, with virtually no overlap. The ApoB-100 elutes essentially as a single peak at fraction 13 (mean density of 1.042 g/ml), whereas most of the apoB-31 elutes as a peak at fraction 41 (mean density of 1.147 g/ml).

with results obtained by nondenaturing gradient gel electrophoresis.[19] However, on density gradient ultracentrifugation, the major portion of plasma apoB-31 floats at a mean density of 1.147 g/ml, clearly separated from the apoB-100-containing LDL but essentially at the identical density as apoA-I-containing HDL (Fig. 5). The high density of apoB-31-containing particles is due almost certainly to the fact that they, like apoA-I containing HDL particles, contain a high protein : lipid ratio. In support of this, Young *et al.*[36] purified apoB-37-containing HDL particles and apoA-I-containing HDL

[36] S. G. Young, F. P. Peralta, B. W. Dubois, L. K. Curtiss, J. K. Boyles, and J. L. Witztum, *J. Biol. Chem.* **262,** 16604 (1987).

particles from the HDL fractions of several apoB-37 heterozygotes and found that the two different types of particles had nearly identical densities and protein : cholesterol ratios, despite having considerably different sizes. Similarly, apoB-32-containing HDL particles were larger than the apoA-I-containing HDL particles, but both types of particles were very similar in terms of chemical composition.

Heparin Binding and Precipitation Characteristics of Truncated Apolipoprotein B Species

Seven heparin binding sites have been demonstrated on apoB-100 using a combination of chemical and proteolytic digestion and heparin-Sepharose affinity chromatography.[37] Truncated apoB's differ in affinities for heparin depending on their lengths, with shorter forms having significantly less affinity for heparin binding than the full-length apoB-100. Similarly, studies on the heparin precipitation characteristics of truncated-apoB-containing lipoproteins have demonstrated that the propensity to be precipitated by heparin and manganese increases with the length of the truncated apoB (S. McCormick and S. Young, unpublished observations, 1992). This finding presumably reflects the increased number of heparin binding sites on the longer apoB proteins. Thus, lipoproteins containing apoB-32, which contains three of the seven apoB-100 heparin binding sites, are only partially precipitated by heparin–manganese; apoB-46 and apoB-61 lipoproteins are also only partially precipitated, although to a greater extent than apoB-32; and longer truncated apoB species such as apoB-83, which has all seven heparin binding sites, are (like apoB-100) completely precipitated by heparin.

Purifying Lipoproteins That Contain Truncated Apolipoprotein B Species

Because truncated apoB's invariably are present in low concentrations in the plasma of heterozygotes (relative to the amount of apoB-100), purification of the truncated apoB-containing lipoproteins from the apoB-100-containing lipoproteins is challenging and has been reported in only a few instances. Nevertheless, the purification of lipoproteins containing truncated apoB's has proved useful for defining structural requirements for the interaction between apoB and the LDL receptor. For example, Young *et al.*[36] used immunoaffinity chromatography to purify apoB-37-containing lipoproteins from the HDL fraction of apoB-37 heterozygotes. To accomplish this, the entire HDL fraction from these individuals was loaded onto an anti-apoB-37 immunoaffinity column. The apoA-I particles did not bind

[37] K. H. Weisgraber and S. C. Rall, Jr., *J. Biol. Chem.* **262,** 11097 (1987).

to the column; the apoB-37-containing particles were eluted from the column with 3 *M* potassium iodide. The purified apoB-37-containing particles (mean diameter, 126 ± 11 Å) did not bind to the LDL receptor of cultured fibroblasts.[36]

Krul *et al.*[32] have reported purification of apoB-89-containing lipoproteins from the plasma of an apoB-40/apoB-89 compound heterozygote. In that instance, sequential ultracentrifugation was used to obtain a lipoprotein fraction that contained predominantly apoB-89-containing lipoproteins. Subsequently, Krul *et al.*[25] purified apoB-75-containing lipoproteins by applying the d = 1.063–1.090 g/ml fraction from an apoB-75 heterozygote to a monoclonal antibody B1B3 immunoaffinity column. Because apoB-75 contains 3386 amino acids but the epitope for B1B3 is located at approximately apoB-100 amino acid 3500, the antibody column bound the apoB-100-containing lipoproteins but allowed the apoB-75-containing lipoproteins to flow through. The apoA-I-containing particles in the "flow through" fraction then were removed by an anti-apoA-I immunoaffinity column, permitting enhanced purification of the apoB-75-containing lipoproteins.

Specific methods for affinity chromatography have been described elsewhere.[25,36] Krul *et al.*[25,32] found that the purified apoB-75- and apoB-89-containing lipoproteins bound to the LDL receptor with 1.2- to 1.5-fold greater affinity than apoB-100-containing LDL.[25,32] Welty and co-workers have used a similar approach to purify the apoB-67-containing particles from the plasma of an apoB-67 heterozygote; they found that the apoB-67 particles did not bind to the LDL receptor (F. Welty, personal communication, 1994). Finally, McCormick and co-workers have used heparin affinity chromatography to separate apoB-32-containing HDL particles from the small amounts of apoB-100-containing particles within the HDL density fraction (S. McCormick, unpublished observations, 1992). Two milliliters of an HDL fraction (d = 1.075–1.291 g/ml) that had been dialyzed against 50 m*M* Tris-HCl (pH 7.4), 10 m*M* citrate were loaded onto a 6-ml heparin-Sepharose column that had been equilibrated in the same buffer. The apoA-I-containing HDL particles did not bind to the column. The apoB-32-containing particles were eluted from the column using a NaCl gradient, eluting at approximately 80 m*M* NaCl.

Transfection of Human Apolipoprotein B cDNA Constructs into Cultured Cells

Further characterization of several of the naturally occurring truncated apoB species has been made possible by the expression of human apoB cDNA constructs in cultured cells. These apoB cDNA (or "minigene") expression studies have yielded important insights into the relationship between apoB length and the buoyant density of lipoproteins,[11] the efficiency of secretion of newly synthesized truncated apoB species,[11] and the molecular mechanisms involved in the secretion of both a truncated apoB-86 and a full-length apoB-100 by a single mutant allele, the apoB-86 allele.[26]

Modifying Apolipoprotein B Expression Vector. The first apoB-100 expression vector (a minigene construct that contained introns 26–28) was reported in 1990 by Blackhart *et al.*[38] Later, Yao *et al.*[11] developed a series of apoB cDNA expression vectors coding for carboxyl-terminally truncated forms of apoB (apoB-18, B-23, B-28, B-31, B-37, B-48, and B-53). Each of the apoB expression vectors was constructed using the pCMV5 plasmid in which the expression of the cDNA or minigene is driven by a cytomegalovirus (CMV) promoter.[39] Our laboratory has developed a number of apoB expression vectors in pRCMV (InVitrogen, San Diego, CA), a mammalian expression vector with a CMV promoter that also contains a neomycin resistance gene within the plasmid.

To express naturally occurring truncated apoB species in cultured cells, it is generally necessary to modify an existing apoB expression vector. The large size of the apoB-100 minigene (>22 kb) and the paucity of unique restriction sites within the apoB cDNA make site-directed mutagenesis of the full-length apoB-100 expression vector challenging. Performing site-directed mutagenesis in 1- to 1.5-kb apoB cDNA on gene fragments and returning the mutagenized fragment to the full-length vector usually is not an attractive option because this approach generally requires rebuilding the entire expression vector, a process that can involve many ligation reactions. Although mutagenesis of very large (>22 kb) expression vectors remains challenging, we have found that the double-stranded mutagenesis technique of Deng and Nickoloff[40] can be used to modify apoB expression vectors that are up to 12 kb in length. Because this technique uses denatured plasmid DNA as the substrate for the mutagenesis reaction, mutagenizing expression vectors of 12 kb or less can be performed simply, without subclolning or shuttling fragments of DNA from one vector to another. Our attempts to mutagenize apoB expression vectors that are larger than 12 kb with the Deng and Nickoloff technique, however, have been unsuccessful. Therefore, for the purpose of modifying the full-length apoB-100 expression vectors, we have assembled a full-length expression vector from various cassettes that are 10 kb or less. Mutagenesis is performed within the 10-

[38] B. D. Blackhart, Z. Yao, and B. J. McCarthy, *J. Biol. Chem.* **265,** 8358 (1990).

[39] S. Andersson, D. L. Davis, H. Dahlbäck, H. Jörnvall, and D. W. Russell, *J. Biol. Chem.* **264,** 8222 (1989).

[40] W. P. Deng and J. A. Nickoloff, *Anal. Biochem.* **200,** 81 (1992).

kb cassettes, which are then returned to the full-length vector with one or two cloning steps (M. Linton and S. Young, unpublished observations, 1993).

Transient Transfection of COS-7 Cells. The pCMV and pRCMV vectors were designed for high levels of expression in COS-7 cells (ATCC).[39] We have found that transient transfection of COS-7 cells is a relatively easy method to determine whether a newly constructed vector yields an apoB protein of the expected size. We find transient transfection useful before embarking on the time-consuming task of developing stably transformed hepatoma cell lines because we have observed that unwanted point mutations occasionally arise in these large plasmid constructs. For transfection, the COS-7 cells are grown in T-75 flasks at 37° in a 5% (v/v) CO_2 incubator in Dulbecco's modified Eagle's medium (DMEM) containing 10% (v/v) fetal bovine serum. Nearly confluent cells are split 1:8 on the day prior to transfection. Each flask is transfected with 10–15 μg of cesium chloride-purified apoB expression vector DNA by the calcium phosphate coprecipitation and glycerol shock method described by Chen and Okayama.[41] Following the glycerol shock, 10 ml fresh medium is added to each flask, and the cells are incubated at 37° for 48 hr.

Although very short apoB's such as apoB-18 are secreted by COS-7 cells,[11] longer truncated apoB proteins do not appear to be secreted. Therefore, we assess apoB expression in the cells by analyzing the apoB content of the cellular homogenates rather than by examining the apoB content of the medium. To accomplish this, the cells are washed twice in 10 ml phosphate-buffered saline (PBS), scraped into 5 ml PBS, and pelleted by low-speed centrifugation. The cells then are resuspended in 150 μl buffer containing 10 m*M* Tris-HCl (pH 8.3) and 3% SDS and sonicated for 15 sec with the Branson Sonifier 450 (1/8-inch-diameter microtip) on constant duty cycle with the output control set at 7. The cellular homogenates then are electrophoresed on 3–12% SDS–polyacrylamide gels, and the apoB content of the samples is assessed by Western blotting with an "amino-terminal" apoB-specific monoclonal antibody, such as 1D1.

Generation of Stably Transformed Cell Lines. We have generated stable cell lines expressing a variety of truncated apoB's using the rat hepatoma cell line McA-RH7777.[11] The McA-RH7777 cells are grown in DMEM containing 10% fetal bovine serum and 10% horse serum. We prefer to grow McA-RH7777 cells in Primaria T-75 flasks (Fisher Scientific, Springfield, NJ) because the cells adhere better to the positively charged surfaces of these flasks and will form better monolayers. Transfections are performed with 10–15 μg cesium chloride-purified plasmid DNA using the calcium

[41] C. Chen and H. Okayama, *Mol. Cell. Biol.* **7,** 2745 (1987).

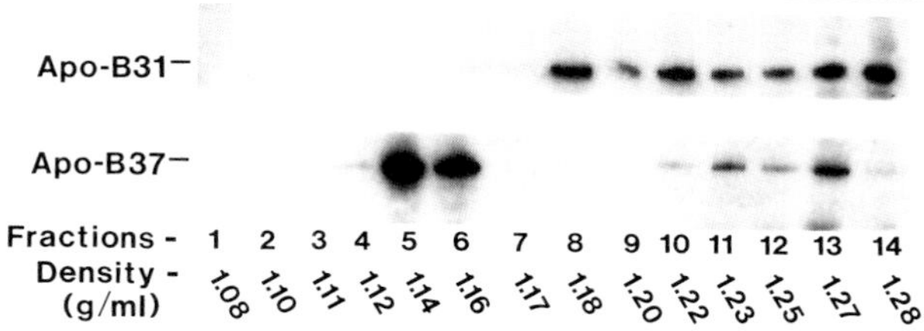

FIG. 6. Density distribution of the apoB-31- and apoB-37-containing lipoproteins secreted by rat hepatoma cells that had been stably transformed with apoB-31 and apoB-37 cDNA expression vectors.[11] Culture medium (DMEM containing 10% fetal calf serum) was placed on the stably transformed rat hepatoma cells for 24 hr. The medium then was concentrated to 1 ml by an Amicon filter and adjusted to a density of 1.21 g/ml. The medium then was fractionated by discontinuous salt gradient ultracentrifugation, and the different density fractions were analyzed by Western blotting using the apoB-specific monoclonal antibody 1D1. [Reproduced with permission from M. F. Linton, R. V. Farese, and S. G. Young, *J. Lipid Res.* **34,** 521 (1993).]

phosphate coprecipitation and glycerol shock method.[41] When using an apoB expression vector lacking a *neo* gene, we cotransfect the human apoB expression construct with pSV2neo at an apoB cDNA:*neo* molar ratio of 20:1. Following the glycerol shock, the cells are grown for 48 hr and then split at various ratios ranging from 1:4 to 1:48 into 100-cm dishes and grown in medium containing 400 μg/ml active G418. After approximately 2 weeks, G418-resistant colonies are visible and can be harvested and transferred to 96-well plates. Individual colonies are harvested by pipetting 5 μl of 0.05% trypsin up and down on the colony, which then is transferred in the trypsin to a 96-well plate containing 200 μl medium. The cells are allowed to grow in the 96-well plates for 1 week in G418-containing medium. The medium from each well is then screened for the presence of human apoB proteins by a radioimmunoassay[42] or by dot blotting.

Dot blots are performed by transferring 200 μl medium from each well onto 0.2-μm nitrocellulose membranes using a dot- or slot-blot apparatus. Dilutions of human LDL are included as positive controls. The blot then is washed twice with 200 μl transfer buffer, blocked with a 5% nonfat milk solution, and incubated with the apoB-specific monoclonal antibody 1D1. Positive clones are identified using an anti-mouse second antibody and an enhanced chemiluminescent detection system (Amersham, Arlington Heights, IL). Clones expressing human apoB are expanded and screened for secretion of the appropriate truncated apoB by SDS–polyacrylamide gel electrophoresis and Western blotting. For the initial Western blot

[42] R. V. Farese, Jr., L. M. Flynn, and S. G. Young, *J. Clin. Invest.* **90,** 256 (1992).

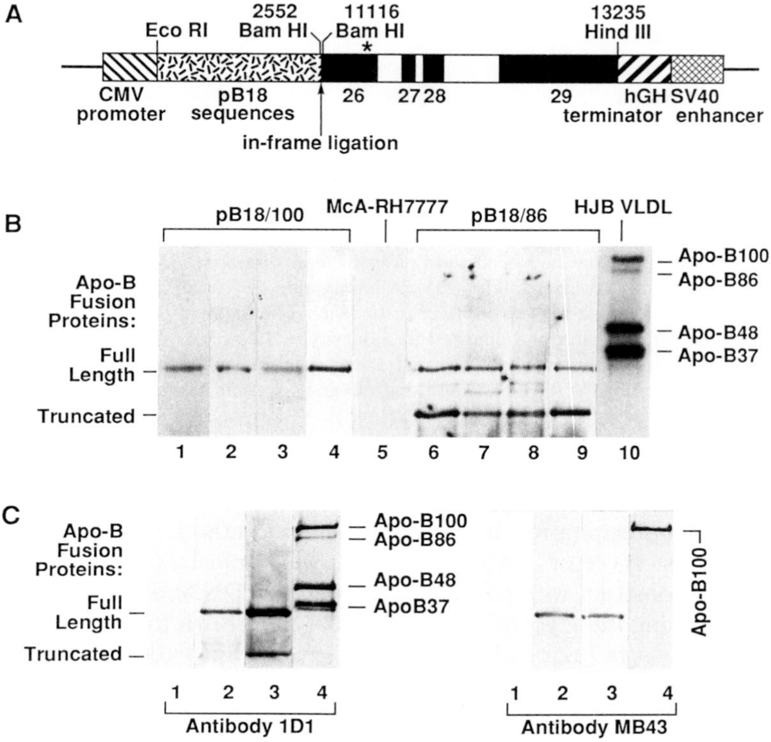

FIG. 7. (A) Schematic diagram of the apoB fusion protein expression vector that was used to investigate how the apoB-86 allele yields both apoB-86 and apoB-100. The location of the apoB-86 mutation (which was contained in pB18/86 but not in pB18/100) is marked by a star. The pB18 portion of the vector codes for apoB amino acids 1–782; the exon 26–29 gene fragment, which is ligated in-frame to the pB18 portion of the vector, codes for apoB amino acids 3636–4343. The full-length apoB fusion protein produced from the pB18/100 vector is predicted to contain 1490 amino acids; the truncated fusion protein that is predicted to be produced from the pB18/86 vector should contain 1044 amino acids. Apolipoprotein B-100 contains 4536 amino acids; apoB-86, 3996 amino acids; apoB-48, 2152 amino acids; and apoB-37, 1728 amino acids. (B) Western blot analysis of the apoB proteins secreted by the pB18/100 and pB18/86 stable transformants, using monoclonal antibody 1D1 (epitope: apoB amino acids 474–539). The apoB proteins in the medium were concentrated by adsorption onto Cab-O-Sil,[11] electrophoresed on 3–12% SDS–polyacrylamide gels, and then transferred to nitrocellulose membranes for Western blotting. Lanes 1–4, human apoB fusion proteins from four different pB18/100 stable transformants; lane 5, proteins from nontransfected McA-RH7777 cells; lanes 6–9, apoB fusion protein from four different pB18/86 stable transformants; lane 10, VLDL of an apoB-86/apoB-37 compound heterozygote. This experiment showed that the cells transformed with the mutant expression vector secreted both a truncated fusion protein and a full-length fusion protein. (C) Western blots of SDS–polyacrylamide gels demonstrating that the full-length apoB fusion protein produced by the pB18/86-transformed cells is detected by monoclonal antibody MB43 (epitope: apoB amino acids 4027–4081). The apoB

screening, the lipoproteins are extracted from 4 ml medium using Cab-O-Sil (250 μg) as described.[43]

For detailed analysis of truncated apoB's secreted by stably transformed McA-RH7777 cells, it is often desirable to isolate the $d < 1.21$ g/ml lipoproteins by ultracentrifugation, as this step concentrates the apoB and separates it from irrelevant secreted proteins. For these studies, subconfluent (~50% confluency) stable transformants are incubated in serum-free medium for 24 hr. The culture medium from four 75-cm^2 flasks (40 ml) is collected and the d < 1.21 g/ml fraction is isolated by ultracentrifugation for 48 hr in a 60 Ti rotor (10°). The lipoprotein samples then are dialyzed and lyophilized prior to analysis by SDS–polyacrylamide gel electrophoresis and Western blotting.

A generation of stably transformed rat hepatoma cell lines expressing human truncated apoB's has been useful in understanding the relationship between apoB length and the buoyant density of lipoproteins.[11] For example, from studies on the lipoproteins of subjects with hypobetalipoproteinemia, an inverse relationship was observed between the density of truncated apoB's in the plasma and their length. Thus, for example, apoB-37 tends to be found in more buoyant lipoprotein fractions than apoB-31.[19] These observations could not determine, however, whether apoB-37 binds more lipid than apoB-31 because apoB-37 contains more lipid-binding domains or because apoB-31-containing lipoproteins undergo more extensive catabolism in the plasma and therefore are found in denser subfractions. To determine whether the inverse relationship between apoB length and lipoprotein density was due to an intrinsic structural property of newly secreted lipoproteins or to differing metabolism of apoB-31- and apoB-37-containing lipoproteins in the plasma, expression vectors for both apoB-31 and apoB-37 were constructed and stably expressed in rat hepatoma cell lines. To examine the densities of newly secreted apoB-31 and apoB-37, the medium of the apoB-31 and apoB-37 stable transformants was subjected to discontinuous density gradient ultracentrifugation,[4,11] and different density fractions than were examined by Western blot analysis. These experiments clearly showed that the density of newly secreted apoB-31 lipoproteins was

[43] D. E. Vance, D. B. Weinstein, and D. Steinberg, *Biochim. Biophys. Acta* **792,** 39 (1984).

---

fusion proteins were partially purified from the medium by isolating the d < 1.21 g/ml lipoproteins by ultracentrifugation. For the 1D1 and MB43 blots: lane 1, lipoproteins of $d <$ 1.21 g/ml from nontransfected cells; lane 2, lipoproteins from cells stably transformed with pB18/86; lane 3, lipoproteins from cells stably transformed with pB18/100; lane 4, VLDL of an apoB-86/apoB-37 compound heterozygote. [Reprinted with permission from M. F. Linton, V. Pierotti, and S. G. Young, *Natl. Acad. Sci. U.S.A.* **89,** 11431 (1992).]

greater than that of newly secreted apoB-37 lipoproteins (Fig. 6) and thus that the lipoprotein density was an intrinsic property determined by the length of the truncation.

The apoB-31- and apoB-37-secreting cell lines also were used in [^{35}S]methionine pulse–chase studies to examine the efficiency of secretion of newly synthesized truncated apoB's.[11] These studies revealed that the newly synthesized apoB-31 and apoB-37 were secreted promptly and efficiently by cultured cells, suggesting that the low levels of apoB-31 and apoB-37 in plasma of affected subjects were not due to a secretory defect.

We have used cell culture expression studies to elucidate a unique mutant apoB allele, the apoB-86 allele, which was originally identified in an apoB-37/apoB-86 compound heterozygote within the H.J.B. kindred.[26,44] The VLDL of this compound heterozygote contained a substantial amount of apoB-100 in addition to apoB-37 and apoB-86 (Fig. 7B,C). Genetic studies in the 41 members of the H.J.B. kindred suggested, but could not prove, that the apoB-86 allele yielded a full-length apoB-100 in addition to apoB-86.[1,2,4,26] The apoB-86 allele contains the deletion of a single cytosine (cDNA nucleotide 11840) within exon 26 of the apoB gene. This deletion results in a frameshift and is predicted to yield a truncated apoB, apoB-86, containing 3896 amino acids. As the mutation did not alter an intron–exon splice junction and did not create an alternative splice site, the explanation for apoB-100 synthesis by this allele was initially not clear.

To investigate the mechanism whereby the apoB-86 allele yielded both a truncated apoB and a full-length apoB-100, we expressed the mutation in McA-RH7777 cells. Wild type (pB18/100) and mutant (pB18/86) fusion expression vectors (Fig. 7A) were constructed as described.[26] The fusion vectors consisted of an exon 26–29 apoB gene fragment that was ligated in-frame to the 3′ end of an apoB-18 cDNA expression vector. Expression of the wild-type vector yielded only the predicted full-length protein (Fig. 7B,C). The expression of the mutant pB18/86 expression vector, however, yielded not only the predicted truncated protein but also the full-length protein (Fig. 7B,C). Thus, these cell culture studies verified human studies that indicated that the apoB-86 allele produced full-length apoB in addition to the predicted truncated apoB protein.

The stably transformed cell lines also enabled us to identify the mechanism of reading frame restoration in the apoB-86 allele. We examined the cDNA derived from the stably transformed pB18/86 cell lines and found no evidence of alternative splicing. Sequencing of cDNA clones derived from pB18/86-transformed cell lines in the region of the apoB-86 mutation

[44] D. Steinberg, S. M. Grundy, H. Y. I. Mok, J. D. Turner, D. B. Weinstein, W. V. Brown, and J. J. Albers, *J. Clin. Invest.* **64,** 292 (1979).

revealed the explanation for the synthesis of a full-length apoB by the apoB-86 allele.[26] The deletion of cytosine-11840 in the apoB-86 allele results in a stretch of eight consecutive adenines. When we examined the cDNA of cells transformed with the mutant apoB-86 expression vector (pB18/86), we found that 11% of the cDNA clones contained the insertion of an additional adenine, thereby creating a stretch of nine consecutive adenines. The insertion of an extra adenine at the site of the 1-bp deletion "corrects" the mutant reading frame caused by the 1-bp deletion. Thus, it appears that the apoB-86 allele yields apoB-100 in addition to apoB-86 because a transcriptional error occurs at a significant rate at this long stretch of adenines and corrects the mutant reading frame by inserting an extra adenine. Transcriptional slippage at long stretches of adenines previously had been observed in *Escherichia coli*,[45] but never before in eukaryotes. The delineation of this mechanism would have been difficult, if not impossible, without cell culture expression studies.

In the future, the expression of other naturally occurring truncated apoB species in stable cell lines will undoubtedly be useful for such purposes as obtaining pure populations of lipoproteins for assessment of binding of truncated apoB's to the LDL receptor, assessing whether truncated apoB's have the capability to interact with apo(a) to form lipoprotein(a) particles, and identifying other functionally distinct domains of the apoB molecule.

Acknowledgments

The authors thank Maurizo Averna, Robert Kitchens, and Tish Kettler for performing the experiments illustrated in Figs. 4 and 5.

[45] L. A. Wagner, R. B. Weiss, R. Driscoll, D. S. Dunn, and R. F. Gesteland, *Nucleic Acids Res.* **18,** 3529 (1990).

[8] Apolipoprotein B-48: Problems Related to Quantification

By DIANA M. LEE

Introduction

Discovery of Intestinal Apolipoprotein B

Krishnaiah *et al.*[1,2] in 1978 and 1980 were the first to identify intestinal apolipoprotein B (apoB). They found this apoB species present in rat lymph lipoproteins and in plasma very-low- and low-density lipoproteins (VLDL and LDL) after the rats were fed a high-fat and high-cholesterol diet. On sodium dodecyl sulfate–polyacrylamide gel electrophoresis (SDS–PAGE), this apoB species showed a smaller molecular weight than that of the major apoB from LDL (i.e., the major form derived from liver). These two apoB forms showed common antigenic determinants with similar amino acid compositions.[2] In the same year (1980), Kane *et al.*[3] found an intestinal apoB species in human chylomicrons which was not present in any other density fractions. They found significant differences in amino acid compositions between the intestinal apoB and the hepatic apoB. The apparent molecular weight of the intestinal apoB species on SDS–PAGE was 48% of that of the hepatic apoB. Because of the inconsistent molecular weight determinations among laboratories, Kane *et al.*[3] proposed the centile nomenclature apoB-100 for the large apoB and apoB-48 for the small apoB. This terminology has been adopted by most investigators in the lipoprotein field not only for human apoB, but also for all species of animal apoB.

Is Apolipoprotein B-48 the Only Form of Intestinal Apolipoprotein B?

Although both forms of apoB are synthesized and secreted by rat liver,[4,5] apoB-48 was initially considered to be the only apoB form in rat intes-

[1] K. V. Krishnaiah, J. Ong, L. F. Walker, J. Borensztain, and G. S. Getz, *Fed. Proc., Fed. Am. Soc. Exp. Biol.* **37,** 257a (1978).

[2] K. V. Krishnaiah, L. F. Walker, J. Borensztain, G. Schonfeld, and G. S. Getz, *Proc. Natl. Acad. Sci. U.S.A.* **77,** 3806 (1980).

[3] J. P. Kane, D. A. Hardman, and H. E. Paulus, *Proc. Natl. Acad. Sci. U.S.A.* **77,** 2465 (1980).

[4] C. E. Sparks and J. B. Marsh, *J. Lipid Res.* **22,** 519 (1981).

[5] J. Elovson, Y. O. Huang, N. Baker, and R. Kannan, *Proc. Natl. Acad. Sci. U.S.A.* **78,** 157 (1981).

tine.[2,6,7] However, analyses of freshly collected rat mesenteric chyle and its chylomicrons by SDS–PAGE and immunoblotting showed the presence of apoB-100.[8]

Secretion of intestinal apoB-100 was also observed in human intestinal Caco-2 cells[9] and in human plasma chylomicrons under well-preserved conditions (i.e., in the presence of antibiotics, antioxidants, and proteinase inhibitors), and with chylomicrons analyzed within a few hours of blood collection.[10] However, when samples were analyzed after 1–2 days, satellite bands smaller than apoB-100, as well as an apoB-48-like band containing the amino-terminal end, appeared.[10] These results suggest that intestinal apoB-100 is very sensitive to proteolytic degradation and may be converted to a smaller molecular form resembling authentic apoB-48.

Synthesis of apoB-100 was reported in human fetal intestine, but it progressively decreased during maturation.[11] Hoeg *et al.*[12] and Levy *et al.*[13] presented evidence that both apoB-48 and apoB-100 are synthesized and secreted by adult intestine. Even in abetalipoproteinemic subjects, local synthesis of both intestinal apoB-48 and apoB-100 was found but without secretion.[14] Taken together, these results show that apoB-100 is synthesized and secreted in the intestine of humans and rats in addition to apoB-48 and that the intestinal apoB-100 degrades rapidly into smaller molecular forms *in vitro.*

Metabolism of Apolipoprotein B-48

Mechanism of Apolipoprotein B-48 Synthesis

Studies on cDNA clones of apoB-100 indicated that the apoB-100 gene is present in both the liver and intestine of humans, monkeys, and

[6] L. L. Swift, P. D. Soulé, M. E. Gray, and V. S. LeQuire, *J. Lipid Res.* **25,** 1 (1984).

[7] F. M. Van't Hoff, D. A. Hardman, J. P. Kane, and R. J. Havel, *Proc. Natl. Acad. Sci. U.S.A.* **79,** 179 (1982).

[8] D. M. Lee, E. Koren, S. Singh, and T. Mok, *Biochem. Biophys. Res. Commun.* **123,** 1149 (1984).

[9] D. M. Lee, N. Dashti, and T. Mok, *Biochem. Biophys. Res. Commun.* **156,** 581 (1988).

[10] D. M. Lee and S. Singh, *Biochim. Biophys. Acta* **960,** 148 (1988).

[11] R. M. Glickman, M. Rogers, and J. N. Glickman, *Proc. Natl. Acad. Sci. U.S.A.* **83,** 5296 (1986).

[12] J. M. Hoeg, D. D. Sviridov, G. E. Tennyson, S. J. Demosky, Jr., M. S. Meng, D. Bojanovski, I. G. Safonova, V. S. Repin, M. B. Kuberger, V. N. Smirnov, K. Higuchi, R. E. Gregg, and H. B. Brewer, Jr., *J. Lipid Res.* **31,** 1761 (1990).

[13] E. Levy, C. Rochette, I. Londono, C. C. Roy, R. W. Milne, Y. L. Marcel, and M. Bendayan, *J. Lipid Res.* **31,** 1937 (1990).

[14] R. P. F. Dullaart, B. Speelberg, H. J. Schuurman, R. W. Milne, L. M. Havekes, Y. L. Marcel, H. J. Geuze, M. M. Hulshof, and D. W. Erkelens, *J. Clin. Invest.* **78,** 1397 (1986).

rats,[15–18] suggesting that the apoB-100 synthesized in the intestine is identical to apoB-100 synthesized in the liver. However, the synthesis of apoB-48 is due to the inframe stop codon in intestinal mRNA of apoB resulting from a C to U substitution in the codon CAA encoding Gln^{2153} to UAA, which is a termination codon for apoB-48.[19–21] Thus, this stop codon results in apoB-48 with 2152 amino acids and isoleucine at the carboxyl terminus. This discovery has facilitated studies on the structure and regulation of apoB-48, and established that apoB-48 is the major apoB form in the intestine.

Modulation of Apolipoprotein B mRNA Editing

In rodents, both apoB-100 and apoB-48 are produced by the liver. The level of production of the two apoB species can be modulated metabolically by thyroid hormone. Hypothyroid rats produce mainly apoB-100, and hyperthyroid rats produce mainly apoB-48. This modulation acts by regulating the level of the stop codon in apoB mRNA.[22] Thyroid hormone analogs, but not growth hormone, also modulate the mRNA editing.[23] Fasting and refeeding a high-carbohydrate diet modulates the editing of mRNA as well.[24,25] Leighton *et al.*[24] found that fasting caused a 60% decrease in the amount of apoB mRNA possessing the stop codon in rat hepatocytes, resulting in a selective decrease in the secretion of apoB-48, whereas apoB mRNA levels remained constant. They concluded that posttranscriptional

[15] A. J. Lusis, R. West, M. Mahrabian, M. A. Reuben, R. C. LeBouef, J. S. Kaptein, D. F. Johnson, V. N. Schumaker, M. P. Yuhasz, M. C. Schotz, and J. Elovson, *Proc. Natl. Acad. Sci. U.S.A.* **82,** 4597 (1985).

[16] T. J. Knott, S. C. Rall, Jr., T. L. Innerarity, S. F. Jacobson, M. S. Urdea, B. Levy-Wilson, L. M. Powell, R. J. Pease, R. Eddy, H. Nakai, M. Byers, L. M. Priestly, E. Robertson, L. B. Rall, C. Betsholta, T. B. Shows, R. W. Mahley, and J. Scott, *Science* **230,** 37 (1985).

[17] S. S. Deeb, C. Disteche, A. G. Motulsky, R. V. Lebo, and W. Y. Kan, *Proc. Natl. Acad. Sci. U.S.A.* **83,** 419 (1986).

[18] S. W. Law, K. J. Lackner, A. V. Hospattankar, J. M. Anchors, A. Y. Sakaguchi, S. L. Naylor, and H. B. Brewer, Jr., *Proc. Natl. Acad. Sci. U.S.A.* **82,** 8340 (1985).

[19] L. M. Powell, S. C. Wallis, R. J. Pease, Y. H. Edwards, T. J. Knott, and J. Scott, *Cell (Cambridge, Mass.)* **50,** 831 (1987).

[20] S.-H. Chen, G. Habib, C.-Y. Yang, Z.-W. Gu, B. R. Lee, S.-A. Weng, S. R. Silberman, S.-J. Cai, J. P. Deslyperé, M. Rosseneu, A. M. Gotto, Jr., W.-H. Li, and L. Chan, *Science* **238,** 363 (1987).

[21] A. V. Hospattankar, K. Higuchi, S. W. Law, N. Meglin, and H. B. Brewer, Jr., *Biochem. Biophys. Res. Commun.* **148,** 279 (1987).

[22] N. O. Davidson, L. M. Powell, S. C. Wallis, and J. Scott, *J. Biol. Chem.* **262,** 13482 (1988).

[23] N. O. Davidson, R. C. Carlos, and A. M. Lukaszewicz, *Mol. Endocrinol.* **4,** 779 (1990).

[24] J. K. Leighton, J. Joyner, J. Zamarripa, M. Deines, and R. A. Davis, *J. Lipid Res.* **31,** 1663 (1990).

[25] C. L. Baum, B.-B. Teng, and N. O. Davison, *J. Biol. Chem.* **265,** 19263 (1990).

mechanisms must play a role in regulating apoB secretion. Baum *et al.*[25] also observed that hepatic apoB-100 synthesis became undetectable in rats subjected to 48 hr of fasting and subsequently refed a high-carbohydrate diet for either 24 or 48 hr. This change was accounted for by an increase in the proportion of edited apoB mRNA from 37% UAA in fasted animals to 79 and 91% UAA at 24 and 48 hr after refeeding, respectively. Lau *et al.*[26] concluded that rat hepatic apoB mRNA editing is not a cotranscriptional event; it is an intranuclear event that occurs posttranscriptionally. Thus, it is clear that in rats the finding of increased apoB-48 in plasma VLDL cannot be interpreted as a reflection of an increased synthesis or secretion of intestinal apoB.

It is generally agreed that, in humans, apoB mRNA editing is developmentally regulated in fetal small intestine, producing parallel changes in the synthesis and secretion of apoB-100 and apoB-48.[27,28] Neither fetal nor adult liver contained detectable levels (<1%) of edited apoB mRNA. By contrast, apoB mRNA is incrementally edited during development in several extraintestinal human fetal tissues, including membrane, kidney, stomach, and colon. Teng *et al.*[27] found that fetal membranes demonstrated editing activity at 8 weeks gestational age (~7% UAA), increasing to 50% UAA at 11–12 weeks. By 13–14 weeks, more than 80% of the apoB mRNA was edited. In kidney, apoB mRNA was 15% UAA at 12 weeks, doubled by 20 weeks, and was more than 40% in adult kidney. In colon, apoB mRNA increased from approximately 10% UAA at 11–12 weeks to 80% UAA by 15 weeks. In adult colon, 50–91% editing was found. Fetal stomach contained more than 50% edited apoB mRNA, whereas adult stomach contained over 90% edited apoB mRNA. However, despite the presence of edited apoB mRNA in several tissues, apoB-48 synthesis and secretion are confined to the small intestine.[27] Hoeg *et al.*,[12] using intestinal cultures from normal adults incubated in the presence of protease inhibitors in medium supplemented with [^{35}S]methionine, found that, although apoB-48 was the principal apoB species newly synthesized, apoB-100 was also synthesized and secreted, accounting for 5% of the intestinal apoB. However, on the basis of Coomassie blue protein staining, they found that apoB-100 was the predominant species secreted into the culture medium, whereas apoB-48 was the primary species within the organ culture; on the basis of quantitation of mRNA, they found 16 ± 3% of the intestinal apoB mRNA coding for apoB-100. As shown later, the methionine content of apoB-48

[26] P. P. Lau, W. Xiong, H.-J. Zhu, S.-H. Chen, and L. Chan, *J. Biol. Chem.* **266,** 20550 (1991).
[27] B.-B. Teng, M. Verp, J. Salomon, and N. O. Davidson, *J. Biol. Chem.* **265,** 20616 (1990).
[28] A. P. Patterson, G. E. Tennyson, J. M. Hoeg, D. D. Sviridov, and H. B. Brewer, Jr., *Arterioscler. Thromb.* **12,** 468 (1992).

is 28% higher than that of apoB-100 (see Table I). Quantitation of apoB synthesis on the basis of radioactivity of methionine incorporation would overestimate apoB-48 by 28%.

Removal of Apolipoprotein B-48

Chylomicron remnants are rapidly taken up by liver cells.[29] Turnover studies of the two apoB species in rats show that the removal of apoB-48 is faster than that of apoB-100[4,5,30] and the injected apoB-48 is recovered mainly in the liver.[4,7] In hyperchylomicronemic subjects the removal of apoB-48 is slower than that of apoB-100.[31]

The receptor responsible for chylomicron remnant removal is still a subject of controversy. Bersot *et al.*[32] suggested that the receptor for β-VLDL is responsible for the removal of chylomicron remnants. Using human monocyte-derived macrophages, Koo *et al.*[33] found that the uptake of β-VLDL is mediated through the apoB/E (LDL) receptor and not by β-VLDL receptor. Hussarn *et al.*[34] reported that apoE-enriched chylomicron remnants are taken up in liver cells by the receptor for activated α_2-macroglobulin (α_2M)/LDL-receptor-related protein (LRP). However, Jäckle *et al.*[35] demonstrated that the removal of β-VLDL and chylomicron remnants is not mediated via LRP and does not compete with α_2M.

Structure of Apolipoprotein B-48

Amino Acid Composition of Apolipoprotein B-48

The amino acid sequence of apoB-100 has been deduced from the sequence of its complementary DNA clone.[36–39] Apolipoprotein B-100 is

[29] A. D. Cooper, *Biochim. Biophys. Acta* **488,** 464 (1977).
[30] A.-L. Wu and H. G. Windmueller, *J. Biol. Chem.* **256,** 3615 (1981).
[31] P. J. Nestel, T. Billington, and N. H. Fidge, *Biochim. Biophys. Acta* **751,** 422 (1983).
[32] T. P. Bersot, T. L. Innerarity, R. E. Pitas, S. C. Rall, Jr., K. H. Weisgraber, and R. W. Mahley, *J. Clin. Invest.* **77,** 622 (1986).
[33] C. Koo, M. E. Wernette-Hammond, Z. Garcia, M. J. Molloy, R. Uany, C. East, D. W. Bilheimer, R. W. Mahley, and T. L. Innerarity, *J. Clin. Invest.* **81,** 1332 (1988).
[34] M. M. Hussarn, F. R. Maxfield, J. Más-Oliva, I. Tabas, Z.-S. Ji, T. L. Innerarity, and R. W. Mahley, *J. Biol. Chem.* **266,** 13936 (1991).
[35] S. Jäckle, C. Huber, S. Moestrup, J. Gliemaun, and U. Beisiegel, *J. Lipid Res.* **34,** 309 (1993).
[36] S.-H. Chen, C. Y. Yang, P.-F. Chen, D. Setzer, M. Tanimura, W.-H. Li, A. M. Gotto, Jr., and L. Chan, *J. Biol. Chem.* **261,** 12918 (1986).
[37] T. J. Knott, S. C. Wallis, L. M. Powell, R. J. Pease, A. J. Lusis, B. Blackhart, B. J. McCarthy, R. W. Mahley, B. Levy-Wilson, and J. Scott, *Nucleic Acids Res.* **14,** 7501 (1986).
[38] S. W. Law, S. M. Grant, K. Higuchi, A. Hospattankar, K. Lackner, N. Lee, and H. B. Brewer, Jr., *Proc. Natl. Acad. Sci. U.S.A.* **83,** 8142 (1986).
[39] C. Cladaras, M. Hadzopoulou-Cladaras, R. T. Nolte, D. Atkinson, and V. I. Zannis, *EMBO J.* **5,** 3495 (1986).

a single polypeptide chain consisting of 4536 amino acid residues. Although all reports agree on the total number of amino acid residues, there are 75 differences found in DNA, 53 of which result in amino acid substitutions.[40] Some of these differences may be due to DNA sequencing errors, but many have been verified by direct amino acid sequence analyses.[40] This leads to heterogeneity in the amino acid sequence of apoB-48. Many of the DNA sequence differences create restriction fragment length polymorphisms in the apoB gene.[41] To date, 13 different mutations in apoB gene associated with hypobetalipoproteinemia have been identified, and they are listed in the review of Young and Linton.[42] Eleven of the mutations result in truncated apoB species detectable in plasma lipoprotein fractions. Among them, apoB-46 (2057 amino acid residues[43]) and apoB-50 (2252 amino acid residues[44]) have a peptide length close to that of apoB-48 (2152 amino acid residues). Their differentiation from apoB-48 on the basis of size only is difficult. Apolipoprotein B-46 was found in VLDL, LDL, and high-density lipoproteins (HDL) of the affected subject,[43] whereas apoB-50 was found only in VLDL.[44,45]

To facilitate the comparison we have calculated the molecular weight and the amino acid composition of apoB-48 and apoB-100 on the basis of the sequence data of Chen *et al.*,[36] most of which has been confirmed with amino acid sequence analyses. The results are listed in Table I. The molecular weights of apoB-100 and apoB-48 are calculated to be 512,906 and 240,858, respectively. The latter turned out to be 47% of that of apoB-100. The value of 48% as estimated by SDS–PAGE, including the mass of carbohydrate,[3] is very close to this value in the absence of carbohydrate. Of 20 amino acid residues there are 10 residues showing a difference more than 6% between the two species (Table I). On an equal mass basis, apoB-48 has lower histidine and higher arginine contents than apoB-100. The most striking differences are in the sulfur-containing amino acids: apoB-48 has a 53% higher content of half-cystine and 28% higher content of methionine than apoB-100. Among aromatic amino acids, both phenylalanine and tryptophan are lower in apoB-48 (16 and 25%, respectively). These differences may affect the dye binding, chromogenicity, and absorption at

[40] C. Y. Yang, Z. W. Gu, S. A. Weng, W. K. Tae, S.-H. Chen, H. J. Pownall, P. M. Sharp, S.-W. Liu, W.-H. Li, A. M. Gotto, Jr., and L. Chan, *Arteriosclerosis (Dallas)* **9,** 96 (1989).

[41] S. G. Young, *Circulation* **82,** 1574 (1990).

[42] S. G. Young and M. F. Linton, *Trends Cardiovasc. Med.* **1,** 59 (1991).

[43] S. G. Young, S. T. Hubl, D. A. Chappell, R. S. Smith, F. Claiborne, S. M. Snyder, and J. F. Terdiman, *N. Engl. J. Med.* **320,** 1604 (1989).

[44] M. J. Malloy, J. P. Kane, A. Hardman, R. L. Hamilton, and K. B. Dalal, *J. Clin. Invest.* **67,** 1441 (1981).

[45] D. A. Hardman, C. R. Pullinger, J. P. Kane, and M. J. Malloy, *Circulation* **80**(Suppl. 2), II-466 (1989).

TABLE I
AMINO ACID COMPOSITION CALCULATED FOR APOLIPOPROTEINS B-100 AND B-48[a]

Amino acid	No. of residues in apoB-100	Total residue weight/mole apoB-100	No. of residues in apoB-48	Total residue weight/mole apoB-48	Mole residues per 10^5 g apoB-100	Mole residues per 10^5 g apoB-48	Change (%) of apoB-48 fro apoB-100[b]
Lys	356	45635.64	166	21279.54	69.41	68.92	
His	115	15776.85	50	6859.50	22.42	20.76	−7.4
Arg	148	23119.08	77	12028.17	28.86	31.97	+10.8
Asp	234	26933.4	107	12315.70	45.62	44.42	
Thr	298	30127.8	144	14558.40	58.10	59.79	
Ser	392	34139.28	192	16721.28	76.43	79.72	
Glu	299	38609.87	132	17045.16	58.30	54.80	−6.0
Pro	168	16317.84	82	7964.66	32.76	34.05	
Gly	207	11813.49	117	6677.19	40.36	48.58	+20.4
Ala	267	18981.03	138	9810.42	52.06	57.30	+10.1
Cys	25	2580.25	18	1857.78	4.87	7.47	+53.4
Val	251	24886.65	115	11402.25	48.94	47.75	
Met	78	10234.38	47	6166.87	15.21	19.51	+28.3
Ile	288	32592.96	109	12335.53	56.15	45.25	−19.4
Leu	524	59301.08	259	29311.03	102.16	107.53	
Tyr	152	24804.88	75	12239.25	29.64	31.14	
Phe	223	32823.37	88	12952.72	43.48	36.54	−16.0
Trp	37	6890.14	13	2420.86	7.21	5.40	−25.1
Asn	244	27845.28	120	13694.40	47.57	49.82	
Gln	230	29474.5	103	13199.45	44.84	42.76	
H_2O	1	18.0	1	18.0			
	Σ 4536	Σ 512,905.77	Σ 2152	Σ 240,858.16 = 47% of apoB-100	Σ 884.39	Σ 893.48	

[a] Calculations were based on the amino acid sequence data of Chen *et al.* [S.-H. Chen, C. Y. Yang, P.-F. Chen, D. Setzer, M. Tanimura, W.-H. Li, A. M. Gotto, Jr., and L. Chan, *J. Biol. Chem.* **261,** 12918 (1986)].
[b] Listed only for differences exceeding 6% of apoB-100.

280 nm of the two apoB species, as discussed later in connection with quantitation of the protein. In addition, apoB-48 contains more hydrophilic glycine (20%) and less hydrophobic isoleucine (19%). Together with the lower aromatic amino acid content, this may result in an enhanced water solubility of apoB-48 compared to apoB-100.

Hydropathy Value of Apolipoprotein B-48

Using the hydropathy index of individual amino acids,[46] we calculated that the average hydropathy value for apoB-48 is −0.34. This negative value reflecting the hydrophilic nature of apoB-48, based solely on the

[46] J. Kyte and R. F. Doolittle, *J. Mol. Biol.* **157,** 105 (1982).

protein moiety, is comparable to those of most water-soluble enzymes.[46] Comparing this value to the hydropathy value of −0.31 for apoB-100,[47] apoB-48 is slightly more hydrophilic than apoB-100. We have suggested that the hydrophobic nature of apoB-100 could be largely attributed to its covalently bound fatty acids.[47–51] Whether apoB-48 is similarly acylated as apoB-100 remains to be determined.

Carbohydrate Moiety of Apolipoprotein B-48

The carbohydrate moiety of human small intestinal apoB-48 was studied by Sasak *et al.*[52] They found that apoB-48 contained 4% carbohydrate, with the monosaccharide composition in mole/mole being the following: *N*-acetylglucosamine, 16.8; mannose, 17.8; galactose, 13.4; fucose, 3.4; xylose, 2.4; and *N*-acetylgalactosamine, 2.3. They proposed that apoB-48 contains 16% high-mannose oligosaccharides with $Man_8GlcNAc_2$ as the predominant species, and probably 6% higher-branched complex-type oligosaccharides. In comparison, no fucose or xylose residues have been found in hepatic apoB-100 from human plasma.[53,54] In chylomicron retention disease, the incorporation of mannose into intracellular chylomicrons was reduced by 80% in patients compared with normal subjects.[55]

Physiological and Pathophysiological Aspects of Apolipoprotein B-48

Apolipoprotein B-48 and Fasting Very-Low-Density Lipoproteins of Normolipidemic and Hyperlipoproteinemic Subjects

Whereas apoB-48 is usually not detected in the LDL of human plasma, it occurs in some fasting VLDL, particularly in subjects with elevated triglycerides. Meng *et al.*[56] reported that among hyperlipoproteinemias,

[47] D. M. Lee, *Adv. Exp. Med. Biol.* **285,** 49 (1991).
[48] G. Huang, D. M. Lee, and S. Singh, *Biochemistry* **27,** 1395 (1988).
[49] V. S. Kamanna and D. M. Lee, *Biochem. Biophys. Res. Commun.* **162,** 1508 (1989).
[50] J. M. Hoeg, M. S. Meng, R. Ronan, S. J. Demosky, Jr., T. Fairwell, and H. B. Brewer, Jr., *J. Lipid Res.* **29,** 1215 (1988).
[51] W. R. Fisher and S. Gurin, *Science* **143,** 362 (1964).
[52] W. V. Sasak, J. S. Lown, and K. A. Colburn, *Biochem. J.* **274,** 159 (1991).
[53] M. Vauhkonen, *Glycoconjugate J.* **3,** 35 (1986).
[54] T. Taniguchi, Y. Ishikawa, M. Tsunemitsu, and H. Fukuzaki, *Arch. Biochem. Biophys.* **273,** 197 (1989).
[55] E. Levy, Y. Marcel, R. J. Deckelbaum, R. Milne, G. Lepage, E. Seidman, M. Bendayan, and C. C. Roy, *J. Lipid Res.* **28,** 1263 (1987).
[56] M. S. Meng, R. E. Gregg, E. J. Schaefer, J. M. Hoeg, and H. B. Brewer, Jr., *J. Lipid Res.* **24,** 803 (1983).

types I and V, apoB-48 was present in VLDL of subjects with plasma triglyceride levels higher than 1200 mg/dl, but not in those with triglyceride levels lower than 1200 mg/dl. Apolipoprotein B-48 was found in $d < 1.006$ g/ml of type III hyperlipoproteinemic subjects, but not in types IIa, IIb, and IV hyperlipoproteinemic subjects. In type III hyperlipoproteinemic subjects with apoE deficiency, apoB-48 was detected not only in $d < 1.006$ g/ml but also in intermediate density lipoproteins (IDL) and LDL fractions.[56] This is the only fasting LDL reported to contain apoB-48. Others also identified apoB-48 in VLDL of types III and IV hyperlipoproteinemic subjects.[57,58] A patient with lecithin–cholesterol acyltransferase (phosphatidylcholine–sterol *O*-acyltransferase) deficiency was also found to have a small quantity of apoB-48 in fasting VLDL.[56] The finding of apoB-48 in fasting VLDL is interpreted as the result of a defect in chylomicron remnant removal.[31] Most investigators do not find apoB-48 in fasting VLDL of normolipidemic subjects. However, Cazzolato *et al.*[58] reported that apoB-48 is a constant finding in fasting VLDL of Italians with normolipidemia. Apolipoprotein B-48 was also found in every fasting VLDL of 14 normolipidemic Japanese subjects studied.[59]

If methodological differences did not contribute to the different findings, the question is raised whether a high-carbohydrate diet (a typical Italian diet contains 55% of calories from carbohydrate) may contribute to this difference.[58] Indeed, diet may have some effect. Rat intestinal apoB-48 synthesis was found unaltered by dietary triglyceride intake[60] but was regulated by bile salt, fatty acid, and phospholipid flux.[61] In the postprandial state, the ratio of apoB-48/apoB-100 is significantly increased in patients with coronary artery disease compared to control subjects, even when age, triglycerides, and cholesterol were included in the analysis.[62] This finding is consistent with the hypothesis that postprandial lipoproteins may be atherogenic.[63]

Apolipoprotein B-48 as Intestinal Marker

In the past, radioactive-labeled retinyl ester has been used frequently as the intestinal marker in metabolic studies designed to trace chylomicron

[57] F. Tercé, R. W. Milne, P. K. Weech, J. Davignon, and Y. L. Marcel, *Arteriosclerosis* (*Dallas*) **5,** 201 (1985).

[58] G. Cazzolato, G. B. Bon, and P. Avogaro, *Arteriosclerosis* (*Dallas*) **5,** 88 (1985).

[59] H. Hidaka, H. Kojima, Y. Nakajima, T. Aoki, T. Nakamura, T. Kawabata, T. Nakano, Y. Harano, and Y. Shigeta, *Clin. Chim. Acta* **189,** 287 (1990).

[60] N. O. Davidson, M. E. Kollmer, and R. M. Glickman, *J. Lipid Res.* **27,** 30 (1986).

[61] N. O. Davidson, M. J. Drewek, J. I. Gordon, and J. Elovson, *J. Clin. Invest.* **82,** 300 (1988).

[62] L. A. Simons, T. Dwyer, J. Simons, L. Bernstein, P. Mock, N. S. Poonia, S. Balasubramaniam, D. Baron, J. Branson, J. Morgan, and P. Roy, *Atherosclerosis* **65,** 181 (1987).

[63] D. B. Zilversmit, *Circulation* **60,** 473 (1979).

remnants.[63] However, Krasinski *et al.*[64] have reported that metabolism of plasma retinyl ester did not parallel those of apoB-48 or plasma triglycerides. Retinyl ester concentration peaked 6 hr after a fat-rich meal, whereas plasma triglycerides and apoB-48 both peaked at 3 hr. In fasting or 9 hr postprandial plasma, a large percentage of retinyl esters was found in LDL (9% in 3 hr; 19% in 9 hr; 44% in fasting), suggesting that apoB-48, but not retinyl ester, is a suitable marker for chylomicrons or chylomicron remnants.

Quantitation of Apolipoprotein B-48

Lack of Specific Antibodies for Immunochemical Quantitation of Apolipoprotein B-48

To date, all polyclonal and monoclonal antibodies produced against apoB-48 have cross-reacted with apoB-100. There is not a single monospecific anti-apoB-48 available for quantitation of this apoB form. Although an antibody produced against a synthetic heptapeptide of the carboxyl-terminal end of apoB-48 has been claimed to be specific to apoB-48,[65] the authors failed to demonstrate the lack of cross-reactivity with apoB-100. Whether this antibody can be used for immunochemical quantitation of apoB-48 remains to be established.

There are several monoclonal antibodies with their determinants located at the carboxyl-terminal portion of apoB-100,[66,67] such as the monoclonal antibodies 3F5 and 4G3[66] and MB47,[67] which do not react with apoB-48. Antibodies produced against synthetic peptides of the carboxyl-terminal half of apoB-100 mostly do not react with apoB-48. Theoretically, these antibodies can be used to quantitate the apoB-100, and the polyclonal antibodies against apoB-100 can be used to quantitate the sum of apoB-100 and apoB-48, providing the calibration is properly done. Then the difference between the two would be apoB-48. This approach is suitable only if the relative content of apoB-48 is not small. Otherwise, the error would be high when quantitation is based on difference. Utilizing this principle, Hoeg *et al.*[12] incorporated immunoblotting and used ^{125}I-labeled second antibody or ^{125}I-labeled protein A to quantitate the two species after electrotransfer of the protein bands. This is more of a qualitative

[64] S. D. Krasinski, J. S. Cohn, R. M. Russell, and E. J. Schaefer, *Metabolism* **39,** 357 (1990).

[65] A. S. Peel, P. Komanduri, C. M. Williams, L. M. Morgan, and B. J. Gould, *Biochem. Soc. Trans.* **20,** 165S, 1992.

[66] Y. L. Marcel, P. K. Weech, P. Milthorp, F. Tercé, C. Vezina, and R. W. Milne, *Prog. Lipid Res.* **23,** 169 (1985).

[67] S. G. Young, J. L. Witztum, D. C. Casal, L. K. Curtiss, and S. Bernstein, *Arterioisclerosis (Dallas)* **6,** 178 (1986).

method because the amount of each protein lost due to electrotransfer is unknown.

In practice, there are three methods used for the quantitation of apoB-48, all based on ratios with respect to apoB-100. In describing these procedures, special attention is given to the potential shortcomings of each method and precautionary measures needed.

Scanning of SDS–Polyacrylamide Gels

The most widely used and the simplest method is the scanning of the stained protein bands after SDS–PAGE of the intact lipoproteins or the delipidized and redissolved proteins. The areas corresponding to the two protein bands are then measured with a planimeter or by automatic integrator. The percentage of each area is calculated with or without correction for chromogenicity difference to represent the percent concentration of each apoB form. The format for SDS–PAGE can involve either a slab gradient polyacrylamide gel from 3 to 20%[68] or a 3.3% polyacrylamide slab gel,[9] or simply a 3.3% polyacrylamide disk gel, according to the method of Weber and Osborn[69] or Fairbanks *et al.*[70] The advantage of using a gradient gel is that, in addition to apoB-100 and apoB-48, all the small molecular weight apolipoproteins present in chylomicrons, VLDL and IDL, etc., can also be included in the quantitation. The major difference between the Weber and Osborn method[69] and the Fairbanks *et al.* method[70] is in the buffer system. Whereas the former uses Na_2HPO_4/NaH_2PO_4 buffer, the latter uses Tris–sodium acetate–EDTA buffer. Because Na_2HPO_4 may contain a relatively high content of iron and heavy metals, it may cause oxidation of apoB during electrophoresis; therefore, we have recommended the use of the Fairbanks method.[8–10] In addition, we have also incorporated the following modifications to minimize the contaminants and oxidative damage.

Precautionary Measures in SDS–PAGE

Recrystallization of Acrylamide. We recommend that the acrylamide be recrystallized from redistilled chloroform when ultrapure acrylamide is not commercially available.

Prerun Gels. Polymerization of acrylamide gels requires the use of an oxidizing agent, ammonium persulfate. The residual ammonium persulfate may cause oxidation of apoB species. Although the concentration may not

[68] U. K. Laemmli, *Nature (London)* **227,** 680 (1970).
[69] K. Weber and M. Osborn, *J. Biol. Chem.* **244,** 4406 (1969).
[70] G. Fairbanks, T. L. Steck, and D. F. H. Wallach, *Biochemistry* **10,** 2606 (1971).

be sufficient to cause cleavage of apoB, it certainly could oxidize tyrosine and tryptophan residues and decrease the intensity of band staining[71]; the aromatic amino acids are the major amino acids responsible for dye binding, and both tyrosine and tryptophan are highly sensitive to oxidation. Inactivation of enzymes or creation of enzymatic artifacts after polyacrylamide gel electrophoresis was also observed by others.[72] To avoid this potential oxidative damage, it is recommended that the gels be prerun prior to the sample application for 2 hr at 2 mA/tube in a buffer containing antioxidant (0.2% v/v thiodiglycol) to remove the ammonium persulfate.[73] Then the sample is electrophoresed in fresh buffer containing 0.02% (v/v) thiodiglycol.[8–10]

Elimination of Stacking Gel. We have always avoided use of a stacking gel because of concern over the possibility of apoB being exposed to free radical formation when the sample is embedded in the gel by photopolymerization.[74] No apparent loss in resolution or alteration in banding patterns have been observed with such a practice.

Precautionary Measures during Delipidization

There are several disadvantages in using delipidized samples for quantitation, in comparison to the intact lipoproteins when using SDS–PAGE. (i) There might be differential losses due to delipidization because the solubility of apoB-48 might be slightly higher than that of apoB-100. (ii) There is a risk of exposure of the samples to organic solvents containing oxidants or contaminants. Ethanol–diethyl ether (3:1, v/v) is the most frequently used solvent for delipidization. Frequently, diethyl ether contains peroxides and needs pretreatment before use[73,75]; otherwise, oxidation of lipids and apoB may result in incomplete solubilization of apoB in SDS solutions, or apoB may undergo oxidative cleavage.[73,75] Any of these damaging reactions may lead to inaccurate results. Most likely, the loss may affect apoB-100 to a greater extent than apoB-48, resulting in a higher apoB-48/apoB-100 ratio. The second most frequently used organic solvent for delipidization is $CHCl_3:CH_3OH$ (3:1 or 2:1, v/v). Chloroform also frequently contains contaminants, such as phosgene and diethyl chloroformate,[76] and needs pretreatment before use. The protocols for purification of diethyl ether and chloroform are included at the end of this chapter.

[71] D. M. Lee and S. Singh, unpublished results (1986).
[72] W. M. Mitchell, *Biochim. Biophys. Acta* **147,** 171 (1967).
[73] D. M. Lee, A. J. Valente, W. H. Kuo, and H. Maeda, *Biochim. Biophys. Acta* **666,** 133 (1981).
[74] S. Hjerten, S. Jerstedt, and A. Tiselius, *Anal. Biochem.* **11,** 219 (1965).
[75] S. S. Huang and D. M. Lee, *Biochim. Biophys. Acta* **577,** 424 (1979).
[76] E. J. Cone, W. F. Buchwald, and W. D. Darwin, *Drug Metab. Dispos.* **10,** 561 (1982).

(iii) Delipidization is the step in which the sample is most vulnerable to oxidation, because of the air space in the delipidization tube. We recommend that ethanol be saturated with N_2 before mixing it with diethyl ether, and that 0.02% (w/v) butylated hydroxytoluene (BHT) be added as an antioxidant to the solvent.[73] After adding the lipoprotein samples, the air space is filled with argon before starting extraction.

A new delipidization procedure has been reported for the preparation of samples for SDS–PAGE.[77] The procedure involves extraction of lipoproteins (1 ml) with diethyl ether (10 ml) in the presence of trichloroacetic acid (0.2 ml of 4.9 *M*) and sodium deoxycholate (0.2 ml of 3.6 m*M*). The precipitated apolipoproteins are then dissolved in Tris buffer (0.125 *M*, pH 6.8) containing SDS (41 mg/ml), bromphenol blue (0.02 mg/ml), 20% (v/v) glycerol, and 10% (v/v) 2-mercaptoethanol. The pH is neutralized with NaOH. The authors claim that this method gives an improved protein recovery, and the procedure requires only 1 hr.

Precautionary Measures in Dialysis

Potassium salts should be avoided during isolation of lipoproteins if direct analysis of samples by SDS–PAGE is desired. Potassium ions precipitate with SDS, and the samples must be dialyzed to eliminate them. Dialysis of lipoprotein samples presents another risk of exposure to oxidation. Amino acid residues including tyrosine, tryptophan, methionine, and cysteine, which play important roles in quantitation of proteins, are prone to oxidation. Therefore, as the first measure, the salt used in the dialysis buffer should be free of Fe^{3+} and Cu^{2+}. Phosphate buffer is used very frequently for dialyzing lipoprotein solutions, but, as mentioned earlier, phosphate has a high content of iron. Even in the presence of EDTA, there is no guarantee that lipid peroxidation would not occur during dialysis. In fact, EDTA is a poor chelator for iron. Deferoxamine mesylate is a better chelating agent for iron. We have established that 1 m*M* EDTA is not sufficient to protect apoB from fragmentation and have routinely used 3 m*M* EDTA in addition to other preservatives.[8,10,73,78] Sodium azide should be avoided because it promotes free radical oxidation.[79] The dialysis buffer should be saturated with N_2 by blowing the gas to the bottom of the buffer for 30 min, and the mouth of the container should be covered with a double layer of Parafilm (not stretched) and kept sealed tightly during dialysis. The sites of oxidative cleavage of apoB-100 have not been identified[73,79];

[77] M. A. Mindham and P. A. Mayes, *J. Lipid Res.* **33,** 1084 (1992).

[78] D. M. Lee and D. Downs, *J. Lipid Res.* **23,** 14 (1982).

[79] J. Schuh, G. F. Fairclough, and R. H. Haschemeyer, *Proc. Natl. Acad. Sci. U.S.A.* **75,** 3173 (1978).

for quantitation of apoB-48/apoB-100 on the basis of intact sizes of the two species, protection against autoxidation cannot be emphasized enough.

Using these precautionary measures, 10–12 μg fresh apoB-100 gives a very intense band when stained with Coomassie Brilliant Blue R.

Dye Binding Capacities of Two Apolipoprotein B Species

Most investigators have assumed that the dye binding capacities are the same for both apoB-100 and apoB-48, and the apoB-48/apoB-100 ratio is calculated directly from the areas under the peaks of the protein bands after gel scanning. This may have resulted in underestimation of apoB-48. The amino acid compositions show that the average aromatic amino acid content of apoB-48 is 12% lower than that of apoB-100 (see Table I). If all aromatic amino acids have equal dye binding capacity (which may not be the case) and if no oxidation occurs during SDS–PAGE, one would expect the chromogenicity of apoB-48 to be 12% lower than that of apoB-100. If the oxidizing agent, ammonium persulfate, is not removed from the polyacrylamide gel, the chromogenicity of apoB-48 would further decrease after electrophoresis. Because apoB-48 moves faster than apoB-100 in the electrophoretic field, the former is exposed to oxidants first. Although the reducing agent 2-mercaptoethanol or dithiothreitol is generally applied with the samples, these agents are neutral and do not migrate with the samples. Poapst *et al.*[80] separated apoB-48 and apoB-100 by preparative SDS–PAGE and quantitated them on analytical SDS–PAGE after staining with Coomassie Brilliant Blue R-250. They found that the chromogenicity of the resulting apoB-48 was 43% lower than that of apoB-100. Such a disparity was not observed by Zilversmit and Shea,[81] who separated the two apoB species by gel filtration.

High-Performance Liquid Chromatography

Hidaka *et al.*[59] used high-performance liquid chromatography (HPLC), monitored at 280 nm, to quantitate apoB-48. They employed a column of Shim-Pack Diol-300 (7.9 mm × 25 cm, Shimadzu, Kyoto, Japan) maintained at 40°. The column was packed with hydrophilic silica gel with pore size 300 Å. Apolipoproteins (5–10 μg) in 10 μl were eluted with 0.1% SDS, 0.2 *M* NaCl, and 10 m*M* sodium phosphate buffer (pH 7.0) at a flow rate of 0.4 ml/min. Apolipoproteins B-100 and B-48 were separated with retention times of 12.1 and 16.9 min, respectively. This very fast quantitative method was validated with the SDS–PAGE scanning method. Although the per-

[80] M. Poapst, K. Uffelman, and G. Steiner, *Atherosclerosis* **65,** 75 (1987).
[81] D. B. Zilversmit and T. M. Shea, *J. Lipid Res.* **30,** 1639 (1989).

centage of apoB-48 determined by HPLC was significantly correlated with percentage of apoB-48 by SDS–PAGE, the quantity was 6–8 times higher by the former. It is not clear why the discrepancy between the two methods is so high. One possibility is that quantitation by absorption at 280 nm might give misleading results because of cystine generated in the samples. When cysteine is autoxidized to cystine, the absorbance at 280 nm increases sharply, as can be demonstrated with reduced and oxidized glutathione. During the delipidization procedure, Hidaka *et al.* stored the lipoproteins in diethyl ether for 1 week in the absence of antioxidants.[59] Peroxides might have been generated in the diethyl ether during a week's time, and it is possible that a large portion of the tyrosine, tryptophan, and cysteine residues might have been oxidized. This would result in decreased tyrosine and tryptophan contents for dye binding, and increased cystine for 280 nm absorption, leading to a dramatic difference in results by the two monitoring systems. By the HPLC method, the authors found 17.1 $\pm$ 5.6% apoB-48 present in every fasting VLDL from all 14 normolipidemic subjects studied. They demonstrated that the percentage of apoB-48 in fasting VLDL was significantly correlated with VLDL triglyceride and cholesterol concentrations.

Separation of apoB species by HPLC was also performed by Schwandt *et al.*[82] using a TSK 4000 SW column, 60 $\times$ 0.75 cm i.d. (LKB, Graefelfing, Germany), based on gel-permeation chromatography. They separated apoB into fractions identified as apoB-100, apoB-48, and apoB-26, but did not find apoB-74. As apoB-26 and apoB-74 are the complementary fragments of apoB-100,[3] there should be three times as much apoB-74 as apoB-26, and it is not clear why the major fraction was not found in their system.

Quantitation by Radioiodination

In the radioiodination method, intact or delipidized lipoproteins are first radioiodinated, then separated by SDS–PAGE, and the radioactivity is determined in stained bands as reported by Zilversmit and Shea.[81] Delipidized lipoproteins of $d < 1.006$ g/ml are separated on a Superose 6 column in phosphate-buffered saline containing 0.01% EDTA and 0.1% SDS monitored at 280 nm. Then the protein concentration is measured by absorbance at 220 nm and labeled in 2% SDS.[81] After SDS slab gel electrophoresis,[68] the gel is stained with Coomassie blue G-250 and scanned on a Bio-Rad (Richmond, CA) Model 620 densitometer at 600 nm. The areas are determined by the 1-D Analyst version 2.01 and 2-D Analyst version 1.01 software from Bio-Rad.

[82] P. Schwandt, W. O. Richter, and P. Weisweiler, *Clin. Chim. Acta* **157,** 249 (1986).

The absorption maxima for both apoB-48 and apoB-100 were found to be the same (600 nm) with Coomassie blue G-250. The specific activities of apoB-100 were 25 ± 1.5% lower than those of apoB-48.[81] However, the amino acid composition showed that the tyrosine content of apoB-100 was only 5% lower than that of apoB-48 (see Table I). Using animal lipoproteins containing high levels of apoB-48, when labeling was done on intact lipoproteins, the specific activities for both apoB-100 and apoB-48 were similar. When labeling was performed after delipidization, the specific activity of apoB-100 was 35% lower than that of apoB-48.[81] These results suggest that delipidization might have damaged tyrosine residues more in apoB-100 than in apoB-48.

Comparison of the gel scanning method with the labeling method and the dye elution method (with 25% pyridine in water) using known mixtures of apoB-48 and apoB-100 showed that the three procedures were comparable, with the iodination method being the most accurate; at low percent apoB-48 concentration, the scanning procedure was less reliable. At low percent apoB-48 concentration, Coomassie Brilliant Blue R-250 was found to be less accurate than Coomassie Brilliant Blue G-250. The authors found that the calibration curves of apoB staining were linear and passed through the origin, in contrast to the findings of Poapst *et al.*,[80] and suggested that the two-dimensional scans gave better accuracy than the one-dimensional results.

Primary Protein Assay

The above three quantitative methods all depend on a primary protein assay to provide the absolute concentration for apoB-48. Zilversmit and Shea[81] found satisfactory results using absorbance at 220 nm against a crystalline bovine serum albumin standard (Miles, Elkhart, IN) that had been lyophilized and dried in a vacuum desiccator over P_2O_5, without correction factor for chromogenicity. Henderson *et al.*[83] compared all protein assays for apoB. They found the Kjeldahl method to be the most accurate. The Lowry method,[84] with the inclusion of SDS to eliminate the turbidity contributed by lipids, gives reasonably good results.[83] The protein used as primary standard, either bovine serum albumin or human serum albumin, should be free of fatty acids and dried. We usually use human serum albumin dried under vacuum in P_2O_5 at 80° (under boiling benzene) for 48–72 hr. The protein solution is then determined by the Kjeldahl method or by amino acid analysis, and that value serves as the primary

[83] L. O. Henderson, M. K. Powell, S. J. Smith, W. H. Hannon, G. R. Cooper, and S. M. Marcovina, *Clin. Chem.* **36,** 1911 (1990).

[84] O. H. Lowry, H. J. Rosebrough, A. L. Farr, and R. J. Randall, *J. Biol. Chem.* **193,** 265 (1951).

standard. Because all lipoprotein samples contain EDTA, the same concentration of EDTA is added to the standard protein. The slope (absorbance/protein mass) of the calibration curve is slightly lower in the presence than in the absence of EDTA.

Precautionary Measures against Proteases

The above reports demonstrated that it is feasible to quantitate apoB-48 accurately if the structural integrity of lipoproteins and apolipoproteins is carefully maintained during isolation, dialysis, delipidization, labeling, chromatography, and electrophoresis. At every step of the process there is a large risk of altering a portion of apoB and, therefore, altering the apoB-48/apoB-100 ratio, because of the susceptibility of apoB to autoxidation,[73,75,79] to alkaline pH (because of the presence of labile thiolester linkages),[47–49,85–88] and to proteases.[8–10,89–96] Each of these conditions could cause cleavage of apoB.

The use of serum should be avoided because thrombin is activated during coagulation, and thrombin cleaves apoB.[91,92,94] Even for the collection of plasma, contact with glass should be avoided because it activates kallikrein, which cleaves apoB-100 into apoB-74 and apoB-26.[94,95] The use of polyanion precipitation of apoB-containing lipoproteins is not recomended because polyanions such as dextran sulfate cause contact activation of factor XII and activation of prekallikrein, resulting in an increased concentration of kallikrein and, possibly, other proteases which cleave apoB-100.[93] Any time cleavage of apoB-100 occurs, the ratio of apoB-48/apoB-100 increases. Whether the same protease would cleave apoB-48 at the same site as apoB-100 is not known. Even if both species are cleaved

[85] S. Singh and D. M. Lee, *Biochim. Biophys. Acta* **876,** 460 (1986).

[86] D. M. Lee, *Prog. Lipid Res.* **30,** 245 (1991).

[87] D. M. Lee and S. Singh, *in* "Atherosclerosis IX: Proceedings of the Ninth International Symposium on Atherosclerosis" (O. Stein, S. Eisenberg, and Y. Stein, eds.), p. 233. R. & L. Creative Communications, Tel Aviv, Israel, 1992.

[88] E. Koren, N. Dashti, P. R. Wilson, and D. M. Lee, *Mol. Cell. Biochem.* **124,** 67 (1993).

[89] R. B. Triplett and W. R. Fisher, *J. Lipid Res.* **19,** 478 (1978).

[90] J. C. H. Steele, *Thromb. Res.* **15,** 573 (1979).

[91] D. M. Lee and S. Singh, *Circulation* **68,** II, III-217 (1983).

[92] W. A. Bradley, A. M. Gotto, D. L. Peavy, and S. H. Gianturco, *Arteriosclerosis* (*Dallas*) **3,** 508a (1983).

[93] R. E. Byrne and A. M. Scanu, *J. Lipid Res.* **30,** 109 (1989).

[94] A. D. Cardin, K. R. Witt, J. Chao, H. S. Margolius, V. H. Donaldson, and R. L. Jackson, *J. Biol. Chem.* **259,** 8522 (1984).

[95] D. A. Hardman, A. Gustafson, J. W. Schilling, V. H. Donaldson, and J. P. Kane, *Biochem. Biophys. Res. Commun.* **137,** 832 (1986).

[96] K. V. Krishnaiah and H. Wiegandt, *FEBS Lett.* **40,** 265 (1974).

at the same site, the apoB-48/apoB-100 ratio would change. ε-Aminocaproic acid is an effective inhibitor for plasmin and kallikrein. In the presence of 10 m*M* ε-aminocaproic acid[73] we have never observed the presence of apoB-74 and apoB-26 in freshly isolated VLDL or LDL from normolipidemic subjects. Other frequently used inhibitors include the serine protease inhibitors leupeptin (1 m*M*) and phenylmethylsulfonyl fluoride (1 m*M*). The latter is soluble in lipids and is difficult to remove by dialysis. Taking precautionary measures against bacterial enzymes is a good practice. Antibiotics (e.g., 500 units/ml penicillin G and 50 μg/ml streptomycin sulfate),[73] covering a wide spectrum of both gram-negative and gram-positive bacteria, have been found satisfactory.

Inhibition of Lipid Peroxidation

As mentioned earlier, oxidation of apoB or its lipoproteins causes not only fragmentation and aggregation of apoB,[73,75,79,97] but also a decrease in the amino acid residues critical in dye binding and in iodine binding. In our experience, the best combination of antioxidants is 0.005% (w/v) BHT plus 0.05% (w/v) reduced glutathione (GSH) plus 0.1% (w/v) EDTA added to plasma before the plasma bag is exposed to air.[73,98] A protocol for preparing a 200× alcoholic BHT solution is described below. The BHT should be added drop by drop to plasma with gentle shaking of the plasma. Otherwise, the lipoproteins and proteins in contact with alcohol may precipitate. Because BHT is insoluble in water, it is useless to add BHT stock solution to aqueous buffers: the BHT will form flakes which float to the surface of aqueous solutions. Once BHT is dissolved in plasma, it remains with the lipoproteins and is not dialyzable.[99] Both GSH and EDTA are water-soluble and dialyzable, and they should be added to all solutions in contact with lipoproteins. The GSH is readily oxidized to oxidized glutathione (GSSG) in air, so the air in water should be displaced with N_2 prior to dissolving GSH. Moreover, because N_2 is lighter than O_2, it exchanges with O_2 any time the container is opened, so the opened GSH container should always be flushed with nitrogen or argon. Reduced glutathione is acidic; the pH needs to be adjusted to neutral with dilute NaOH solution. It is convenient to prepare a 100-fold GSH solution, divided in small aliquots and saturated under argon stored at −10°. For monitoring lipid peroxide formation in lipoprotein samples, the presence of GSH and BHT does not

[97] L. G. Fong, S. Parthasarathy, J. L. Witztum, and D. Steinberg, *J. Lipid Res.* **28,** 1466 (1987).
[98] D. M. Lee and P. Alaupovic, *Biochim. Biophys. Acta* **879,** 126 (1986).
[99] D. M. Lee, unpublished observations (1981).

interfere with the thiobarbituric acid (TBA) reagent[100,101]; the antioxidants suppress only the generation of TBA reactive substances.

Protocols for Preparation of Reagents to Prevent Lipid Peroxidation

Alcoholic Butylated Hydroxytoluene Solution. Prepare daily fresh 1% BHT in 75% ethanol. Dissolve 0.1 g BHT in 7.5 ml ethanol. Add 2.5 ml double-distilled water drop by drop to the BHT solution with gentle shaking. Blow argon to the top and cover it to avoid contact with air and keep at room temperature.

Testing for Presence of Peroxides in Diethyl Ether with Potassium Iodide. Pour equal volumes of diethyl ether and 1 *M* freshly prepared KI into a glass-stoppered tube. Shake vigorously, then let stand to allow phase separation. If peroxide is present in ether, the KI develops a yellow tint or a pink color, depending on the amount of peroxides present. If no peroxides are present, the KI should remain colorless.

Purification of Diethyl Ether with Ferrous Sulfate Solution. Prepare a fresh solution of 30% (w/v) $FeSO_4$ in double-distilled water just prior to use. To 500 ml diethyl ether add 100 ml $FeSO_4$ and shake vigorously in a separatory funnel. The peroxides cause the $FeSO_4$ solution to change color from aqua to a greenish or muddy color. Repeat several times with fresh $FeSO_4$ until the last solution remains aqua, then wash with 100 ml distilled water 3–4 times. The diethyl ether is dried overnight with anhydrous $CaCl_2$ (67 g/liter) and redistilled. The fraction with boiling point 34° is to be collected for diethyl ether.[73] (*Note:* Do not let the flask run dry during distillation. An explosion may occur if a trace amount of peroxide is present). If the diethyl ether is not used within 2 days, it should be tested again with KI.

Purification of Chloroform. To 2 liters of chloroform add about 1 liter of double-distilled water. Tighten the cap and shake vigorously. Remove the water phase by siphoning. Repeat several times until the water phase does not show yellowish color. Add anhydrous $CaCl_2$ (67 g/liter) and let stand overnight. Redistill the washed chloroform and collect the fraction with boiling point 60°. Alternatively, purification of chloroform may be achieved by elution through activated alumina (neutral, 80–200 mesh, 2.5 × 20 cm column) or by standing over calcium hydroxide (10 g powder/200 ml, 18 hr).[76]

[100] D. M. Lee, *Biochem. Biophys. Res. Commun.* **95,** 1663 (1980).
[101] K. Kameda-Takemura, C. N. Corder, and D. M. Lee, *Artery* **20,** 189 (1993).

Conclusions

It is still not clear what metabolic necessity prompts a gene to make two proteins. Some speculate that the formation of apoB-48, which lacks an LDL-receptor binding domain, favors a faster removal of chylomicron remnants and, thus, prevents their possible accumulation in the circulation. Although apoB-48 is structurally identical to nearly half of apoB-100, the properties of the two proteins are quite different. In humans, a number of tissues are capable of editing apoB mRNA, yet only intestine is found to synthesize and secrete apoB-48. Thus, the editing mechanism is not fully in use by most tissues. What, then, is the signal commanding the packaging of lipoproteins with apoB-48 and their secretion into circulation? Does secretion parallel gene expression? How do diet and hormones affect the production and secretion of apoB-48?

To investigate the regulation of secretion or removal of apoB-48, quantitation of apoB-48 is imperative. So far, the methods available for quantitation of apoB-48 are indirect methods—all depend on separation of apoB-48 from apoB-100 and simultaneous measurement of both proteins. Separation of the proteins is achieved either by SDS–PAGE or by HPLC. Quantitation is based on (1) dye binding, (2) radioactivity associated with or incorporated into the protein bands, or (3) absorption at 280 or 220 nm of the separated proteins. Basically, SDS–PAGE is an oxidizing system. Apolipoprotein B-48 will be preferentially oxidized, leading to variable underestimation of this apoB species when dye binding is the method of quantitation. Only if the oxidation of samples can be avoided during electrophoresis does this simple method have merit. The HPLC approach is fast but expensive and is not available in every laboratory. Finally, the main source of errors for all quantitative procedures lies in the preparation of samples. It is hoped that by taking all the necessary precautionary measures, results from different laboratories or by different methods may converge.

Acknowledgments

The author would like to express appreciation to Dr. P. Alaupovic for valuable discussion, and to Ms. J. Pilcher for typing the manuscript.

[9] Immunochemical Separation of Apolipoprotein B-48- and B-100-Containing Lipoproteins

By ROSS W. MILNE

Introduction

Apolipoprotein (apo) B exists in two forms, apoB-100 and apoB-48, that have apparent molecular weights of 549,000 and 264,000, respectively.[1] The two species are encoded by the same gene. Apolipoprotein B-48 is the product of a tissue-specific mRNA editing mechanism that converts codon 2153 of the apoB transcript from CAA, specifying glutamine, to the translational stop codon UAA. Apolipoprotein B-48 thus corresponds to the amino-terminal 2152 residues of apoB-100. In humans, apoB mRNA editing is restricted to enterocytes of the small intestine, whereas, in certain other species, such as the rat, apoB-48 is produced in both small intestine and liver.

Apolipoprotein B-100 is necessary for the hepatic assembly of very-low-density lipoproteins (VLDL) and apoB-48 for the formation of chylomicrons in intestinal enterocytes. A large proportion of the triglyceride-rich apoB-100-containing VLDL secreted by the liver are transformed within the plasma compartment to cholesteryl ester-rich low-density lipoproteins (LDL) by the action of lipases and by lipid and apolipoprotein exchanges with other lipoprotein particles. Chylomicrons also undergo metabolic processing in the plasma. Chylomicrons and their remnants are rapidly cleared by the liver, a process mediated by cell surface receptors specific for the apoE that is present on the particles. Because of the rapid clearance, apoB-48 levels are very low in plasma from fasting normal subjects but can be at higher concentrations in patients with hyperlipoproteinemias that are associated with impaired clearance of remnants of triglyceride-rich lipoproteins (TRL). It has been proposed that the postprandial lipoproteins may directly or indirectly promote atherogenesis.[2]

Immunoaffinity chromatography has been used to isolate apoB-48 and apoB-100 TRL in order to determine their respective lipid and apolipopro-

[1] L. Chan, *J. Biol. Chem.* **267,** 25621 (1992).
[2] D. B. Zilversmit, *Circulation* **60,** 473 (1979).

tein compositions,[3,4] abilities to bind to cell surface receptors,[5] and contributions to postprandial lipidema[6,7] and as a purification step in experiments designed to establish the carboxy terminus of apoB-48.[8] The separations have been based on using monoclonal antibodies (MAb) specific for epitopes situated in the carboxy-terminal 52% of apoB-100. As this portion of apoB-100 primary structure is not shared with apoB-48, these mAb, when immobilized, will specifically retain apoB-100-containing particles. Apolipoprotein B-48 lipoproteins can be recovered in the nonretained fraction, and the apoB-100 lipoproteins can be released from the immunoaffinity matrix under conditions that disrupt the antibody–antigen interaction. Several antisera, prepared against synthetic peptides representing the carboxy terminus of apoB-48, have been found to be apoB-48-specific.[9,10] In principle, these antisera could be used to specifically retain lipoproteins containing apoB-48. To date, however, utilization of these antisera for the separation of apoB-48 and apoB-100-containing TRL has not been reported, and it is not clear whether the antisera react with lipoprotein-bound, nondenatured apoB-48, a property that would be essential, if the antisera were to be used as reagents for immunoaffinity chromatography of TRL.

Problems Associated with Immunoaffinity Separation of Apolipoprotein B-48- and B-100-Containing Lipoproteins

We have described in detail our methodology for preparation of immunoaffinity columns and their use in the isolation of lipoprotein subfractions.[11] The remainder of this chapter is therefore devoted to a discussion of problems particular to the preparation of apoB-48- and apoB-100-containing TRL.

[3] R. W. Milne, P. K. Weech, L. Blanchette, J. Davignon, P. I. Alaupovic, and Y. L. Marcel, *J. Clin. Invest.* **73,** 816 (1984).

[4] F. R. Tercé, R. W. Milne, P. K. Weech, J. Davignon, and Y. L. Marcel, *Arteriosclerosis (Dallas)* **5,** 201 (1985).

[5] D. Y. Hui, T. L. Innerarity, R. W. Milne, Y. L. Marcel, and R. W. Mahley, *J. Biol. Chem.* **259,** 15060 (1984).

[6] B. O. Schneeman, L. Kotite, K. M. Todd, and R. J. Havel, *Proc. Natl. Acad. Sci. U.S.A.* **90,** 2069 (1993).

[7] J. S. Cohn, E. J. Johnson, J. S. Millar, S. D. Cohn, R. W. Milne, Y. L. Marcel, R. M. Russel, and E. J. Schaefer, *J. Lipid Res.* **34,** 2033 (1993).

[8] T. L. Innerarity, S. G. Young, K. S. Poksay, R. W. Mahley, R. S. Smith, R. W. Milne, Y. L. Marcel, and K. H. Weisgraber, *J. Clin. Invest.* **80,** 1794 (1987).

[9] Z. Yao, B. D. Blackhart, D. F. Johnson, S. M. Taylor, K. W. Haubold, and B. J. McCarthy, *J. Biol. Chem.* **267,** 1175 (1992).

[10] A. S. Peel, A. Zampelas, C. M. Williams, and B. J. Gould, *Clin. Sci.* **85,** 521 (1993).

[11] R. W. Milne, P. K. Weech, and Y. L. Marcel, *in* "Lipoprotein Analysis; A Practical Approach" (C. A. Converse and E. R. Skinner, eds.), p. 61. Oxford Univ. Press, Oxford, 1992.

Because of their complex structure, apoB-100- and apoB-48-containing TRL have a number of physical and immunochemical characteristics that have practical implications in the application of immunoaffinity chromatography to their separation. Apolipoprotein B-containing lipoproteins are heterogeneous in terms of size, hydrated density, surface charge, and lipid and apoliproprotein composition. The physical and chemical heterogeneity influences the conformation of apoB and this is, in turn, manifested in immunochemical heterogeneity of the apoB. We have observed that when a single MAb, 5E11, specific for an epitope situated between apoB-100 residues 3441–3569, was used as an immunoadsorbent, it retained only a portion of the apoB-100-containing lipoproteins present in the VLDL fraction of plasma from a type III dyslipoproteinemic subject.[3] Repassage of the sample on the same column did not eliminate the residual apoB-100-containing TRL. A second MAb, 4G3, specific for an epitope between apoB-100 residues 2980–3084 was even less efficient. However, when the sample was passed sequentially over the 5E11 and 4G3 immunoaffinity columns, virtually all of the apoB-100-containing lipoproteins could be retained.

Thus, there appeared to be heterogeneity among the lipoproteins in the expression of the two epitopes. It is therefore recommended that individual anti-apoB-100 MAb or combinations of MAb be evaluated for the capacity to immunoprecipitate radiolabeled apoB-100-containing TRL quantitatively[11] to assure their suitability as reagents for the preparation of immunoadsorbents. The capacity of the immunoadsorbents should be established by application of increasing amounts of TRL and analysis of the retained and nonretained fractions by immunoassay using anti-apoB-100 and anti-apoB-100/B-48 antibodies and by sodium dodecyl sulfate (SDS)–polyacrylamide gel electrophoresis.[3]

We have observed that recovery of the combined apoB-48 (nonretained) and apoB-100 (retained) fractions is routinely about 70% of the lipoproteins applied to the column. Repassing of the two fractions on the immunoaffinity column indicated that the losses are predominantly restricted to the retained fraction. The poor recovery of the retained fraction is only partially rectified by using more drastic elution conditions. We have proposed[11] that lipoproteins in the size range of intermediate-density lipoproteins and LDL are of a diameter that is close to the exclusion limit of the affinity gel, and, as a consequence, particles of this size tend to get physically trapped within the matrix of the gel. We have found that recovery of the retained fraction could be improved if the flow of the column is reversed during the elution step. On the other hand, we did not observe an appreciable improvement when Sepharose 4B was replaced by Sepharose 2B as the matrix of the affinity column. Higher recoveries may also be achieved by using batch rather than column immunoadsorption. Recoveries of 85 to 95% have been reported following batch immunoadsorption of postprandial TRL with immobilized anti-apoB-100 MAb.[7] In this case, the recovery may have also been improved through the use of the chaotropic agent sodium thiocyanate to elute the retained apoB-100-containing lipoproteins.

Another potential problem in the isolation of apoB-48 and apoB-100 TRL is maintaining the integrity of the particles. We have reported the physical, chemical, and functional characterization of the respective fractions isolated from the plasma of patients with type III dyslipoproteinemia. The particles in both retained and nonretained fractions appeared to be intact by electron microscopy[3] and showed normal binding to cell surface receptors.[5] For these experiments, the retained apoB-100 containing lipoproteins were eluted in a low pH/high salt buffer (0.1 *M* citric acid/1 *M* NaCl). We have not evaluated retained fractions eluted with chaotropic agents. We have, on occasion, observed a small secondary peak that absorbed at 280 nm and that eluted with phosphate-buffered saline after the primary nonretained peak containing the apoB-48 lipoproteins. This peak contained immunoreactive apoE as measured by radioimmunoassay, although no Coomassie blue-staining protein was found when the peak was concentrated and subjected to SDS–polyacrylamide gel electrophoresis nor was apoE detected after Western blotting. Although we have not further characterized the components of this fraction, it may represent apolipoproteins or fragments that have detached from the lipoproteins during washing of the columns. Even if this were the case, however, given the low protein concentration, it is unlikely that the apolipoprotein loss from the particles, represented by this minor peak, would significantly change the apolipoprotein composition of the separated subpopulations.

Conclusions

Immunoaffinity separation of apoB-100 and apoB-48 TRL has been used to determine the lipid and apolipoprotein compositions of the respective fractions[3,4] and to demonstrate that, although both apoB-48- and apoB-100-containing TRL could bind to cell surface receptors, in the case of the apoB-48-containing fraction, binding was mediated by apoE, whereas, in the case of apoB-100 fraction, both apoE and apoB contributed to the binding.[5] The technique has been applied to the study of postprandial lipoprotein metabolism, and it has been shown that apoB-100-containing lipoproteins make a major contribution to postprandial lipemia.[6,7] These studies have further suggested that retinyl esters may not be a good marker of intestinally derived lipoproteins.[7] Aside from the analysis of postprandial lipoproteins, the immunoaffinity isolation of apoB-48- and apoB-100-containing TRL could be potentially used to study the metabolism of lipid-soluble vitamins and serve as the basis for development of an immunoassay for measuring apoB-48.

[10] Quantitation of Apolipoprotein E

By Elaine S. Krul and Thomas G. Cole

Introduction

Human apolipoprotein E (apoE) is a single-chain protein of M_r of 34,200 which is found to be associated with most lipoprotein classes in plasma. Apolipoprotein E is evolutionarily related to apoA-I, apoA-II, apoC-I, apoC-II, and apoC-III in that the genes have very similar structures, each consisting of four exons and three introns. Apolipoprotein E is unique among the apolipoproteins in that it is synthesized and secreted by a wide variety of tissues in the body. Despite this, the liver has been shown to be the primary source for the apoE in plasma.[1] The gene for apoE resides on the long arm of chromosome 19 (19q13.2)[2,3] and encodes for one of three common allelic forms of apoE: apoE3 (most common), apoE2 ($Arg^{158} \rightarrow$ Cys), and apoE4 ($Cys^{112} \rightarrow$ Arg). Davignon *et al.*[4] have estimated that 60% of the variation in plasma cholesterol concentrations is genetically determined and that apoE heterogeneity accounts for roughly 14% of that variation.

Apolipoprotein E is the best characterized apolipoprotein in terms of structure and function (for a review, see Weisgraber[5]). It is initially synthesized as a prepeptide of 317 amino acids, and intracellular, cotranslational cleavage of an 18-residue signal peptide gives rise to the mature 299-amino acid protein. O-Linked glycosylation of apoE occurs as a late posttranslational event at Thr^{194}, and the importance of this glycosylation step in apoE function is not understood. However, as a result of the glycosylation and

[1] W. März, B. Peschke, V. Ruzicka, R. Siekmeier, W. Gross, W. Schoeppe, and E. Scheuermann, *Transplantation* **55,** 284 (1993).

[2] B. Olaisen, P. Teisberg, and T. Gedde-Dahl, Jr., *Hum. Genet.* **62,** 233 (1982).

[3] H. K. Das, J. McPherson, G. A. P. Bruns, S. K. Karathanasis, and J. L. Breslow, *J. Biol. Chem.* **260,** 6240 (1985).

[4] J. Davignon, R. E. Gregg, and C. F. Sing, *Arteriosclerosis* (*Dallas*) **8,** 1 (1988).

[5] K. H. Weisgraber, *Adv. Protein Chem.* **45,** 249 (1994).

subsequent sialylation, plasma apoE exhibits chemical heterogeneity in addition to the genetic polymorphism.

Functions of Apolipoprotein E

The functional importance of apoE in lipoprotein metabolism was realized soon after its initial description in the early 1970s. Havel and Kane[6] described the presence of elevated amounts of apoE in the cholesterol ester-rich lipoproteins, designated as β-VLDL, which accumulated in the plasma of subjects with the lipoprotein disorder type III hyperlipoproteinemia or familial dysbetalipoproteinemia (FD). At about the same time, Shore and Shore[7] observed that similar apoE-containing β-migrating lipoproteins accumulated in the plasma of rabbits fed high-cholesterol diets. These and many subsequent studies suggested that apoE plays an important role in the transport and metabolism of cholesterol esters and/or unesterified cholesterol. In the early 1980s when it was observed that apoE was synthesized and secreted by a wide variety of tissues and organs of the body,[8] it was speculated that apoE may have an important role in the transport of cholesterol between tissues or cells. In fact, some of the proposed esoteric functions of apoE may be related to its role in the local transport and metabolism of cholesterol; for example, at least part of the role of apoE in nerve regeneration is its ability to scavenge cholesterol-enriched lipid from degenerating nerve cells and to make the cholesterol available to growing cells via its ability to bind to lipoprotein receptors.[9]

Ligand for Lipoprotein Receptors

Apolipoprotein E is a key mediator in the clearance of triglyceride-rich lipoproteins from plasma. These include chylomicrons and their remnants[10] and very-low-density lipoproteins (VLDL) from hypertriglyceridemic subjects.[11] Apolipoprotein E, as well as apoB, mediates the binding and uptake of lipoproteins by the low-density lipoprotein (LDL) or apoB/E receptor, and at least some of the apo-E mediated clearance of the triglyceride-rich lipoproteins from plasma can be accounted for by this pathway. The apoE-

[6] R. J. Havel and J. P. Kane, *Proc. Natl. Acad. Sci. U.S.A.* **70,** 2015 (1973).

[7] B. Shore and V. Shore, *Biochem. Biophys. Res. Commun.* **58,** 1 (1974).

[8] D. M. Driscoll and G. S. Getz, *J. Lipid Res.* **25,** 1368 (1984).

[9] G. E. Handelmann, J. K. Boyles, K. H. Weisgraber, R. W. Mahley, and R. E. Pitas, *J. Lipid Res.* **33,** 1677 (1992).

[10] Y. Hui, T. L. Innerarity, and R. W. Mahley, *J. Biol. Chem.* **256,** 5646 (1981).

[11] S. H. Gianturco, A. M. Gotto, Jr., S.-L. C. Hwang, J. B. Karlin, A. H. Y. Lin, S. C. Prasad, and W. A. Bradley, *J. Biol. Chem.* **258,** 4526.

containing lipoproteins bind with significantly higher affinity to LDL receptors than do apoB-containing lipoproteins. The higher affinity of apoE-containing lipoproteins is due to the interaction of four apoE molecules with four LDL receptors.[12] In addition, the LDL receptor-related protein/activated α_2-macroglobulin receptor (LRP/α_2-MR) can bind apoE-enriched β-VLDL, and it has been postulated that this receptor may be involved in apoE-mediated chylomicron clearance.[13] A third receptor, the VLDL receptor, which binds specifically to apoE-containing lipoproteins, has been identified recently in several tissues including heart, muscle, brain, and adipose tissue, but not in liver. The VLDL receptors may be responsible for delivery of lipoprotein-derived fatty acids to these tissues.[14]

The genetic polymorphism of apoE is important with regard to the ability of apoE to mediate lipoprotein binding to the LDL receptor. Apolipoprotein E2 binds to LDL receptors with approximately 1% of the ability of apoE3 or apoE4. Binding to LRP/α_2-MR or VLDL receptors is unaffected by the allelic form of apoE.[15,16]

Role in Reverse Cholesterol Transport

High-density lipoproteins (HDL) are thought to play an important and antiatherogenic role in transporting cholesterol from peripheral tissues to the liver.[17] In many species lacking cholesteryl ester transfer protein (CETP), such as dogs, rats, and mice, and in CETP-deficient human subjects,[18] HDL with apoE or HDL_1 is a prominent lipoprotein carrying the majority of the plasma cholesterol. The HDL_1 pool expands after animals are fed a high fat, high cholesterol diet, and this fraction is referred to as HDL_c. This lipoprotein class, by virtue of the high amounts of apoE associated with these particles, has very high affinity for LDL receptors and provides a pathway for reverse cholesterol transport.[19]

In species that possess CETP, two pathways of reverse cholesterol trans-

[12] T. Funahashi, S. Yokoyama, and A. Yamamoto, *J. Biochem.* (*Tokyo*) **105,** 582 (1989).

[13] R. W. Mahley and M. M. Hussain, *Curr. Opin. Lipidol.* **2,** 170 (1991).

[14] T. Yamamoto, S. Takahashi, J. Sakai, and Y. Kawarabayasi, *Trends Cardiovasc. Med.* **3,** 144 (1993).

[15] U. Beisiegel, W. Weber, G. Ihrke, J. Herz, and K. K. Stanley, *Nature* (*London*) **341,** 162 (1989).

[16] K. Oida, S. Takahashi, J. Suzuki, M. Kohno, S. Miyabo, M. Okubo, T. Murase, T. Yamamoto, and T. Nakai, *Circulation* **90,** I-2 (1994).

[17] A. R. Tall, *J. Clin. Invest.* **86,** 379 (1990).

[18] S. Yamashita, D. L. Sprecher, N. Sakai, Y. Matsuzawa, S. Tarui, and D. Y. Hui, *J. Clin. Invest.* **86,** 688 (1990).

[19] R. E. Pitas, T. L. Innerarity, K. S. Arnold, and R. W. Mahley, *Proc. Natl. Acad. Sci. U.S.A.* **76,** 2311 (1979).

port are possible, a pathway via HDL as described in CETP-deficient species and another pathway via LDL. The transfer protein transfers cholesterol esters from HDL to VLDL, intermediate-density lipoproteins (IDL), and LDL in exchange for triglyceride, so CETP-containing species such as humans, primates, and rabbits transport the bulk of plasma cholesterol in LDL, leaving the burden of cholesterol removal on the LDL pathway. The role of apoE in this pathway may involve a function of enhancing the transfer of HDL cholesterol to VLDL by increasing the affinity of VLDL for CETP.[20] Reverse cholesterol transport via the HDL pathway can still operate in CETP-containing species as suggested by the following studies. Cole *et al.*,[21] observed an increase in HDL-apoE in human subjects with familial hypercholesterolemia fed a high-fat, high-cholesterol diet in the absence of an increase in total plasma apoE concentrations, indicating that even in the presence of CETP, apoE-enriched HDL_1 or HDL_c particles can accumulate after cholesterol feeding. Thus, the apoE present on the HDL_c particles which binds to LDL receptors can function in a direct pathway for reverse cholesterol transport. Apolipoprotein E may play another role in the HDL pathway as suggested in studies by Leblond and Marcel[22] who demonstrated that apoE has a function in the selective uptake of HDL cholesterol esters by human hepatoma Hep G2 cells.

Activation of Plasma Enzymes

Apolipoprotein E can modulate the activity of lipoprotein lipase[23] and can activate hepatic lipase[24] and lecithin–cholesterol acyltransferase (LCAT; phosphatidylcholine–sterol *O*-acyltransferase).[25] Apolipoprotein E2 is less effective at increasing lipoprotein lipase and hepatic lipase activity compared to apoE3, whereas all polymorphic forms of apoE are equally stimulatory for LCAT.

Other Functions

Apolipoprotein E may function in immunoregulation,[26] nerve regeneration,[9] modulation of steroidogenesis,[27] in several aspects of reproductive

[20] M. Kinoshita, H. Arai, M. Fukasawa, T. Watanabe, K. Tsukamoto, K. Inoue, K. Kurokawa, and T. Teramoto, *J. Lipid Res.* **34,** 261 (1993).

[21] T. G. Cole, B. Pfleger, O. Hitchens, and G. Schonfeld, *Metabolism* **34,** 486 (1985).

[22] L. Leblond and Y. Marcel, *J. Biol. Chem.* **268,** 1670 (1993).

[23] C. Ehnholm, R. W. Mahley, D. A. Chappell, K. H. Weisgraber, E. Ludwig, and J. L. Witztum, *Proc. Natl. Acad. Sci. U.S.A.* **81,** 5566 (1984).

[24] T. Thuren, K. H. Weisgraber, P. Sisson, and M. Waite, *Biochemistry* **31,** 2332 (1992).

[25] A. Steinmetz, H. Kaffarnik, and G. Utermann, *Eur. J. Biochem.* **152,** 747 (1985).

[26] M. G. Pepe and L. K. Curtiss, *J. Immunol.* **136,** 3716 (1986).

[27] M. E. Reyland and D. L. Williams, *J. Biol. Chem.* **266,** 21099 (1991).

tissue function,[28,29] and as a pathological chaperone protein that induces β-sheet conformation in amyloidogenic peptides in a variety of systemic and cerebral amyloidoses.[30] Apolipoprotein E4 may play a significant role in the latter function, as this allele of apoE is associated with Alzheimer's disease and, compared to apoE3, binds to synthetic β-amyloid peptide with much higher affinity.[31] In addition, having the apoE2 allele appears to confer protection against Alzheimer's disease.[32]

Structure of Apolipoprotein E

The apoE molecule consists of two independent structural domains (amino acids 20–165 and 225–299) connected by a central, protease-susceptible region (165–210).[5] The amino-terminal domain is extremely stable, containing two highly conserved sequences, one being the LDL receptor binding region of apoE (residues 136–150) and another the region between residues 29–61. The specific function of the latter region has not been determined. The carboxyl-terminal domain mediates the binding of apoE to lipid by means of three amphipathic helical domains.

The 22-kDa thrombin-derived fragment of apoE (residues 1–191) was used in the first crystallographic studies of the protein.[33] In solution, the structure contains five helices, four of which are arranged in an antiparallel four-helix bundle. Hydrophobic side chains are sequestered in the interior of the bundle, and the hydrophilic side chains are exposed to solvent. The receptor binding region is located in helix 4.

Apolipoprotein E binding to LDL receptors only occurs when apoE is lipid-associated. Under these conditions, the four-helix bundle structure of apoE "opens" up without disrupting the helices, allowing the hydrophobic inner faces of the bundle to interact with lipid.[5]

[28] M. Nicosia, W. H. Moger, C. A. Dyer, M. M. Prack, and D. L. Williams, *Mol. Endocrinol.* **6,** 978 (1992).

[29] L. M. Olson, X. Zhou, and J. R. Schreiber, *Biol. Reprod.* **50,** 535 (1994).

[30] T. Wisniewski and B. Frangione, *Neurosci. Lett.* **135,** 235 (1992).

[31] W. J. Strittmatter, K. H. Weisgraber, D. Y. Huang, L.-M. Dong, G. S. Slavesen, M. Pericak-Vance, D. Schmechel, A. M. Saunders, D. Goldgaber, and A. D. Roses, *Proc. Natl. Acad. Sci. U.S.A.* **90,** 8098 (1993).

[32] C. Talbot, C. Lendon, N. Craddock, S. Shears, J. C. Morris, and A. Goate, *Lancet* **343,** 1432 (1994).

[33] C. Wilson, M. R. Wardell, K. H. Weisgraber, R. W. Mahley, and D. A. Agard, *Science* **252,** 1817 (1991).

Heterogeneity of Apolipoprotein E

Genetic Polymorphism

The three common allelic forms of human apoE (apoE3, apoE2, and apoE4) can be distinguished by isoelectric focusing. Apolipoprotein E2 is the most acidic, having a p*I* of approximately 5.7. Apolipoproteins E3 and E4 differ from apoE2 by one and two positive charge units due to the presence of one and two arginine residues, respectively, and migrate with correspondingly higher p*I* values. The average apoE allelic frequencies are as follows: $\varepsilon 2$, ~0.073; $\varepsilon 3$, ~0.783; and $\varepsilon 4$, ~0.143. Six common phenotypes are therefore present in the general population: E2/E2, E2/E3, E2/E4, E3/E3, E3/E4, and E4/E4. It should be noted that other, more rare variants of apoE have been identified that migrate in isoelectric focusing gels in the positions of the common apoE alleles. These apoE variants may or may not have significant LDL receptor binding activity; for example, $Arg^{136} \rightarrow$ Ser and $Arg^{228} \rightarrow$ Cys both focus at the E2 position, but only the former has defective receptor binding activity.[5]

Apolipoprotein E2, having cysteine residues at positions 112 and 158, demonstrates a very low capacity to bind to LDL receptors compared to apoE3 or apoE4. The E2/E2 phenotype is associated with type III hyperlipoproteinemia, a condition characterized by the accumulation of cholesterol-rich VLDL particles having β mobility (β-VLDL) on agarose electrophoresis. However, not all individuals with the E2/E2 phenotype develop hyperlipoproteinemia. It has been demonstrated that an additional factor in addition to the homozygous E2/E2 state has to be present for the disorder to be expressed (recessive transmission with low penetrance). The receptor-binding activity of the E2 ($Arg^{158} \rightarrow$ Cys) variant can be influenced by the lipoprotein lipid composition.[34] This is not the case for some of the rare variants of apoE whose receptor binding activity is permanently defective.[5]

The three major isoforms of apoE significantly influence plasma cholesterol concentrations.[4] Apolipoprotein E4 is associated with the highest plasma cholesterol levels, apoE2 with the lowest, and apoE3 with intermediate levels. Knowledge of the structures of the three forms of apoE allows some interpretation of this observation. In the presence of apoE2, remnant lipoproteins are removed from the plasma at a slower than normal rate, and the conversion of VLDL to LDL is retarded. As a consequence, LDL

[34] T. L. Innerarity, D. Y. Hui, T. P. Bersot, and R. W. Mahley, *in* "Lipoprotein Deficiency Syndromes" (A. Angel and J. Frohlich, eds.), p. 273. Plenum, New York, 1986.

receptors are upregulated, leading to a further lowering of plasma LDL concentrations. Even E2/E2 individuals with normal cholesterol concentrations accumulate β-VLDL and have elevated plasma apoE concentrations compared to E3/E3 individuals.

Elevated cholesterol concentrations in apoE4 compared to apoE3 individuals are largely due to the preferential association of apoE4 with the triglyceride-rich lipoproteins compared to apoE3. this is due to two factors: (a) the presence of an arginine or positive charge at residue 112 alters salt bridge interactions within apoE, affecting its association and distribution among the lipoprotein classes, and (b) the lack of a cysteine in apoE precludes the formation of homo- or heterodimers (with apoA-II) which preferentially associate with HDL.[5] The increased association of apoE4 with triglyceride-rich remnant particles, then, leads to their efficient removal from plasma and a consequent downregulation of LDL receptors. The end result is an elevation of plasma LDL concentrations and a concomitant decrease in plasma apoE concentrations compared to the apoE3 phenotype.[35] Also, Kesäniemi *et al.*[36] demonstrated that the apoE phenotype may influence intestinal cholesterol absorption directly in the order apoE2 < apoE3 < apoE4.

For apoE quantitation, it is important that the antibodies used in immunodetection bind to all isoforms of apoE with equal affinity. Strategies to circumvent this problem are to immunize with a mixture of the common apoE isoforms or use monoclonal antibodies (MAb) with epitopes beyond the critical amino acid residue differences. Although a monoclonal antibody specific to apoE2 ($Art^{158} \rightarrow Cys$) has been generated, this required immunizations with specific peptide sequences of apoE.[37] Anti-apoE monoclonal antibodies generated by two independent groups showed that most of the these MAb reacted similarly with the three common apoE isoforms.[38,39]

Chemical Heterogeneity

Between 80 and 85% of human plasma apoE is in the asialo form, whereas the remainder is present as monosialo and disialo isoforms.[40] Newly

[35] M. Smit, P. DeKnijff, M. Rosseneu, J. Bury, E. Klasen, R. Frants, and L. Havekes, *Hum. Genet.* **80,** 287 (1988).

[36] Y. A. Kesäniemi, C. Ehnholm, and T. A. Miettinen, *J. Clin. Invest.* **80,** 578 (1987).

[37] K. Gerritse, P. de Knijff, G. van Ierssel, L. M. Havekes, R. R. Frants, M. M. Schellekens, N. D. Zegers, E. Claassen, and W. J. A. Boersma, *J. Lipid Res.* **33,** 273 (1992).

[38] R. W. Milne, P. Douste-Blazy, Y. L. Marcel, and L. Retegui, *J. Clin. Invest.* **68,** 111 (1981).

[39] E. S. Krul, M. J. Tikkanen, and G. Schonfeld, *J. Lipid Res.* **29,** 1309 (1988).

[40] V. I. Zannis and J. L. Breslow, *Biochemistry* **20,** 1033 (1981).

secreted apoE, however, is highly sialylated,[41] as is cerebrospinal fluid apoE.[42] The effects of sialylation on apoE function are not well understood, but the sialic acid residues may prevent immediate cellular reuptake of newly secreted apoE. Different degrees of sialylation of apoE should be kept in mind when assaying culture fluids, cerebrospinal fluids, or plasma apoE.

Conformational Heterogeneity

Several lines of evidence indicate that apoE function is modulated by its conformation on lipoprotein surfaces. (1) Apolipoprotein E does not bind to the LDL receptor in a lipid-free state but only when complexed with phospholipid.[43] (2) Conformational changes induced in apoE2 by addition of a positive charge at residue 158 and cleavage of the carboxyl-terminal portion of apoE results in an improvement of binding to the cellular LDL receptor.[44] (3) Using enzymatic probes, Bradley *et al.*[45] demonstrated the presence of accessible and inaccessible forms of apoE on hypertriglyceridemic VLDL, and showed that the accessible form was responsible for apoE-mediated uptake of lipoproteins by cells. (4) Krul *et al.*[39] demonstrated that increased accessibility of apoE on the surface of lipoprotein particles, as determined by MAb binding, was associated with increased lipoprotein receptor recognition. (5) Rubinstein *et al.*[46] demonstrated that apoE in VLDL existed in an exchangeable and nonexchangeable pool; however, after *in vitro* incubation of VLDL with lipoprotein lipase, conformational changes occurred which increased the proportion of exchangeable apoE. (6) The uptake of β-VLDL by the LRP/α_2-MR requires incubation with exogenous apoE to achieve binding, suggesting that the large amount of apoE already present on the β-VLDL is inaccessible to the receptor.[47] The latter hypothesis is supported by the observation that β-VLDL apoE is relatively resistant to enzymatic cleavage (S. R. Thorpe and A. Daugherty, personal communication, 1989).

[41] V. I. Zannis, J. L. Breslow, T. R. SanGiacomo, D. P. Aden, and B. B. Knowles, *Biochemistry* **20,** 7089 (1981).

[42] R. E. Pitas, J. K. Boyles, S. H. Lee, D. Y. Hui, and K. H. Weisgraber, *J. Biol. Chem.* **262,** 14352 (1987).

[43] T. L. Innerarity, R. E. Pitas, and R. W. Mahley, *J. Biol. Chem.* **254,** 4186 (1979).

[44] T. L. Innerarity, K. H. Weisgraber, K. S. Arnold, S. C. Rall, Jr., and R. W. Mahley, *J. Biol. Chem.* **259,** 7261 (1984).

[45] W. A. Bradley, S.-L. C. Hwang, J. B. Karlin, A. H. Y. Lin, S. C. Prasad, A. M. Gotto, Jr., and S. H. Gianturco, *J. Biol. Chem.* **259,** 14728 (1984).

[46] A. Rubinstein, J. C. Gibson, H. N. Ginsberg, and W. V. Brown, *Biochim. Biophys. Acta* **879,** 355 (1986).

[47] R. C. Kowal, J. Herz, J. L. Goldstein, V. Esser, and M. S. Brown, *Proc. Natl. Acad. Sci. U.S.A.* **86,** 5810 (1989).

Differences in apoE conformation on the surface of lipoprotein particles probably occur as a result of different degrees of apoE self-association,[48,49] association with other apolipoproteins,[50,51] and differences in lipoprotein lipid composition.[34] These factors must be kept in mind when quantitating apoE, and methods should be employed to fully expose all apoE epitopes in samples used in immunoassay. Because apoE readily forms tetramers in solution,[48] efforts must be taken to ensure that all epitopes of pure apoE used as standard in the immunoassay are available for binding to antibody. Furthermore, when using cell homogenates in apoE immunoassays, it should be noted that apoE can exist in multimeric form in such samples (E. S. Krul, unpublished observation, 1992).

Apolipoprotein E Quantitation

Clinical Significance

Apolipoprotein E deficiency is associated with type III hyperlipoproteinemia, tuberoeruptive and palmar cutaneous xanthomas, and premature atherosclerotic vascular disease.[52,53] Individuals with this rare genetic defect primarily suffer from a delayed clearance of triglyceride-rich intestinal remnants from plasma as a result of the lack of plasma apoE protein. Animal models of this condition, mice in which the apoE gene has been mutated to a null allele through homologous recombination (apoE knockout mice), develop spontaneous hypercholesterolemia and atherosclerosis.[54,55] These human and animal observations point to a critical role of apoE in maintaining normal cholesterol metabolism and in preventing atherosclerosis. This was convincingly demonstrated by Yamada *et al.*,[56] who showed that apoE administered intravenously to Watanabe heritable hyperlipidemic rabbits (WHHL) for 8.5 months reduced atherosclerotic lesion areas by 50%. The antiatherogenic role of apoE may be partly due to its

[48] S. Yokoyama, Y. Kawai, S. Tajima, and A. Yamamoto, *J. Biol. Chem.* **260,** 16375 (1985).
[49] A. D. Dergunov and M. Rosseneu, *Biol. Chem. Hoppe-Seyler* **375,** 485 (1994).
[50] Y. Ishikawa, C. J. Fielding, and P. E. Fielding, *J. Biol. Chem.* **263,** 2744 (1988).
[51] W. Windler and R. J. Havel, *J. Lipid Res.* **26,** 556 (1985).
[52] G. Ghiselli, E. J. Schaefer, P. Gascon, and H. B. Brewer, Jr., *Science* **214,** 1239 (1981).
[53] M. S. Meng, R. E. Gregg, E. J. Schaefer, J. M. Hoeg, and H. B. Brewer, Jr., *J. Lipid Res.* **24,** 803 (1983).
[54] S. H. Zhang, R. L. Reddick, J. A. Piedrahita, and N. Maeda, *Science* **258,** 468 (1992).
[55] A. S. Plump, J. D. Smith, T. Hayek, K. Aalto-Setala, A. Walsh, J. G. Verstuyft, E. M. Rubin, and J. L. Breslow, *Cell (Cambridge, Mass.)* **71,** 343 (1992).
[56] N. Yamada, I. Inoue, M. Kawamura, K. Harada, Y. Watanabe, H. Shimano, T. Gotoda, M. Shimada, K. Kohzaki, T. Tsukada, M. Shiomi, Y. Watanabe, and Y. Yazaki, *J. Clin. Invest.* **89,** 706 (1992).

ability to accelerate the plasma clearance of cholesterol, as shown in studies of WHHL or cholesterol-fed New Zealand White rabbits injected intravenously with apoE.[57,58] Mutant forms of apoE that are defective in binding to LDL receptors may pose an atherogenic risk comparable to apoE deficiency when a high-cholesterol, high-fat diet is consumed. Such a diet fed to transgenic mice expressing high levels of $apoE3_{Leiden}$, a defective binding mutant of apoE, developed hypercholesterolemia, high plasma levels of the mutant apoE, and atherosclerosis.[59] Mice overexpressing normal human apoE, in contrast, are resistant to diet-induced hypercholesterolemia.[60] The antiatherogenic role of apoE may involve more than its role in plasma cholesterol clearance as cells in atherosclerotic lesions have been shown to synthesize apoE,[61] suggesting a role in local cholesterol and/or cellular metabolism.

Measurement of total plasma apoE is not predictive of coronary heart disease despite its documented antiatherogenic role described above. There may, however, be clinical utility in measurements of ratios of apoE/apoB or in the distribution of apoE among lipoprotein classes (see below). Commercial assay kits and reagents make the establishment of apoE immunoassays readily available to most laboratories.

Quantitating both apoE and apoB can provide a simple means to screen subjects for type III hyperlipoproteinemia without having to perform the more sophisticated methods of ultracentrifugation (to isolate VLDL cholesterol), agarose electrophoresis (to confirm the presence of β-VLDL), and apoE phenotyping or genotyping.[62] März *et al.*[62] demonstrated that the determination of apoE/apoB ratios in subjects discriminated between type III and other types of hyperlipoproteinemia with 88% specificity. For many laboratories not equipped to perform the assays described above, the availability of automated apoB and apoE immunoassays can permit the apoE/apoB ratio test as a first-line screening parameter for identifying those patients in need of additional testing.

[57] N. Yamada, H. Shimano, H. Mokuno, S. Ishibashi, T. Gotohda, M. Kawakami, Y. Watanabe, Y. Akanuma, T. Murase, and F. Takaku, *Proc. Natl. Acad. Sci. U.S.A.* **86,** 665 (1989).

[58] R. W. Mahley, K. H. Weisgraber, M. M. Hussain, B. Greenman, M. Fisher, T. Vogel, and M. Gorecki, *J. Clin. Invest.* **83,** 2125 (1989).

[59] B. J. van Vlijmen, A. M. van den Maagdenberg, M. J. Gijbels, H. van der Boom, H. HogenEsch, R. R. Frants, M. H. Hofker, and L. M. Havekes, *J. Clin. Invest.* **93,** 1403 (1994).

[60] H. Shimano, N. Yamada, M. Katsuki, M. Shimada, T. Gotoda, K. Harada, T. Murase, C. Fukazawa, F. Takaku, and Y. Yazaki, *Proc. Natl. Acad. Sci. U.S.A.* **89,** 1750 (1992).

[61] M. E. Rosenfeld, S. Butler, V. A. Ord, B. A. Lipton, C. A. Dyer, L. K. Curtiss, W. Palinski, and J. L. Witztum, *Arterioscler. Thromb.* **13,** 1382 (1993).

[62] W. März, G. Feussner, R. Siekmeier, B. Donnerhak, L. Schaaf, V. Ruzicka, and W. Gross, *Eur. J. Clin. Chem. Clin. Biochem.* **31,** 743 (1993).

Bittolo Bon *et al.*[63] observed that normolipidemic survivors of myocardial infarctions had lower HDL-apoE/apoAI ratios compared to controls. The data support the concept that subjects with atherosclerosis can be characterized by normal lipid concentrations and that other factors, such as distributions of lipoproteins, may play a role in atherogenesis. The clinical utility of measuring HDL-apoE/total apoE is underscored by the finding that gemfibrozil, which was shown to be efficacious in the prevention of coronary heart disease in the Helsinki Heart Study,[64] increases the proportion of plasma apoE that is associated with HDL (HDL-apoE).[65]

Rifai *et al.*[66] initially reported that changes in cerebrospinal fluid (CSF) apoE/serum apoE ratios permitted discrimination between multiple sclerosis patients in remission (higher ratios) and those in exacerbation (lower ratios). Combining the apoE index (ratio of CSF apoE/albumin : serum apoE/albumin) with the conventional immunoglobulin G (IgG) index allowed for maximum discrimination between controls and patients undergoing exacerbation. In a subsequent report measuring CSF apoE alone, Carlsson *et al.*[67] reported that 60% of subjects with inflammatory central nervous system (CNS) disease had apoE concentrations two standard deviations above the mean of the control group. These studies indicate that the determination of CSF apoE can be a useful, additional parameter in predicting the course and prognosis of inflammatory CNS disease.

Research Applications

Much of what is known about apoE metabolism has been derived from studies employing methods to measure apoE concentations in human and animal plasma and plasma isolates, biological fluids, cell homogenates, and culture media. Measurement of apoE in this context continues to be an important research tool. Clinical trials will prove to be the bridge between research and clinical utility of apoE quantitation. For example, administration of the 3-hydroxy-3-methylglutaryl-coenzyme A (HMG-CoA) reductase inhibitor pravastatin to non-insulin-dependent diabetics reduced their

[63] G. Bittolo Bon, G. Cazzalato, M. Saccardi, G. M. Kostner, and P. Avogaro, *Atherosclerosis* **53,** 69 (1984).

[64] M. H. Frick, O. Elo, K. Haapa, O. P. Heinonen, P. Heinsalmi, P. Helo, J. K. Huttunen, P. Kaitaniemi, P. Koskinen, V. Manninen, H. Mäenpää, M. Mälkönen, M. Mänttäri, S. Norola, A. Pasternack, J. Pikkarainen, M. Romo, T. Sjöblom, and E. A. Nikkilä, *N. Engl. J. Med.* **317,** 1237 (1987).

[65] P. Gambert, M. Farnier, G. Girardot, J.-M. Brun, and C. Lallemant, *Atherosclerosis* **89,** 267 (1991).

[66] N. Rifai, R. H. Christenson, B. B. Gelman, and L. M. Silverman, *Clin. Chem.* **33,** 1155 (1987).

[67] J. Carlsson, V. W. Armstrong, H. Reiber, K. Felgenhauer, and D. Seidel, *Clin. Chim. Acta* **196,** 167 (1991).

elevated plasma apoE concentrations to normal control levels.[68] However, a report comparing the effects of lovastatin and niacin treatment of nondiabetic, hypercholesterolemic subjects[69] demonstrated that total plasma apoE concentrations changed little after treatment with either of these two commonly used drugs, despite a marked improvement in the lipoprotein profiles of the subjects. Given the important role of apoE in plasma cholesterol clearance, the latter results suggest that apoE quantitation should be combined with a method to assess apoE distribution among the lipoproteins to assess the efficacy of hypolipidemic drug therapy.[63,65]

Methods of Quantitation of Apolipoprotein E

Many published immunoassays for apoE exist in the literature, and readers are referred to a review by Wu *et al.*[70] For measurements of apoE in human plasma, commercial kits are available (PerImmune, Inc., Rockville, MD (ELISA kit); Daiichi Chemicals, Tokyo, Japan; Kamiya Biomedical Co., Thousand Oaks, CA) and can be used for such applications. Unfortunately, commercial kits, particularly those intended for use in automated instrumentation, are generally designed for quantitating proteins in plasma. Thus the range of sample volumes and dilutions may be limited, and it may be impractical to assay research samples such as isolated plasma lipoproteins or cell culture fluids where concentrations of apoE may be below the range assayable by the commercial kits. In such cases, the researcher is faced with developing an in-house assay. Some commercial sources of reagents are as follows: human apoE (BioDesign International, Kennebunk, ME; Chemicon, Temecula, CA; Fitzgerald Industries International, Inc., Concord, MA; IMMUNO AG, Vienna, Austria; Organon Teknika, Durham, NC), anti-human apoE polyclonal antibodies (American Qualex, La Mirada, CA: BioDesign International; Chemicon; Fitzgerald Industries International, Inc.; IMMUNO AG, and anti-human apoE monoclonal antibodies (Accurate Chemical & Scientific Corp., Westbury, NY; BioDesign International; Biosys, Compiegne, France; Boehringer Mannheim, Indianapolis, IN; Chemicon; Fitzgerald Industries International).

Quantitating apoE in different lipoprotein fractions may be a useful tool to assess the effects of a dietary or drug treatment on apoE metabolism.

[68] F. Umeda, J. Watanabe, K. Inoue, A. Hisatomi, K. Mimura, T. Yamauchi, Y. Sako, M. Kunisaki, Y. Tajiri, and H. Nawata, *Endocrinol. Jpn.* **39,** 45 (1992).

[69] D. R. Illingworth, E. A. Stein, Y. B. Mitchel, C. A. Dujovne, P. H. Frost, R. H. Knopp, P. Tun, R. V. Zupkis, and R. A. Greguski, *Arch. Intern. Med.* **154,** 1586 (1994).

[70] L. H. Wu, J. T. Wu, and P. N. Hopkins, *in* "Laboratory Measurement of Lipids, Lipoproteins and Apolipoproteins" (N. Rifai and G. R. Warnick, eds.), p. 279. AACC Press, Washington, D.C., 1994.

However, ultracentrifugation "strips" apoE from lipoprotein particles and causes apoE to sediment in the density greater than 1.21 g/ml fraction of plasma in an essentially lipid-free form.[10] For this reason, it is recommended that nondisruptive methods of fractionating lipoproteins be used, such as precipitation methods or gel filtration, to avoid this artifact when fractionating plasma, serum, or culture medium for apoE quantitation.

In the ELISA (enzyme-linked immunosorbent assay) method described below, β-octylglucopyranoside (β-OGP) is used as a detergent to solubilize lipoprotein particles and to expose all apoE epitopes present. This nonionic detergent was chosen because of its suitability as a detergent for apoE crystallographic studies;[33] moreover, it maintains the helical structure of apoE and is compatible with immunoassays. It was determined that 0.025 to 0.05% (w/v) β-OGP is the optimal concentration to fully expose all apoE epitopes in the ELISA method (data not shown). Other nonionic detergents such as Triton X-100 or Tween 20 were found to be effective in exposing apoE epitopes and can also be used. The use of different detergents should be evaluated with the specific antibodies employed in the ELISA.

The sensitive ELISA method described here is applicable to any research laboratory, and the availability of commercial sources of apoE and anti-apoE antibodies allows the assay to be set up, with minor modifications, in most laboratories. A similar, noncompetitive apoE ELISA employing commercial reagents has been published.[71]

ELISA Method for Apolipoprotein E

In the apoE ELISA, a partially purified polyclonal rabbit anti-human apoE antibody is coated onto the surface of microtiter wells (capture antibody). Standards or samples are added to the wells and incubated overnight. The mass of apoE in the standards or samples is quantitated by the addition of a second antibody which is a purified monoclonal antibody to human apoE. After the incubation, the amount of MAb bound (which is proportional to the amount of apoE in the standards or samples) is detected with a horseradish peroxidase-conjugated goat anti-mouse IgG. The antibodies selected for this assay were determined to be "pan" antibodies, that is, antibodies which bind equally to all lipoprotein forms of apoE.

Materials

Purified human apoE primary standard: Apolipoprotein E is most stable when stored frozen at concentrations ≥1 mg/ml in phosphate-buffered saline (PBS) in aliquots at −70°. Avoid repeated freeze–

[71] V. Gracia, C. Fiol, I. Hurtado, X. Pintó, J. M. Argimon, and M. J. Castiñeiras, *Anal. Biochem.* **223,** 212 (1994).

thawing of apoE. It is recommended for the sake of economy and convenience that a secondary standard such as plasma or serum be calibrated against the primary standard. The secondary standard can then be used in the immunoassay. Periodically check the secondary standard and every new batch of secondary standard against the primary standard

Partially purified rabbit polyclonal anti-human apoE antibody solution: The specific antibody R224-4 was generated at the Lipid Research (Washington University, St. Louis, MO) Center by immunizing rabbits with purified human apoE. The polyclonal antibody is partially purified according to the method of Voller *et al.*[72] and stored frozen in aliquots

Purified monoclonal anti-human apoE: WU E-14 was generated at the Lipid Research Center by immunization of mice with human VLDL. The monoclonal antibody is purified according to the method of Temponi *et al.*[73] and is stored frozen in aliquots

10 m*M* Phosphate-buffered saline, pH 7.4 (PBS)

3% (w/v) Bovine serum albumin (BSA), 0.05% (w/v) β-octylglucopyranoside in PBS, pH 7.4 (3% BSA–β-OGP–PBS): Use radioimmunoassay grade BSA or equivalent (Sigma, St. Louis, MO). Prepare 5% (w/v) stock of β-octylglucopyranoside (β-OGP) (Sigma) in Milli-Q water; add 1 ml per 100 ml 3% BSA–PBS. Adjust pH to 7.4 after adding BSA and β-OGP to PBS; 40–50 ml 3% BSA, β-OGP–PBS is need per 96-well assay plate

1% BSA, 0.05% β-octylglucopyranoside in PBS, pH 7.4 (1% BSA–β-OGP–PBS): Prepare as above

96-well microtiter plates: NUNC Maxisorp plates (Marsh Biomedical Products, Inc., Rochester, NY) are recommended

Humidified plate chambers: Sealed plastic containers lined with wetted paper towels are suitable

Enzyme-conjugated second antibody: Goat anti-mouse IgG coupled to horseradish peroxidase (Boehringer Mannheim) is diluted 1 : 1000 in 1% BSA–PBS (or appropriate dilution as determined for specific lots of secondary antibodies)

Substrate solution: Add 2 ABTS tablets [2,2′-azinobis(3-ethylbenzothiazoline sulfonate), Boehringer Mannheim] per 100 ml substrate buffer solution (Boehringer Mannheim)

[72] A. Voller, D. Bidwel, and A. Bartlett, *in* "Manual of Clinical Immunology" (N. R. Rose and H. Friedman, eds.), p. 506. American Society of Microbiology, Washington, D.C., 1976.

[73] M. Temponi, T. Kageshita, F. Perosa, R. Ono, H. Okada, and S. Ferrone, *Hybridoma* **8,** 85 (1989).

30% Hydrogen peroxide: For ELISA use, H_2O_2 should be stored tightly capped at 4° and should never come in contact with metal

20 m*M* Sodium azide: This is the stop solution for the enzyme reaction (0.65 g NaN_3/500 ml) and can be stored at room temperature

Procedure

1. Determine the number of 96-well plates to be used for the assay. A standard curve and a minimum of two blank wells (background) should be included on each plate.

2. Prepare NUNC plates. Check bottoms of the wells for defects. Rinse the wells with deionized water, flick the plates to ensure that all the water has been removed, and place in the humidified chambers.

3. Dilute the purified polyclonal antibody (R224-4) to 10 μg/ml in PBS. Dispense 100 μl per well, seal the plates with microplate adhesive sealing film (Packard Instruments, Meriden, CT), and incubate overnight at room temperature (in humidified chambers).

4. Wash wells three times with PBS using a plate washer.

5. Block each well with 300 μl 3% BSA–β-OGP–PBS. (Do not allow plates to sit for long periods between additions to plate.) Incubate 1 hr at room temperature.

6. Make dilutions of standards and samples at this point. For the primary standard, it is convenient to prepare a 10–50 μg/ml solution in 1% BSA–β-OGP–PBS and make serial dilutions from this over an approximate range of 0–1 μg/ml or 0–100 ng/well apoE. The range required to achieve saturation of apoE binding sites should be evaluated for each set of antibodies used in the ELISA. Dilutions of approximately 1:500 are appropriate for plasma samples. All sample dilutions should be made in 1% BSA–β-OGP–PBS.

7. Wash the plates once with the plate washer.

8. Pipette 100 μl of the standards and samples into replicate wells. To replicate background wells, add 100 μl buffer alone. Seal plates.

9. Incubate plates overnight at room temperature.

10. Wash plates three times with PBS.

11. Add 100 μl of the purified monoclonal antibody (10 μg/ml in 1% BSA–β-OGP–PBS) to all wells. Seal the plates. Incubate at room temperature for 4 hr.

12. Wash plates three times with PBS.

13. Add 100 μl of the appropriately diluted secondary antibody conjugate in 1% BSA–β-OGP–PBS to all wells. Seal the plates. Incubate for 1–2 hr.

14. Prepare the substrate reagent (add ABTS tablets to substrate buffer). Just before addition to the plates add 10 μl 30% hydrogen peroxide to the substrate reagent.

15. Wash plates three times with PBS, making sure that the wells are free of liquid, and immediately add 100 μl substrate buffer to each well. Add in rows at 5- to 15-sec intervals. Agitate plates (Titer Plate Shaker, LabLine, Melrose Park, IL) to ensure even color development. Stop the reaction when the darkest wells are almost at the upper threshold of detectability for the ELISA reader (absorbance at 410 nm should be greater than

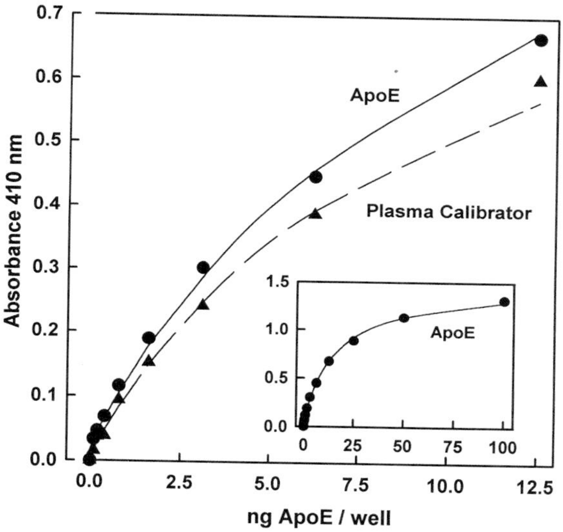

FIG. 1. Apolipoprotein E ELISA standard curves using purified human apoE (E2) (●) and a plasma calibrator (E2/E3) (▲). The apoE concentration in the plasma calibrator was determined by repeated radioimmunoassay, using a purified human apoE standard, according to the method of J. M. Falko, G. Schonfeld, J. L. Witztum, J. B. Kolar, S. W. Weidman, and R. Steelman, *J. Clin. Endocrinol. Metab.* **50,** 521 (1980). The data for the standard curves was analyzed by the program Immunosoft Version 3.4 (Dynatech Laboratories, Inc., Chantilly, VA) which determines the best fit for the four-parameter logistic equation:

$$Y = \frac{A - D}{1 + (x/C) \wedge B} + D$$

where A, B, C, and D correspond to the background absorbance, slope, ED_{50} value, and upper limit of absorbance, respectively. The inset shows the standard curve obtained with purified apoE over the range 0–100 ng apoE/microtiter well. Binding of the apoE to the microtiter wells is saturating at concentrations above 25 ng apoE/well; therefore, for assay purposes the lower part of the curve is employed (0–12.5 ng apoE/well). Note that the purified apoE and plasma calibrator yielded standard curves with similar slopes.

0.300 but less than 1.9). Optimal color development usually occurs within 2 to 5 min.

16. Stop the reaction by adding 50 μl 20 mM sodium azide. Add the azide at 5- to 15-sec intervals as for the addition of the substrate buffer. Agitate to mix reagents. Record the time the color development was stopped for future reference.

17. Read the ELISA plates within 30 min after the addition of sodium azide.

18. Read ELISA plates in the dual mode, at a test absorbance of 410 nm and reference absorbance of 570 nm. Use the reagent blank (no sample) or air blank for the plates.

Additional Notes

1. Incubation times can be reduced; however, this should be evaluated with the specific antibodies used for ELISA.

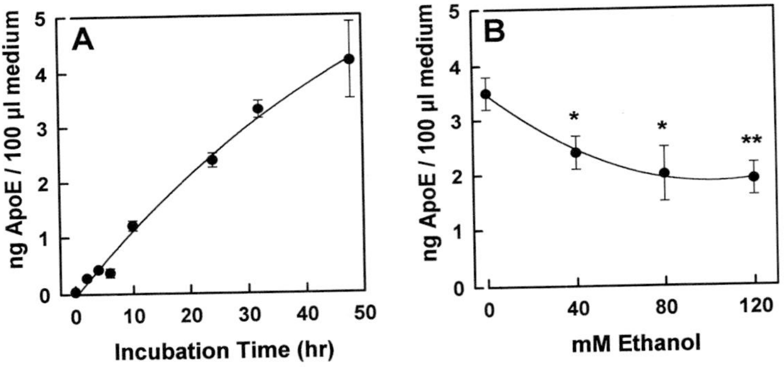

FIG. 2. Secretion of apoE by THP-1 cells in culture. The THP-1 cells were seeded into 12-well cluster culture dishes in DME/F-12 medium with HEPES containing 10% fetal bovine serum and 10^{-7} M phorbol 12-myristate 13-acetate (PMA). After 1 day the adherent cells were washed with serum-free DME/F-12 medium and then incubated in this medium containing PMA for 24 hr. (A) The medium above the cells was replaced with fresh DME/F12. The medium was collected from duplicate wells at the indicated time points for apoE quantitation. The ApoE concentrations represent the mean of duplicate determinations ($\pm$ standard deviations) for each of the duplicate culture wells. Apolipoprotein E secretion into the medium increased linearly for up to 48 hr. (B) In this experiment, the medium was replaced with serum-free medium containing PMA and the indicated amounts of ethanol in triplicate wells. The ApoE concentrations represent the mean of triplicate determinations ($\pm$ standard deviations) for each of the triplicate wells after 24 hr. Apolipoprotein E secretion into the medium was significantly reduced by all the concentrations of ethanol tested ($p <$ 0.05 for 40 and 80 mM ethanol and p = 0.005 for 120 mM ethanol).

2. After each wash step, the microtiter plates should be inverted over a paper towel and tapped to ensure that all liquid has been removed from the wells before the next step.

3. Make sure the microtiter wells are free of scratches, condensation, or air bubbles before reading the plates on the ELISA reader.

Assay Performance. Standard curves using purified apoE and a plasma calibrator are shown in Fig. 1. The slopes of the curves are very similar in the range of apoE concentrations shown. Purified human apoA-I gave background binding when tested in this assay (data not shown).

The mean apoE concentration for a group of human plasma samples ($N = 26$) was 3.1 ± 0.6 mg/dl (mean cholesterol concentrations of 280 ± 37 mg/dl). Intraassay and interassay coefficients of variation for this assay are approximately 6 and 15%, respectively.

The sensitivity of the assay permitted measurement of apoE secreted into the medium of cultured THP-1 cells (a human monocytic cell line). The results of two such experiments are summarized in Fig. 2. A time course experiment was conducted in Fig. 2A demonstrating that the secretion of apoE by THP-1 cells in culture was linear over a 48-hr period. The effects of physiological concentrations of ethanol on apoE secretion were then tested. Our hypothesis was that ethanol may have a deleterious effect on the peripheral synthesis and secretion of apoE. As shown in Fig. 2B, doses of ethanol ranging from 40 to 120 m*M* decreased apoE secretion into the medium of THP-1 cells over 24 hr by approximately 40% ($p < 0.05$).

Summary

Quantitation of apoE has proved to be extremely useful in studies of the regulation of apoE synthesis and metabolism. Measurement of serum apoE and/or its distribution among the lipoprotein classes may have clinical utility, although this remains to be established. Some of the unique properties of apoE such as its genetic, chemical, and structural heterogeneity, its propensity to self-associate, and its ability to freely exchange on the surfaces of a wide variety of lipoprotein classes are factors that should be considered in measurements of apoE. The availability of commercial kits and reagents for human apoE quantitation make the development of apoE immunoassays readily achievable in most research and clinical laboratories.

Acknowledgments

The authors acknowledge the expert technical assistance of Angela Crisci and Tish Kettler in the development of the apoE ELISA.

[11] Quantification of Apolipoprotein C-II by Immunochemical and Chromatographic Methods

By PHILIP W. CONNELLY, CAMILLA VEZINA, and GRAHAM F. MAGUIRE

Introduction

Korn[1] demonstrated that plasma activates lipoprotein lipase (LPL). Apolipoprotein C-II (apoC-II) is the major apolipoprotein activator of LPL.[2] Hyperchylomicronemia due to a deficiency of LPL was first described by Havel and Gordon.[3] Familial apoC-II deficiency was first described by this laboratory[4] in a family with 14 homozygotes and 26 obligate heterozygotes. These studies proved that apoC-II is of key importance in lipoprotein metabolism. Apolipoprotein C-II is a 79-residue apoprotein with a molecular weight of 8915. The sequence of apoC-II has been reviewed in a previous volume of this series.[5] The purpose of this chapter is to describe the methods of quantification of apoC-II currently used in our laboratory.

Purification of Apolipoprotein C-II

Immunoassay of apoC-II requires a pure apoC-II for immunization and for calibration of the assay. In addition, authentic apoC-II is required for positive identification of apoprotein peaks obtained by chromatographic methods. To maximize the yield, apoC-II is isolated from apochylomicrons, if present, and very-low-density lipoprotein apoproteins (apoVLDL) of hypertriglyceridemic subjects fasted for 12–16 hr. Blood samples are collected into Na_2EDTA tubes and the plasma centrifuged in a Beckman 80Ti or 50.2 rotor at 20,000 rpm for 30 min at 20° to obtain the chylomicron fraction. The remaining plasma is centrifuged at 45,000 rpm for 16.5 hr at 10° to obtain VLDL. Both lipoprotein fractions are washed by centrifugation under the conditions of isolation, dialyzed against 0.01% Na_2EDTA, pH 8.2, lyophilized, and delipidated using ethanol–diethyl ether (3 : 1, v/v).[6]

It should be noted that anhydrous solvents should be used for the

[1] E. D. Korn, *J. Biol. Chem.* **215,** 15 (1955).

[2] R. J. Havel, *Circ. Res.* **27,** 595 (1970).

[3] R. J. Havel and R. S. Gordon, *J. Clin. Invest.* **39,** 1777 (1960).

[4] W. C. Breckenridge, J. A. Little, G. Steiner, A. Chow, and M. Poapst, *N. Engl. J. Med.* **298,** 1265 (1978).

[5] R. L. Jackson and G. Holdsworth, this series, Vol. 128, p. 288.

[6] A. M. Scanu and C. Edelstein, *Anal. Biochem.* **44,** 576 (1971).

delipidation step because the C apolipoproteins are slightly soluble in a "wet" organic phase. This decreases recovery of the apolipoproteins and invalidates any quantitative information. Also, to avoid formation of artifacts in apoproteins, it is important to ensure that the ether used is peroxide-free. It is best to store the ether for short periods, in small quantities, and at 4° in a refrigerator designed for solvent storage.

Preparative Isoelectric Focusing

Our first step in the purification of apoC-II from apochylomicrons or apoVLDL is preparative isoelectric focusing on Ultradex gel with pH 4–6 ampholines (Pharmacia, Piscataway, NJ), using a procedure essentially as described by Havel *et al.*[7] The Ultradex gel, containing 4% Ultradex (Pharmacia), 4.5 *M* urea, 1.2% ampholine, pH 4–6 (Pharmacia), and 15 m*M* dithiothreitol (DTT), is prefocused at 300 V for 1 hr at 12°. Apoprotein (50–75 mg) is dissolved in 3 ml of 10 m*M* Tris-HCl, pH 8.2, containing 8 *M* urea and 75 m*M* DTT. After 30 min at 37°, the sample is mixed with gel, removed from the cathodic area of the plate, and electrophoresed at 8 W for 18 hr at 12° in a Hoefer Isobox (Hoefer, San Francisco, CA). The separated proteins are visualized against a dark background using a fluorescent light, and gel containing apoC-II is transferred to an elution column (Pharmacia) and mixed with 1 ml of 10 m*M* Tris-HCl, pH 8.2, 0.5% sodium decyl sulfate. After gentle rocking for 10 min at 10°, the eluent is spun off using an Omnifuge RT centrifuge (Heraeus, American Hospital Supply, McSaw Park, IL) at 1000 rpm for 5 min at 10°. This is repeated and the eluents pooled. Apolipoprotein C-II protein is precipitated using 9 volumes of ice-cold acetone and is washed with more acetone. Ampholine, which is also precipitated, is removed by extraction with 2–3 ml $CHCl_3$–methanol (1.5 : 1, v/v). The isolated apoC-II is dried under a stream of N_2 at 37°.

Evaluation of Purity of Apolipoprotein C-II by Analytical Isoelectric Focusing

Analytical isoelectric focusing gels of 0.75 mm thickness are made using 7.5% polyacrylamide with an acrylamide : bisacrylamide ratio of 36 : 1. The gel contains 8 *M* urea and 2% ampholine. A pH range of 4–6.5 is obtained by mixing ampholines pH 4–6 and 5–7 in a ratio of 4 : 1. Gels are prefocused for 1 hr at 100 V. For sample preparation, 75 μg apoVLDL or 1–2 μg isolated apoC-II are dissolved for 30 min at 37° in 10 m*M* Tris-HCl, pH 8.2, containing 8 *M* urea and 20 m*M* DTT. The samples are electrofocused

[7] R. J. Havel, L. Kotite, and J. P. Kane, *Biochem. Med.* **21,** 121 (1979).

for 20 hr at 200 V. The gel is fixed for 1 hr in 12% trichloroacetic acid and stained for 1 hr in 0.03% Coomassie Blue G-250 in 3.5% perchloric acid.[8]

Western blotting[9] provides a more sensitive assessment of purity of isolated apoC-II. Half the amount of sample is electrophoresed. Following focusing, proteins are transferred to either nitrocellulose or polyvinylidene difluoride (PVDF; Millipore, Bedford, MA) membranes using a Hoefer Transfor cell (Hoefer). The buffer chosen for transfer is either (a) 0.7% acetic acid containing 20% methanol or (b) 25 m*M* Tris-HCl, 192 m*M* glycine, pH 8.3. When using buffer b, the gel is presoaked for 30 min in buffer b containing 20% methanol. Proteins are transferred at 500 mA for 3 hr at 10°. Following transfer, apoC-II is detected by Western blot according to the manufacturer's protocol for their immunoblot assay kit. Anti-apoC-II is used as first antibody at a 1/2000 dilution, while goat anti-rabbit immunoglobulin G (IgG)–alkaline phosphatase conjugate (1/2000) serves as the second antibody.

Separation by Anion-Exchange Chromatography

Anion-exchange high-performance liquid chromatography (HPLC) offers an alternative to preparative isoelectric focusing for the semipreparative and preparative isolation of apoC-II. In addition, the protocol that we use eliminates urea and thus any potential artifactual carbamylation. We isolate apoC-II from apoVLDL using Aquapore AX-300 anion exchange columns (Applied Biosystems, Inc., San Jose, CA).

For separation of apoC-II from large quantities (5–20 mg) of apoVLDL, we have used a Prep-10 (Applied Biosystems, San Jose, CA) anion-exchange column (10 cm × 10 mm column, packed with 300 Å, wide pore, spherical silica of 20 μm particle size). This procedure can also serve as a second step in the purification of apoC-II, after an initial isolation from apoVLDL on Ultradex IEF. For this purpose the column is equilibrated in 99% buffer A (30 m*M* Tris-HCl, pH 7.45, 40% 2-propanol), 1% buffer B (0.3 *M* Tris-HCl, pH 7.45, 50% v/v acetonitrile). Apolipoprotein C-II, partially purified by preparative isoelectric focusing, is dissolved in 0.1 *M* NH_4HCO_3, 0.2% sodium decyl sulfate (5 mg/ml), and 200 μl is injected onto the column. Because the apoproteins are retained on the column until the gradient is initiated, the volume injected does not typically affect resolution. The apoproteins are eluted with a linear gradient from 1 to 100% of buffer B in 30 min, at a flow rate of 5 ml/min with detection at 280 nm, range 0.1 absorbance units (AU). Apolipoprotein C-II is pooled from several runs and the solvent removed with N_2 at 37°. The remaining

[8] A. H. Reisner, P. Nemes, and C. Bucholtz, *Anal. Biochem.* **64,** 509 (1975).

[9] H. Towbin, T. Staehelin, and J. Gordon, *Proc. Natl. Acad. Sci. U.S.A.* **76,** 4350 (1979).

aqueous solution is then dialyzed against 0.1 *M* NH_4HCO_3 and stored at −70°.

We more frequently use a small AX-300 guard column (3 cm × 4.6 mm, packed with 7 μm particle size) sufficient for the semipreparative isolation of apoproteins from 1–2 mg apoVLDL. We use slightly different conditions for separating apoproteins on this smaller anion-exchange column. For sample preparation, apoVLDL is dissolved in buffer A at 1 mg protein/ml. One milliliter is injected with a programmed gradient of 8 min at 1% B, followed by stepwise gradients of 1–20% B in 4 min and 20–55% B in 35 min at a flow rate of 1 ml/min, with detection at 280 nm using a range of 0.1 AU.

Analytical Separation

The AX-300 guard column is also used for the analytical separation of C apolipoproteins in apoVLDL. Typically 150 μg apoVLDL is dissolved in 100 μl buffer A at 37° for 30 min. This is then injected onto the column under the following HPLC conditions: 8 min at 1% B, and 1–100% B for 15 min, at a flow rate of 1 ml/min, with detection at 280 nm using a range of 0.05 AU.

Semipreparative Isolation of Apolipoprotein C-II by Reversed-Phase Chromatography

To obtain apoC-II of the highest purity for use as an antigen, the preparative isoelectric focusing step can be followed by reversed-phase HPLC. The general principles in column selection that we have followed are to choose the shortest column that will give the required resolution to maximize recovery of protein, and the largest pore size to maximize interaction of the protein with the hydrophobic phase. One approach to this has been previously described.[5] The method used in our laboratory is significantly different and will be described. Apolipoprotein C-II obtained by preparative isoelectric focusing is redissolved in 0.1 *M* NH_4HCO_3 buffer (1 mg/ml) and up to 250 μg injected onto the reversed-phase column. We typically use a Supelcosil LC-308 column, 5 cm by 4.6 mm, 5 μm particle size, 300 Å pore diameter (Supelco Canada, Oakville, ON), equipped with an LC-308 guard column, 2 cm by 4.6 mm (Supelco). Elution buffer A is 0.0675% trifluoroacetic acid (TFA) in water. Buffer B is 0.0675% TFA in 60% acetonitrile. It is important to use the highest grade of TFA available (Pierce, Rockford, IL). To remove background due to contamination, the TFA is passed through a Sep-Pak Plus C_{18} solid-phase cartridge (Waters, Milford, MA). The apoC-II is eluted using a program of 42% buffer B at time 0, increasing to 97% buffer B over 33 min, at a flow rate of 1 ml/min.

Apoprotein is detected at 226 nm with a range of 2.0 AU. The apoC-II peak is collected, and solvent is removed under N_2 at 37°. The preparation is then lyophilized and stored at −70°.

Analytical Separation of Apolipoprotein VLDL by Reversed-Phase Chromatography

The identical chromatographic conditions can be used to analyze apoC-II in whole apoVLDL. Use of a recording integrator or a computerized data acquisition system (Varian GC Star, Varian Canada, Toronto, ON), allows expression of the absorbance in terms of percent of apoC's. We have also used this method to analyze the relative amounts of normal and mutant apoC-II.

Immunoassay of Apolipoprotein C-II

Production of Polyclonal Antisera to Apolipoprotein C-II

Immunization Protocol. Methods for production of polyclonal antisera in a variety of species are readily available in the literature.[10,11] We have found the rabbit to be a useful species in which to raise antisera directed against apoC-II within a reasonable time frame and with adequate titer for use in quantitative immunoassay.

New Zealand White, female, rabbits (2.5–3 kg) are acclimatized and their health monitored for 1 to 2 weeks prior to injection. In our experience, different rabbits will respond individually to apoC-II, and therefore at least two animals are immunized. All immunization protocols must be performed by trained personnel, follow acceptable immunological procedures, and be covered by an appropriate license.

Purified apoC-II that has been stored at −70° as aliquots in 0.1 *M* NH_4HCO_3, 0.2% sodium decyl sulfate, is suitable as an immunogen. A small preimmunization blood sample (5–10 ml) is taken from the ear vein of each animal. An aliquot of 100 μg apoC-II is diluted to 0.5 ml with sterile isotonic saline and emulsified with 1.0 ml of Freund's complete adjuvant (Difco Laboratories Ltd., Detroit, MI) by working the solution through two syringes, attached together with a dual female end Teflon Luer adapter. Using sterile techniques and a 19-gauge needle, 0.5 ml of the inoculate is injected subcutaneously at one site in the interscapular region and 0.5 ml intramuscularly at two sites in the thigh muscle. Ten days later,

[10] B. A. L. Hurn and S. M. Chantler, this series, Vol. 70, p. 104.

[11] D. Catty and C. Raykundalia, *in* "Antibodies Volume 1: A Practical Approach" (D. Catty, ed.), p. 19. IRL Press, Oxford, 1988.

the rabbit is given a booster immunization of apoC-II (50 μg), emulsified as above with Freund's incomplete adjuvant, at three sites, subcutaneously. Ten days later, a second booster (50 μg) is given.

Twenty-seven days after the initial immunization, 2–5 ml blood is drawn from the marginal ear vein to assess antibody response. If there is adequate titer, the rabbit can be bled from the central ear artery with a 19-gauge needle. About 45 ml blood can be collected at monthly intervals. The antibody titer should be assessed by solid-phase immunoassay at each sampling. Booster injections should be given at appropriate intervals, either monthly or, if the antigen is in short supply, when the titer begins to fall. Rabbits can be maintained and bled for as long as 18 months if they are healthy and responding. About 60 ml antiserum can be harvested when the animals are sacrificed by exsanguination under anaesthetic.

The blood is allowed to clot for 30 min at 37°, and the clot is freed from the walls of the container. The blood is then left to sit at least 30 min or overnight at 4°. The major clot is carefully removed from the vessel, and the serum is spun at 1500 rpm for 20 min to remove remaining red cells. Serum samples are not pooled before testing and are best stored frozen as 5- or 10-ml aliquots, with several smaller aliquots of 0.5 ml set aside for testing of titer and specificity. Tubes of appropriate size with good seals should be used for storage. Antisera are stored at −70°.

Test of Antiserum Titer

Antibody titer can be easily monitored by solid-phase immunoassay. Either VLDL or apoVLDL is a convenient source of apoC-II for this purpose.

Materials

10 m*M* phosphate, 150 m*M* NaCl (PBS), pH 7.4
Bovine serum albumin [BSA; radioimmunoassay (RIA) grade, Sigma, St. Louis, MO]
VLDL or apoVLDL (about 250 μg protein/ml)
96-well microtiter plates (NUNC Maxisorp, Gibco/BRL, Grand Island, NY)
Tween 20 (enzyme immunoassay purity, Bio-Rad, Richmond, CA)
Anti-rabbit IgG–horseradish peroxidase conjugate (Bio-Rad)
TMB Peroxidase Substrate (Bio-Rad)
1 *N* H_2SO_4
Multichannel pipettor
Microtiter plate photometer (e.g., Multiskan MCC/340, Flow/ICN, Costa Mesa, CA)
Microtiter plate shaker (e.g., Minishaker, Dynatech, Chantilly, VA)

Procedure. Microtiter wells are coated with 100 μl VLDL or apoVLDL, diluted to 5 μg protein/ml in PBS. The plates are sealed and left at room temperature for 2 hr. The coating solution is then discarded, and the wells are washed once with PBS. After tapping out excess liquid, the remaining protein binding sites are blocked by incubation for 30 min at room temperature with 300 μl of 0.5% BSA in PBS. Aliquots of the antiserum to be tested are diluted in 0.1% BSA in PBS. Usually 1 ml of a 1/1000 dilution is prepared in a 12 by 75 mm test tube, and then 10 serial doubling dilutions of 100 μl are made directly in the coated plate. This titration step is most easily done with a multichannel pipettor. Control wells might include dilutions of preimmunization serum and/or previous bleeds from the same animal. After incubation at room temperature for 1 hr, the wells are emptied and washed three times with PBS containing 0.05% Tween 20 (PBST). An appropriate dilution of anti-rabbit IgG–horseradish peroxidase is prepared, usually 10 ml of a 1/20,000 or a 1/50,000 dilution in 0.1% BSA–PBS, and 100 μl is added to the wells. The plate is incubated at room temperature for 1 hr, and the wells are emptied and washed three times with PBST. TMB peroxidase substrate (10 ml) is freshly prepared according to the manufacturer's instructions, and 100 μl is added to the wells. The plate is placed on a shaker for 10–30 min, and when color has developed sufficiently, the reaction is stopped with 100 μl of 1 *N* H_2SO_4. The plate is read at 450 nm on the plate reader, and the optical density is plotted against antibody dilution. An example of this type of titration assay is shown in Fig. 1.

Test of Antiserum Specificity

Once a reasonable titer is established, the antiserum specificity must be validated. Analytical isoelectric focusing of apoVLDL and/or high-density lipoprotein apoprotein (apoHDL), followed by immunoblotting with the test antiserum, will indicate if it is monospecific for apoC-II.

Solid-phase immunoassay should also be used to establish that the antiserum is monospecific for apoC-II, since the sensitivity is different from that of the immunoblotting techniques. Microtiter wells can be coated with purified apoproteins (e.g., apoC-III, apoA-II) instead of apoVLDL in the titer assay described above.

Immunoaffinity Purification of Polyclonal Anti-Apolipoprotein C-II

Methods for coupling proteins to insoluble matrices are readily available in the literature.[12] Commercially available activated affinity supports such

[12] S. C. March, I. Parikh, and P. Cuatrecasas, *Anal. Biochem.* **60,** 149 (1974).

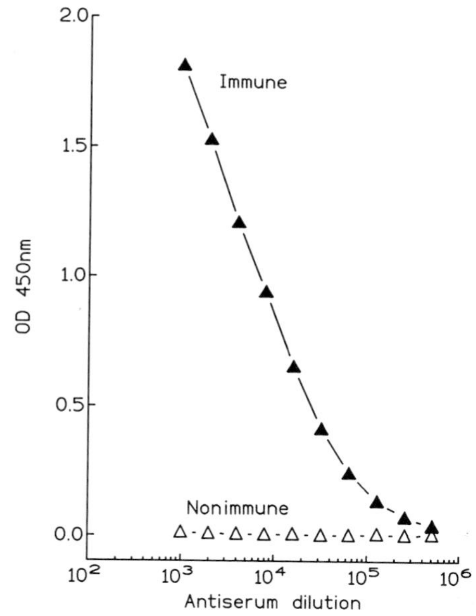

FIG. 1. Titration curve of anti-apoC-II antiserum. Rabbit serum taken before immunization (nonimmune) and the test anti-apoC-II antiserum (immune) were diluted and incubated in microtiter wells coated with VLDL, at a concentration of 5 μg protein/ml, as described in the text. Anti-rabbit IgG–horseradish peroxidase, at a dilution of 1/50,000, was used to detect bound antibody.

as CNBr-Sepharose (Pharmacia) or Affi-Gel 15 (Bio-Rad) are suitable for this purpose and are easy to use.

Materials

Purified apoC-II (at least 2 mg protein)
Dialysis tubing of low molecular weight cutoff (e.g., Spectra/Por3, Spectrum Medical Industries, Los Angeles, CA)
Affi-Gel 15 (Bio-Rad)
0.1 *M* HEPES, pH 7.5
2-Propanol
1 *M* Ethanolamine hydrochloride, pH 8.0
10 m*M* Tris-HCl, 0.15 *M* NaCl, 1 m*M* EDTA, pH 7.4 (TBS–EDTA)
TBS–EDTA containing 0.02% (w/v) Merthiolate (thimerosol)
0.1 *M* Tris-HCl, pH 8.5
0.2 *M* Glycine, pH 2.8
3 *M* NaSCN in 10 m*M* phosphate, pH 7.0

Econo-Columns (Bio-Rad)

10 mM phosphate buffer, 0.15 M NaCl, pH 7.4 (PBS)

Procedure. Apolipoprotein C-II (2 mg), stored in 0.1 M NH_4HCO_3, is dialyzed overnight at 4° in 0.1 M HEPES, pH 7.5. About 3–5 ml Affi-Gel 15 is washed quickly at 4° and added to the dialyzed apoC-II. After coupling according to the manufacturer's suggestions, the apoC-II-Affi-Gel is stored in TBS–EDTA containing 0.02% Merthiolate.

Prior to each use, the apoC-II affinity column is washed with 3–5 column volumes of the elution buffers to be used in subsequent procedures and then reequilibrated by washing with about 10 volumes of TBS–EDTA. The gel is also primed with nonimmune rabbit serum by an initial batch incubation of 2–5 ml of rabbit nonimmune serum, diluted with an equal volume of TBS–EDTA for 4 hr at 4°. After washing with the elution buffers, the gel is reequilibrated with TBS–EDTA.

The washed gel is packed into a small Econo-Column. Polyclonal antiserum to apoC-II (3–5 ml) is diluted with an equal volume of TBS–EDTA and applied to the gel at a slow flow rate (3–5 ml/hr). The unadsorbed proteins are washed through with TBS–EDTA. When the absorbance at 280 nm is less than 0.02, the flow rate is increased to 20 ml/hr, and the bound immunoglobulins are eluted with 2–3 column volumes of 0.2 M glycine, pH 2.8, into a tube containing enough 0.1 M Tris-HCl, pH 8.5, to neutralize the acid. High-affinity antibody is eluted with 2 column volumes of 3 M NaSCN, pH 7.0. The eluted fractions are dialyzed immediately against PBS, pH 7.4, containing 0.02% sodium azide at 4°. Solid-phase immunoassay, as described in tests for antibody titer, can be used to test the capacity of the column and to monitor the eluted fractions for anti-apoC-II. The gel is reequilibrated in TBS–EDTA containing 0.02% Merthiolate and stored at 4°.

The protein content of the eluted fractions is measured by the Lowry procedure[13] using an IgG standard (Bio-Rad). The immunopurified anti-apoC-II is stored at −70° at a concentration of 2 mg/ml. If necessary, adequate concentration can be accomplished by coating the outside of a dialysis bag containing the antiserum fraction with either Aquacide II (Calbiochem, San Diego, CA) or polyvinylpyrrolidone (BDH). The concentrated fraction is then dialyzed in PBS, containing 0.02% (w/v) sodium azide, prior to storage.

Preparation of Anti-Apolipoprotein C-II–Enzyme Conjugate

Immunopurified, monospecific anti-apoC-II (the glycine fraction or the NaSCN fraction) is covalently linked to horseradish peroxidase or alkaline

[13] O. H. Lowry, N. J. Rosebrough, A. L. Farr, and R. J. Randall, *J. Biol. Chem.* **193,** 265 (1951).

phosphatase by standard techniques.[14] We have found the alkaline phosphatase conjugate to be very stable over time.

Materials

Immunopurified anti-apoC-II, 0.5 mg
Alkaline phosphatase, 1.25 mg (Boehringer Mannheim, Indianapolis, IN)
25% Glutaraldehyde (v/v) (Fisher, Springfield, NJ)
Dialysis tubing of small diameter
PBS, containing 0.02% (w/v) sodium azide
50 m*M* Tris, pH 8.0, containing 0.02% sodium azide
2% (w/v) BSA (RIA grade, Sigma) in 50 m*M* Tris, pH 8.0, containing 2 m*M* $MgCl_2$ and 0.02% sodium azide

Procedure. The anti-apoC-II is mixed with the alkaline phosphatase in a 1.5-ml conical tube. The incubation volume is adjusted to 500 μl with PBS, and glutaraldehyde is added to a final concentration of 0.2%. The solution is vortexed and then incubated for 2 hr at room temperature in a dark container. The solution is then dialyzed in 4 liters of PBS (2 changes of 2 liters), followed by 4 liters of 50 m*M* Tris, pH 8.0, overnight. The mixture is transferred to a 1.5-ml conical tube with an O-ring seal and the volume completed to 1 ml with 2% BSA in 50 m*M* Tris, pH 8.0, containing 2 m*M* $MgCl_2$ and 0.02% sodium azide. The anti-apoC-II–alkaline phosphatase conjugate is stable for at least 6 months when stored in a dark container at 4°.

Sandwich ELISA for Apolipoprotein C-II Quantitation

Sandwich ELISA (enzyme-linked immunosorbent assay) techniques for apoC-II quantitation have been described in the literature.[15,16] The protocol in use in our laboratory is essentially the technique of Bury *et al.*,[15] with modifications to buffers and incubation times to allow the assay to be completed within one working day.

Materials

Immunopurified anti-apoC-II
Anti-apoC-II–alkaline phosphatase conjugate
Purified apoC-II standard
96-well microtiter plate (NUNC Maxisorp)

[14] A. Voller, D. E. Bidwell, and A. Bartlett, "The Enzyme-Linked Immunosorbent Assay." Flowline Press, Guernsey, Channel Islands, U.K., 1979.

[15] J. Bury, G. Michiels, and M. Rosseneu, *J. Clin. Chem. Clin. Biochem.* **24,** 457 (1986).

[16] N. Alsayed, R. Rebourcet, and J. Chapman, *Clin. Chem.* **36,** 2047 (1990).

Microtubes in 96-well array (Bio-Rad)
50 m*M* Sodium carbonate buffer, pH 9.6, containing 0.02% sodium azide
10 m*M* phosphate, 0.15 *M* NaCl, pH 7.4 (PBS)
PBS, containing 0.05% Tween 20 (PBST)
0.5% BSA (RIA grade, Sigma) in PBST
0.1% BSA in PBST
Alkaline phosphatase substrate kit (Bio-Rad)
0.4 *N* NaOH

Equipment

Multichannel pipettor and/or repeating pipettor
Diluter/dispenser (e.g., Hamilton Microlab, Hamilton Company, Reno, NV)
Microtiter plate shaker
Microtiter plate washer or squeeze bottles for manual washing
Microtiter plate reader
Computer with curve-fitting software designed to accept and analyze data from plate reader

General Procedure. A microtiter plate, of the highest optical quality available, is coated with capture antibody by incubating the wells with 100 μl affinity-purified anti-apoC-II (diluted appropriately in 10 ml of 50 m*M* sodium carbonate buffer, pH 9.6), for 2 hr, at room temperature. The coating solution is discarded and the wells are washed once with PBS. After tapping out excess fluid, any remaining binding sites are blocked by incubation with 300 μl of 0.5% BSA in PBST, for 30 min, at room temperature. The blocking solution is discarded, the wells are washed once with PBS and emptied. Purified apoC-II, plasma or plasma lipoprotein fractions are diluted in 0.1% BSA–PBST, usually in duplicate or triplicate. Final dilutions can be made into 1-ml microtubes arranged in a 96-well array from which transfers by a multichannel pipettor are convenient. A semiautomated diluter is desirable for improved precision, as dilutions of serum in the assay are in the 1000-fold range. Aliquots (100 μl) of the dilutions are transferred to the coated wells, and the plate is incubated for 1 hr at room temperature. The solution is discarded, and the wells are washed three times with PBST and blotted dry. The anti-apoC-II–alkaline phosphatase conjugate is diluted appropriately with 0.1% BSA in PBST. It is added to the wells (100 μl), and the plate is incubated at room temperature for 1 hr. After washing the wells three times with PBST, 100 μl of the substrate, *p*-nitrophenyl phosphate, is added. The plate is placed on the shaker at medium speed, and color is allowed to develop for 30 min. The reaction is stopped with 100 μl of 0.4 *N* NaOH, and the color intensity is read on the plate reader at 405 nm.

Establishment of Optimal Assay Conditions. Optimal concentrations of coating antibody and enzyme-linked detection antibody are determined by a dilution series of the respective antibodies in a checkerboard design. Coating antibody is serially diluted, in doubling dilutions, with 100 μl carbonate buffer, across the columns of the plate, usually starting at a concentration of 20 μg protein/ml. The plate is incubated and washed as described above. Purified apoC-II is diluted to 250 ng/ml, a relatively high concentration, in PBST. Enough solution is prepared for the full plate. Aliquots (100 μl) are added to the wells, and the plate is incubated as above. After washing, the anti-apoC-II alkaline phosphatase is serially diluted, in doubling dilutions of 100 μl, in PBST, down the rows of the plate, usually starting at a dilution ratio of 1/62.5. Concentrations yielding an optical density (OD) in the range of 1.5 to 2.0 OD units are appropriate. Ideal coating antibody concentration is usually in the range of 5 μg/ml, whereas the optimal dilution of the anti-apoC-II alkaline phosphatase is in the range of 1000- to 2000-fold with these assay conditions (Fig. 2).

Once the optimal coating and conjugate concentrations have been set, a titration curve of purified apoC-II can be attempted, to determine the working range of the assay. Initially, serial doubling dilutions of apoC-II in PBST can be based on the Lowry protein estimation, starting at a concentration of 250 ng/ml. The standard curve consists of at least six points measured in duplicate or triplicate. The protein concentration must be verified by quantitative amino acid analysis and absorbance measurements of the pure protein at 280 nm as discussed below. A typical standard curve is shown in Fig. 3. With these antibodies and assay conditions, our assay is sensitive down to 5 ng/ml with a working range of 10–200 ng/ml.

Validation of Assay for Plasma and Isolated Lipoprotein Fractions. Plasma or serum, and isolated lipoprotein fractions, are serially diluted to demonstrate that the curves generated from the dilution series are parallel to that of the standard apoC-II and therefore that the assay is a valid method to quantitate apoC-II in the samples. An example of such a titration experiment is shown in Fig. 4. Plasma and isolated lipoprotein samples should be delipidated or exposed to detergent and retested, to ensure that the apoC-II epitopes are fully exposed in the assay. We have found that including 0.05% Tween 20 in the assay buffer increases the response to plasma apoC-II about 2.5-fold, consistent with previous reports.[15,17] The assay accuracy can be determined by adding purified apoC-II to plasma

[17] S. I. Barr, B. A. Kottke, J. Y. Chang, and J. T. Mao, *Biochim. Biophys. Acta* **633,** 491 (1981).

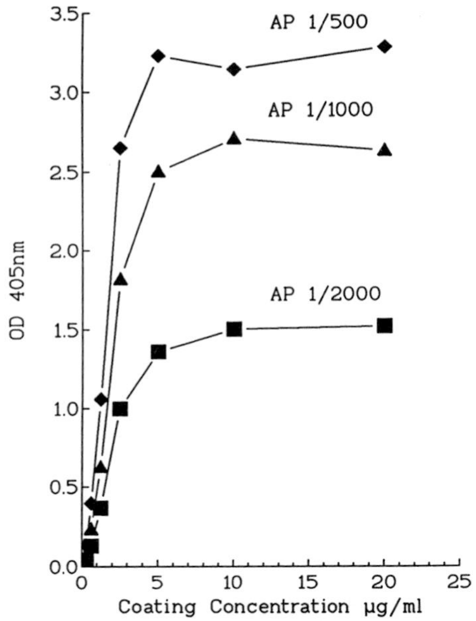

FIG. 2. Titration of coating and enzyme-linked detection antibodies. The immunopurified anti-apoC-II coating antibody was serially diluted in carbonate buffer, from 20 to 0.3 μg protein/ml, in microtiter wells across the plate. Purified apoC-II was diluted to 250 ng/ml in PBST and added to the coated wells. The incubation and washing procedures were as described in the text. Anti-apoC-II, linked to alkaline phosphatase, was serially diluted in PBST from 1/62.5 to 1/2000, in wells down the plate. Shown are the coating antibody dilutions at three alkaline phosphatase dilutions.

samples at several concentrations and comparing the measured apoC-II concentration with the expected apoC-II concentration.

Standardization of Assay

Assignment of Mass to Purified Standard. The protein concentration of purified apoC-II used as a standard must be accurately determined. The principal chromogenic groups in proteins to which the Lowry method[13] is sensitive are tyrosine and tryptophan, with some contribution by cystine, cysteine, and histidine. The extent to which color will develop in the Lowry assay cannot be predicted on the basis of structure, and therefore one would expect the values obtained, using BSA as a standard, to differ from the actual protein mass and thus should be considered as relative protein values.

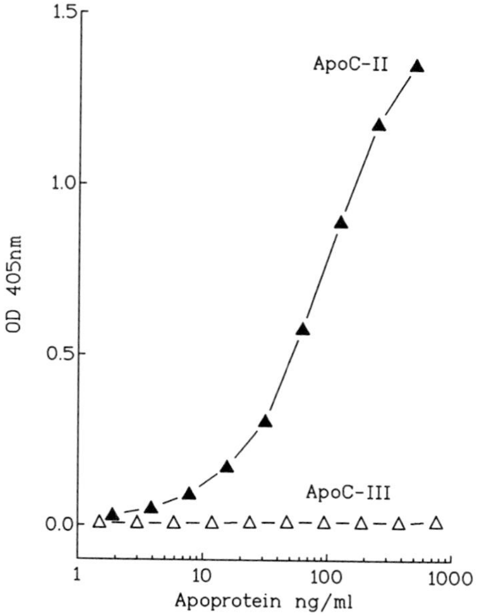

FIG. 3. Standard curve of purified apoC-II. Microtiter wells were coated with 100 μl immunopurified anti-apoC-II, at a concentration of 5 μg protein/ml. Serial doubling dilutions of HPLC-purified apC-II were made in PBST, starting at 500 ng/ml, and 100 μl was added to the coated wells. Anti-apoC-II, linked to alkaline phosphatase and diluted 1500-fold, was used for detection. Incubation and washing procedures are described in the text. Apolipoprotein C-III had no affinity for anti-apoC-II in this assay.

To obtain more accurate values for the mass of apoC-II present in a sample, an amino acid analysis can be performed. For this purpose 2–10 μg purified protein is placed in a thick-walled test tube, and to it is added 250 μl constant-boiling 5.8 *N* HCl. Norleucine (5 nM) is added as an internal standard. The tube is sealed under vacuum and heated at 110° for 19–24 hr. After hydrolysis the contents of the tube are dried and immediately dissolved in 100–120 μl of pH 2.2 buffer. Hydrolyzate from 2–3 μg protein is injected onto a single column system (e.g., Beckman 121M, Beckman Instruments, Fullerton, CA). A correction is made in the calculated protein mass to account for tryptophan, which is not determined with this method.

The mass of apoC-II can also be estimated by use of molar extinction coefficients. For apoC-II a value of 12,300 is used, based on tyrosine and tryptophan being 1350 and 5550, respectively. Solutions of apoC-II are made in either 0.1 *M* NH_4HCO_3 buffer or 3 *M* guanidinium hydrochloride and the absorbances measured at 280 nm.

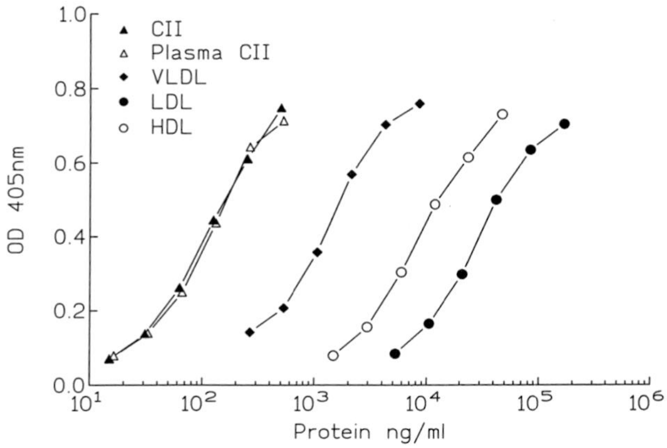

FIG. 4. Dilution series of purified apoC-II standard, plasma, and isolated lipoprotein fractions. Microtiter wells were coated with 100 μl immunopurified anti-apoC-II, at a concentration of 5 μg protein/ml. Six serial, doubling dilutions of HPLC-purified apoC-II, plasma, VLDL ($d < 1.006$ g/ml), low-density lipoprotein (LDL, $1.006 < d < 1.063$ g/ml), and high-density lipoprotein (HDL, $d < 1.21$ g/ml) were made in PBST, starting at a dilution estimated to contain about 500 ng apoC-II/ml. Anti-apoC-II, linked to alkaline phosphatase and diluted 2000-fold, was used for detection. Curves for apoC-II and plasma are expressed as apoC-II concentration; those for the isolated lipoprotein fractions are expressed as protein concentration.

The Lowry protein method (modified to include 0.2% sodium dodecyl sulfate) consistently overestimated the amount of apoC-II by factors of 2 and 1.4 when compared to values obtained by amino acid analysis and from the molar extinction coefficient, respectively. A typical preparation of apoC-II standard would be assigned a value of 0.87 mg/ml by the modified Lowry method, 0.45 mg/ml based on amino acid analysis, and 0.62 mg/ml based on the molar extinction coefficient. Owing to these differences, it is important to provide the details and reproducibility of the method used to assign mass values.

Calibration of Standardized Serum and Quality Control Serum. Once an accurate mass of the standard has been established, a serum pool can be calibrated and used as a secondary standard. A 200-ml serum pool, filtered and mixed well, can be divided into aliquots in 0.5-ml O-ring seal tubes (Sarstedt, Princeton, NJ) and stored at −70° for at least 1 year. A target value can be obtained after 20 determinations. The use of a secondary standard results in improved between-run precision. Similarly, a quality control pool can be prepared and used to monitor interassay variation. The intraassay and interassay variation of our control pools is 4 and 9%, respectively. The use of curve-fitting software to fit an equation to the standard data and to interpolate concentrations of samples improves precision and reproducibility.

Linearity of Response. The National Committee for Clinical Laboratory Standards protocol EP-6P[18] is a useful exercise to demonstrate linearity of the assay in the range of expected apoC-II serum concentrations. A serum of low apoC-II concentration is mixed at different ratios, with a serum of high apoC-II concentration, to produce five pools, equally spaced in concentrations, that are measured in quadruplicate. The dose–response curve should be linear.

Impact of Heterogeneity of Apolipoprotein C-II on Quantification and Separation

Heterogeneity of Plasma Apolipoprotein C-II

Proteolytic Processing. Human plasma apoC-II exists in two forms that differ by the N-terminal hexapeptide Thr-Gln-Gln-Pro-Gln-Gln. The major form of apoC-II in human plasma is the full-length 79-residue protein containing this hexapeptide, with the processed form being a minor component. Fojo *et al.*[19] were the first to suggest that apoC-II could be processed by a mechanism similar to that for the proteolytic processing of apoA-I and apoA-II. Studies by others have identified species variation in the amino-terminal sequence of apoC-II, and in cynomolgus monkeys processing of apoC-II is relatively efficient due to a sequence polymorphism.[20] The only reliable method for distinguishing the two forms of apoC-II consists of two-dimensional electrophoresis.[21] Both forms of apoC-II have been reported to be equivalent activators of lipoprotein lipase.[19] These can be theoretically distinguished by the use of an antipeptide antibody directed against this sequence, as has been reported for apoA-I.[22]

[18] National Committee for Clinical Laboratory Standards. Evaluation of the linearity of quantitative analytical methods; Proposed Guideline. NCCLS publication EP6-P. Vol. 6, No. 18. NCCLS, Villanova, Pennsylvania, 1986.

[19] S. S. Fojo, L. Taam, T. Fairwell, R. Ronan, C. Bishop, M. S. Meng, J. M. Hoeg, D. L. Sprecher, and H. B. Brewer, Jr., *J. Biol. Chem.* **261,** 9591 (1986).

[20] P. N. Herbert, L. L. Bausserman, K. M. Lynch, A. L. Saritelli, M. A. Kantor, R. J. Nicolosi, and R. S. Shulman, *Biochemistry* **26,** 1457 (1987).

[21] D. L. Sprecher, L. Taam, and H. B. Brewer, Jr., *Clin. Chem.* **30,** 2084 (1984).

[22] A. Barkia, C. Martin, P. Puchois, J. C. Gesquiere, C. Cachera, A. Tartar, and J. C. Fruchart, *J. Lipid Res.* **29,** 77 (1988).

Oxidation. Hughes *et al.*[23] have reported the presence of two forms of apoC-II, separated by reversed-phase HPLC. Chemical characterization was consistent with oxidation of methionine, with the more oxidised form eluting earlier (i.e., as a more hydrophilic protein). von Eckardstein *et al.*[24] analyzed two apoC-II isoforms, isolated by preparative isoelectric focusing, followed by reversed-phase HPLC. The two isoforms were shown to be native apoC-II and apoC-II with methionine sulfoxide, consistent with modification of both Met-9 and Met-60. Apolipoprotein C-II with only a single modified Met residue has not been identified. Separation of the apolipoproteins by reversed-phase HPLC of delipidated apoHDL or apoVLDL allows estimation of the amount of native and oxidized apoC-II. We typically see minor amounts of oxidized apoC-II; however, the amount can be altered during solvent delipidation and may be artifactually increased by the use of poor-quality solvents or conditions that promote oxidation.

Immunoblot and Reaction Disproportionate to Mass. We have made extensive use of Western blotting in our studies of structural mutations of apoC-II.[25,26] We have consistently noted that the minor isoforms of normal apoC-II, as judged by protein stain, are much more intense when judged by immunoblot. The size and isoelectric points of these minor isoforms are consistent with oxidized forms of apoC-II. We conclude that the immunostaining of oxidized apoC-II, when using an unfractionated polyclonal antiserum, is disproportionately greater than would be expected.

Genetically Determined Structural Heterogeneity of Apolipoprotein C-II

Rare Mutations in Apolipoprotein C-II. The mutants apoC-II$_{\text{Toronto}}$[25] (apoC-II-T) (residues Asp^{69}–Glu^{79} mutated to Thr^{69}–Cys^{74}) and apoC-II$_{\text{St. Michael}}$[26] (apoC-II-S) (residues Gln^{70}–Glu^{79} mutated to Pro^{70}–Pro^{96}) have a cysteine residue not present in normal C-II. We have used the chemical reactivity of cysteine to make charged derivatives of the mutants and chromogenic derivatives, such as pyridylethylcysteine. These modifications allow distinction of the mutant from the normal apoC-II by their behavior in isoelectric focusing and their chromatographic properties and enhanced UV absorbance.

[23] T. A. Hughes, M. A. Moore, P. Neame, M. F. Medley, and B. H. Chung, *J. Lipid Res.* **29,** 363 (1988).

[24] A. von Eckardstein, M. Walter, H. Holz, A. Benninghoven, and G. Assman, *J. Lipid Res.* **32,** 1465 (1991).

[25] P. W. Connelly, G. F. Maguire, T. Hoffman, and J. A. Little, *Proc. Natl. Acad. Sci. U.S.A.* **84,** 270 (1987).

[26] P. W. Connelly, G. F. Maguire, and J. A. Little, *J. Clin. Invest.* **80,** 1597 (1987).

Chromatography to Detect Pyridylethyl Cysteine Residues. The presence of cysteine in both apoC-II-T and apoC-II-S was useful in elucidating the structure of both proteins. In the case of apoC-II-T, 50 mg acetone-soluble apoVLDL from a homozygous subject was reduced and reacted with 25

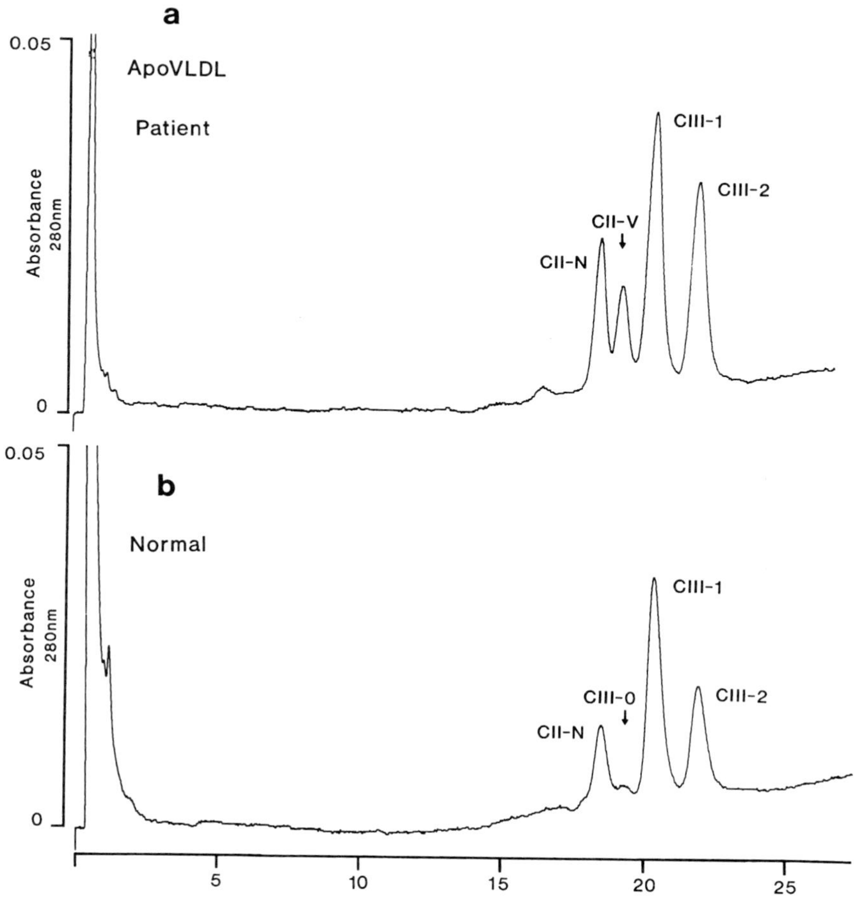

FIG. 5. Anion-exchange HPLC of apoVLDL of (a) subject heterozygous for the C-II variant, C-II-Thr-19, and (b) subject with normal C-II, C-II-Lys-19. For sample preparation, 150 μg apoVLDL was dissolved in 100 μl buffer A (30 m*M* Tris-HCl, pH 7.45, containing 40% 2-propanol), at 37° for 30 min. The HPLC conditions were as follows: column, AX-300 (3 cm × 4.6 mm); sample, 150 μg apoVLDL injected; elution, 8 min at 1% buffer B (0.3 *M* Tris-HCl, pH 7.45, containing 50% acetonitrile) followed by a gradient from 1 to 100% B in 15 min at a flow rate of 1 ml/min; detection, 280 nm at a range of 0.05 AU. Peak identification: CII-N, apoC-II-Lys-19; CII-V, apoC-II-Thr-19; CIII-0, apoC-III-0; CIII-1, apoC-III-1; CIII-2, apoC-III-2.

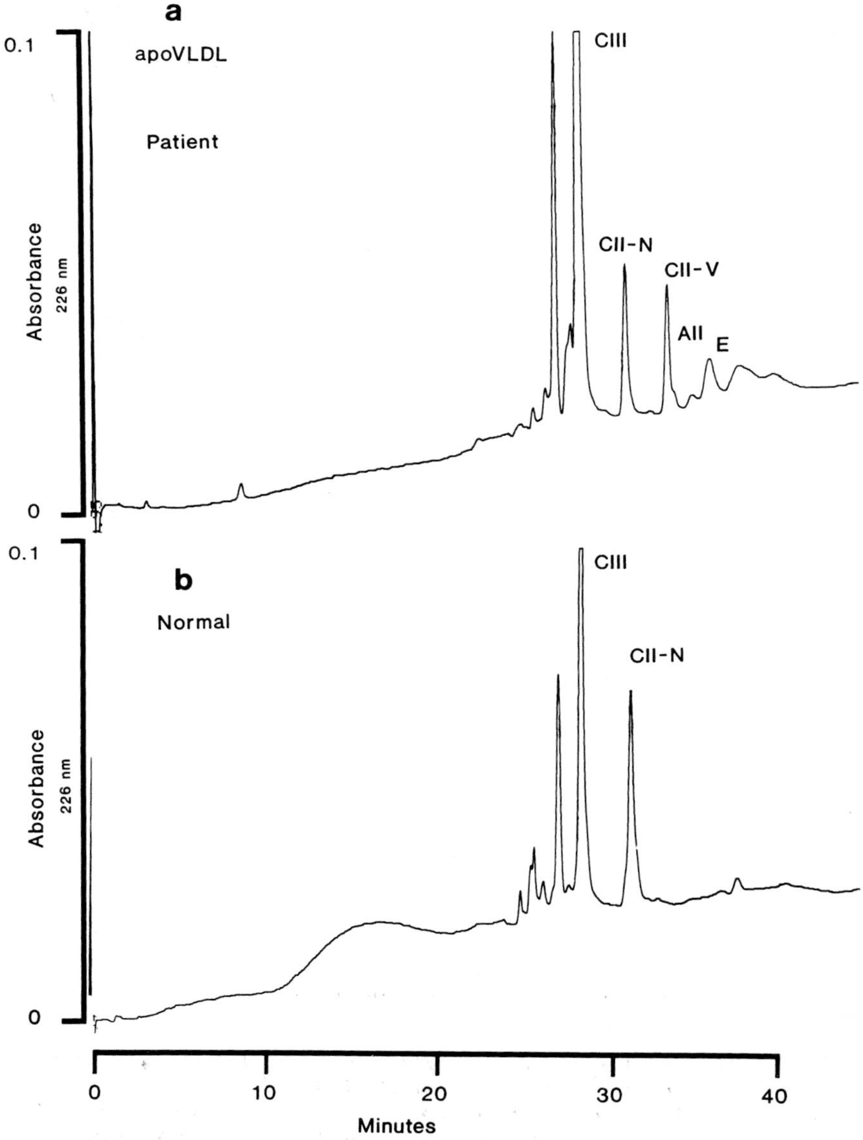

FIG. 6. Reversed-phase HPLC of apoVLDL of (a) subject heterozygous for the C-II mutant, C-II-Thr-19, and (b) subject with normal CII, C-II-Lys-19. For sample preparation, 150 μg apoVLDL was dissolved in 150 μl of 0.1 M NH_4HCO_3, 0.2% sodium decyl sulfate, pH 8.0. The HPLC conditions were as follows: column, Supelcosil LC-308 (5 cm $\times$ 4.6 mm); sample, 20 μg apoVLDL injected; elution, stepwise gradient of 15–60% buffer B (0.067%

m*M* 4-vinylpyridine. Pyridylethylated apoC-II-T was then isolated using preparative isoelectric focusing. Normal apoC-II, apoC-II-T, and pyridylethyl-apoC-II-T were digested with trypsin and the tryptic peptides separated on reversed-phase HPLC.[25] The pyridylethylated peptide, containing cysteine, was identified by the increased absorbance at 280 nm, relative to 226 nm.

Neither of these mutants has been detected in our normal ELISA for apoC-II. We have shown that most of apoC-II-T is covalently attached to other cysteine-containing apoproteins (apoB-100, apoE3, apoA-II, and itself) and a relatively small amount exists in a monomeric form.[27]

Rocket electroimmunoassay for apoC-II in heterozygotes for apoC-II-T can result in a "blasted" rocket, that is, an immunoprecipitin arc without a peak.[28] This suggests that the mutant interferes with the complete immunoprecipitation of normal apoC-II.

Analysis of Rare Variants of Apolipoprotein C-II

Two structural polymorphisms of apoC-II have been documented. These are apoC-II-Thr-19 and apoC-II-Gln-55. Both variants of apoC-II have been reported to activate LPL as effectively as normal apoC-II. In each case, the variant amino acid has been substituted for a lysine residue, which has resulted in a change in the isoelectric point relative to normal apoC-II. The isoelectric points of both variants are very similar and indistinguishable from that of oxidized apoC-II. We have quantified the amount of apoC-II-Thr-19 in apoVLDL by isoelectric focusing and densitometry of the protein-stained gel, by anion-exchange HPLC (Fig. 5) and reversed-phase HPLC (Fig. 6). The ratio of normal to variant was 1.5 : 1 by all methods. Our initial observations with regard to apoC-II-Thr-19 suggest that this variant may have an altered distribution between VLDL and HDL relative to normal apoC-II.[29] We have studied a family with subjects

[27] P. W. Connelly, G. F. Maguire, C. Vezina, R. A. Hegele, and J. A. Little, *J. Lipid Res.* **34,** 1717 (1993).

[28] W. C. Breckenridge, personal communication (1983).

[29] M. W. Huff, A. J. Evans, B. M. Wolfe, P. W. Connelly, G. F. Maguire, and W. L. P. Strong, *J. Lipid Res.* **31,** 385 (1990).

TFA in 60% acetonitrile) in buffer A (0.0675% TFA in distilled water) over 15 min, followed by 60–95% B in 25 min, flow rate of 1 ml/min; detection, 226 nm, at a range of 0.1 AU. Peak identification: CII-N, apoC-II-Lys-19; CII-V, apoC-II-Thr-19; CIII, apoC-III-0, apoC-III-1, and apoC-III-2; AII, apoA-II; E, apoE. The major peak immediately preceding C-III coelutes with authentic apoC-I.

heterozygous for the apo-II-Gln-55 variant and have found that there is complete concordance of identification of heterozygotes using detection by densitometry of apoVLDL and genomic DNA sequence analysis (P. W. Connelly, C. Vezina, and G. F. Maguire, unpublished observations, 1993). We have not applied reversed-phase HPLC to the analysis of the apoC-II-Gln-55 variant; however, it would be anticipated that it could be separated from normal apoC-II by HPLC.

The unequivocal identification of the structure of variant apoC-II is now most simply made by polymerase chain reaction-based analysis of genomic DNA.[30]

[30] R. A. Hegele, P. W. Connelly, G. F. Maguire, M. W. Huff, L. Leiter, B. M. Wolfe, A. J. Evans, and J. A. Little, *Dis. Markers* **9,** 73 (1991).

[12] Immunochemical Methods for Quantification of Human Apolipoprotein C-III

By MOTI L. KASHYAP

Introduction

The C group of apolipoproteins is mainly found in the triglyceride-rich lipoproteins [very-low-density lipoprotein (VLDL) and chylomicrons] and high-density lipoproteins (HDL). Apolipoprotein C (apoC) consists of three peptides (apoC-I, C-II, and C-III) with distinct differences in structure and physical and chemical properties. Apolipoprotein C-III is a peptide (molecular weight 9300) that exists in at least three forms. The species contain 0, 1, or 2 mol of sialic acid and are known as apoC-III_0, C-III, and C-III_2, respectively.

Although the exact function of apoC-III is unclear, current evidence indicates that it is intimately involved in normal and abnormal triglyceride metabolism. Early experiments indicated that apoC-III inhibited apoC-II activation of lipoprotein lipase. However, subsequent experiments using perfused rat liver indicated that whereas apoE or its isoforms promote the recognition by hepatic receptors for chylomicron remnants or intermediate-density lipoproteins, apoC-III inhibited uptake of the apoE-mediated removal of triglyceride-rich particles.[1] Studies indicate that transgenic mice

[1] E. Windler, Y. Chao, and R. J. Havel, *J. Biol. Chem.* **255,** 8303 (1980).

with overexpression of human apoC-III have severe hypertriglyceridemia.[2,3] The apoC-III-rich particles behave normally as substrates for lipoprotein lipase, but uptake of such particles by hepatocytes is impaired, supporting the earlier findings that the major mechanism for hypertriglyceridemia in disorders characterized by excess apoC-III may be by inhibition of uptake by the liver of apoE-mediated clearance of triglyceride-rich particles.[4] Thus, there is a need for useful methods for measurement of apoC-III and its subspecies in the evaluation of hypertriglyceridemic states in humans and also in experimental work at various levels that may require sensitive and accurate methods for measurement of this important apolipoprotein, which may underlie the pathophysiology of certain hyperlipidemic states.

Apolipoprotein C-III, like the other C apoproteins, is distributed in both the triglyceride-rich lipoprotein and in HDL. High-density lipoprotein acts as a reservoir for the C apoproteins, and during VLDL production or chylomicron production (from exogenous fat intake) apoC's transfer from HDL to chylomicrons for subsequent metabolism.[5] Thus, with increasing triglyceride levels and, hence, increased concentrations of triglyceride-rich particles, greater proportions of apoC-II and apoC-III are present in the triglyceride-rich lipoproteins with reciprocal decline in the amount of C apoproteins in HDL.[6–8] During triglyceride-rich lipoprotein catabolism in which apoC-II activates lipoprotein lipase, the C apoproteins are transferred back to HDL, and the catabolic products of triglyceride-rich lipoproteins are taken up by the liver. Thus, it appears that apoC-II activates lipoprotein lipase and exposes apoE for clearance of such particles by the liver, and apoC-III appears to modulate this process. Excessive production of apoC-III therefore can lead to hypertriglyceridemic states, which are characterized by increased contents of apoC-III relative to other apoproteins in such particles.[4]

[2] Y. Ito, N. Azrolan, A. O'Connel, A. Walsh, and J. L. Breslow, *Science* **249,** 790 (1990).

[3] H. V. Desilva, S. J. Lauer, R. W. Mahley, K. H. Weisgraber, and J. M. Taylor, *Biochem. Soc. Trans.* **21,** 483 (1993).

[4] K. Aalto-Setala, E. A. Fisher, X. Chen, T. Chajek-Shaul, T. Hayek, R. Sechner, A. Walsh, R. Ramakrishnan, H. N. Ginsberg, and J. L. Breslow, *J. Clin. Invest.* **90,** 1889 (1992).

[5] R. J. Havel, J. P. Kane, and M. L. Kashyap, *J. Clin. Invest.* **52,** 32 (1973).

[6] M. L. Kashyap, L. S. Srivastava, B. A. Hynd, P. S. Gartside, and G. Perisutti, *J. Lipid Res.* **22,** 800 (1981).

[7] M. L. Kashyap, L. S. Srivastava, C. Y. Chen, G. Perisutti, M. Campbell, R. F. Lutmer, and C. J. Glueck, *J. Clin. Invest.* **60,** 171 (1977).

[8] M. L. Kashyap, L. S. Srivastava, B. A. Hynd, G. Perisutti, D. W. Brady, P. S. Gartside, and C. J. Glueck, *Lipids* **13,** 993 (1978).

Requirements of Valid Immunoassay for Apolipoprotein C-III

In the development of an assay in the laboratory, the most important qualities include specificity, accuracy, sensitivity, and precision. The sensitivity varies for different methods discussed below, and precision also varies inherently with different methodologies. The most important and difficult aspects of the assay include tests for monospecificity of the antibody and, even more importantly, the accuracy of the method. For monospecificity, it is important that the particular monoclonal or polyclonal antibody have no cross-reactivity with any other apolipoprotein or have homology with any other protein in a biological sample such as plasma. Methods for testing specificity are discussed below.

The question of accuracy is more difficult in apolipoprotein quantification because, unlike free proteins in solution, apoproteins are bound to other constituents of the lipoprotein particle; thus, certain epitopes may not be accessible to the antibody. For this reason the antibody chosen should have the ability to measure an epitope that is universally expressed in all lipoprotein particles, or the sample may have to be pretreated to expose all immunoreactive epitopes. Thus, a given antibody should be capable of measurement of total mass of the antigen bound to lipoprotein particles in samples. Various approaches to assess accuracy for apoC-III are also discussed in the description of the methodology.

The antibody should also be capable of recognizing all isoproteins of apoC-III identically. In hypertriglyceridemic states in humans, the distribution of these isoforms can be abnormal. Hence, there is a need for the absolute quantitation of individual isoforms in a given sample. To achieve this objective, the apoC-III isoforms can be separated by analytical isoelectric focusing with staining of the individual band and quantitation of each apoprotein by quantitative densitometry. If this approach is taken, it is important that the area under the scan for each band be in the rectilinear portion of the curve describing the relationship between apoC-III mass and densitometric area.[6] Alternatively, anion-exchange fast protein liquid chromatographic (FPLC) methods can also be used for quantitation of the individual isoforms.[9,10]

[9] H. Mezdour, V. Clavey, I. Kora, M. Koffigan, A. Barkia, and J. Fruchart, *J. Chromatogr.* **414,** 35 (1987).

[10] P. Weisweiler, C. Friedl, and P. Schwandt, *Clin. Chem.* **32,** 992 (1986).

Overview of Immunochemical Methods for Apolipoprotein C-III Quantitation

The principal methods that are used include radioimmunoassay (RIA),[6,11,12] electroimmunoassay (EIA),[13] radial immunodiffusion,[14] enzyme-linked immunoassay (ELISA),[15–20] immunonephelometric assay,[21] and immunoturbidometric methods.[22,23] A summary of the pros and cons of these methods follows.[24] The radioimmunoassay is sensitive, useful for dilute solutions and *in vitro* experimental work, etc. It spares antibody and the primary apoC standards, which are sometimes difficult to isolate in sufficient quantities. High throughput is possible with automation, and with appropriate antibodies, no plasma pretreatment may be required. The limitations are that it is an expensive method to set up and maintain. The electroimmunoassay method is relatively rapid, simple, and less expensive, but its disadvantage is that large amounts of polyclonal antibodies are necessary. It has a low throughput, and delipidation of lipemic samples may be necessary for accurate determination. It is also not as sensitive as the radioimmunoassay and may be of limited use with dilute lipoprotein subfractions or experimental work.

The radioimmunodiffusion method is the simplest; however, limitations are as for the electroimmunoassay, and it may also be less sensitive than EIA. The ELISA can be performed as a competitive or noncompetitive (sandwich) assay. This method has several advantages over radioimmunoassay but may have lesser sensitivity and in most laboratories has a lower

[11] G. Schonfeld, P. K. George, J. Miller, P. Reilly, and J. Witztum, *Metabolism* **28,** 1001 (1979).

[12] S. J. T. Mao, P. K. Bhatnager, A. M. Gotto, Jr., and J. T. Sparrow, *Biochemistry* **19,** 315 (1980).

[13] M. D. Curry, W. J. McConathy, J. D. Fesmire, and P. Alaupovic, *Biochim. Biophys. Acta* **617,** 503 (1980).

[14] K. Itakura, T. Matsudate, T. Sakurai, S. Hashimoto, K. Ito, H. Kanno, M. Hirata, and K. Nakamura, *Clin. Chim. Acta* **161,** 275 (1986).

[15] L. Holmquist, *J. Immunol. Methods* **34,** 243 (1980).

[16] D. Parsy, V. Clavey, C. Fievet, I. Kora, P. Duriez, and J. C. Fruchart, *Clin. Chem.* **31,** 1632 (1985).

[17] J. Bury and M. Rosseneu, *J. Clin. Chem. Clin. Biochem.* **23,** 63 (1985).

[18] M. Fu, B. Liu, W. Liu, and D. Fang, *Clin. Chim. Acta* **167,** 339 (1987).

[19] W. F. Riesen and E. Sturzenegger, *Ann. Clin. Biochem.* **24,** 66 (1987).

[20] N. Alsayed, R. Rebourcet, and J. Chapman, *Clin. Chem.* **36,** 2047 (1990).

[21] J. Bury, W. De Keersgieter, M. Rosseneu, F. Belpaire, and J. Christophe, *Clin. Chim. Acta* **145,** 249 (1985).

[22] N. Rifai and L. M. Silverman, *Clin. Chem.* **32,** 1969 (1986).

[23] A. Noma, Y. Hata, and Y. Goto, *Clin. Chim. Acta* **199,** 147 (1991).

[24] M. L. Kashyap, *in* "Proceedings of the Workshop on Apolipoprotein Quantification" (K. Lippel, ed.), p. 373. NIH Publication No. 83, Bethesda, Maryland, 1983.

precision than the RIA. The immunonephlometic assay is relatively simple, rapid, and can be automated, but it requires a nephelometer, is influenced by lipoprotein particle size, and therefore requires a preliminary pretreatment step to the sample. Relatively large amounts of antibody may be required, and it has a lower sensitivity than the enzyme-linked immunoassay or the radioimmunoassay. Immunoturbidometric methods can be adapted to spectrophotometric analyzers and may be of slight advantage compared to immunonephelometric methods which require a nephelometer. This may be a useful technique where high throughput is needed, and where sensitivity is not an important issue.

Quantification of Apolipoprotein C-III and Subspecies by Combined Radioimmunoassay and Isoelectric Focusing

The detailed method to be described here is the radioimmunoassay,[6] which is a comprehensive method for measurement of this apoprotein precisely, accurately, and with high sensitivity. References to other immunological methods are given after a detailed description of the radioimmunoassay method. Several steps in the preparation of the reagents are similar, and the principles about validity of the assay as discussed above should be kept in mind for adaptation of the particular method chosen for a given purpose.

Antigen Isolation and Characterization

Because apoC-III_1 is the predominant isoform of apoC-III, this should be isolated from pooled human plasma. Ideal patients for such collections are patients with severe endogenous hypertriglyceridemia with fasting chylomicronemia (type V lipoprotein phenotype) who are subjected to a plasma exchange of several units. The blood should be collected in tubes containing the disodium salt of ethylenediaminetetraacetic acid (EDTA) to give a final concentration of 1.0 mg/ml. Plasma is then separated by slow-speed centrifugation at 4°. The triglyceride-rich lipoproteins (TRL) are isolated by preparative ultracentrifugation. The TRL should be ultracentrifuged at least once and possibly twice to get rid of any contaminating lipoproteins or albumin after layering with 0.15 *M* sodium chloride in 1 m*M* disodium salt of EDTA (EDTA–saline) for further purification.

Apolipoprotein C-III_1 is isolated from the delipidated TRL according to Brown *et al.*[25] This method involves extensive dialysis of VLDL against EDTA–saline, lyophilization and delipidation with cold (4°) ethanol–

[25] W. V. Brown, R. I. Levy, and D. S. Fredrickson, *J. Biol. Chem.* **244,** 5687 (1969).

diethyl ether at least twice, followed by two washes with dimethyl ether alone. The ethyl ether is evaporated under a gentle stream of nitrogen. The apoVLDL is dissolved in a 0.2 *M* Tris-HCl buffer, pH 8.2, containing 0.1 *M* sodium decyl sulfate. Approximately 40–60 mg apoTRL are applied to a Sephadex G-150 (or Sephacryl) column (1.5 × 100 cm). Two major peaks of apoB and apoC will be obtained. The C apoproteins are isolated, then dialyzed extensively against 5 m*M* Tris-HCl, pH 8.2, at 4°. This material is applied to a column of DEAE-cellulose equilibrated with 5 m*M* Tris-HCl, pH 8.2, at 4°. Individual C apoproteins are separated by means of a sodium chloride gradient to a concentration to 0.15 *M*. Apolipoproteins C-I, C-II, C-III_1, and C-III_2 are resolved. The protein in tubes containing apoC-III_1 is subjected to analytical isoelectric focusing for assessment of purity. The tubes from the top and descending slope of the apoC-III_1 peak, which show a single band on analytical isoelectric focusing, are pooled and dialyzed extensively against 50 m*M* ammonium bicarbonate and lyophilized. The amino acid composition of the protein is then determined and should be used to confirm the purity of the antigen.

Mezdour and associates[9] have also described a method using an anion-exchange fast protein liquid chromatographic method for separation and purification of apolipoproteins from human plasma. This method is rapid and may be used where facilities are available. Nevertheless, the purity of the antigen used should be confirmed.

Antibody Preparation

Female albino New Zealand rabbits weighing approximately 2.5 kg are immunized with approximately 100 μg apoC-III_1 (in 0.5 ml of 0.15 *M* sodium chloride) emulsified with 0.15 ml Freund's complete adjuvant. Injections are given directly into the popliteal lymph nodes and also intradermally in multiple sites on the back. Booster injections (100 μg in Freund's complete adjuvant) on the back are given at approximately 3-week intervals and the animals bled 14 days after the last injection. The serum can be kept at −70°. Depending on the titer of the antibody, experiments will have to be done to determine the exact dilution required for the assay development.

Labeling Apolipoprotein C-III_1 with Iodine-125

Basically, the method of Greenwood *et al.*[26] is used. To 5 μg lyophilized apoC-III_1, 0.1 ml of 0.5 *M* phosphate buffer, pH 7.6, is added, and the tube is gently tapped to dissolve the apoprotein. This is followed by the addition of 0.5 mCi ^{125}I and 87.5 μg chloramine-T in 25 μl of 50 m*M* phosphate

[26] F. C. Greenwood, W. M. Hunter, and J. S. Glover, *Biochem. J.* **89,** 114 (1963).

buffer, pH 7.6. After 30 sec, 250 μg sodium metabisulfite, in 0.1 ml of 50 mM phosphate buffer, pH 7.6, is added to stop the reaction. Fifty microliters of 7% bovine serum albumin is then added. The mixture is subjected to gel chromatography using Sephadex G-100 or Sephacryl. A 1 $\times$ 100 cm column previously coated with 3% bovine serum albumin can be used. The radioiodinated apoC-III is then eluted with 70 mM barbital buffer containing 1% bovine serum albumin, pH 8.6, to separate labeled apoC-III from [^{125}I]iodine. The elution profile should show an initial peak of high molecular weight aggregates and a second larger peak of highly purified labeled apoC-III_1. The homogeneity of the radiolabeled apoC-III_1 should be checked by validating that it coelutes in the same peak on gel chromatography with unlabeled apoC-III_1 and also that it comigrates with unlabeled apoC-III_1 by electrophoretic techniques such as isoelectric focusing.

Assay

To a 10 $\times$ 75 mm glass tube are added (a) 0.4 ml of 1% bovine serum albumin in 0.2 M barbital buffer (pH 8.5), (b) 0.1 ml plasma or unknown sample diluted 1:100–1:1200 (or greater with barbital buffer), or apoC-III_1 standard, (c) 0.1 ml antibody diluted to an appropriate level with barbital buffer containing normal rabbit serum, and (d) 0.1 ml of ^{125}I-labeled apoC-III_1 containing approximately 10,000 counts/min (cpm). The assay is performed with tubes at 4°. After incubation of the mixture for 72 hr at 4°, 0.1 ml goat anti-rabbit gammaglobulin is added. The samples are incubated for 48 hr at 4° and then centrifuged at slow speed for 20 min at 4°. The supernatant is aspirated then washed (with 1.0 ml barbital buffer), and radioactivity in the precipitate is measured in a autogamma scintillation spectrometer. All measurements should be done at least in duplicate.

Analytical Isoelectric Focusing of Apolipoprotein C-III Subspecies in Triglyceride-Rich Lipoproteins

As indicated above, in order to measure the absolute concentration of apoC-III subspecies in the plasma, the suggested approach is to isolate VLDL and to scan, preferably with a laser densitometer, the individually separated apoC-III isoforms to obtain the relative ratios. These ratios are then applied to the total plasma concentration of apoC-III measured by the immunoassay. The assumption made in this approach is that the distribution of apoC-III isoforms in VLDL and HDL are identical. A similar approach (of separating apoC-III isoforms in HDL) is not practical because other apoproteins (e.g., apoA-II) complicate the band picture and make it less amenable for precise quantitation of the specific apoC-III isoforms.

One method for quantitative determination of the distribution of apoC-

III isoforms in VLDL is as follows. An aqueous acrylamide/bisacrylamide stock solution is made using 30 g acrylamide, 0.8 g *N,N*-bisacrylamide (Eastman Kodak, Rochester, NY), and glass-distilled water to give a final solution of 100 ml. Six milliliters acrylamide/bisacrylamide stock solution is placed in a flask together with 20 μl N,N,N^1,N^1-tetramethylethylenediamine (TEMED; Eastman Kodak) and 10.24 g ultrapure urea. To this mixture is added 7.8 ml glass-distilled water and an ampholine mixture consisting of 0.8 ml of 40% (v/v) ampholines pH 3.5–5.0, 0.2 ml of 40% ampholines pH 4.0–6.0, and 0.2 ml of 40% ampholines pH 5.0–8.0. The flask is stirred slowly until all the urea is dissolved and then degassed for 15 min. A solution of 14 mg/ml aqueous ammonium persulfate is prepared while the gel is degassing, and 0.8 ml of this solution is added to the flask immediately before the gels are cast. Glass gel tubes are filled with the gel mixture and allowed to polymerize for 30 min. The gels are then prefocused for 1 hr with an anode buffer of 0.1 *M* aqueous phosphoric acid and a cathode buffer of 20 m*M* aqueous sodium hydroxide.

Samples are prepared by placing an appropriate amount of TRL protein in a test tube and incubating at room temperature with an equal volume of 8.6 *M* tetramethylurea (TMU; Sigma, St. Louis, MO), for 15 min. The mixture is then centrifuged at slow speed for 10–15 min, after which half the total volume of the TMU-soluble protein is aspirated and placed in a separate tube together with an appropriate amount of 80% aqueous sucrose solution. The extract–sucrose mixture is then applied directly onto the gel surfaces by layering under the buffer in the tube. The gels are then focused for approximately 75 min and finally focused at a higher voltage for at least 3.5 hr.

Gels are removed from the tubes and stained overnight at room temperature in 0.04% Coomassie Blue G-250 (Sigma) in 3.5% aqueous perchloric acid. Destaining and subsequent gel storage are possible in 7% aqueous acetic acid at room temperature. The gels can be scanned by a laser densitometer, which can be automated to give accurate values of the percent of each peak of the whole. For quantitative work, it is important that the method be validated to show that rectilinearity between the mass of apoC-III applied and the densitometric area is appropriate for the sample to be quantitated. Because apoC-II isoforms are also present in the apoC-III range, it is important to be certain of the bands which represent apoC-III isoforms. To determine this, desialylation of apoC-III with neuraminidase will remove the sialyated forms and result in a single band of apoC-III_0. Apolipoprotein C-II and its isoforms will not behave in this way.

The method for desialylation is as follows. Neuraminidase obtained commercially (free of protease activity) is incubated with TRL isolated by preparative ultracentrifugation. A ratio of 2 units of the enzyme per milligram TRL protein can be used. Incubation is carried out for 24 hr at 37° in 10 m*M* Tris-HCl (pH 6.9), 0.15 *M* sodium chloride, 1 m*M* EDTA, 2 m*M* calcium chloride. The mixture is dialyzed against 0.15 *M* sodium chloride in 1 m*M* EDTA and then ultracentrifuged. The TRL is isolated and then delipidated with tetramethylurea, and apoC-III subspecies and apoC-II are separated by isoelectric focusing. Control tubes contain only normal saline and no enzymes. After isoelectric focusing, the bands are stained as described above.

Sample Preparation

Ideally, a fresh sample should be used. However no extensive studies have been conducted on frozen plasmas. In general, the experience we have indicates that in moderate hypertriglyceridemic plasmas a single thawing of frozen samples does not significantly affect measurement of total apoC-III. We have also found that it is important to prewarm the sample to 37° for 15–20 min. This prevents aggregation of triglyceride-rich particles, particularly those containing saturated fatty acids. The coefficient of variation of the immunoassay when this step is introduced is much reduced. In some assays the sample may have to be pretreated with Tween 20, a detergent, to expose all antigenic sites. Additionally, in very lipemic samples, delipidation may be necessary to do the same.

Standards

The primary standard used initially should be apoC-III_1 or apoC-III_2. The concentration of a secondary standard or calibrator can then be assigned to a given plasma sample, which can be subsequently used as the calibrator. For controls a plasma from a normal triglyceridemic and a hypertriglyceridemic patient should be chosen to monitor quality control.

A valid assay is one that is highly specific for apoC-III and is accurate in the measurement of this apoprotein. A major difference between validation of these two criteria for proteins in free solution and for apoproteins is that apoproteins are part of a lipoprotein particle, thereby giving rise to two difference sources of variation. First, the apoprotein conformation in different lipoprotein particles may be such as to give different affinities for a given antibody. Thus, it is important to validate not only that the antibody is highly specific for apoC-III, but also that the affinity for the antibody to the apoprotein is uniform in all lipoprotein particles measured. The second problem is that of accuracy, in which the apoprotein epitopes may not be available to the antibody in all lipoprotein particles.

Monospecificity of the antibody is ascertained by making certain that there is no cross-reactivity with other apolipoproteins using individually isolated apolipoproteins and, for apoB, using narrow-cut low-density lipoprotein (LDL; i.e., density 1.030–1.050 g/ml). The cross-reactivity can be assessed by double immunodiffusion (Ouchterlony method), immunoelectrophoresis, or radioimmunoassay via measurement of the lack of significant displacement by all other apoproteins and lipoprotein-free plasma components in the assay. The question of affinity of anti-apoC-III to apoC-III in different lipoprotein particles is assessed by determining the displacement curves obtained by different particles of lipoproteins containing apoC-III. Lipoproteins can be separated by column chromatography or by ultracentrifugation, and individual fractions, namely, VLDL, intermediate-density lipoproteins (IDL), and HDL, can be assessed for parallelism with the standard curve using the pure apoprotein as standard. Likewise, parallelism between the different isoforms of apoC-III should also be assessed to confirm that a particular antibody does not have different affinities which would be reflected in nonparallelism with the primary standard curve.

As for accuracy, the following approaches are recommended. (a) Quantitative recovery of apoC-III in separated lipoprotein samples can be measured. Plasma samples with a wide range of plasma triglyceride values should be ultracentrifuged at density 1.006 g/ml and the apoC-III content in the supernatant and infranatant measured. The sum of the subfractions should approximately equal to the plasma level. (b) Pure apoC-III can be added to a plasma sample and the recovery calculated. (c) Plasma samples can be delipidated using ethanol–diethyl ether or ethanol–acetone, and the total mass measurement of the delipidated apoprotein should be similar to that in the native plasma. (d) The use of Tween 20 under certain circumstances has facilitated measurement of apoproteins because this detergent results in exposure of epitopes that may be hidden from the antibody. Thus plasma apoC-III should be measured with and without Tween 20, and if the values are similar the detergent can be used as part of a pretreatment of the sample prior to measurement.

Other Immunochemical Methods of Apolipoprotein C-III Measurement

Two enzyme-linked immunosorbent assay (ELISA) methods are in common use. These are the noncompetitive (sandwich) enzyme-linked immunoassays[16,17,19,20] and the competitive immunoassays.[15,18] In a noncompetitive sandwich ELISA, Parsy *et al.*[16] use affinity-purified antibodies to apoC-III absorbed to the surface of microtiter plates. After washing, the solid-phase antibody is incubated with antigen (serum from fasting subjects), washed, and then incubated with peroxidase-labeled purified antibodies to apolipoprotein C-III. After a final wash, the bound label is assayed, providing a direct measurement of the antigen. A similar assay has also been described by Bury and Rosseneu.[17] In the competitive assays, the apoC-III for apoC-III-containing lipoprotein (e.g., VLDL) can be used to coat microtiter plates. The binding of antibody to antigen is measured by use of second antibody labeled with alkaline phosphatase or with peroxidase-labeled purified antibodies to apoC-III. Inhibition of binding is measured by addition of unlabeled apoC-III_1 for development of the standard curve. Unknowns are measured against the standard curve using the primary or a secondary standard. For other assays, readers are referred to earlier papers which give details of the turbidometric,[22,23] immunonephelometric,[21] radial immunodiffusion,[14] and electroimmunoassay methods.[13] The pros and cons of the assays have been discussed earlier, and a detailed description of the methods is given in these publications.

An interlaboratory, national or international, standardization program for the measurement of apoprotein C-III is not yet available. If future research indicates the utility of this method beyond use in basic and clinical research, standardization of the principles of isolation and characterization of reagents for the development of such an assay in different laboratories is possible.

[13] Electrophoretic Methods for Quantitation of Lipoprotein [a]

By JOHN W. GAUBATZ, PAVAN MITAL, and JOEL D. MORRISETT

Introduction

Lipoprotein (Lp) [a] is a lipoprotein whose plasma concentration is highly correlated with cardiovascular disease. Electrophoretic methods have been used widely for many years for qualitative detection and quantitative measurement of plasma lipoproteins, especially Lp[a]. The movement of a lipoprotein or apoprotein in an electrophoretic field is determined by several physiochemical properties such as charge, size, shape, and composition. When electrophoretic migration is used with immunochemical reaction, the combined properties can be used for positive identification of specific migrating species. However, migration and immunoreactivity may be significantly altered by variations in specific structural features. In the case of apolipoprotein (apo) [a], variations in the number of kringle repeats, carbohydrate content, and disulfide pairing can have profound effects. For

Lp[a], lipid content and association with other apoB-containing lipoproteins can be important determinants of migratory and immunoreactive behavior.

In this chapter, the topic of electrophoretic methods for quantification has been expanded beyond concentration determination to include other properties of apo[a] or Lp[a] amenable to measurement. These include apo[a] polymorph classification and phenotyping for individual and population analyses, molecular weight (M_r) determination of apo[a] polymorphs, and calculation of relative electrophoretic mobility. Investigations that combine various aspects of these analytical electrophoretic methods expand the opportunities for assessing structure and function relationships. Some of the features of apo[a] and Lp[a] and how these impact on the determinations are briefly reviewed.

Properties of Apolipoprotein [a] and Lipoprotein [a] Affecting Electrophoretic Behavior

Physical and Apoprotein Compositional Heterogeneity of Lipoprotein [a]

In plasma Lp[a], apo[a] is complexed to apoB-100 most likely by a single disulfide linkage between one molecule of apo[a] and one molecule of apoB-100. Native, cholesteryl ester-rich Lp[a] is by far the most abundant physical form in which apo[a] appears. However, a less dense triglyceride-rich Lp[a] particle has been reported whose contribution to the total apo[a] plasma pool is dependent on donor prandial state and the method used for Lp[a] isolation.[1-3] In some cases this particle contributed more than 20% of the total plasma apo[a] content, and, when ultracentrifugation was avoided, apoE and the apoC's were also associated with the particle.[1] Hence, the use of ultracentrifugation during isolation may cause the dissociation of triglyceride-rich Lp[a], making it difficult to study this association phenomenon and assess its physiological significance.

Trieu *et al.*[4] have provided evidence that Lp[a] can associate with other apoB-containing lipoproteins such as low-density lipoprotein (LDL) and very-low-density lipoproteins (VLDL) by an interaction that involves lysine

[1] J.-M. Bard, S. Delattre-Lestavel, V. Clavey, P. Pont, B. Derudas, H.-J. Parra, and J.-C. Fruchart, *Biochim. Biophys. Acta* **1127,** 124 (1992).

[2] J. S. Cohn, C. W. K. Lam, D. R. Sullivan, and W. J. Hensley, *Atherosclerosis* **90,** 59 (1991).

[3] C. Sandholzer, G. Feussner, J. Brunzell, and G. Utermann, *J. Clin. Invest.* **90,** 1958 (1992).

[4] V. N. Trieu, T. F. Zioncheck, R. M. Lawn, and W. J. McConathy, *J. Biol. Chem.* **266,** 5480 (1991).

residues of apoB and kringle domains of apo[a].[5] Lipoprotein [a] association with LDL is also suggested by electrophoretic analyses of mixtures of Lp[a] and LDL_2 showing, instead of the characteristic pre-β_1 and β bands for these two lipoproteins, a single band with intermediate mobility. Binding of VLDL and LDL to Lp[a] may be mediated by Ca^{2+}, known to form bridges involving sialic acid residues, which are present on both apoB-100 and apo[a].[6] Apolipoprotein [a] has also been found as a lipid-poor complex containing apoB-100[3,7,8] representing from 1.8 to 27.2% of the total apo[a][9–12] in plasma. One laboratory has indicated that apo[a] in triglyceride-rich Lp[a] and in normal Lp[a] is recognized identically by their antibody.[3] Nevertheless, the presence of apo[a] in lipoprotein particles of different lipid, carbohydrate, and protein composition should be carefully considered when using conventional electrophoretic and immunologic techniques for quantitation.

Kringle Unit Heterogeneity of Apolipoprotein [a]

Apolipoprotein [a] consists of two distinct polypeptide domains: one containing tandemly linked kringle units and the other containing a serine protease-like region. The kringles are triloop structures stabilized by three disulfide bridges and contain 78–80 amino acid residues. Within the kringle domain of the 530-kDa human apo[a] polymorph for which the amino acid sequence has been inferred,[13] there are 11 different kringle types.[14] Apolipoprotein [a] kringle types 1–10 are highly homologous but not identical to plasminogen kringle 4, and apo[a] kringle type 11 is homologous to plasminogen kringle 5. Apo[a] kringle type 2 is repeated a varying number

[5] V. N. Trieu, T. F. Zioncheck, R. M. Lawn, and W. J. McConathy, *Biochemistry* **266,** 5480 (1991).

[6] A. Yashiro, J. O'Neil, and H. F. Hoff, *J. Biol. Chem.* **268,** 4709 (1993).

[7] H.-J. Menzel, H. Dieplinger, C. Lackner, F. Hoppichler, J. K. Lloyd, D. R. Muller, C. Labeur, P. J. Talmud, and G. Utermann, *J. Biol. Chem.* **265,** 981 (1990).

[8] M. L. Koschinsky, J. E. Tomlinson, T. F. Zioncheck, K. Schwartz, D. L. Eaton, and R. M. Lawn, *Biochemistry* **30,** 5044 (1991).

[9] J. W. Gaubatz, C. Heideman, A. M. Gotto, J.D. Morrisett, and G. H. Dahlen, *J. Biol. Chem.* **258,** 4582 (1983).

[10] A. Gries, J. Nimpf, M. Nimpf, H. Wurm, and G. M. Kostner, *Clin. Chim. Acta* **164,** 93 (1987).

[11] C. R. Duvic, G. Smith, W. E. Sledge, L. T. Lee, M. D. Murray, P. S. Roheim, W. R. Galloner, and J. J. Thompson, *J. Lipid Res.* **26,** 540 (1985).

[12] G. M. Fless, M. Snyder, and A. M. Scanu, *J. Lipid Res.* **30,** 651 (1989).

[13] J. W. McLean, J. E. Tomlinson, W.-J. Kuang, D. L. Eaton, E. Y. Chen, G. M. Fless, A. M. Scanu, and R. M. Lawn, *Nature (London)* **330,** 132 (1987).

[14] J. Guevara, A. Y. Jan, R. Knapp, A. Tulinsky, and J. D. Morrisett, *Arterioscler. Thromb.* **13,** 758 (1993).

of times, giving rise to a wide range of apo[a] molecular weight polymorphs, of which 34 have been detected by sodium dodecyl sulfate–polyacrylamide gel electrophoresis (SDS–PAGE).[15]

All monoclonal and polyclonal antibodies against apo[a] described thus far show reactivity for all polymorphs. However, with certain monoclonals,[16] differential immunoreactivity has been observed among the various polymorphs, and this difference has been influenced by whether apo[a] was in the solid phase (nitrocellulose or microtiter plate) or the liquid phase. Although this problem was not noted for polyclonal antibodies, these reagents generally present a greater degree of cross-reactivity with plasminogen,[16,17] although in actual assays cross-reactivity has not been a frequent obstacle. The authors have noted that treatments such as absorption with a $d > 1.21$ g/ml plasma fraction, affinity chromatography over plasminogen-Sepharose, and absorption with plasminogen all reduce the cross-reactivity of a polyclonal antiserum to plasminogen. Monoclonals reactive to a unique epitope in a kringle that occurs only once in apo[a], or to a specific region of the serine protease domain, would be insensitive to apo[a] polymorph size differences resulting from variations in the number of kringle type 2 repeats. Even in the case of a polyclonal antibody, a technique such as rocket electrophoresis, which involves the formation of an immunoprecipitin complex, may be rather insensitive to differences in the size of apo[a] polymorphs because each particle may contribute equally to the complex regardless of the number of kringle repeats in its apo[a] protein. The sensitivity of an assay to apo[a] polymorph size and hence the number of kringle 2 repeats is partially dependent on the selected property utilized for the reference standard and sample measurement. Total protein or apo[a] protein content varies with polymorph size, whereas components such as cholesterol or apoB-100 do not.

Carbohydrate Heterogeneity

Apolipoprotein [a] is a glycoprotein containing 25–45% carbohydrate by weight.[8,18] Regions between the apo[a] kringles contain from 26 to 36 amino acid residues. These interkringle regions are rich in serine and threonine, suggesting they may be principal glycosylation sites. The size

[15] S. M. Marcovina, Z. H. Zhang, V. P. Gaur, and J. J. Albers, *Biochem. Biophys. Res. Commun.* **191,** 1192 (1993).

[16] H.-C. Guo, V. W. Armstrong, G. Luc, C. Billardon, S. Goulinet, R. Nustede, D. Seidel, and M. J. Chapman, *J. Lipid Res.* **30,** 23 (1989).

[17] N. Vu-Dac, H. Mezdour, H. J. Parra, G. Luc, I. Luyeye, and J. C. Fruchart, *J. Lipid Res.* **30,** 1437 (1989).

[18] G. M. Fless, M. E. ZumMallen, and A. M. Scanu, *J. Biol. Chem.* **261,** 8712 (1986).

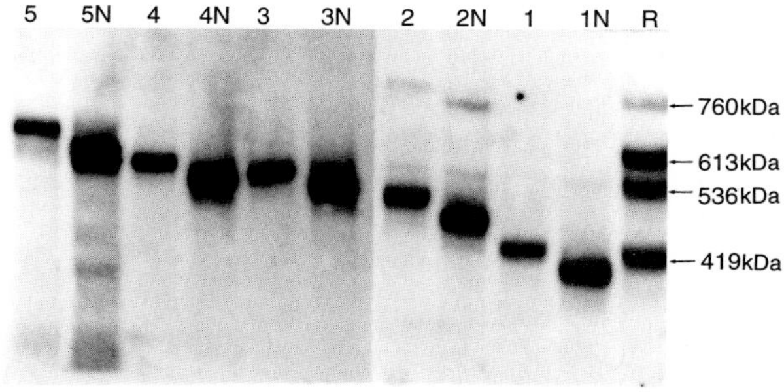

FIG. 1. Immunoblot analysis of apo[a] polymorphs 1–5 before and after neuraminidase treatment (N). Lane R contains 4 reference polymorphs of apo[a]. (Reproduced from Morrisett *et al.*[19] with permission.)

and composition of oligosaccharides attached at these sites can vary widely, providing another source of chemical and electrophoretic heterogeneity. The protein component of Lp[a] is richer than that of LDL in galactosamine, glucosamine, galactose, and sialic acid, whereas the lipid components of these two lipoproteins have these sugars in approximately equal abundance. Notably, glucose is entirely absent from Lp[a] glycoproteins,[18] and mannose is absent from its glycolipids. Apolipoprotein [a] contains substantially more glucosamine, galactose, and sialic acid than apoB-100. The amount of these sugars, especially sialic acid, can vary widely, resulting in subpopulations of Lp[a] differing in carbohydrate content. The high sialic acid content of apo[a] is largely responsible for the slow pre-β mobility imparted to Lp[a] on agarose electrophoresis, a property used to detect the lipoprotein in early studies. The carbohydrate in apo[a] contributes markedly to the mobility as observed on SDS–PAGE (Fig. 1). The mobility is increased after desialyation with neuraminidase and the apparent molecular mass decreased by about 50 kDa,[19] whereas a more complete deglycosylation leads to an apparent molecular mass reduction of 200 kDa.[8] Although glycosylation can alter the conformation of a protein, the removal of sialic acid with neuraminidase does not necessarily affect the binding of antibodies.[16] However, other biological properties could be modified, as has been shown to be the case for LDL,[20] even if the antibody recognition is unaffected.

[19] J. D. Morrisett, J. W. Gaubatz, R. D. Knapp, and J. G. Guevara, Jr., *in* "Lipoprotein [a]" (A. M. Scanu, ed.), p. 53. Academic Press, San Diego, 1990.

[20] A. N. Orekhov, V. V. Tertov, I. A. Sobenin, V. N. Smirnov, D. P. Via, A. M. Gotto, Jr., and J. D. Morrisett, *J. Lipid Res.* **33,** 805 (1992).

Oxidative and Reductive Modification of Lipoprotein [a]

A disulfide bond provides the covalent linkage between apoB-100 and apo[a]. Additional intramolecular disulfide bonds play a critical role in maintaining the native structure of kringles in apo[a]. Virtually all methods for separating apo[a] from apoB-100 involve treatments with reducing agents such as 1,4-dithiothreitol (DTT) or 2-mercaptoethanol (ME), which also reduce intrakringle disulfides and profoundly affect kringle structure. Apolipoprotein [a] without disulfide linkage to apoB-100 but with intact kringles can be obtained from a few cell lines and from one animal model.[8,21,22] Marked differences in molecular weight, fibrin binding, and immunological reactivity are observed after complete reduction of apo[a].[8,23,24] When the susceptibility of Lp[a] to reduction by sulfhydryl compounds of physiological significance such as homocysteine, glutathione, and cysteine was studied, all of these properties were affected although to a lesser degree than observed with DTT and ME.[23,24] Treatment with reducing agents produced no immunochemically detectable changes in apoB of Lp[a].

Lipoprotein [a] in an oxidized form has been isolated from human atheromatous lesions,[25] and *in vitro* oxidative modification of Lp[a] with Cu^{2+} or human mononuclear cells has been shown to influence scavenger receptor recognition in addition to producing polypeptide cleavage, increasing negative charge, and causing aggregate formation.[26,27] The polypeptide fragments produced by oxidation dissociate from Lp[a] only under denaturing conditions, such as with SDS–PAGE, and retain immunoreactivity in immunoblots.[26,28] Because oxidation of Lp[a] may result in a more atherogenic particle, extensive interest has been generated in this modification and the protection against it afforded by certain antioxidants such as vitamin E, β-carotene, ascorbate, and probucol.[26,27,29,30]

[21] G. Chiesa, H. H. Hobbs, M. L. Koschinsky, R. M. Lawn, S. D. Maika, and R. E. Hammer, *J. Biol. Chem.* **267,** 24369 (1992).

[22] A. L. White, D. L. Rainwater, and R. E. Lanford, *J. Lipid Res.* **34,** 509 (1993).

[23] P. C. Harpel, V. T. Chang, and W. Borth, *Proc. Natl. Acad. Sci. U.S.A.* **89,** 10193 (1992).

[24] A. M. Scanu, D. Pfaffinger, G. M. Fless, K. Makino, J. Eisenbart, and J. Hinman, *Arterioscler. Thromb.* **12,** 424 (1992).

[25] H. F. Hoff, J. O'Neil, and A. Yashiro, *J. Lipid Res.* **34,** 789 (1993).

[26] W. Sattler, G. M. Kostner, G. Waeg, and H. Esterbauer, *Biochim. Biophys. Acta* **1081,** 65 (1991).

[27] M. Naruszewicz, E. Selinger, and J. Davignon, *Metabolism* **41,** 1215 (1992).

[28] U. P. Steinbrecher, J. L. Witztum, S. Parthasarathy, and D. Steinberg, *Arteriosclerosis (Dallas)* **7,** 135 (1987).

[29] I. Jialal, D. A. Freeman, and S. M. Grundy, *Arterioscler. Thromb.* **11,** 482 (1991).

[30] I. Jialal and S. M. Grundy, *J. Clin. Invest.* **87,** 597 (1991).

TABLE I
EFFECT[a] OF PROPERTIES/TREATMENTS OF LIPOPROTEIN [a]/APOLIPOPROTEIN [a] ON ELECTROPHORETIC ANALYSES

Electrophoretic procedure	Property measured	Increased no. of kringle repeats	Desialylation/ deglycosylation	Disulfide reduction	Oxidation
Rocket	Antigen–antibody reactivity (precipitin height or area)	—	— (16)	↑ (9) — (31)	— (26, 28)
SDS–PAGE	Apparent molecular weight (M_r)	↑	↓ (8, 22, 32)	[b] ↑ (8, 22, 23) ↓ (9, 23, 24, 32, 33, 34)	↓ (26, 27, 28)
Western blot	Antigen–antibody reactivity (no. of antibody bound)	↑ ↓ (16)	— (32)	[c] ↑ (16) ↓ (23, 24)	— (26, 28)
Standard 0.5% agarose	Charge (electronegativity)	—	↓ (6, 33)	[d] ↑ (33) ↓	↑ (25, 26, 27)

[a] Measured on a per particle or molar basis. ↑, Increase; ↓, decrease; —, no change.
[b] Free apo[a] not disulfide-linked to apoB increases in apparent M_r on reduction, whereas there is a decrease in apparent M_r of the apo[a] if it is present as an apo[a]–apoB complex before reduction.
[c] Response depends on immunogen (reduced or nonreduced) and antibody, especially with regard to monoclonals.
[d] The Lp[a$^-$] lipid-staining component remaining as a result of apo[a] dissociation decreases in mobility. Apolipoprotein [a] itself increases due to negative net charge concentration.

Oxidative and reductive modifications can result in marked changes in biological properties and structure; possible effects on immunoreactivity should be evaluated when these modifications have been produced. Table I summarizes the effects of various properties or modifications of Lp[a] or apo[a] on different electrophoretic procedures.[31–34] Some of the observed effects differ among studies, and they probably reflect differences in responses of various antibodies, illustrating the need to evaluate these on an individual basis.

Rocket Electroimmunoassay

Quantitation of Apolipoprotein [a] in Plasma

Over the years, the microplate enzyme-linked immunoassay (ELISA) has surpassed the electroimmunoassay (EIA) as the most widely used

[31] J. M. Pepin, J. A. O'Neil, and H. F. Hoff, *J. Lipid Res.* **32,** 317 (1991).
[32] G. F. Grinstead and R. D. Ellefson, *Clin. Chem.* **34,** 1036 (1988).
[33] A. Gries, C. Fievet, S. Marcovina, J. Nimpf, H. Wurm, H. Mezdour, J. C. Fruchart, and G. M. Kostner, *J. Lipid Res.* **29,** 1 (1988).
[34] G. M. Fless, C. A. Rolih, and A. M. Scanu, *J. Biol. Chem.* **259,** 11470 (1984).

method for Lp[a] quantitation.[15,35–37] The ELISA affords higher sensitivity, greater capacity for processing a large number of samples, and easier automation in terms of sample handling, data collection and reduction, and final result display. Nevertheless, a number of researchers continue to use EIA for quantitation, probably because the assay is simple to set up and run, and requires no sophisticated instrumentation.[10,26,38,39] Additionally, EIA has often been used as an independent reference method for Lp[a] measurement when a new assay was being validated.[17,40,41] Indeed, Kostner[42] has indicated that an electroimmunoassay using polyclonal antibodies is an excellent reference method for Lp[a] measurement because it is rather insensitive to particle size, apo[a] isoform molecular weight, and Lp[a]–LDL complexes. However, EIA is sensitive to the reductive treatment of Lp[a], an effect also seen with a microplate ELISA.[24] With the ELISA, reductive cleavage of apo[a] from apoB results in a decreased apparent concentration for apo[a]. With the EIA, reduction causes a higher but weaker staining immunoprecipitin rocket, resulting in an overestimation of apo[a].[9] This significant alteration in apo[a] immunoreactivity on treatment with reducing agents such as 1,4-dithiothreitol or 2-mercaptoethanol probably arises not only from cleavage of the disulfide(s) joining apo[a] to apoB, but also from a scission of intrakringle disulfides, thereby allowing the kringles to unfold and alter the structure of immunologic epitopes. Unlike apo[a], apoB does not undergo alteration in immunoreactivity when treated with reductants.[24,33] Accordingly, caution must be exercised in generalizing results from one assay system to another.

Although many laboratories currently utilize a microplate ELISA for Lp[a] quantitation, the EIA can be an important part of the analysis protocol as a semiquantitative tool. For example, for our ELISA,[35] human Lp[a]

[35] J. W. Gaubatz, G. L. Cushing, and J. D. Morrisett, this series, Vol. 129, p. 167. (1986).

[36] H. C. Guo, J.-B. Michel, Y. Blouquit, and M. J. Chapman, *Arterioscler. Thromb.* **11,** 1030 (1991).

[37] H. G. Kraft, H. Dieplinger, E. Hoye, and G. Utermann, *Arteriosclerosis* (*Dallas*) **8,** 212 (1988).

[38] C. Sandholzer, D. M. Hallman, N. Saha, G. Sigurdsson, C. Lackner, A. Csaszar, E. Boerwinkle, and G. Utermann, *Hum. Genet.* **86,** 607 (1991).

[39] H. G. Kraft, C. Sandholzer, H. J. Menzel, and G. Utermann, *Arterioscler. Thromb.* **12,** 302 (1992).

[40] G. L. Cushing, J. W. Gaubatz, M. L. Nava, B. J. Burdick, T. M. A. Bocan, J. R. Guyton, D. Weilbaecher, M. E. DeBakey, G. M. Lawrie, and J. D. Morrisett, *Arteriosclerosis* (*Dallas*) **9,** 593 (1989).

[41] W. Marz, R. Siekmeier, E. Grob, and W. Grob, *Clin. Chim. Acta* **214,** 153 (1993).

[42] G. M. Kostner, *Clin. Chim. Acta* **211,** 191 (1992).

plasma concentrations span a greater than 100-fold range,[43,44] and because the slope of our ELISA concentration–response curve allows only an 8-fold range of measurements, a single dilution of plasma would leave many samples outside the range of the standard curve. Hence, semiquantitative determination of plasma samples by EIA allows the selection of an appropriate single dilution from 1:2500 to 1:200,000 for the ELISA so that more than 95% of the samples assayed fall within the range of the standard curve. This enables the development of an ELISA with a steep slope where most plasma samples can be assigned a numerical value rather than threshold limits, without reassay.

The following modifications of our EIA procedure were made for this application[35]: (1) 4 times the usual antibody volume is added to the gel; (2) the Lp[a] standard is shifted from the usual range of 5–50 μg/ml to 20–400 μg/ml; and (3) 2.5 μl undiluted plasma samples is added. Samples not giving a visible immunoprecipitin rocket are diluted 1:2500 (v/v) for the ELISA. Other samples are diluted from 1:5000 to 1:200,000 depending on the concentration estimated from the rocket height of the standard.

Quantitation of Apolipoprotein [a] in Tissue

The presence of apo[a] in human atherosclerotic plaque was initially demonstrated by Walton *et al.*,[45] using a fluorescein-labeled antiserum. Relatively few studies have quantified apo[a] in the vascular bed,[31,40,46] and only one of these[40] utilized electrophoresis for the quantitation. The EIA procedure is based on the early work of Smith and Slater,[47] who developed it to quantify LDL in arterial tissue. In initial studies, they observed large losses in immunoreactivity of apoB following tissue homogenization. Subsequently, a procedure was developed in which fine minces of aortic intima were inserted directly into large rectangular wells of an antibody-containing agarose gel.[48] The lipoprotein, along with other negatively charged species, is electrophoresed from the arterial tissue by an electric field and subsequently migrates through the gel toward the anode. When the concentration of apoprotein reaches equivalence with the antibody in the gel, a rocket-shaped precipitin line forms. The area under the rocket is proportional to the concentration of the precipitating antigen.

[43] G. Utermann, *Science* **246,** 904 (1989).

[44] C. Labeur, J. Shepherd, and M. Rosseneu, *Clin. Chem.* **36,** 591 (1990).

[45] K. W. Walton, J. Hitchens, H. N. Magnani, and M. Khan, *Atherosclerosis* **20,** 323 (1974).

[46] M. Rath, A. Niendorf, T. Reblin, M. Dietel, H. Krebber, and U. Beisiegel, *Arteriosclerosis (Dallas)* **9,** 579 (1989).

[47] E. B. Smith and R. S. Slater, *Atherosclerosis* **11,** 417 (1970).

[48] C. B. Laurell, *Anal. Biochem.* **15,** 45 (1966).

Hoff *et al.* improved on this technique by using homogenates of tissue instead of actual tissue minces, but still quantitated LDL by electroimmunoassay.[49,50] In preliminary studies to determine if the same technique could be used for quantitation of tissue Lp[a], control experiments were performed to compare the recoveries of apo[a] and apoB immunoreactivity after homogenization under the same conditions that Hoff used for LDL (i.e., with a Polytron tissue disrupter). Only a 10% loss of immunoreactivity for apoB was measured by EIA, but a 70% loss of reactivity toward apo[a] was determined.[51] When samples were pulverized with a small Kontes hand homogenizer, the results were similar to those obtained with the Polytron.

Another technical difficulty encountered in quantitating tissue apo[a] is the limitation of tissue mass available. In the case where tissue samples are small, such as with those obtained from human surgery or small animals,[52] the tissue mince technique may be preferred. The use of homogenates or some form of solubilized tissue extract other than tissue minces may be preferred when the above problems can be overcome.

Comparison of the immunoreactivity of apo[a] or Lp[a] in tissue to that in plasma, which is often used as a secondary standard, can best be evaluated in solubilized extracts of tissue apo[a]. The validity of an immunoassay requires that antibody recognition of the antigen in the standard and the test sample be identical. The most commonly used test for this is parallelism. If the linear regression lines generated by appropriate dilutions of the standard and tissue samples are parallel, then the requirement is met. Even when this condition is satisfied, the accuracy of the assay can be determined only if the apo[a] or Lp[a] can actually be isolated and purified from the tissue, and then compared to the standard by the same independent measure used for its standardization. Most often, then, the values reported for quantification in tissue must be regarded as apparent values.

The following protocol has been used for quantitation of Lp[a] protein in resected vein graft tissue,[40] where immunoreactivity of Lp[a] is expressed in terms of Lp[a] total protein. Goat anti-human polyclonal antibody is added to 25 ml molten (55°) 1% agarose (type C, Behring Diagnostics, La Jolla, CA) at a 1:200 dilution. The mixture is poured into the opening of a 18 × 9.5 × 1.5 mm glass sandwich form and allowed to solidify. Wells are then made in the gel with a 4 mm diameter punch used in conjunction with an electrophoresis template (Bio-Rad, Richmond, CA). The gel is then

[49] H. F. Hoff, C. L. Heideman, J. W. Gaubatz, A. M. Gotto, E. E. Erickson, and R. L. Jackson, *Circ. Res.* **40,** 56 (1977).

[50] H. F. Hoff, C. L. Heideman, A. M. Gotto, and J. W. Gaubatz, *Circ. Res.* **41,** 684 (1977).

[51] J. W. Gaubatz, G. L. Cushing, and J. D. Morrisett, unpublished results, 1987.

[52] H. F. Hoff and M. G. Bond, *Atherosclerosis* **43,** 329 (1982).

placed on the cooled plate (10°) of an electrophoresis chamber (Behring Diagnostics, Somerville, NJ) containing barbiturate buffer. Ultrawicks are used to complete contact between gels and the buffer chamber. Tissue segments are loaded into larger wells made by multiple stabs with the 4.0-mm punch, and molten agarose is used to fill the residual volume not occupied by the segment. Dilutions of plasma secondary standards containing 100 to 1000 ng apo Lp[a] are added to the smaller wells. This secondary standard is calibrated against purified Lp[a] whose total protein content had been determined. Samples are electrophoresed at 3.5 V/cm for 18 hr followed by 4.5 V/cm for 4 hr. These times and voltages are necessary to electrophorese out of the tissue the maximum amount of apo[a], which may be present as free apo[a], as a disulfide complex, as intact Lp[a], or as some other form. Tissues are sliced from the gels, and the gels are gently agitated in 150 m*M* NaCl for 18 hr then in deionized water for 4 hr. This washing regimen also lowers background staining, thereby allowing visualization of rockets in cases where these might otherwise not be observed. Electrophoresis gels are dried and stained with Coomassie Blue R-250 (Bio-Rad). The areas under all rockets are measured by triangulation.

Tissue values of Lp[a] are calculated relative to the plasma standards, and are usually expressed as nanograms apoprotein per milligram tissue. Representative apo Lp[a] rockets are shown in Fig. 2. As in the case with apoB, not all of the tissue apo[a], apo[a]–apoB, or Lp[a] can be removed by electrophoresis in this manner; the fraction of apoB removed has been designated as "loosely bound" by Smith[47] and corresponds to the "buffer-extractable" fraction indicated by Hoff.[49,50] Smith obtained an additional "tightly bound" fraction of apoB which could be removed on additional electrophoresis of plasmin-treated, minced tissue. Hoff found a similar fraction liberated by elastase treatment after buffer extraction of the homogenates. Both plasmin and elastase treatment of Lp[a] caused losses of apo[a] immunoreactivity. No rockets for apo[a] could be resolved in tissue treated with these enzymes, and hence attempts to use this technique to quantitate a tightly bound apo[a] fraction in tissue minces have not been successful.

Electrophoretic Procedures

Development and Applications

The size heterogeneity of apo[a] was first described in 1983[11] with the recognition of three different M_r polymorphs demonstrable by 3.25% sodium dodecyl sulfate–polyacrylamide gel electrophoresis (SDS–PAGE)

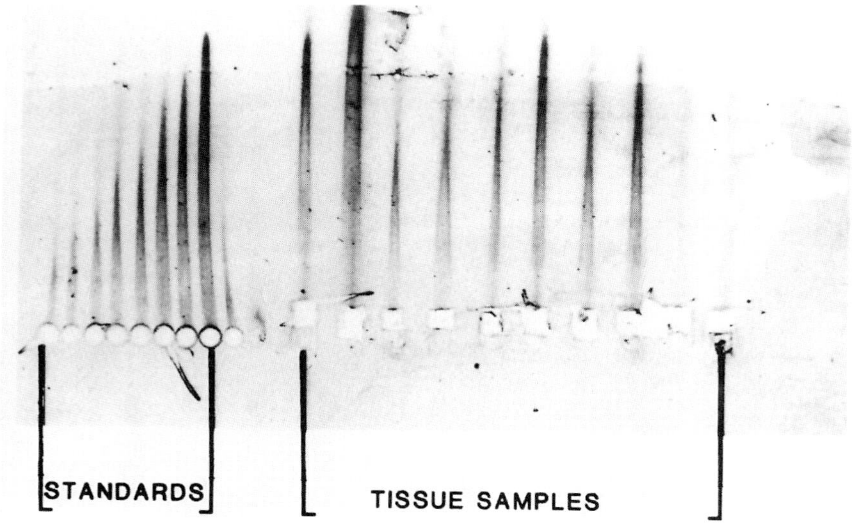

FIG. 2. Apolipoprotein [a] in vein graft tissue was quantified by comparing the rocket areas of a tissue sample with known standards. (Reproduced from Cushing *et al.*[40] with permission.)

followed by immunoblotting. Utermann *et al.*[53] developed a discontinuous SDS–PAGE system using a 6.6% slab gel in which up to six apo[a] M_r polymorphs were resolved among the study population, but only one or two polymorphs were present in any given individual plasma (Fig. 3).

Subsequent changes in the gel and running conditions have led to resolution of an increasing number of polymorphs, with up to 11 in human[54–56] and baboon[55] plasma separated by various polyacrylamide systems (Fig. 4). In 1991, Kamboh *et al.*[57] introduced the use of an agarose electrophoretic system in which 23 polymorphs were demonstrable. An additional improvement in sensitivity was also obtained which revealed bands in all 270 individuals examined, with 77.4% exhibiting two bands and 22.6% showing one band. This advance led to a population estimation of the number of apo[a] polymorphs approximately coinciding with the number of genomic alleles

[53] G. Utermann, H. Menzel, H. G. Kraft, H. C. Duba, H. G. Kemmler, and C. Seitz, *J. Clin. Invest.* **80,** 458 (1987).

[54] J. W. Gaubatz, K. I. Ghanem, J. Guevara, M. L. Nava, W. Patsch, and J. D. Morrisett, *J. Lipid Res.* **31,** 603 (1990).

[55] D. L. Rainwater, G. S. Manis, and J. L. VandeBerg, *J. Lipid Res.* **30,** 549 (1989).

[56] W. Y. Craig, S. E. Poulin, N. R. Forster, L. M. Neveux, T. B. Ledue, and R. F. Ritchie, *Arterioscler. Thromb.* **11,** 1483a (1991).

[57] M. I. Kamboh, R. E. Ferrell, and B. A. Kottke, *Am. J. Hum. Genet.* **49,** 1063 (1991).

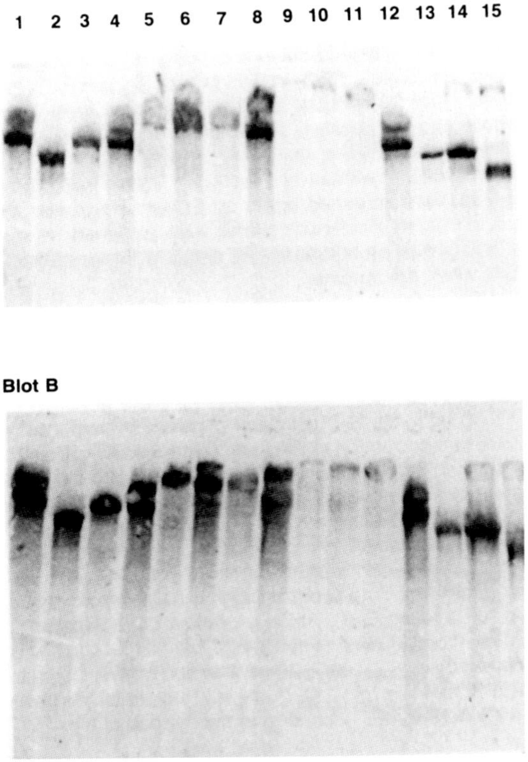

FIG. 3. Resolution of six polymorphs of apo[a] from individual plasma samples by immunoblot analysis following 6.6% SDS–PAGE. (Reproduced from Kraft *et al.*[37] with permission.)

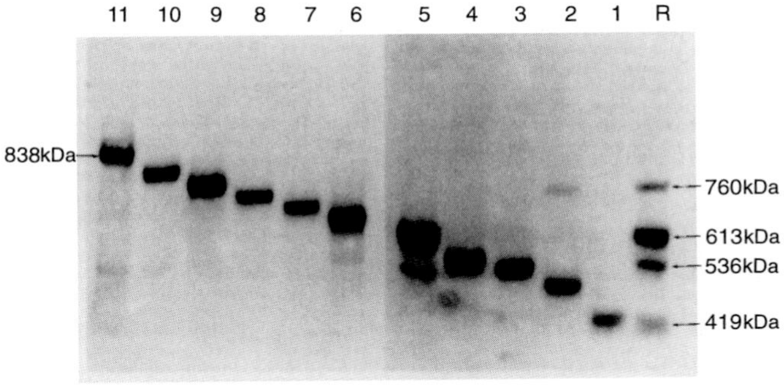

FIG. 4. Resolution of 11 polymorphs of apo[a] from individual plasma samples by immunoblot analysis following 3.75% SDS–PAGE. (Reproduced from Gaubatz *et al.*[54] with permission.)

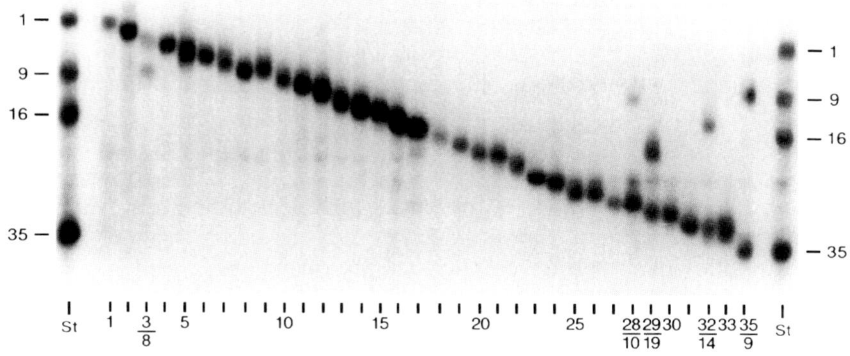

FIG. 5. Resolution of 34 polymorphs of apo[a] from individual plasma samples by immunoblot analysis following 1.5% SDS–agarose gel electrophoresis. (Reproduced from Marcovina *et al.*[15] with permission.)

which could be demonstrated in 102 individuals by restriction fragment analysis followed by pulsed-field gel electrophoresis.[58] The latter study demonstrated two alleles in 94% of those tested. Both studies, in contrast to previous results,[54,59] found that the population distribution of apo[a] polymorphs conformed to the Hardy–Weinberg equilibrium.This discrepancy was probably due to increased detection of bands in individuals with a double-band phenotype (81.4%) compared to previous studies in which single-band phenotypes were most abundant.[54,55,59,60]

The agarose electrophoretic system has been modified so as to distinguish between 34 apo[a] size polymorphs[15] (Fig. 5). Concurrent with this development was the recent report of 34 size alleles identified by pulsed-field gel electrophoresis.[61] The occurrence of 34 polymorphs is consistent with a model of single incremental repeats of a kringle unit, beginning with 11 kringles in the lowest apparent M_r form, increasing to 44 in the highest form. Each polymorph corresponds to a 2 mm distance between adjacent bands on the agarose gels, with a total span of about 7 cm between the largest and smallest polymorphs.[15]

The SDS–PAGE or SDS–agarose electrophoresis technique has the potential for providing valuable quantitative information about apo[a] such

[58] C. Lackner, E. Boerwinkle, C. C. Leffert, T. Rahmig, and H. H. Hobbs, *J. Clin. Invest.* **87,** 2153 (1991).

[59] E. Boerwinkle, H. J. Menzel, H. G. Kraft, and G. Utermann, *Hum. Genet.* **82,** 73 (1989).

[60] M. Farrer, F. L. Game, P. C. Adams, M. F. Laker, and K. G. M. M. Alberti, *Clin. Chim. Acta* **207,** 215 (1992).

[61] J. C. Cohen, G. Chiesa, and H. H. Hobbs, *J. Clin. Invest.* **91,** 1630 (1993).

as (i) frequencies of the apo[a] polymorphs, allowing statistical comparisons; (ii) apparent M_r determinations, allowing the assessment of polymorph size changes *in vitro*[22] or *in vivo*[62] resulting from various experimental or pharmacological interventions (this parameter can also be used to determine if the same sized allele in different individuals invariably results in expression of the same sized apo[a] product); and (iii) actual concentrations of apo[a] determined from the intensities of bands quantified on the immunoblot by densitometry. Several potential technical difficulties associated with this method of quantitation have been discussed elsewhere[63] and include the following: (a) differential reactivity of the apo[a] polymorphs to the antibody used, (b) incomplete and/or differential transfer of apo[a] polymorphs during Western blot analyses, (c) nonlinear signal response of the film in autoradiography,[64] and (d) differential losses in immunoreactivity attending heating and reduction during sample preparation.

At present, a reliable system for quantitating apo[a] polymorphs has not been published. Densitometric scanning of autoradiographs of SDS–PAGE gels has been used to determine the contribution of various tissues to apo[a] expression in transgenic mice.[21] This method would also allow an assessment of the percentage distribution of apo[a] polymorphs between two bands in double-band phenotype individuals. A qualitative determination of these distributions has been reported by several investigators.[15,54,57] A quantitative method would allow comparison of the allele transcript concentration with corresponding apo[a] polymorph expression, two variables that do not always change synchronously.[65,66]

Sample Preparation

In preparing samples for SDS gels, some investigators add plasma or serum directly to the SDS sample buffer,[37,57,60] but we recommend initial lyophilization of an appropriate plasma aliquot (e.g., 50 μl). To this is added 200–500 μl SDS sample buffer (10 m*M* Tris, 1 m*M* EDTA, 2% (w/v) SDS, 10% (v/v) glycerol, 5% (v/v) 2-mercaptoethanol, pH 8.0). The volume of sample buffer added can be altered depending on the apo[a] concentration in the plasma, but volumes less than 500 μl may lead to gel formation following boiling. After the sample buffer mixture is heated for 3–5 min

[62] H. G. Kraft, H. J. Menzel, F. Hoppichler, W. Vogel, and G. Utermann, *J. Clin. Invest.* **83,** 137 (1989).

[63] A. M. Scanu, *J. Int. Med.* **231,** 679 (1992).

[64] M. D. Wygant and J. W. Nelson, *Am. Lab.* **40,** RRR (1993).

[65] J. E. Hixson, M. L. Britten, G. S. Manis, and D. L. Rainwater, *J. Biol. Chem.* **264,** 6013 (1989).

[66] D. P. Wade, B. L. Knight, K. Harders-Spengel, and A. K. Soutar, *Atherosclerosis* **91,** 63 (1991).

at 100°, it is cooled to room temperature before loading on the gel or is frozen until later use. The apo[a] concentration is determined by ELISA, and the sample load is normalized. The actual load depends on the sensitivity of the detection system used; the 50- to 250-ng range is sufficient, but higher loads are recommended for increased detectability of both polymorphs in the case of subjects with double bands.

Electrophoresis Conditions

Agarose electrophoresis is performed essentially according to Kamboh *et al.*[57] and Marcovina *et al.*[15] Agarose gels (15 × 25 cm) are cast in a Hoefer (San Francisco, CA) submarine gel unit using a solution of 150 ml of 1.5% ultrapure agarose (Bethesda Research Laboratories, Bethesda, MD) dissolved in 90 m*M* Tris, 90 m*M* borate, 2 m*M* EDTA, and 0.1% SDS. Sample volumes up to 40 μl can be applied to the gel in 3-mm-wide and 2-mm-deep wells located 3 cm from the cathode. Electrophoresis is performed in tank buffer (45 m*M* Tris, 45 m*M* borate, 2 m*M* EDTA, and 0.1% SDS) at constant 100 V for 16 hr at 4°. The two outermost lanes of each gel contain a reference cocktail prepared by mixing plasma samples from individuals with different apo[a] polymorphs; the molecular weight reference is used to control for gel-to-gel variation in apo[a] migration.

Western Blot Analysis

After electrophoresis, proteins are transferred to nitrocellulose (0.45 μm) using a Hoefer Transphor Cell at 60 V for 3–18 hr in 25 m*M* Tris, 86 m*M* glycine containing 5% methanol (pH 8.3). After transfer, the membrane is incubated for 1 hr in a solution of 0.5% bovine serum albumin in Tris-buffered saline (TBS: 100 m*M* Tris, 10 m*M* NaCl, 1 m*M* EDTA, 1 m*M* sodium azide, pH 7.4) to block any remaining protein binding sites. Alternatively, 2–5% nonfat milk can be substituted for bovine serum albumin in buffers. All additional incubations are performed in the same TBS buffer containing 0.1% Tween 20 detergent. If peroxidase is used in the detection system, azide should be omitted from the buffers since it inhibits the enzyme.

Following the blocking step, the filter is incubated for 2 hr at room temperature with antibody. A rabbit polyclonal immunoglobulin fraction (0.5 mg/100 ml) purified by Lp[a] affinity chromatography provides good results. Although serum fractions can be used,[57,60] less background staining is obtained with a purified immunoglobulin fraction. Marcovina *et al.*[15] used a monoclonal antibody directed to an epitope expressed on kringle 4 of apo[a] with no cross-reactivity to plasminogen. Additionally, this antibody had the same affinity for apo[a] whether the disulfide bonds were intact or reductively cleaved. Other investigators have found that their antibodies exhibited a decreased immunoreactivity toward apo[a] after its reductive release from apoB,[16,23,24] a necessary procedure for apo[a] polymorph analysis with this method. A polyclonal antibody with increased reactivity to reduced apo[a] for use with this procedure can be produced by using reduced Lp[a] or reduced apo[a] separated from Lp[a] by ultracentrifugation[67] as the immunogen.

Following incubation with antibody, the membrane is washed with TBS and Tween 20 followed by incubation for 2 hr with ^{125}I-labeled protein A (ICN Biomedicals, Inc., Irvine, CA), having specific activity >30 μCi/μg giving about 10 μCi per membrane. Radioactive labeled bands are visualized after overnight exposure at −70° to X-Omat XAR5 film (Eastman Kodak Co., Rochester, NY) using a Lightning Plus intensifying screen (DuPont, Wilmington, DE). Protein A has a high affinity for antibodies prepared in rabbit, but protein G has a greater affinity for those prepared in rat or goat.

The separation of 34 apo[a] polymorphs as a composite of two agarose gels is depicted in Fig. 5.[15] Number 1 is the highest M_r polymorph in this system and 35 the lowest. The polymorph number in a test sample was assigned using the equation $[(34 \times B) \div A] + 1$ where A is the distance in millimeters between isoforms 1 and 35 in the reference standard and B is the distance between isoform 1 and the isoform in the test sample. If the gels are scanned with a laser densitometer, the distance of migration can be taken as the midpoint of the peak, and the technique is adaptable to computer-assisted calculations. A weak anomalous band, which appears across all lanes of the gel and whose intensity usually corresponds to the sample volume loaded rather than Lp[a] concentration, has been noted by many investigators at the position corresponding to apoB-100.[53,54,55,57] Accordingly, a band at this position should be ignored unless it is intense and corresponds to a high Lp[a] plasma concentration. The addition of LDL to the buffer during incubation of the membrane with antibody has been observed to decrease this anomalous staining. Numerous authors have also observed some faint, low molecular weight bands which may be due to partial degradation of apo[a][57] or cross-reactivity to some structurally related lower molecular weight proteins such as plasminogen (94 kDa) unless these have migrated off the gel.

With the 34-polymorph classification system, a total of 595 phenotypes are predicted for the number of possible double and single band combinations of polymorphs in the population. The value of the additional information provided by this system compared to the original 6-polymorph system developed by Utermann *et al.*[53] has been questioned. Certain statistical tests, such as chi-square, used for population quantitative analyses require

[67] V. W. Armstrong, A. K. Walli, and D. Seidel, *J. Lipid Res.* **26,** 1314 (1985).

cells to contain at least five elements ($n = 5$). In the study of Marcovina *et al.*[15] involving 1507 individuals, combining of cells was required for certain analyses. Nevertheless, the resolution of a greater number of polymorphs more accurately reflects the expression of the actual allele size differences and, as such, is currently the preferred method for apo[a] phenotyping. Further improvements in the separation distance between polymorphs on the gels should still be sought. With the 11-polymorph system using an 8-cm-long agarose–acrylamide gel,[54] the polymorphs were spread over a total distance of 4 cm, compared to 34 polymorphs in the described 1.5% agarose system spread over a 7-cm distance in a 25-cm-long gel (Figs. 4 and 5). A longer agarose–acrylamide gel may give better separation than that obtained with the agarose alone system. We have found that 2% agarose (Sea Plaque, FMC, Rockland, ME) run at constant 12.5 W for 20 hr gives a separation distance of 10 cm.

Molecular Weight Determination of Apolipoprotein [a] Polymorphs

A number of investigators have determined molecular weights (M_r) of the various apo[a] polymorphs resolved in their SDS electrophoretic systems[9,32,53,54,56,68,69] essentially by the method of Weber and Osborn. This method uses the relative mobility of the polymorphs to determine the apparent M_r from semilog plots of molecular weight standards versus migration distance relative to the gel length. The two principal difficulties with this method have been (i) the scarcity of suitable M_r standards in the 250- to 1000-kDa size range and (ii) the anomalous mobility of glycoproteins on SDS–PAGE, especially those with high sialic acid content. Carbohydrate-rich glycoproteins can show large deviations in apparent M_r from values based on actual chemical composition.[70,71] Several of the most commonly used standard cocktails have been the Pharmacia (Piscataway, NJ) high molecular weight standard mixture consisting of thyroglobulin (330 kDa), ferritin (220 kDa), albumin (67 kDa), catalase (60 kDa), and lactate dehydrogenase (36 kDa),[17,53,69] and cross-linked phosphorylase *b* (97–584 kDa; Sigma Chemical Co., St. Louis, MO).[32,54] Some of these mixtures have included apoB-100 with a designated molecular mass of 500–550 kDa.[8,9,53] The use of haptoglobin 2-2 polymers in the 172- to 859-kDa range has been suggested as an improved high molecular weight standard mixture.[56]

The newest SDS system used for resolving 34 polymorphs of apo[a]

[68] L. J. Seman and W. C. Breckenridge, *Biochem. Cell Biol.* **64,** 999 (1986).

[69] P. M. Laplaud, L. Beaubatie, S. C. Rall, G. Luc, and M. Saboureau, *J. Lipid Res.* **29,** 1157 (1988).

[70] J. F. Poduslo, *Anal. Biochem.* **114,** 131 (1981).

[71] J. P. Segrest, R. L. Jackson, E. P. Andrews, and V. T. Marchesi, *Biochem. Biophys. Res. Commun.* **44,** 390 (1971).

TABLE II
APPARENT MOLECULAR WEIGHT OF APOLIPOPROTEIN [a] POLYMORPHS[a]

	Molecular weight ($\times 10^{-3}$)	
Apo[a] band	3.75% Gel	2.75% Gel
1	419	565
2	489	658
3	536	687
4	553	724
5	613	760
6	680	798
7	705	838
8	742	881
9	760	925
10	796	970
11	838	1023

[a] Dependence of apparent molecular weight of apo[a] polymorphs on polyacrylamide gel system used.

involves agarose electrophoresis, and the validity of molecular weight determinations in agarose has not been established. The deviation between the apparent M_r determined using SDS–PAGE (2.5–15% gradient) and the mass calculated from the amino acid composition has been determined for a recombinant apo[a] expressed in a human embryonic kidney cell line.[8] The engineered protein has a calculated protein mass of 250 kDa (307.5 kDa including carbohydrate) but an apparent molecular weight based on SDS–PAGE of 500,000 under reducing conditions. Shift to a larger apparent M_r is observed after reduction,[8] probably a result of kringle unfolding.

The comparison of apo[a] M_r determined on one gel system to values determined on different systems may not be valid. For example, changing the acrylamide concentration from 2.75 to 3.75% produces a large shift in apo[a] apparent M_r relative to apo B-100 and the phosphorylase *b* oligomers used as a standard (Table II). This acrylamide change produced a shift in M_r of about 150,000 for most of the apo[a] polymorphs. On the 2.75% gel, no polymorphs migrated faster than apo B-100, but on the 3.75% gel, three of these polymorphs did migrate faster. Hence, when different investigators have used different gels and different buffer systems for SDS–PAGE analysis, the comparison of polymorphs has been difficult even when the same molecular weight standards have been used. Nevertheless, the assignment of an apparent M_r to a specific apo[a] polymorph on the basis of SDS–PAGE migration would allow investigators to identify apo[a] polymorphs in their systems as being identical and would allow an integration of information from these different sources. Hence, the assignment of M_r is helpful as long as these assignments are comparable from one system to another, regardless of whether they represent the true molecular mass.

The use of a reference polymorph(s) of apo[a] for the calculation of relative electrophoretic mobilities is recommended.[15,60] The use of apo B-100 as one of the reference species is inadvisable since the mobilities of apo[a] polymorphs can shift relative to that of apoB-100.

Standard Agarose Electrophoresis

The use of agarose electrophoresis as a qualitative tool for identifying individuals with elevated Lp[a] levels by detection of the pre-β_1 band has been described.[35] Studies on the oxidative modification of LDL[28,29] and Lp[a][25,27] and the assessment of the protective capacities of certain antioxidants[27,30,72] as well as the discovery of electrophoretically altered lipoproteins in human atherosclerotic lesions[25] have revived interest in this type of electrophoresis as a method for quantifying the electrophoretic mobilities of these lipoproteins. Electrophoresis is performed at pH 8.6 in 50 m*M* barbiturate buffer on 0.5% agarose gels. Bovine serum albumin is added to a final concentration of 20 mg/ml to ensure reproducible migration distances.[28] The gels are stained with Sudan Black B. Alternatively, the lipoproteins can be transferred to nitrocellulose by simple pressure blotting and then immunolocalized.[27] The relative electrophoretic mobility (REM) is then calculated by dividing the migration distance of the test lipoprotein by the distance migrated by native, unmodified LDL. A REM as high as 4.3 has been reported for oxidized LDL.[30] A statistical evaluation of the technique showed that repeated *in vitro* oxidation of LDL from a single subject was characterized by a coefficient of variation (*CV*) of 3.5%, whereas a thiobarbituric acid-reacting substances assay (TBARS) gave a *CV* of 10%.[29] Moreover, linear regression analysis showed that the time–course alteration in electrophoretic mobility and the oxidation of sterols gave a regression coefficient of 0.932, further suggesting the use of this technique as a quantitative analytical tool. When immunoblotting is used for detection of apo[a] instead of Sudan Black B staining of the lipid, alterations in the mobility of the free apo[a] can be studied, where apo[a] has been released from Lp[a] by reduction.[27]

[72] L. S. Cristol, I. Jialal, and S. M. Grundy, *Atherosclerosis* **97,** 11 (1992).

[14] Quantitation of Lipoprotein (a) after Lysine-Sepharose Chromatography and Density Gradient Centrifugation

By GUNTHER M. FLESS and MARGARET L. SNYDER

Introduction

Lipoprotein (a) [Lp(a)] is a novel low-density lipoprotein (LDL)-like lipoprotein whose plasma concentration can vary over 100-fold among different individuals.[1] These concentration differences are inherited and controlled to greater than 90% by the apolipoprotein (a) [apo(a)] gene.[2] Lipoprotein (a) is polymorphic in nature because apo(a) isoforms vary in the number of kringle 4 domains,[3] and the abundance of kringle 4 repeats in the apo(a) gene is inversely correlated to the concentration of Lp(a).[4,5] Compared to other lipoproteins such as LDL and high-density lipoprotein (HDL), plasma levels of Lp(a) are relatively small, which creates a problem in the purification of Lp(a) from the more abundant lipoproteins.

The presence of the disulfide-linked, heavily glycosylated apolipoprotein (a) shifts the density distribution of the LDL like Lp(a) particles intermediate between that of LDL and HDL. The size of apolipoprotein (a) together with Lp(a) lipid are major determinants of Lp(a) density.[6] Individuals that express two apo(a) isoforms will have a density profile in which Lp(a) particles may be well separated when the difference in size between the two apo(a) isoforms is large, but may not have well-separated bands on the density gradient when size differences are small. Separation is also affected by the relative concentration of the two Lp(a) species. Because the density distribution of Lp(a) overlaps those of LDL and HDL, one cannot isolate Lp(a) encompassing its complete density spectrum without including isopycnic LDL and HDL particles. Similarly, when separation methods are used that utilize size, such as gel filtration, in the purification

[1] G. M. Fless, M. L. Snyder, and A. M. Scanu, *J. Lipid Res.* **30,** 651 (1989).

[2] E. Boerwinkle, C. C. Leffert, J. P. Lin, C. Lackner, G. Chiesa, and H. H. Hobbs, *J. Clin. Invest.* **90,** 52 (1992).

[3] C. Lackner, J. C. Cohen, and H. H. Hobbs, *Hum. Mol. Genet.* **2,** 933 (1993).

[4] G. Utermann, H. J. Menzel, H. G. Kraft, H. C. Duba, H. G. Kemmler, and C. Seitz, *J. Clin. Invest.* **80,** 458 (1987).

[5] C. Lackner, E. Boerwinkle, C. Leffert, T. Rahmig, and H. Hobbs, *J. Clin. Invest.* **87,** 2153 (1991).

[6] G. M. Fless, *in* "Lipoprotein (a)" (A. M. Scanu, ed.), p. 41. Academic Press, San Diego, 1990.

METHODS IN ENZYMOLOGY, VOL. 263

of Lp(a) from LDL and HDL, Lp(a) cannot be completely separated from LDL because of overlapping size distributions.

A method that is useful in overcoming these problems is chromatography of plasma on lysine-Sepharose.[7] This support is extensively used in the purification of plasminogen from plasma because plasminogen has two lysine binding domains that are situated in kringles 1 and 4.[8] Apolipoprotein (a) has extensive homology to plasminogen and like this protein exhibits lysine binding properties.[7,9] These most likely reside in kringle 4_{37} of apo(a), which is one of the many kringle 4 domains that constitute the bulk of this molecule.[9-11] Lysine-Sepharose chromatography of plasma allows the analysis of Lp(a) exhibiting the complete density spectrum of all particles without arbitrary cutoffs imposed by the problems described above. By passing small volumes of plasma over a lysine-Sepharose minicolumn, the Lp(a)-containing eluate can be incorporated in a discontinuous NaBr density gradient and placed in a single tube of the Beckman (Fullerton, CA) SW-40 swinging-bucket rotor. On centrifugation to equilibrium, Lp(a) fractions of different density can then be analyzed by enzyme-linked immunosorbent assay (ELISA). One may also quantitate apo(a) bound to triglyceride-rich particles located at the top of the gradient, as well as free unconjugated apo(a) present in the bottom fraction.

Lysine-Sepharose Chromatography

Principle

Lipoprotein (a) has an affinity for lysine-Sepharose by virtue of lysine binding kringle 4 domain(s) located on apo(a). The most important domain appears to be kringle 4_{37}, which has the greatest homology to kringle 4 of plasminogen, although there may be other kringles with lesser affinity for lysine which also contribute to the interaction of Lp(a) with lysine-Sepharose.[9-11] Plasminogen and Lp(a) have similar affinities for lysine-

[7] D. L. Eaton, G. M. Fless, W. J. Kohr, J. W. McLean, Q.-T. Xu, C. G. Miller, R. M. Lawn, and A. M. Scanu, *Proc. Natl. Acad. Sci. U.S.A.* **84,** 3224 (1987).

[8] L. Sottrup-Jensen, H. Claeys, M. Zajdel, T. E. Petersen, and S. Magnusson, *in* "Progress in Chemical Fibrinolysis and Thrombolysis" (J. F. Davidson, R. M. Rowan, M. M. Samama, and P. C. Desnoyers, eds.), p. 191. Raven, New York, 1978.

[9] J. McLean, J. Tomlinson, W. Kuang, D. Eaton, E. Chen, G. Fless, A. Scanu, and R. Lawn, *Nature (London)* **300,** 132 (1987).

[10] J. Guevara, R. D. Knapp, S. Honda, S. R. Northup, and J. D. Morrisett, *Proteins Struct. Funct. Genet.* **12,** 188 (1992).

[11] A. M. Scanu, L. A. Miles, G. M. Fless, D. Pfaffinger, J. E. Eisenbart, E. Jackson, J. L. Hoover-Plow, T. Brunck, and E. F. Plow, *J. Clin. Invest.* **91,** 283 (1993).

Sepharose; however, Lp(a) species with different apo(a) isoforms may have affinities that are slightly greater or weaker than that of plasminogen.[12]

The buffer of choice in the isolation of plasminogen from plasma by lysine-Sepharose affinity chromatography has been 0.1 *M* phosphate buffer, pH 7.4. When the same buffer system is used in the chromatography of Lp(a), not all the lipoprotein is found to bind to the lysine-Sepharose.[13] For example, we discovered that approximately 80% of Lp(a) contained in plasma had the capacity to interact with lysine-Sepharose, whereas the rest was incompetent.[11] Purified Lp(a) had the same ratio of competent to incompetent particles. This ratio may also be affected by the quantity of lysine coupled to the Sepharose, because we found that less Lp(a) is bound to commercially available lysine-Sepharose which has about 3- to 4-fold less lysine per milliliter gel than lysine-Sepharose prepared in the laboratory. The percentage of Lp(a) binding to lysine-Sepharose can be increased by lowering the ionic strength of the buffer medium.[14] However, binding is increasingly nonspecific in nature at lower ionic strength, as LDL and HDL also start binding. Certain triglyceride-rich particles also bind nonspecifically, especially when unfasted plasma is passed over lysine-Sepharose, although this binding cannot be eliminated by changing the ionic strength of the buffer medium.

Besides ionic strength, and the density of lysine groups on the gel, temperature is also an important variable affecting lysine-Sepharose chromatography. Lipoprotein (a) species with large apo(a) isoforms tend to self-associate in the cold, and this process is concentration dependent.[12] As plasma percolates through the column, Lp(a) is concentrated at the top of the gel bed because it binds to the lysine groups coupled to the gel matrix. If the column is run at 4°, Lp(a) may cryoprecipitate in a thin zone at the top of the column from which it cannot be eluted with ε-aminocaproic acid (EACA). For this reason it is best to perform the chromatographic isolation at room temperature.

Materials

Sepharose 4B
L-Lysine monohydrochloride
Cyanogen bromide (5 *M* solution in acetonitrile)

[12] G. M. Fless and M. L. Snyder, *Chem. Phys. Lipids* **67/68,** 69 (1994).
[13] V. W. Armstrong, B. Harrach, H. Robenek, M. Helmhold, A. K. Walli, and D. Seidel, *J. Lipid Res.* **31,** 429 (1990).
[14] M. Helmhold, J. U. Wieding, and V. W. Armstrong, 59th European Atherosclerosis Society Congress, Nice, France, 1992 (Abstract 53).

Preparation of Lysine-Sepharose 4B

We prepare lysine-Sepharose using a procedure slightly modified from the one described by Deutsch and Mertz.[15] Packed Sepharose 4B (250 ml) is washed with 8 liters of water on a coarse sintered glass funnel and activated with 25 g CNBr dissolved in 50 ml acetonitrile. The reaction is carried out in a well-ventilated hood, on ice, and the pH is maintained with 6 *N* NaOH at pH 11. After approximately 15 to 30 min, the activated Sepharose 4B is washed with 8 liters of 0.1 *M* $NaHCO_3$, pH 8.1. The agarose is then packed by filtration, diluted with 250 ml of 0.1 *M* $NaHCO_3$, pH 8.1, containing 50 g lysine, and stirred gently overnight at 4°. The freshly conjugated lysine-Sepharose is then washed with 6 to 10 liters of 1 m*M* HCl followed by 8 liters of 0.1 *M* $NaHCO_3$, pH 8.1, and an aliquot is saved for determination of the concentration of immobilized lysine residues using the method of Wilkie and Landry.[16] The concentration of coupled lysine varies from 15 to 25 μmol per milliliter packed gel when prepared by this method, which is approximately 3- to 4-fold higher than lysine-Sepharose 4B obtained from Pharmacia (Piscataway, NJ). The lysine-Sepharose is stored in 0.1 *M* phosphate, pH 7.4, containing 0.01% Na_2EDTA and 0.01% NaN_3.

Chromatography

Bio-Rad (Richmond, CA) Econo-Pac columns (1 × 12 cm) are packed with 5 ml lysine-Sepharose which is preequilibrated with column buffer (e.g., 0.1 *M* phosphate, 0.01% Na_2EDTA, and 0.01% NaN_3, pH 7.4). A porous polymer frit is placed on top of the lysine-Sepharose gel bed to prevent the column from running dry. Plasma samples smaller than 3 ml are applied to the column and allowed to run through by gravity at room temperature. Larger volumes (up to 50 ml) should be applied with a pump or by gravity, but at flow rates that should not exceed 20 ml/cm^2/hr. The samples are washed into the column with four 0.5-ml aliquots of column buffer to be followed with four 5-ml aliquots, before Lp(a) is eluted with 0.2 *M* EACA in 10 m*M* phosphate, pH 7.4. One-milliliter aliquots are applied at a time, and 1-ml fractions are collected in separate tubes. Lipoprotein (a)- and plasminogen-containing fractions (tubes 4 through 10) are located by their absorbance at 280 nm. The volume of applied plasma depends on the Lp(a) content and on the sensitivity of the absorbance monitor that is part of the density gradient fractionating system.

[15] D. G. Deutsch and E. T. Mertz, *Science* **170,** 1095 (1970).
[16] S. D. Wilkie and D. Landry, *BioChromatography* **3,** 205 (1988).

Density Gradient Centrifugation

Principle

The mass of Lp(a) recovered from the minicolumns is small, being less than 200 μg Lp(a) protein for a 1-ml plasma sample with an Lp(a) protein content of 20 mg/dl or 100 μg for a 50-ml sample with an Lp(a) protein content of 0.2 mg/dl. This fact dictates the choice of the density gradient. A steep gradient covering the density range of 1.0 to 1.21 g/ml is preferred because this tends to concentrate the Lp(a) sample to a small portion of the gradient. We chose to use a step gradient made with only two solutions (e.g., the elution buffer containing the sample and a solution of 20% NaBr), because of its ease of preparation. Representative Lp(a) density profiles of six individuals with different plasma levels of Lp(a) are given in Fig. 1.

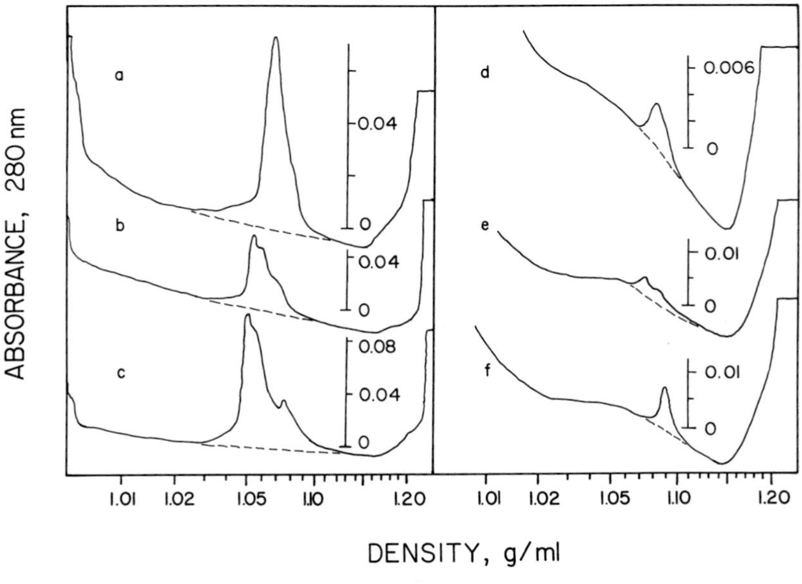

FIG. 1. Lipoprotein (a) density profiles of six individuals. Two milliliters plasma was applied to 5-ml lysine-Sepharose minicolumns, and the bound Lp(a) was eluted with 200 m*M* EACA, 10 m*M* phosphate, pH 7.4. Step gradients were constructed in SW-40 centrifuge tubes by layering the bound Lp(a) on top of 5 ml of 20% NaBr and topping the tubes with 200 m*M* EACA, 10 m*M* phosphate. The SW-40 rotor was run at 39,000 rpm, 20°, for 66 hr before the gradients were monitored at 280 nm with an ISCO density gradient fractionation system. Densities of a control gradient were measured with a Mettler/Paar density meter. Plasma Lp(a) protein levels were as follows: subject a, 16.3 mg/dl; b, 14.6 mg/dl; c, 25.3 mg/dl; d, 1.6 mg/dl; e, 2.1 mg/dl; f, 4.2 mg/dl. By Western blot analysis, the Lp(a) of subjects a, d, and e was homozygous, whereas that of b, c, and f was heterozygous.

Lipoprotein (a) contained in 2 ml plasma was applied to the gradients, which were monitored at 280 nm at different sensitivity levels.

The drawback of step gradients is that 2 to 3 days of centrifugation are required in order to reach equilibrium. When knowledge of Lp(a) density is not needed, a step gradient such as the one proposed by Chung *et al.*[17] (Table II, procedure 6) for use in the vertical rotor (Beckman VTi 80), may be more suitable because of the short centrifugation time (44 min) required to achieve separation.

Equipment

Preparative ultracentrifuge (Beckman, Palo Alto, CA, L7-55 or equivalent)
Swinging-bucket rotor (Beckman SW-40)
Density gradient fractionation (ISCO, Lincoln, NE, Model 640 or 185) and monitoring system (ISCO Model UA-5)
Density meter (Mettler/Paar, Hightstown, NJ, Model DMA46) or refractometer (Zeiss/Abbe, Thornwood, NY, Model A)

Procedure

Place 5 ml of 20% (w/w) NaBr into a SW-40 ultracentrifuge tube (ultraclear). Carefully layer the eluate from the lysine-Sepharose column (up to 8 ml) on top of the NaBr solution and, if necessary, top off the tube with 0.2 *M* EACA, 10 m*M* phosphate, pH 7.4. Place the tubes in the buckets of the swinging-bucket rotor and centrifuge 64 hr at 39,000 rpm and 20°. After centrifugation is completed, the tubes are carefully removed from the buckets and placed in the density gradient fractionating system. The tubes are pierced at the bottom, and the gradient is pushed out the top at a flow rate of 1 ml/min with a dense fluorocarbon oil, Fluorinert FC-40 (ISCO), that has a density of 1.85 g/ml. The chart speed is 1 cm/min, and the fraction collector is set to 0.5 ml/tube. The gradient is monitored at 280 nm, and the sensitivity of the chart recorder is adjusted according to the Lp(a) content of the eluate. Densities of the various fractions can be measured with a density meter by established techniques.[18] In case a density meter is not available, the eluate from the lysine-Sepharose 4B column can be dialyzed against 0.1 *M* NaBr, 10 m*M* phosphate, 0.01% Na_2EDTA, 0.01% NaN_3, pH 7.4, so that densities can be measured by refractometry. Because the introduction of a dialysis step would decrease the recovery of Lp(a), it is recommended that two density gradient tubes be used for each sample:

[17] B. H. Chung, J. P. Segrest, M. J. Ray, J. D. Brunzell, J. E. Hokanson, R. M. Krauss, K. C. Beaudrie, and J. T. Cone, this series, Vol. 128, p. 181.
[18] O. Kratky, H. Leopold, and H. Stabinger, this series, Vol. 27, p. 98.

one tube to which the sample in eluate buffer is added directly for quantitation by ELISA, and a second to which the dialyzed sample is added for density determination by refractometry.

Enzyme-Linked Immunosorbent Assay

Principle

Density gradient centrifugation of the lysine-Sepharose eluate has the potential to separate three different apo(a)-containing fractions. Located at the top of the density gradient, in one or two tubes, may be triglyceride-rich particles containing apo(a). In the density gradient itself, usually between densities 1.05 to 1.10 g/ml, is located Lp(a) containing predominantly cholesteryl esters, whereas free apo(a) or fragments of apo(a) may be present at the bottom of the tube.

Two different ELISA formats are necessary to quantitate the different apo(a)-containing materials. Apolipoprotein (a) present in lipoproteins can be measured in an ELISA that employs capture antibodies specific to apo(a) and detecting antibodies specific to apoB.[1] There are two reasons for choosing this format. First, if the antiserum to apo(a) is cross-reactive to plasminogen, by using antibodies specific to apoB only molecules containing both apo(a) and apoB are quantitated. Even if plasminogen were bound to the plate, it would not be measured because it does not contain apoB. Second, the choice of anti-apoB as detecting antibody also avoids problems arising due to heterogeneity of apo(a). Because apo(a) consists mainly of kringle 4 domains that are highly homologous to one another, the potential exists that it contains many repetitive epitopes. The number of identical epitopes would be greater in large apo(a) polymorphs and smaller in apo(a) molecules of low molecular weight. A detecting antibody specific to a repetitive domain would therefore amplify or exaggerate the Lp(a) content of samples containing large apo(a) isoforms. This problem is avoided by using anti-apoB as the detecting antibody.

The measurement of the free apo(a) content of bottom fraction, which also contains plasminogen, requires the use of an anti-apo(a) capture antibody with a specificity exclusive to apo(a) and no cross-reactivity with plasminogen. The detecting antibody must also be specific to apo(a).

Reagents

Buffers

Coating buffer: 10 m*M* Tris, 0.15 *M* NaCl, pH 7.6 (Tris–saline)

Blocking buffer: Tris-saline containing 1% bovine serum albumin (BSA)

Wash/diluent buffer: 0.1 *M* $NaHCO_3$, 0.5 *M* NaCl, pH 8.1 containing 0.1% Tween 20 and 1% BSA

Substrate buffer: 10% diethanolamine (97 ml DEA/1 liter buffer) in 0.01% $MgCl_2$, pH 9.8, with 0.02% NaN_3, to which *p*-nitrophenyl phosphate is added to a final concentration of 1 mg/ml just before use

Immunochemicals and Other Biologicals

Polyclonal anti-human Lp(a) antibody

Affinity-purified anti-human apoB antibody conjugated to alkaline phosphatase

Affinity-purified anti-human apo(a) antibody

Affinity-purified anti-human apo(a) antibody, nonreactive with plasminogen

Affinity-purified anti-human apo(a) antibody conjugated to alkaline phosphatase

Affinity columns: Lp(a)-Sepharose, LDL-Sepharose, and plasminogen-Sepharose

Lp(a) free of LDL

LDL free of Lp(a)

Plasminogen free of Lp(a) and very-low-density lipoprotein (VLDL)

Alkaline phosphatase from bovine intestinal mucosa (Sigma, St. Louis, MO)

Equipment

Bio-Rad Econo-System or other low-pressure liquid chromatography setup

Microtiter plate reader (Dynatech, Chantilly, VA), or other comparable photometer

Production of Antibodies

Purified preparations of Lp(a) and plasminogen (0.2–1.0 mg in a volume of 0.2–0.5 ml) are mixed with Freund's complete adjuvant (2:1 Freund's:antigen) and injected intramuscularly in the animal of choice (usually rabbit or goat). Six weeks later, a booster is given with the antigen mixed 2:1 with Freund's incomplete adjuvant and again injected intramuscularly. Ten to 14 days after the booster, 1–2 ml blood should be drawn to determine the titer of the antisera. If the titer is low, another booster may be administered until the titer is high enough to allow collection of blood for antiserum production.

A note of caution is in order. It appears that apolipoprotein B is a much better antigen than apolipoprotein(a). When an antiserum to Lp(a) raised in the goat was analyzed with respect to its reactivity to apoB and apo(a), it was found that 85% of the total immunoglobulins recovered were specific to apoB and only 15% to apo(a). Furthermore, two-thirds of the antibodies specific to apo(a) were also cross-reactive with plasminogen, leaving only 5% of the anti-Lp(a) serum truly specific to apo(a). It is advisable then to use besides Lp(a) also apo(a) or, if available, recombinant apo(a) or even plasminogen to determine antibody titer. This could provide the information necessary in order to decide whether to continue or stop the booster injections. Relying only on titrations with Lp(a) to determine antibody titer may lead to premature bleeding of the animal of choice, because the percentage of antibodies directed to apoB may be very high. Alternatively, apo(a) prepared from Lp(a)[19,20] or recombinant apo(a) can be used as antigen in order to get better yields.

Affinity Columns

Conjugation of LDL, Lp(a), and plasminogen to CNBr-activated Sepharose 4B (Pharmacia) to achieve a concentration of 1 to 2 mg protein per milliliter gel is carried out according to the manufacturer's instructions. To prevent oxidation and bleaching of the immobilized LDL or Lp(a), the affinity columns are covered with aluminum foil to keep out light and are stored in 10 m*M* phosphate, 150 m*M* NaCl, 0.01% Na_2EDTA, 0.01% NaN_3, pH 7.4 (phosphate-buffered saline, PBS), containing 50 μM butylated hydroxytoluene (BHT). To maintain the integrity of the immobilized plasminogen, the column is stored in PBS containing 1 m*M* benzamidine and 0.01% soybean trypsin inhibitor.

Purification of Antibodies

The size of the affinity columns needed to purify the antiserum depends on the titer of antibodies specific to apoB, apo(a), and plasminogen and should be determined empirically. In our hands a 40-ml LDL column and 20-ml columns of Lp(a) and plasminogen were adequate to isolate all specific antibodies in a single pass of 15 to 20 ml high-titer goat anti-Lp(a) serum.

In sequence, join the LDL-, plasminogen-, and Lp(a)-Sepharose 4B columns (1.5-cm-diameter Bio-Rad Econo columns equipped with flow adapters). Wash the columns with PBS to remove the BHT and benzamidine, which strongly absorb at 280 nm. Load the antiserum at a flow rate

[19] G. M. Fless, M. E. ZumMallen, and A. M. Scanu, *J. Lipid Res.* **26,** 1224 (1985).
[20] G. M. Fless, M. E. ZumMallen, and A. M. Scanu, *J. Biol. Chem.* **261,** 8712 (1986).

of 1 ml/hr, and monitor the absorbance at 280 nm. Wash with PBS until the absorbance starts to decrease toward baseline values and follow with 20 ml of 0.1 *M* $NaHCO_3$, 0.5 *M* NaCl, pH 8.1, to remove nonspecifically bound proteins. Continue washing with PBS until the absorbance has returned to baseline values. Disconnect the three columns and elute the bound antibodies separately with 0.1 *M* glycine, pH 2.5. Immediately titrate the eluted antibodies back to neutral pH with 1 *M* Tris and dialyze them against 0.1 *M* phosphate, pH 6.8. Wash the affinity columns with 0.1 *M* phosphate, 0.01% Na_2EDTA, 0.01% NaN_3, pH 7.4, and store the LDL and Lp(a) columns in the presence of 50 μg/ml BHT and the plasminogen column in 1 m*M* benzamidine and 0.01% soybean trypsin inhibitor.

Conjugation of Antibodies

A simple method of conjugating enzymes to antibodies is the one-step procedure of Avrameas[21] as described by Tijssen.[22] Concentrate, if necessary, the affinity-purified antibody (Amicon, Danvers, MA, Centriprep 30) until the concentration is greater than 5 mg/ml. Transfer 1 mg immunoglobulin G (IgG) and 2 mg alkaline phosphatase (APase) to a small dialysis bag that has been extensively rinsed with double-distilled water (e.g., Milli-Q water) and close with a clamp. Dialyze overnight at 4° against 4 liters of 0.1 *M* phosphate, pH 6.8, made with double-distilled water. In the morning, change the dialyzate and continue dialyzing at room temperature for several more hours. Open the dialysis bag and add directly to the bag 1% glutaraldehyde to a final concentration of 0.025% (e.g., 1 μl of 1% v/v glutaraldehyde per 40 μl reaction volume). Reclose the dialysis bag and mix thoroughly. Incubate the bag in a 50-ml tube containing 0.1 *M* phosphate with 0.025% glutaraldehyde for 3 hr at room temperature. At the end of the incubation period, open the bag and add 1/40 volume of 2 *M* lysine, pH 7.4, to block excess glutaraldehyde. Reclose bag and continue incubating for 2 hr at room temperature in a closed 50-ml tube with 0.5 ml phosphate buffer (no glutaraldehyde) to provide humidity. Dialyze against several changes, 4 liters each, of 0.15 *M* NaCl, 50 m*M* Tris, 2 m*M* $MgCl_2$, 0.02% NaN_3, pH 8.0. To store, add BSA to a final concentration of 10 mg/ml and keep at 4°. Some important notes to remember are (1) use only double-distilled water to make up all the buffers; (2) the weight ratio of APase to IgG should be 2 : 1; and (3) the total protein concentration should be 7.5 μg protein (enzyme plus IgG) per microliter.

[21] S. Avrameas, *Immunochemistry* **6,** 43 (1969).

[22] P. Tijssen, *Lab. Tech. Biochem. Mol. Biol.* **15,** 242 (1985).

Production of Standard Lipoprotein (a)

Lipoprotein (a)-containing plasma [preferably from a donor with only one apo(a) isoform] is treated with 0.2 m*M* phenylmethylsulfonyl fluoride, 0.01% Na_2EDTA and adjusted to a density of 1.21 g/ml with solid NaBr.[23] The Lp(a)-containing top fraction, obtained on centrifugation in the Ti 50.2 rotor for 20 hr, at 45,000 rpm and 20°, is dialyzed against 0.1 *M* phosphate, 0.01% Na_2EDTA, 0.01% NaN_3, 1 m*M* benzamidine, pH 7.4. This lipoprotein fraction is then applied to a lysine-Sepharose column (2.5 × 20 cm) equilibrated in the same buffer at a flow rate of 50 ml/hr at room temperature. The column is monitored at 280 nm and washed with column buffer until the absorbance has returned to baseline values. Lipoprotein (a) is then eluted with 200 m*M* EACA containing 10 m*M* phosphate, 0.01% Na_2EDTA, 0.01% NaN_3, 1 m*M* benzamidine, pH 7.4. The eluate is made 7.5% (w/w) with CsCl and centrifuged 20 hr at 49,000 rpm at 20° in the 50.2 Ti rotor. The CsCl forms a self-generating gradient on which Lp(a) species with different apo(a) isoforms may be separated, and which also serves to remove contaminating plasminogen and nonspecifically bound triglyceride-rich particles. Lipoprotein (a) is dialyzed exhaustively against 0.15 *M* NaCl, 0.01% Na_2EDTA, 0.01% NaN_3, checked for purity by sodium dodecyl sulfate–polyacrylamide gel electrophoresis (SDS–PAGE), and stored filter sterilized (0.45 μm) in screw-capped Sarstedt (Newton, NC) vials that are filled to allow no air space.

Preparation of Plasminogen

An easy and inexpensive method of preparing plasminogen is the classic procedure of Deutsch and Mertz[15] as modified by Castellino and Sodetz.[24] Fasted normolipidemic plasma is treated with 0.2 m*M* phenylmethylsulfonyl fluoride and passed over lysine-Sepharose. The column is washed with 0.5 *M* NaCl, 10 m*M* phosphate, pH 7.4, to remove nonspecifically bound proteins. Plasminogen is eluted with a linear gradient of 0 to 20 m*M* EACA. Early fractions contain Lp(a) and should not be used. The pooled plasminogen should be monitored for purity by SDS–PAGE. If it is contaminated with large molecular weight proteins that could arise from the presence of small quantities of Lp(a) or triglyceride-rich lipoproteins, it should be further purified by gel filtration over Sepharose 6B.

[23] M. L. Snyder, D. Polacek, A. M. Scanu, and G. M. Fless, *J. Biol. Chem.* **267,** 339 (1992).
[24] F. J. Castellino and J. M. Sodetz, this series, Vol. 45, p. 273.

Procedure for Estimation of Lipoprotein (a) Using Antiapolipoprotein B as Detecting Antibody

Polystyrene microtiter plates are coated with 400 ng/well affinity-purified anti-human apo(a) antibody (100 μl of a 4 μg/ml solution in coating buffer), sealed with a thin adhesive-coated plastic sheet, and incubated overnight at room temperature.[1] The amount of antibody coated is the minimum required to obtain a good dose–response curve with low variability in absorbance values. The coating concentration has to be determined empirically and may depend on the affinity of the antibody for apo(a). The following day, the plates are washed three times, 5 min/wash, with blocking buffer to remove the unbound antibodies and then incubated for 2 hr with the blocking buffer to block any remaining binding sites. Next, the plates are dried by blotting, sealed, and stored at $-20°$ until used. The plates may be stored for up to 6 months this way with no loss of reactivity nor increase in background noise if properly sealed.

Allow plates to equilibrate to room temperature before use. Dilute the Lp(a) standard and the samples in 0.1 *M* $NaHCO_3$, 0.5 *M* NaCl, 1% BSA, 0.1% Tween 20, pH 8.1, and add 100 μl of each dilution to the wells in triplicate and seal the plates. Incubate for 2 hr at room temperature and follow with three washes of wash buffer, 5 min/wash. Add 100 μl per well anti-apoB–alkaline phosphatase conjugate diluted in wash/diluent buffer and incubate 1 hr at room temperature. Wash plates three times with wash/diluent buffer, 5 min/wash, and develop color with 1 mg/ml *p*-nitrophenyl phosphate in diethanolamine buffer (100 μl/well). Protect the plate from light for 30 min, after which time the color development is stopped by adding 100 μl of 1 *N* NaOH to each well. The absorbance is read at 410 nm on an ELISA plate reader.

The choice of the proper dilution of the alkaline phosphatase–antibody conjugate necessary for good color development is determined by checkerboard titration. Serial dilutions of Lp(a) standard and enzyme conjugate are incubated in the same plate [coated with anti-apo(a)], and the dilution of conjugate that gives the best color response is chosen for the assay.

The assay is subject to fluctuations in temperature which, when not controlled, can give rise to edge effects. Positioning plates under air conditioning vents or next to windows, hot plates, or other sources of heat or cold can introduce tremendous variability in results. These may be prevented by performing the coating, incubation, and color developing steps in a draft-free box easily constructed from polystyrene sheets. In addition, we place the microtiter plates on 6-mm thick aluminum plates cut to the exact size of the bottom of the microtiter plates. These serve to conduct heat uniformly to all 96 wells of the microtiter plate.

Procedure for Estimation of Apolipoprotein (a) Using Antiapolipoprotein (a) as Detecting Antibody

The plate is coated with 400 ng/well affinity-purified anti-apo(a) that is nonreactive to plasminogen. The rest of the procedure is exactly similar to the method described above except that anti-apo(a) conjugated to alkaline phosphatase is used as detecting antibody. Because the capture antibody does not react with plasminogen, it is not necessary to remove the immunoglobulins specific to plasminogen from the detecting antibody.

Plate Setup and Calculation of Results

To obtain a standard curve it is necessary to add at least five concentrations of the Lp(a) standard (serially diluted 2-fold) to the plate. One dilution is used to determine the absorbance maximum, and the other four are used to establish the straight-line portion of the dose–response curve. In addition, one set of wells is left Lp(a)-free and is filled only with buffer in order to determine the absorbance minimum. These should be sufficient to form a symmetrical dose–response curve. The unknown samples are also serially diluted, and multiple dilutions have to be added to the plate in order to obtain precision and accuracy in the calculated Lp(a) content. The concentration of Lp(a) in the unknown samples can be evaluated manually; however, the reliability of the calculated results can be improved through suitable curve-fitting procedures for which software is readily available.

There are several problems in reporting results that are peculiar to Lp(a). First, it is common to report Lp(a) values in terms of lipoprotein concentration. This practice is cumbersome because it requires knowledge of the lipid composition of Lp(a). It also does not lend itself to standardization because the lipid content of Lp(a) is not fixed and can vary even in Lp(a) particles with the same apo(a) isoform. It would be much easier to report Lp(a) concentration in terms of Lp(a) protein. For example, concentrations of LDL are reported in terms of apoB. Second, because Lp(a) is polymorphic in nature, the molecular weight of Lp(a) protein of different Lp(a) polymorphs will vary. Therefore, true concentrations of Lp(a) will be obtained only if the concentrations of the standard Lp(a) particle or Lp(a) protein are expressed on a molar basis. In practice, however, it is difficult to obtain accurate molecular weights of apo(a) isoforms. Molecular weight analysis by analytical centrifugation is not practical, and SDS–PAGE gives unreliable results because current gel systems are not optimal either for the analysis of large molecular weight proteins or for the analysis of heavily glycosylated proteins that have anomalous electrophoretic mobilities. In light of these difficulties, it is probably best to report results on a weight basis with the understanding that values of unknown

Lp(a) samples with apo(a) isoforms different from the standard will be either higher or lower than the true value.

Acknowledgments

We gratefully acknowledge Dr. Angelo M. Scanu for interest in and support of this work. This work was supported by a Program Project Grant from the U.S. Public Health Service (HL-18577).

[15] Two-Dimensional Nondenaturing Electrophoresis of Lipoproteins: Applications to High-Density Lipoprotein Speciation

By CHRISTOPHER J. FIELDING and PHOEBE E. FIELDING

Introduction

High-density lipoproteins (HDL) are small lipoproteins (6–12 nm diameter) present in intercellular fluid, lymph, and plasma. The HDL are formed extracellularly by the interaction of cellular and medium lipids with lipid-poor apolipoprotein. The major HDL apolipoprotein (apolipoprotein A-I, apoA-I) is secreted from liver and intestine, mainly in the form of a loose complex with triglyceride-rich lipoproteins. Once reaching the plasma, apoA-I dissociates rapidly independently of lipolysis. Lipid-poor apoA-I reacts most avidly with phospholipids and unesterified cholesterol. Continued addition of these lipids leads to the formation of a unique discoidal HDL, in which apoA-I assumes a largely helical form. Convincing biophysical analysis by several laboratories makes it very likely that 22-amino acid repeating sequences evident in the apoA-I primary sequence alternate across the 35-Å width of the disk bilayer, stabilizing its periphery. Subsequent reaction of such discoidal HDL with lecithin–cholesterol acyltransferase (LCAT; phosphatidylcholine–sterol *O*-acyltransferase) converts part of the free cholesterol and phospholipid of the disk to cholesteryl ester, which is retained in the HDL, and lysolecithin, which is lost to the medium. This cholesteryl ester is likely localized mainly in the core of the now spherical mature HDL particle. The apoA-I remains in a highly helical conformation on the surface of the sphere. Later in metabolism, when HDL core lipids have been reduced by neutral lipid transfer or lipase activities, free or lipid-poor apoA-I is regenerated. As a result of these changes there are three main physical forms of HDL: lipid-poor particles, disks, and spheres.[1]

The residence time of apoA-I in human plasma is 4–5 days,[2,3] while the time required for the conversion of lipid-poor apoA-I to a mature spheroidal HDL has been estimated at about 15 min[4] from studies of the conversion of lipid-poor HDL *in vitro*. Similar estimates can be obtained by comparing the turnover number of LCAT in native plasma (5–10 min^{-1}) and the number of cholesteryl ester molecules per spherical HDL particle (30–110).[5] These data imply an extensive recirculation of apoA-I as this protein picks up phospholipid and free cholesterol from cell membranes or other lipoproteins, transesterifies them to cholesteryl ester, and returns them, either on HDL or (after transfer) on other lipoproteins, to the liver for further metabolism.

Historically plasma HDl were defined in terms of their flotation within the density limits 1.063–1.21 g/ml.[6] Almost without exception, lipoproteins within these limits do include apoA-I in their protein moiety; more recently, however, it has become clear that in some important ways centrifugally isolated HDL does not fully reflect the composition or properties of HDL in native plasma or lymph. Some HDL (particularly the lipid-poor forms) disappear completely from the density 1.063–1.21 g/ml fraction.[7] Further apoA-I is stripped off spheroidal HDL particles, and is recovered, along with the native lipid-poor HDL species, in the centrifugal infranatant solution.[8] Finally, during flotation, LCAT and lipid transfer activities continue at low but measureable rates, modifying the HDL lipid composition. These findings have led to a search for alternative isolation procedures that more accurately reflect the heterogeneity of HDL.

In addition to apoA-I, many HDL contain other apolipoproteins. The second most abundant protein (apolipoprotein A-II, apoA-II) is secreted with apoA-I from the liver. Isolated apoA-I and apoA-II both exchange among lipid vesicles *in vitro*,[9] yet metabolic data support a kinetic model

[1] P. E. Fielding and C. J. Fielding, *in* "Biochemistry of Lipids, Lipoproteins and Membranes" (D. E. Vance and J. Vance, eds.), p. 427. Elsevier, Amsterdam, 1991.

[2] G. L. Vega, H. Gylling, A. V. Nichols, and S. M. Grundy, *J. Lipid Res.* **32,** 867 (1991).

[3] K. Ikewaki, D. J. Rader, J. R. Schaefer, T. Fairwell, L. A. Zech, and H. B. Brewer, *J. Lipid Res.* **34,** 2207 (1993).

[4] O. L. Francone and C. J. Fielding, *Eur. Heart J.* **11E,** 218 (1990).

[5] S. Eisenberg, *J. Lipid Res.* **25,** 1017 (1984).

[6] R. J. Havel, H. A. Eder, and J. H. Bragdon, *J. Clin. Invest.* **34,** 1345 (1955).

[7] B. F. Asztalos, C. H. Sloop, L. Wong, and P. S. Roheim, *Biochim. Biophys. Acta* **1169,** 291 (1993).

[8] S. T. Kunitake and J. P. Kane, *J. Lipid Res.* **23,** 936 (1982).

[9] J. A. Ibdah, C. Smith, S. Lund-Katz, and M. C. Phillips, *Biochim. Biophys. Acta* **1081,** 220 (1991).

in which plasma HDL containing only apoA-I, and HDL containing both apoA-I and apoA-II, maintain a largely separate existence in the circulation.[10] Other, less abundant apoproteins probably redistribute passively according to the lipid composition of the HDL particles present.

The apoprotein heterogeneity of HDL provides other approaches for the fractionation of the particles. Immunoaffinity chromatography has been used to determine HDL subfraction distributions in plasma according to particle protein composition.[11,12] High-density lipoproteins containing only apoA-I only (LpA-I only) and HDL with both apoA-I and apoA-II (LpAI, AII) have been estimated in normal plasma and in plasma from patients experiencing several disease states.[13–15] Immunoaffinity chromatography does not distinguish HDL with the same protein composition differing in size, apoA-I conformation, or lipid composition, and recovery of lipoprotein fractions from the antibody matrix involves denaturing solvents such as 3 *M* NaSCN (pH 7.0) or 0.1 *M* acetic acid (pH 3.0).

Another approach based on HDL subfraction protein composition involves two-dimensional nondenaturing electrophoresis.[16–18] Different proteins at the same pH usually have significantly different net charges. In HDL the number of charges exposed can vary with protein conformation on particles with different lipid composition, allowing further differentiation among species. In native plasma, HDL containing only apoA-I usually have a slower (pre-β) electrophoretic migation than α-migating HDL containing both apoA-I and apoA-II.

In this technique first-dimensional separation is in agarose, on the basis of particle net charge. Second-dimensional fractionation is in a gradient of polyacrylamide, and separation here is on the basis of particle diameter. The method is convenient for kinetic studies, because the pattern of protein or lipid tracer can be followed as a function of time from gel fragments containing different HDL subfractions. Several of the fractions best sepa-

[10] D. J. Rader, G. Castro, L. A. Zech, J.-C. Fruchart, and H. B. Brewer, *J. Lipid Res.* **32,** 1849 (1991).

[11] C. J. Fielding and P. E. Fielding, *Proc. Natl. Acad. Sci. U.S.A.* **78,** 3911 (1981).

[12] M. C. Cheung, A. C. Wolf, K. D. Lum, J. H. Tollefson, and J. J. Albers, *J. Lipid Res.* **27,** 1135 (1986).

[13] P. Puchois, A. Kandoussi, P. Fievet, J. L. Fourrier, M. Bertrand, E. Koren, and J.-C. Fruchart, *Atherosclerosis* **68,** 35 (1987).

[14] M. C. Cheung, B. G. Brown, A. C. Wolf, and J. J. Albers, *J. Lipid Res.* **32,** 383 (1991).

[15] P. Puchois, N. Ghalim, G. Zylberberg, P. Fievet, C. Demarquilly, and J.-C. Fruchart, *Arch. Intern. Med.* **150,** 1638 (1990).

[16] G. R. Castro and C. J. Fielding, *Biochemistry* **27,** 25 (1988).

[17] W. D. Stuart, B. Krol, S. H. Jenkins, and J. A. K. Harmony, *Biochemistry* **31,** 8552 (1992).

[18] Y. Huang, A. von Eckardstein, and G. Assmann, *Arterioscler. Thromb.* **13,** 445 (1993).

rated by this procedure represent interesting intermediates in the genesis of spheroidal HDL (Fig. 1, left-hand side).

Pre-β-1 High-Density Lipoprotein. The small (60–75 kDa) preβ-1 HDL fraction has a protein moiety containing only apoA-I. It represents 2–5% of total plasma apoA-I. This HDL is an effective early acceptor of cell-derived cholesterol. Particles similar or identical to pre-β-1 HDL are generated when lipid-free apoA-I is incubated with cultured monolayers of mammalian cells.

Pre-β-2 High-Density Lipoprotein. Pre-β-2 HDL is a large (325 kDa) HDL similar or identical to synthetic discoidal HDL containing three apoA-I per particle, generated from mixtures of apoA-I, phospholipid, and unesterified cholesterol. Such HDL are the most effective substrates for

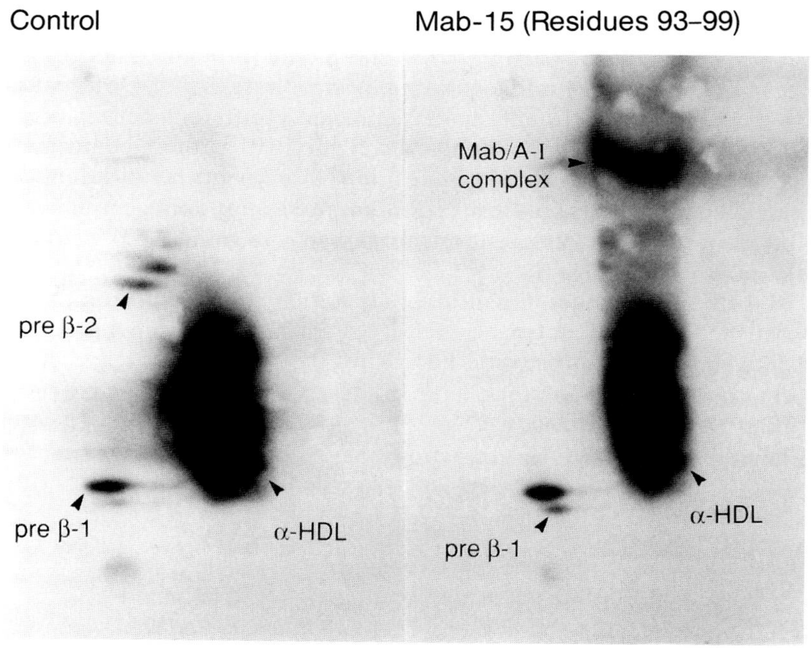

FIG. 1. Two-dimensional electrophoresis of native plasma HDL visualized with ^{125}I-labeled polyclonal rabbit anti-human apoA-I IgG. [Reprinted with permission from P. E. Fielding, M. Kawano, A. L. Catapano, A. Zoppo, S. Marcovina, and C. J. Fielding, *Biochemistry* **33,** 6981 Copyright (1994) American Chemical Society.] *Left:* Plasma in the absence of mouse anti-human apoA-I monoclonal antibody (MAb). *Right:* Plasma passed through an agarose layer containing MAb-15, an antibody recognizing residues 93–99 of the 243-amino acid sequence of mature human apoA-I. This monoclonal antibody is reactive with pre-β-2 HDL but not pre-β-1 HDL. Some reactivity is also seen with α-migrating HDL.

LCAT activity. Their concentration in plasma is comparable to that of the lipid-poor, pre-β-1 HDL fraction.

Pre-β-3 High-Density Lipoprotein. The pre-β-3 particles contribute only a trace amount of total plasma apoA-I but include about 50% of total plasma LCAT. They probably represent complexes of pre-β-2 HDL with LCAT.

α-High-Density Lipoprotein. The major part of HDL has α-electrophoretic migration. It is separated into fractions comparable to the HDL_3, HDL_{2a}, and HDL_{2b} particles demonstrated earlier by sedimentation equilibrium or single-dimensional polyacrylamide gradient electrophoresis.[19,20] The HDL_3 form includes most of the remainder of plasma LCAT.[21] The major function of α-HDL_3 in plasma appears to be the processing of cholesterol and phospholipid derived from low- and very-low-density lipoproteins. The larger HDL_{2a} and HDL_{2b} fractions are formed as the result of continuing LCAT activity. The metabolism of cell-derived cholesterol ("reverse cholesterol transport") mainly takes place in the pre-β-migrating HDL fractions.

The distribution of HDL subfractions following two-dimensional electrophoresis is normally visualized by autoradiography following electrotransfer of the original gel to nitrocellulose[22] and reaction of apoA-I-containing lipoprotein fractions with ^{125}I-labeled polyclonal goat or rabbit anti-human apoA-I immunoglobulin G (IgG).[23] Because of the very different concentrations of α- and pre-β-HDL, autoradiography exposure times need to be adjusted to optimize visualization of the fractions of interest.

A modification of the two-dimensional electrophoretic technique allows study of the interaction of monoclonal antibodies with HDL fractions. This procedure allows all HDL fractions in plasma to be screened simultaneously for the expression of any apolipoprotein epitope.[24] A quantitative procedure for determining HDL subfractions from nondenaturing electrophoresis gels has been described, using the Hitachi phosphoimager.[7] This represents an advance in convenience and reproducibility over earlier manual quantitation techniques.

Two-dimensional electrophoresis is very effective in separating the pre-β-migrating HDL fractions of plasma from the major, α-migrating fractions.

[19] F. Lindgren, A. Nichols, and N. K. Freeman, *J. Phys. Chem.* **59,** 930 (1955).

[20] D. W. Anderson, A. V. Nichols, T. M. Forte, and F. T. Lindgren, *Biochim. Biophys. Acta* **493,** 55 (1977).

[21] P. E. Fielding, T. Miida, and C. J. Fielding, *Biochemistry* **30,** 8551 (1991).

[22] H. Towbin, T. Staehelin, and J. Gordon, *Proc. Natl. Acad. Sci. U.S.A.* **76,** 4350 (1979).

[23] M. A. K. Markwell, *Anal. Biochem.* **125,** 427 (1982).

[24] P. E. Fielding, M. Kawano, A. L. Catapano, A. Zoppo, S. Marcovina, and C. J. Fielding, *Biochemistry* **33,** 6981 (1994).

Because the lipid-poor, discoidal, and spheroidal fractions appear to represent the main stages of HDL synthesis, the two-dimensional procedure can provide metabolic information unavailable from other procedures.

Nondenaturing Two-Dimensional Electrophoresis

Preparation of Plasma Samples

The pattern of HDL subfractions is quite labile, in particular that of pre-β-HDL particles. Major determinants of pre-β-HDL levels in biological fluids include the presence or absence of nucleated cells as cholesterol donors, as well as the activity of the enzyme LCAT.[25] In the absence of nucleated cells pre-β-HDL levels decrease, when LCAT activity continues. Pre-β-HDL completely disappears after 90 min at 37°, but it is little changed after 24 hr at 0°. Inhibition of LCAT activity in plasma, by anti-LCAT antibody or sulfhydryl reagents such as 5,5′-dithiobis(2-nitrobenzoic acid) (DTNB), maintains or modestly increases pre-β-HDL levels. Pre-β-HDL levels are not affected by citrate or EDTA used in normal concentrations as anticoagulants. Protein anticoagulants such as a cocktail of protease inhibitors[16] or streptokinase (final concentration 100 U/ml)[26] are useful when plasma is incubated with cell monolayers prior to electrophoresis. Normal fasting plasma has been quick-frozen and maintained at −70° for periods of several months without change in pre-β-HDL levels on subsequent electrophoresis.[27] Conditions for hyperlipidemic plasma samples have not been established. Plasma from freshly drawn whole blood rapidly cooled in ice water is optimal for HDL fractionation. Samples can also be kept as whole blood for up to 24 hr at 0°, and plasma obtained by centrifugation immediately before use. Failing this, plasma should be quick frozen and stored at −70°.

First-Dimensional Separation

Reagents

50 m*M* Barbital buffer, pH 8.6
Agarose (standard low M_r), electrophoresis grade (Bio-Rad, Richmond, CA)
Gelbond (FMC, Rockland, ME)

Procedure. Two-tenths gram agarose is added to 26.7 ml barbital buffer and the mixture heated in a boiling water bath until the agarose dissolves.

[25] M. Kawano, T. Miida, C. J. Fielding, and P. E. Fielding, *Biochemistry* **32,** 5025 (1993).
[26] T. Miida, C. J. Fielding, and P. E. Fielding, *Biochemistry* **29,** 10469 (1990).
[27] B. Y. Ishida, J. Frolich, and C. J. Fielding, *J. Lipid Res.* **28,** 778 (1987).

Then 3.5 ml agarose solution is added to the hydrophilic side of each Gelbond strip (2.5 × 12.5 cm). After gelling (10 min, room temperature) a 1 × 17 mm well is cut in the gel layer 3 cm from one end with a razor blade and the loose gel fragment removed. The strips are laid laterally on the cooling block (4°) of a flatbed electrophoresis unit (Pharmacia-LKB, Piscataway, NJ). Wicks of blotting paper at each end of the strip connect it to the two electrode reservoirs of barbital buffer. Twenty microliters sample is added to each well. A marker of albumin–bromphenol blue is added at the side of the well. Electrophoresis is then carried out at 200 V until the marker dye has migrated 7.5 cm (1.5–2 hr).

By running the first-dimensional an additional 1.0 cm, Azstalos *et al.* were able to separate an additional, fast-migrating pre-α HDL fraction.[7] Its composition has not yet been determined.

Second-Dimensional Separation: Standard Conditions

Reagents

2% (w/v) Polyacrylamide gel solution: 15 ml distilled water plus 5.5 ml lower buffer (1.5 *M* Tris-HCl, pH 8.8) and 1.5 ml of 30% (w/v) acrylamide solution (29.2 *g* acrylamide plus 0.8 g bisacrylamide in distilled water to 100 ml); add 60 μl ammonium persulfate (0.1 g/ml distilled water)

15% (w/v) Polyacrylamide gel solution: 4.5 ml distilled water plus 4.5 ml lower buffer, 9 ml of 30% acrylamide solution, and 45 μl ammonium persulfate

Procedure. Immediately before pouring the gradient, 8 μl TEMED (*N,N,N′,N′*-tetramethylethylenediamine; Bio-Rad) is added to the 2% solution and 4 μl to the 15% solution. To form the gradient, 18 ml of 2% solution is added to the outside chamber of a two-chamber gradient maker; then 14 ml of the 15% solution is added to the inside chamber, and the gradient is poured. A stacking gel (1 cm) of 3% polyacrylamide in 5 ml upper buffer (0.125 *M* Tris-HCl, pH 6.8) with ammonium persulfate (1 mg/ml) and TEMED (0.7 μl/ml) is then formed over the gradient.

Two longitudinal strips, each 0.5 cm wide, are cut from the first-dimension agarose strip. These are laid end-to-end over the stacking gel. The space above the strips is filled with 0.75% agarose in barbital buffer to keep the strips in place. The tank buffer is 25 m*M* Tris–glycine, pH 8.3. Electrophoresis is then carried out at 200 V, 4°, for 1 hr; and then at 400 V for 3 hr.

For a more complete separation of α-migrating HDL species, electrophoresis can be continued for an additional 2 cm (1 h).

Electrotransfer

Electrotransfer is carried out according to the standard conditions described by Towbin *et al.*[22] In brief, the gels are laid on blotting paper, covered with nitrocellulose paper (Sartorius, Edgewood, NY) and then a further layer of blotting paper, and placed in the cassette of a electrotransfer apparatus (Hoefer, San Francisco, CA). The transfer buffer is 25 m*M* Tris–glycine, pH 8.3, containing 20% (v/v) methanol. Electrotransfer is carried out at 30 V for 16 hr, and then at 40 V for a further 2 hr. The nitrocellulose is sponged clean under distilled water of any adhering particles of gel, and stored at 4°.

Immunoblotting

Polyclonal antiserum to human apoA-I or other apolipoprotein is purified on Protein G-agarose (Bio-Rad). The bound IgG, eluted with Desorbing Buffer according to the manufacturer's instructions, is dialyzed against phosphate-buffered saline (PBS, pH 7.4) and kept at 4°. One hundred micrograms purified antibody is labeled with 1 mCi ^{125}I.[23] For individual experiments, the nitrocellulose blots from electrotransfer are incubated in 2% nonfat powdered milk in PBS (50 ml per sheet) for 2 hr at room temperature to block nonspecific binding sites. The blots are then removed and incubated with antibody solution [1.5×10^6 counts/min (cpm)/ml in powdered milk–PBS solution, 0.25–0.5 μg IgG/ml]. Incubation is carried out with gentle shaking at room temperature for 1.5 hr. The labeled solution is then removed, and the blot is washed with 1% (w/v) powdered milk–PBS solution four times for 30 min each at room temperature, with vigorous shaking. The blots are finally washed with PBS and dried in air. Autoradiography is carried out using Kodak (Rochester, NY) X-OMAT film at −70° for 0.5–3 days.

Following autoradiography, the film can be used to identify the locations and proportions of individual HDL fractions on the nitrocellulose paper. The relevant areas are cut out and the radioactivity determined by gamma scintillation spectrometry.

Further Analysis of High-Density Lipoprotein Subfractions from Gel

Recovery of intact lipoprotein particles from gel fragments can be achieved by electroelution. To locate HDL fractions in the gel, first-dimensional electrophoresis is carried out as described above. The agarose strip is then divided longitudinally, and the pieces laid end-to-end on the polyacrylamide gradient. Following second-dimensional electrophoresis as described above, two identical patterns are generated. The gel is divided down the midline. One pattern is electrotransferred, blotted, and autoradiographed as described above, while the other is maintained at 0°. The autoradiograph is then used as a template to locate HDL subfractions in the reserved gel half. Gel fragments of interest are cut out and extracted as described above. This approach can be used for lipid analysis of HDL fractions, or for measurement of isotopic lipids in kinetic experiments.

Identification of High-Density Lipoprotein Protein Epitopes

First-dimensional electrophoresis is run according to the standard conditions described above.[24] For second-dimensional separation, the gradient gel and stacking gel are cast exactly as described above. One milliliter of a 2% solution of low-melting agarose (SeaPlaque, FMC) in lower buffer is mixed at 37° with 20–100 μg mouse monoclonal IgG prepared against human HDL protein. A comparable solution of nonimmune IgG is made at the same time. A plastic insert is placed vertically, midway across the stacking gel. The solution of anti-lipoprotein IgG is added to one side of the insert, the solution of nonimmune IgG to the other. After the antibody solutions have gelled at room temperature, strips from the first-dimensional agarose gel are added as above and secured with 0.75% agarose solution.

Electrophoresis, electrotransfer, immunoblotting with ^{125}I-labeled anti-apoprotein IgG, and autoradiography are then carried out as described above. When antibody interacts with one or more HDL species of those contained in plasma, this species is displaced vertically, owing to the formation of an IgG–HDL complex (Fig. 1, right-hand side). The extent of reactivity at any concentration of antibody can be determined by comparing the radioactivity in the nitrocellulose transfer from the anti-apoprotein antibody side with that from the nonimmune antibody side.

Summary

No single technique is able to separate each of the many HDL species present in native plasma. Some are present in only trace proportions. Some HDL have no obvious independent metabolic role, beyond perhaps serving as reservoirs of apoproteins active in metabolic events in other lipoproteins. The choice of HDL analytical technique depends mainly on the problem under study. Two-dimensional nondenaturing electrophoresis has been useful in studies of plasma cholesterol metabolism and cholesterol transport from cells, because it separates intermediates in these processes.

Acknowledgment

Original research of the authors cited in this chapter was supported by the National Institutes of Health through Arteriosclerosis SCOR HL 14237.

[16] Heterogeneity of High-Density Lipoproteins and Apolipoprotein A-I as Related to Quantification of Apolipoprotein A-I

By STEVEN T. KUNITAKE, PATRICIA O'CONNOR, and JOSEFINA NAYA-VIGNE

Introduction

Apolipoprotein (apo) A-I is the major protein constituent of human plasma high-density lipoproteins (HDL). It is present on the bulk of HDL particles isolated from the fasting plasma of normolipidemic individuals.

Accurate measurement of apoA-I is central to the study of HDL. Plasma levels of apoA-I correlate well with plasma HDL cholesterol levels,[1] and it has been suggested that apoA-I may actually be a more accurate predictor of cardiovascular risk than HDL cholesterol.[2,3] In addition, HDL complexes are heterogeneous with respect to protein constituents,[4] and therefore an accurate determination of the apoA-I will aid in the elucidation of the apolipoprotein stoichiometry of the various complexes and the distribution of apoA-I among them.

Direct quantitation of apoA-I by immunoassay is complicated by variability in the epitopes of apoA-I. This variability may arise either from structural differences in the apoA-I protein itself or from conformational differences owing to the association of apoA-I with distinct HDL species.

Variability in Chemical Structure of Apolipoprotein A-I

Structural variation in apoA-I can arise by a number of different mechanisms. Apolipoprotein A-I exists, in circulation, as a series of isoforms. Apolipoprotein A-I is apparently secreted into the circulation in the form of a proprotein.[5] The mature protein, the most common form, is generated by cleavage of a hexapeptide from the N terminus of the proprotein.[6,7] The

[1] M. C. Cheung and J. J. Albers, *J. Clin. Invest.* **60,** 43 (1977).
[2] N. Hamada, *Jpn. Circ. J.* **44,** 487 (1980).
[3] G. de Backer, M. Rosseneu, and J. P. Deslypere, *Atherosclerosis* **42,** 197 (1982).
[4] Y. L. Marcel, P. K. Weech, T.-D. Nguyen, R. W. Milne, and W. J. McConathy, *Eur. J. Biochem.* **143,** 467 (1984).
[5] V. I. Zannis, J. L. Breslow, and A. J. Katz, *J. Biol. Chem.* **255,** 8612 (1980).
[6] J. I. Gordon, H. F. Sims, S. R. Lentz, C. Edelstein, A. M. Scanu, and A. W. Strauss, *J. Biol. Chem.* **258,** 4037 (1983).
[7] W. Stoffel, K. Knyrim, and C. Bode, "Physiological Chemistry" p. 1631. de Gruyter, New York, 1983.

remaining isoforms appear to be produced by deamidation of the pro and mature forms of apoA-I.[8] These structural differences in the isoforms of apoA-I would definitely alter its primary structure; however, a number of monoclonal antibodies appear to have the ability to bind to all of the isoforms.[9]

Apolipoprotein A-I can undergo a number of other posttranslational modifications. Apolipoprotein A-I can be glycosylated nonenzymatically[10–12] with glucose, forming a stable linkage to the lysine groups of apoA-I. This modification of apoA-I may pose significant problems in the quantitation of apoA-I derived from the plasma of diabetics, where glycosylated HDL have been found *in vivo*.[10] Apolipoprotein A-I may also undergo fatty acid acylation[13] and phosphorylation[14] with unknown effects on antibody recognition.

Apolipoprotein A-I may be posttranslationally modified during storage, presumably by oxidation.[15,16] Oxidation can lead to the formation of methionine sulfoxide, altering the structure and properties of apoA-I.[16,17] These changes during storage result in dramatic alterations in the immunoreactivity of apoA-I.[18,19]

In addition to posttranslational modifications, many genetic variants of apoA-I exist. Comprehensive listings of such genetic variants have been tabulated.[20] Amino acid substitutions appear to occur throughout the entire length of apoA-I. Many of the genetic variants affect the binding affinity

[8] G. Ghiselli, M. F. Rohde, S. Tanenbaum, S. Krishman, and J. A. M. Gotto, *J. Biol. Chem.* **260,** 15662 (1985).

[9] Y. L. Marcel, X. Collet, E. Raffai, D. Jewer, P. K. Weech, N. V. Dack, J. C. Fruchart, B. Perret, P. Provost, and E. Kassart, *in* "Biotechnology of Dyslipoproteinemias: Applications in Diagnosis and Control," p. 227. Raven, New York, 1990.

[10] E. Schleichler, T. Deufel, and O. H. Wieland, *FEBS Lett.* **129,** 1 (1981).

[11] J. L. Witztum, M. Fisher, T. Poetro, U. P. Steinbrecher, and R. L. Elam, *Diabetes* **31,** 1029 (1982).

[12] P. S. Caines, R. J. Thibert, T. F. Draisey, C. C. Foreback, and J. W. Chu, *Clin. Biochem.* **22,** 285 (1989).

[13] J. M. Hoeg, M. S. Meng, R. Roman, T. Fairwell, and J. H. B. Brewer, *J. Biol. Chem.* **261,** 3911 (1986).

[14] Z. H. Beg, J. A. Stonik, J. M. Hoeg, J. S. J. Demossky, T. Fairwell, and J. H. B. Brewer, *J. Biol. Chem.* **264,** 6913 (1989).

[15] X. L. Wang, N. P. B. Dudman, J. Wang, and D. E. L. Wilcken, *Clin. Chem.* **35,** 2082 (1989).

[16] A. von Eckardstein, M. Walter, H. Holz, and A. Benninghoven, *J. Lipid Res.* **32,** 1465 (1991).

[17] G. M. Anantharamaiah, T. A. Hughes, M. Iqbal, A. Gawish, P. J. Neame, M. F. Medley, and J. P. Segrest, *J. Lipid Res.* **29,** 309 (1988).

[18] X. L. Wang, N. P. B. Dudman, B. L. Blades, and D. E. L. Wicken, *Clin. Chim. Acta* **179,** 285 (1989).

[19] P. Johnson, R. A. Muirhead, and T. Deegan, *Ann. Clin. Biochem.* **18,** 308 (1981).

[20] G. Assmann, A. von Eckardstein, and H. Funke, *Circulation* **87**(Suppl. 3), 28 (1993).

of selected monoclonal antibodies. One variant that has been well characterized is apoA-I_{Milano} with a substitution of cysteine for arginine at residue 173.[21,22] This apoA-I variant can affect an epitope on apoA-I that alters antibody recognition by at least one monoclonal antibody.[23]

All of the structural changes described above can significantly alter specific epitopes on apoA-I. Such alterations could dramatically affect the ability of antibodies to recognize those epitopes and may interfere with accurate immunoquantitation.

Variability in Conformation of Apolipoprotein A-I

All HDL are not identical. In circulation, HDL exist as a heterogeneous collection of discrete particles varying in chemical composition and apolipoprotein constituents. This particle heterogeneity has been demonstrated by analyzing a variety of physiochemical parameters: particle density, particle size, apolipoprotein composition, lipid composition, and electrophoretic mobility.[4,24–27] Structural differences in the distinct HDL complexes apparently affect the conformation of apoA-I residing on them and change the exposure of certain epitopes. Several studies utilizing monoclonal antibodies have shown that a number of antibodies bind to apoA-I with varying affinity depending on its association with lipid[9,23,28–30] and also on the HDL subfraction on which the apoA-I resides.[31–36]

[21] V. Gualandiri, G. Franceschini, C. R. Sitori, G. Gianfranceschi, G. B. Orsini, A. Cerrone, and A. Menotti, *Am. J. Hum. Genet.* **37,** 1083 (1985).

[22] K. H. Weisgraber, T. Bersot, R. W. Mahley, G. Franceschini, and C. R. Sirtori, *J. Clin. Invest.* **66,** 901 (1980).

[23] S. Marcovina, S. Fantappie, A. Zoppo, G. Franceschini, and A. L. Catapano, *J. Lipid Res.* **31,** 375 (1990).

[24] D. W. Anderson, A. V. Nichols, T. M. Forte, and F. T. Lindgren, *Biochim. Biophys. Acta* **493,** 55 (1977).

[25] P. J. Blanche, E. L. Gong, T. M. Forte, and A. V. Nichols, *Biochim. Biophys. Acta* **665,** 408 (1981).

[26] M. Okazaki, N. Hagiwara, and I. Hara, *J. Biochem. (Tokyo)* **92,** 517 (1982).

[27] B. F. Asztalos, C. H. Sloop, L. Wong, and P. S. Roheim, *Biochim. Biophys. Acta* **1169,** 3 (1993).

[28] C. M. Allan, N. H. Fidge, J. R. Morrison, and J. Kanellos, *Biochem. J.* **290,** 449 (1993).

[29] L. K. Curtiss and R. S. Smith, *J. Biol. Chem.* **263,** 13779 (1988).

[30] F. Pio, H. D. Loof, N. V. Dac, V. Clavey, J. C. Fruchart, and M. Rosseneu, *Biochim. Biophys. Acta* **959,** 160 (1988).

[31] L. K. Curtiss and T. S. Edgington, *J. Biol. Chem.* **260,** 2982 (1985).

[32] M. Ayrault-Jarrier, E. Bekaert, E. Petit, D. Pastier, J. Polpnovski, B. Pau, F. Paolucci, E. Hervaud, and M. Laprade, *Adv. Exp. Med. Biol.* **243,** 149 (1988).

[33] E. Petit, M. Aryrault-Jarrier, D. Pastier, H. Robin, J. Polpnovski, I. Aragon, E. Hervaud, and B. Pau, *Biochim. Biophys. Acta* **919,** 287 (1987).

To illustrate this point we describe a subfraction of HDL that we have studied in depth, pre-β HDL. Human HDL (apolipoprotein A-I-containing lipoproteins), in plasma or when isolated by selected affinity immunosorption, can be electrophoretically separated into two main subfractions. The bulk of HDL have α mobility, but a portion of the complexes migrate in the pre-β zone during agarose electrophoresis.[37] Pre-β HDL have been found in the plasma of monkeys,[38,39] mice,[40] and dogs.[41] Moreover, there is evidence derived from two-dimensional gel analysis of human plasma that within these two main subfractions there are more than one type of apoA-I containing complex.[42,43]

When isolated by a combination of selected affinity immunosorption[44,45] and starch block electrophoresis,[37] the pre-β HDL are composed predominantly (>90%) of small particles, approximately 65 kDa in mass, although larger sized components are present (S. T. Kunitake, P. O'Connor, and J. Naya-Vigne, unpublished results). These 65-kDa pre-β HDL are protein-rich (only 10% of mass is lipid), apoA-I is the predominant protein, and no apoA-II is detectable. In contrast, α HDL are significantly larger, are composed of approximately 65% protein, and contain apolipoproteins other than apoA-I. These significant differences in particle composition are reflected in great structural differences between pre-β HDL and α HDL. The apoA-I in pre-β HDL has significantly less helical content than does the apoA-I in α HDL (52 versus 64%) as determined by circular dichroism (CD) analysis.[46] Pre-β HDL are dramatically more sensitive to proteolytic attack than are α HDL,[47] indicating a major difference in apoA-I conformation or exposure.

[34] S. R. Silberman, F. Bernini, J. T. Sparrow, J. A. M. Gotto, and L. C. Smith, *Biochemistry* **26,** 5833 (1987).

[35] N. Fidge, M. J. T. Nugent, and M. Tozuka, *Biochim. Biophys. Acta* **1003,** 84 (1989).

[36] Y. L. Marcel, D. Jewer, C. Vezina, P. Milthorp, and P. K. Weech, *J. Lipid Res.* **28,** 768 (1987).

[37] S. T. Kunitake, K. J. La Sala, and J. P. Kane, *J. Lipid Res.* **26,** 549 (1985).

[38] C. K. Castle, M. E. Pape, K. R. Marotti, and G. W. Melchior, *J. Lipid Res.* **32,** 439 (1991).

[39] L. L. Rudel, J. A. Lee, M. D. Morris, and J. M. Felts, *Biochem. J.* **139,** 89 (1974).

[40] B. Y. Ishida, D. Albee, and B. Pagien, *J. Lipid Res.* **31,** 227 (1990).

[41] M. Lefevre, C. H. Sloop, and P. S. Roheim, *J. Lipid Res.* **29,** 1139 (1988).

[42] G. R. Castro and C. J. Fielding, *Biochemistry* **27,** 25 (1988).

[43] O. L. Francone, A. Gurakar, and C. Fielding, *J. Biol. Chem.* **264,** 7066 (1989).

[44] S. Kunitake, J. McVicar, R. Hamilton, and J. Kane, *Circulation* **66**(Suppl. 2), 240 (1982).

[45] J. P. McVicar, S. T. Kunitake, R. L. Hamilton, and J. P. Kane, *Proc. Natl. Acad. Sci. U.S.A.* **81,** 1356 (1984).

[46] S. Kunitake, K. La Sala, C. Mendel, G. Chen, and J. Kane, *in* "Proceedings of the workshop on lipoprotein heterogeneity," (Lippel, K., ed.) Washington, D.C.: NIH publication no. 87, 2646, 419 (1987).

[47] S. Kunitake, G. Chen, S. Kung, J. Schilling, D. Hardman, and J. Kane, *Arteriosclerosis (Dallas)* **10,** 25 (1990).

As expected, the degree of recognition of apoA-I epitopes by monoclonal antibodies also differs between α and pre-β HDL (S. T. Kunitake and L. K. Curtiss, unpublished results).[46] Some of the apoA-I epitopes are poorly recognized on pre-β HDL, whereas others are more exposed. Such differences can create great difficulties in attempts to quantify the levels of this specific HDL subfraction in plasma.

Quantitation of Apolipoprotein A-I

Several immunoassays developed for the quantitation of apoA-I have been published and tabulated.[48] Considerations of methodology and standards in the immunoassays have been discussed.[49] The method described in this chapter appears to quantitate apoA-I equally well when the apolipoprotein is present in the purified delipidated state, in plasma, on α HDL, and on pre-β HDL.

This method possesses the following salient features: (1) The assay utilizes affinity-purified, monospecific, polyclonal antibodies. Polyclonal antibodies recognize many different epitopes, minimizing the effect that modification of one epitope might have on quantitation. (2) The samples are incubated in a detergent cocktail containing sodium decyl sulfate and polyoxyethelene 9 lauryl ether prior to analysis. The intent of the incubation is not to delipidate the HDL proteins completely but to create a uniform exposure of epitopes on all apoA-I molecules. (3) The samples are isolated and stored at 4° in the presence of preservatives, including antioxidants and protease inhibitors, to minimize the modification of apoA-I prior to analysis. (4) A sandwich type enzyme-linked immunosorbent assay (ELISA) format was adopted to maximize the sensitivity of the assay. High sensitivity is not necessary to quantitate the apoA-I contents of plasma but may be required to determine the levels of quantitatively small subfractions of HDL. (5) Purified apoA-I is used as the reference standard.

Preparation of Materials

Apolipoprotein A-I-containing lipoproteins [Lp(A-I)] are isolated from the plasma of fasting, normolipidemic donors by selected affinity immunosorption.[44,45,47] Pre-β HDL and α HDL are isolated from Lp(A-I) by starch block electrophoresis.[37] Ultracentrifugally isolated HDL are isolated from the plasma of fasting normolipidemic donors by sequential ultracentrifuga-

[48] G. R. Cooper, S. J. Smith, D. A. Wiebe, M. Kuchmak, and W. H. Hannon, *Clin. Chem.* **31,** 233 (1985).

[49] K. K. Steinberg, G. R. Cooper, S. R. Graiser, and M. Rosseneu, *Clin. Chem.* **29,** 415 (1983)

tion.[50] Purified apoA-I is isolated from delipidated apoHDL by Sephacryl S-200 gel-permeation chromatography in the presence of 6 *M* urea.

Monospecific anti-human apoA-I antisera are produced in goats injected subcutaneously with purified apoA-I. Monospecific anti-apoA-I antibodies are isolated by affinity chromatography using an apoA-I bound Sepharose column. The purity and monospecificity on the antibodies are verified by analytical sodium dodecyl sulfate–polyacrylamide gel electrophoresis (SDS–PAGE)[51] and Western blotting.[52] The monospecific anti-apoA-I antibodies are covalently coupled to horseradish peroxidase by reaction with sodium periodate.[53]

To prevent modification of apoA-I, blood is drawn from fasting normolipidemic individuals (males and females) and immediately mixed with the following preservatives: 0.08% (w/v) ethylenediaminetetraacetic acid (EDTA), 0.1% (w/v) sodium azide, 50 μg/ml benzamidine, 10 μg/ml phenylmethylsulfonyl fluoride, 300 mg/ml ε-aminocaproic acid, and 10 μg/ml gentamicin sulfate (final concentrations). Plasma is separated by low-speed centrifugation at 1000 *g* for 45 min at 4°C. α_2-Macroglobulin (10 μg/ml) and the above-listed preservatives are added at all steps in lipoprotein isolation to all isolated lipoproteins, and samples are maintained at 4°.

Enzyme-Linked Immunosorbent Assay

The apoA-I contents in purified apoA-I, plasma, α HDL, and pre-β HDL are quantitated by a sandwich ELISA.

ELISA Solutions

Phosphate-buffered saline (PBS), pH 7.5: 1.22 g Na_2HPO_4, 0.138 g $NaH_2PO_4 \cdot H_2O$, 9.0 g NaCl, and water to 1 liter

Wash buffer: 1 liter PBS, pH 7.5, 0.05% Tween 20, and 0.5% bovine serum albumin (BSA), radioimmunoassay (RIA) grade

Blocking buffer: 3% BSA (RIA grade, Sigma, St. Louis, MO) in PBS, pH 7.5

Detergent mixture: 30 m*M* sodium decyl sulfate and 0.8% (v/v) polyoxyethelene 9 lauryl ether

Antibody dilution buffer: wash buffer with 3 m*M* sodium decyl sulfate and 0.13% polyoxyethelene 9 lauryl ether

[50] R. J. Havel, H. A. Eder, and J. H. Bragdon, *J. Clin. Invest.* **34,** 1345 (1955).

[51] U. K. Laemmli, *Nature* (*London*) **227,** 680 (1970).

[52] H. Towbin, T. Staehelin, and J. Gordon, *Proc. Natl. Acad. Sci. U.S.A.* **76,** 4350 (1979).

[53] P. Nakane, *in* "Immunoassays in the Clinical Laboratory" (R. M. Nakamura, W. R. Dito, and E. S. Tucker III, eds.), p. 81. Liss, New York, 1979.

TMB (tetramethylbenzidine) reagent (Kirkegaard & Perry, Gaithersburg, MD)

TMB stop solution (Kirkegaard & Perry)

Procedure. All samples and standards are pretreated prior to analysis. Samples are prediluted in PBS, pH 7.5, to approximately 0.1 mg/ml apoA-I, then incubated 15 min at 37° with an equal volume of detergent mixture containing 30 m*M* sodium decyl sulfate and 0.8% (v/v) polyoxyethelene 9 lauryl ether. The detergent is subsequently diluted to a final concentration of 5 m*M* sodium decyl sulfate and 0.13% polyoxyethelene 9 lauryl ether. Detergent treatment attempts to create uniform exposure of all apoA-I epitopes.

In the first step of the ELISA, microtiter plates (NUNC flat bottom, Roskilde, Denmark) are coated with purified monospecific polyclonal antiserum to apoA-I at 50 μl/well (1 μg/ml immunoglobulin G diluted in PBS, pH 7.5) and incubated overnight at 4°. The plate is then washed three times with ELISA wash buffer. Second, the plate is blocked with blocking buffer for 1 hr at room temperature while shaking. The plate is washed three times with wash buffer. Third, detergent-treated antigen is applied to the plate in a volume of 50 μl per well for 1 hr at 37°. After incubation the

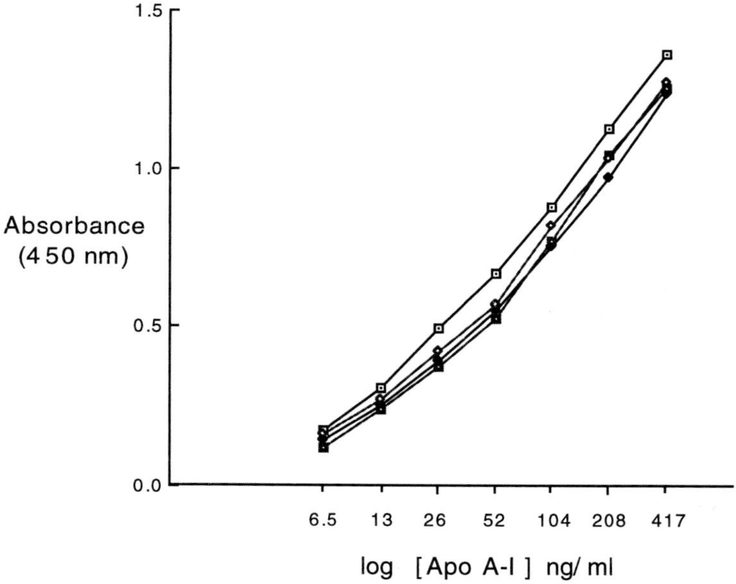

FIG. 1. Quantitation by sandwich ELISA of apoA-I in serial dilutions of detergent-treated samples: apoA-I (⊡), plasma (◆), HDL (◼), and pre-β HDL (◇). Curves are reasonably linear over the range 10–400 ng/ml. The apoA-I in all samples is quantitated equally well.

plate is washed three times with wash buffer. Fourth, bound antigen is subsequently detected by addition of 50 μl of a monospecific polyclonal antibody to apoA-I which has been affinity-purified and conjugated to horseradish peroxidase (HRP). The conjugated antibody is added at a dilution of 1:2000 in antibody dilution buffer (1.3 μg/ml immunoglobulin G). The incubation with conjugated antibody lasts 1 hr at 37°. The plate is washed another three times to remove unbound antibody. Finally, the presence of bound HRP-linked antibody is detected by color generation with 50 μl/well of one-component TMB reagent for 15 min. The reaction is stopped using TMB stop solution at 50 μl/well. The optical density of the reaction mixture in each well is measured at 450 nm and analyzed with a microtiter plate reader (Molecular Devices, Mountain View, CA).

Results. The ELISA was used to generate the curves shown in Fig. 1, using serial dilutions of detergent-treated purified apoA-I, plasma, and electrophoretically isolated α and pre-β HDL. The assay is reasonably linear over a range of 10.0 to 400 ng/ml. The slopes of all the curves are parallel. The assay appears to detect apoA-I equally well in each group.

[17] Chromatographic Methods for Quantitation of Apolipoprotein A-I

By G. M. ANANTHARAMAIAH and DAVID W. GARBER

Introduction

The need for measurement of apolipoprotein A-I (apoA-I) levels in plasma stems from the fact that levels of high-density lipoprotein (HDL) and its major protein component, apoA-I, are inversely correlated with coronary artery disease (CAD). A family history of CAD is commonly associated with low levels of apoA-I and HDL. High-density lipoprotein is a better predictor of CAD than any of the other lipoprotein levels.[1,2] Studies which evaluated the effects of lipoprotein modification on CAD showed that raising HDL levels is associated with regression of atherosclerotic lesion growth and reductions in fatal and nonfatal myocardial infractions.[3,4] Transgenic animal models have allowed the direct effects of human

[1] D. L. Sprecher, H. S. Geigelson, and P. M. Laskarzwski, *Atheroscler. Thromb.* **13,** 495 (1993).
[2] N. R. Phillips, D. Waters, and R. J. Havel, *Circulation* **88,** 2762 (1993).
[3] The Lipid Research Clinics Coronary Prevention Trial results, part 1, *J. Am. Med. Assoc.* **251,** 351 (1984).
[4] J. J. Badimon, V. Fuster, and L. Badimon, *Circulation* **86**(Suppl. 2), 86 (1992).

apoA-I and apoA-II to be tested in mice susceptible to atherosclerosis. These studies have shown that both human apoA-I and apoA-II are associated with circulating HDL particles that are comparable to human HDL in size and density, but the protective effect of these HDL against atherosclerosis resides largely, if not exclusively, with human apoA-I rather than apoA-II.[5-8] It is therefore important to explore a method that unequivocally determines the circulating plasma levels of apoA-I accurately.

Several methods are available for the quantitation of apoA-I. Most involve either electrophoretic or immunochemical techniques. The electrophoretic methods are often used in combination with other methods. Immunochemical techniques used for the estimation of apoA-I are summarized in Table I. There are many inherent problems in the quantitation of apoA-I by these methods, generally related to specificity of antibodies against apoA-I. Apolipoprotein A-I conformation may change depending on the local environment or protein concentration. For instance, antibodies produced against HDL may not have the same specificity toward apoA-I as antibodies produced against free apoA-I. Another problem particularly associated with apoA-I is the presence of different isoforms of the protein. These may arise from deamidation or oxidation, and may have an effect on the quantitation of apoA-I.

Gel-filtration and ion-exchange chromatographic methods have been used for the isolation of apoA-I.[9,10] However, for quantitation, these methods are very slow. A reliable, sensitive, and relatively rapid method for quantitation of apoA-I has been developed using reversed-phase high performance liquid chromatography (HPLC).[11] This involves the selective adsorption of apolipoproteins present in HDL onto a reversed-phase HPLC matrix and sequential release using a suitable organic solvent gradient. Sequential release of adsorbed apolipoproteins depends on the differential hydrophobic interaction between the HPLC matrix and the apolipoproteins. As the organic solvent gradient increases to the point where the solubility

[5] J. R. Shultz, J. G. Verstuyft, E. L. Gong, A. V. Nichols, and E. M. Rubin, *Nature* (*London*) **365,** 762 (1993).

[6] E. M. Rubin, R. M. Krauss, E. A. Spangler, J. G. Verstuyft, and S. M. Clift, *Nature* (*London*) **353,** 265 (1991).

[7] E. M. Rubin, B. Y. Ishida, S. M. Clift, and R. M. Krauss, *Proc. Natl. Acad. Sci. U.S.A.* **88,** 434 (1992).

[8] J. R. Schultz, E. L. Gong, M. R. McCall, A. V. Nichols, S. M. Clift, and E. M. Rubin, *J. Biol. Chem.* **267,** 21630 (1992).

[9] H. B. Brewer, Jr., R. Ronnan, M. Meng, and C. Bishop, this series, Vol. 128, p. 228.

[10] H. Mezdour, V. Clavey, I. Kora, M. Koffigan, A. Barkia, and J. C. Fruchart, *J. Chromatogr.* **414,** 35 (1987).

[11] T. A. Hughes, M. A. Moore, P. Neame, M. F. Medley, and B. H. Chung, *J. Lipid Res.* **29,** 363 (1988).

TABLE I
IMMUNOCHEMICAL METHODS FOR QUANTITATION OF APOLIPOPROTEINS[a]

Method	Advantages	Disadvantages
Radial immunodiffusion (RID)	Easy to perform No dilution required Can be used with polyclonal or monoclonal antibodies Does not require radioisotopes	Cannot be automated Time consuming (2–3 days) Sensitive to particle size Limited sensitivity Between-assay *CV* of 6–8%
Electroimmunodiffusion (EID)	More rapid than RID Greater sensitivity than RID Does not require radioisotopes	Technically more difficult than RID Requires electrophoresis equipment Cannot be automated Somewhat sensitive to size and charge differences in antigen
Immunonephlometric assay (INA)	Highly automated Easy to perform High throughput Does not require radioisotopes	Sensitive to differences in size of analyte
Immunoturbidometric assay (ITA)	Similar to INA	Similar to INA
Radioimmunoassay (RIA)	Highly sensitive Can be automated Can be used with polyclonal or monoclonal antibodies	Requires radioisotopes Adequate precision requires duplicate or triplicate measurements Reagents have short shelf life
Enzyme-linked immunosorbent assay (ELISA)	Same as RIA except does not require radioisotopes Reagents have longer shelf life	Same as RIA except adsorption to plastic can alter epitope expression Competitive ELISA can have artifacts due to diminished epitope recognition

[a] Summarized from J. J. Albers and S. M. Marcovina, *in* "Plasma Lipoproteins and Coronary Artery Disease" (R. A. Kreisberg and J. P. Segrest, eds.), p. 265. Blackwell, Cambridge, MA, 1992.

of an apolipoprotein in the organic solvent is greater than its affinity to the HPLC matrix, that apolipoprotein is eluted from the column. The amount of apoA-I (and other apolipoproteins) can be determined by comparing the eluted peak area with that of an internal standard of known concentration. Whereas in other methods the estimation of each of the apolipoproteins has to be done by a separate analysis devoted to that particular apolipoprotein, in the HPLC method most of the exchangeable apolipoproteins can be estimated simultaneously. A disadvantage of the method is that it is not suitable for the quantitation of apoB because of the insolubility of this protein in the lipid-free state.

The HPLC method is attractive for two main reasons: (1) there have been tremendous improvements in reversed-phase supports, and (2) because of improvements in computer applications to chromatography-based estimation of proteins and other organic compounds, quantitation of peaks in a given HPLC profile has been very precise. High-performance liquid chromatography has been used for amino acid analysis, estimation of certain organic compounds in biological fluids and tissues, and protein sequencing. Such studies have been possible mainly because of high reproducibility, an essential criterion in any quantitation procedure. Despite this, HPLC has not been extensively used for the quantitation of apolipoproteins.

Various forms of HPLC have been used for the isolation of apoA-I.[9] Lipoproteins have been isolated using HPLC by Okazaki and Hara.[12] Edlestein and Scanu applied molecular sieve HPLC to the separation of human plasma apolipoproteins.[13] These methods involved the use of aqueous buffer eluents exclusively. Hancock *et al.* have used ion-pair reversed-phase HPLC for the separation of apolipoproteins A-I and A-II and apolipoproteins from very-low-density lipoproteins (VLDI).[14,15]

Our laboratory has utilized the reversed-phase HPLC method for the quantitative isolation and quantitation of apolipoproteins, particularly apoA-I. The experience gained in the separation of amphipathic peptide analogs of apolipoproteins using HPLC[16,17] has been very helpful for the quantitative isolation and quantitation of exchangeable apolipoproteins. The quantitation of apoA-I using HPLC involves (a) separation of HDL from plasma by gradient ultracentrifugation, (b) delipidation, (c) addition of internal standard that can be separated from the other exchangeable apolipoproteins during the HPLC analysis, and (d) separation and quantitation of apolipoproteins. Although earlier investigators observed that each of the exchangeable apolipoproteins elutes as a single peak using ion-exchange or gel-filtration methods, HPLC analysis of exchangeable apolipoproteins in our laboratory has identified that many of the apolipoproteins can occur as multiple peaks.

[12] M. Okazaki and I. Hara, *in* "Handbook of HPLC for the Separation of Amino Acids, Peptides and Proteins" (W. S. Hancock, ed.), Vol. 2, p. 393. CRC Press, Boca Raton, Florida, 1984.

[13] C. Edlestein and A. Scanu, *in* "Handbook of HPLC for the Separation of Amino Acids, Peptides and Proteins" (W. S. Hancock, ed.), Vol. 2, p. 405. CRC Press, Boca Raton, Florida, 1984.

[14] W. S. Hancock, H. J. Pownall, A. M. Gotto, and J. T. Sparrow, *J. Chromatogr.* **216,** 285 (1981).

[15] W. S. Hancock, C. A. Bishop, A. M. Gotto, D. R. Harding, S. M. Lamplugh, and J. T. Sparrow, *J. Lipids* **16,** 250 (1981).

[16] G. M. Anantharamaiah, J. L. Jones, C. G. Brouillette, C. F. Schmidt, B. H. Chung, T. A. Hughes, A. S. Bhown, and J. P. Segrest, *J. Biol. Chem.* **260,** 10248 (1985).

[17] G. M. Anantharamaiah, this series, Vol. 128, p. 627.

Apart from the existence of apolipoprotein isoforms arising from deamidation,[18] there are also different oxidized isoforms.[19] Methionine is susceptible to oxidation. This is particularly true if ether, which usually contains peroxide that is difficult to eliminate completely, is one of the delipidation solvents. It is also clear that storage of the lipoprotein sample also produces multiple oxidized forms of apolipoproteins.[20] There are three methionines in apoA-I and one in apoA-II. Therefore, one could expect eight peaks for apoA-I and three peaks for apoA-II (which in humans is a dimer). However, both in our hands[19] and in the laboratory of Assman,[21] only two apoA-I isoforms have been seen. These have been termed apoA-Ia (methionine oxidized to methionine sulfoxide) and apoA-Ib (methionine not oxidized). It has been shown in the laboratory of Assman that out of three methionines present in apoA-I, only two (Met-112 and Met-142) are susceptible to oxidation. To oxidize methionine-86, harsh conditions are necessary, such as exposing apoA-I in 3 *M* guanidine hydrochloride solution to hydrogen peroxide overnight. Completely oxidized apoA-I elutes readily from a reversed-phase HPLC column using a gradient of acetonitrile and water in the presence of 0.1% trifluoroacetic acid. As plasma is not subjected to such harsh conditions, this peak is not seen in the usual plasma samples and can therefore be neglected under the usual conditions.

In humans, apoA-II exists as a dimer. As the sequence of apoA-II contains one methionine, one can expect three peaks if methionines in the two strands are involved in oxidation to different extents.[19] The three peaks observed for apoA-II belong to (a) two methionines in the two strands of apoA-II monomers oxidized, (b) one of the two methionines oxidized, and (c) none of the two methionines in the two apoA-II monomers oxidized. If apoHDL is treated with mercaptoethanol, apoA-II dimer is converted to monomer. In this case, apoA-II shows only two peaks: one for apoA-II monomer, and the other for apoA-II monomer with methionine oxidized to methionine sulfoxide. The methionine-oxidized apoA-II monomer and the completely oxidized apoA-II dimer elute from a reversed-phase HPLC column close after the apoA-Ib peak. However, in our experience, oxidation of apoA-I and apoA-II goes hand in hand. If an appreciable amount of apoA-Ia is not present, it is likely that the HPLC-isolated apoA-I does not

[18] G. Ghiselli, M. F. Rohde, S. Tanenbaum, S. Krishnan, and A. M. Gotto, Jr., *J. Biol. Chem.* **260,** 15662 (1985).

[19] G. M. Anantharamaiah, T. A. Hughes, M. Iqbal, A. Gawish, P. J. Neame, M. F. Meddley, and J. P. Segrest, *J. Lipid Res.* **29,** 309 (1987).

[20] P. Milthrop, P. K. Weech, R. W. Milane, and Y. L. Marcel, *Arteriosclerosis* (*Dallas*) **6,** 285 (1986).

[21] A. von Eckardstein, M. Walter, H. Holz, A. Benninghoven, and G. Assman, *J. Lipid Res.* **32,** 1465 (1991).

possess any apoA-II contamination. Therefore, if oxidation is carefully controlled using antioxidants, relatively pure peaks of unoxidized apoA-I and apoA-II can be obtained. Based on this, during delipidation, one should avoid mercaptoethanol.

Hughes *et al.*[11] have conducted a detailed analysis of plasma from many subjects and have demonstrated reproducible values for apoA-I that are in agreement with other methods of apoA-I quantitation. Because the isolation of HDL is a prerequisite for the HPLC estimation of apoA-I in the plasma, the possible presence of about 4% free apoA-I that may not circulate as a part of HDL but may exist as pre-β HDL[22] cannot be quantitated by this method. We now describe the conditions used in our laboratory for the isolation of exchangeable apolipoproteins, particularly apoA-I, and quantitation of these using analytical reversed-phase HPLC columns.

Preparative Isolation of Apolipoprotein A-I

A number of procedures have been used to isolate pure apoA-I, including size-exclusion column chromatography combined with ion-exchange chromatography,[23,24] and reversed-phase HPLC. However, as methods of assessment of purity become more sensitive and standards of purity become more stringent, preparative methods have become more complicated and difficult. Sensitive silver staining of sodium dodecyl sulfate (SDS)–polyacrylamide gels after electrophoresis often shows contaminants which are not apparent with Coomassie blue staining. Use a C_{18} reversed-phase analytical HPLC can demonstrate oxidized and reduced apoA-I.[19] Thus, we have attempted to refine the C_{18} reversed-phase preparative procedure[11] to eliminate trace contaminants and oxidation products.

Sample Preparation

Because apoA-I is known to undergo changes due to either deamidation or methionine oxidation, it is important to take the following precautions: (1) use freshly collected plasma, (2) use reducing agents during the isolation of HDL and apoA-I to avoid methionine oxidation, (3) avoid the use of buffers with pH values higher than 8.0 which are known to cause deamidation, (4) use protease inhibitors, and (5) store the isolated apoA-I as a lyophilized powder at −70° and not in solution, frozen or otherwise.

Plasma is acquired from normolipidemic donors or from the American

[22] B. Y. Ishida, J. Frolich, and C. J. Fielding, *J. Lipid Res.* **26,** 549 (1987).

[23] P. Weisweiler, *Clin. Chim. Acta* **169,** 249 (1987).

[24] H. B. Brewer, Jr., R. Ronan, M. Meng, and C. Bishop, this series, Vol. 128, p. 223.

TABLE II
GRADIENT FOR PREPARATIVE ISOLATION OF APOLIPOPROTEIN A-I BY C_{18} REVERSED-PHASE HPLC[a]

Step	Time (min)	Final %A	Final %B	Gradient rate (%A/min)	Cumulative time (min)
0	1	25	75	0	1
1	38	44	56	0.5	39
2	28	47.5	52.5	0.125	67
3	37	66	34	0.5	104
4	5	90	10	4.8	109
5	10	90	10	0	119
6	5	25	75	−13	124
7	20	25	75	0	144

[a] Flow rate is 4.8 ml/min. Solvent A is acetonitrile plus 0.1% trifluoroacetic acid; solvent B is HPLC-grade water plus 0.1% trifluoroacetic acid. The column is a Vydac C_{18} preparative column (22 × 250 mm). The exclusion time (as determined by the elution time of the salt peak) in this system is 11 min; step 2 must be adjusted accordingly if system exclusion times differ.

Red Cross, and a complete lipoprotein profile[25] is determined. Profiles taken from fresh plasma compared with plasma provided by the American Red Cross show that the Red Cross plasma has been diluted approximately 30% during processing; thus, standards regarding adequate HDL levels in the plasma must be adjusted. As sodium azide may have proteolytic properties, the formerly used preservative cocktail of sodium azide, phenylmethylsulfonyl fluoride, and EDTA has been replaced with PSEEG,[26] which contains (in stock solution) penicillin G (50,000 U/ml), streptomycin sulfate (5 mg/ml), ε-amino-*n*-caproic acid (1 *M*), 10% (v/v) EDTA, and 5% glutathione. PSEEG is added to plasma at a proportion of 1 : 100; all aqueous solutions used during the preparation also contain PSEEG at the same proportion.

Isolation of High-Density Lipoproteins by Ultracentrifugation

Two ultracentrifugation steps are used to isolate HDL relatively free from albumin contamination. First, plasma is adjusted to a density of 1.21 g/ml with solid KBr, placed in screw-top bottles, and centrifuged for 24 hr at 4° and 50,000 rpm in a Beckman 70Ti or 50.2Ti rotor. After centrifugation, the total lipoprotein layer is removed from the top of each tube using a

[25] K. R. Kulkarni, D. W. Garber, S. M. Marcovina, and J. P. Segrest, *J. Lipid Res.* **35,** 159 (1994).
[26] D. M. Lee, A. J. Valente, W. U. Kuo, and H. Maeda, *Biochim. Biophys. Acta* **666,** 133 (1981).

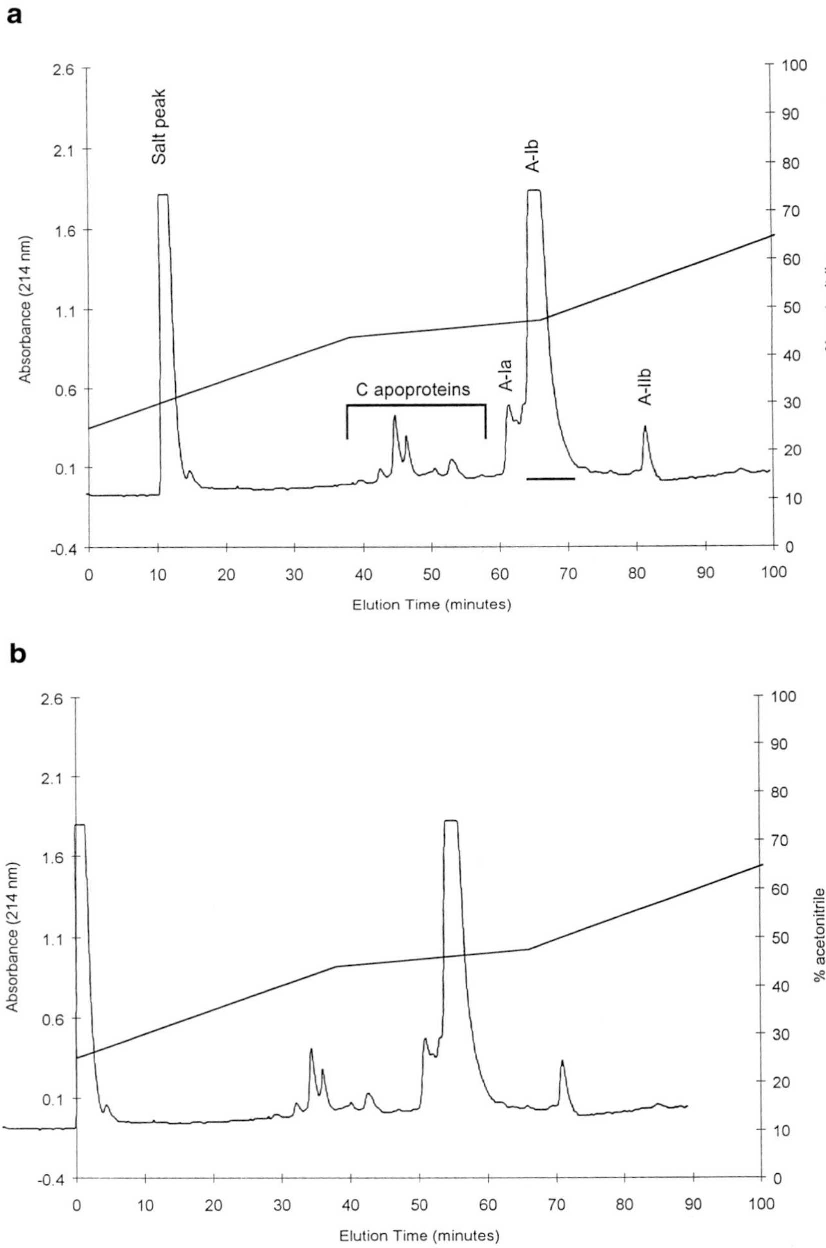

FIG. 1. (a) Preparative HPLC profile used for the purification of human apoA-I. Delipidated HDL was dissolved in guanidine hydrochloride and injected on a C_{18} preparative column (Vydac) equilibrated with solvent containing 25% acetonitrile (with 0.1% trifluoroacetic acid); protein was eluted using the gradient procedure described in Table II. The acetonitrile gradient

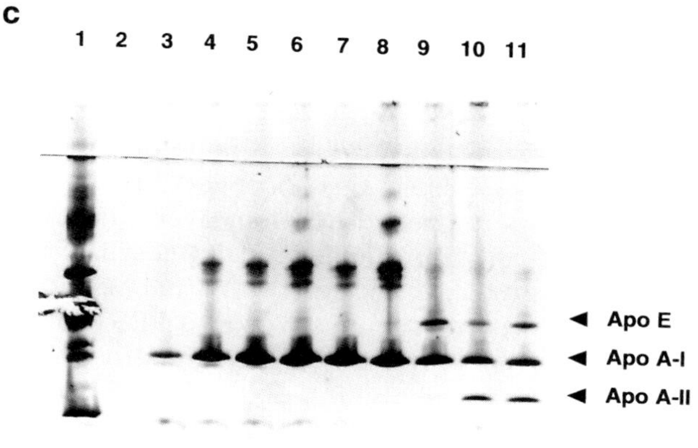

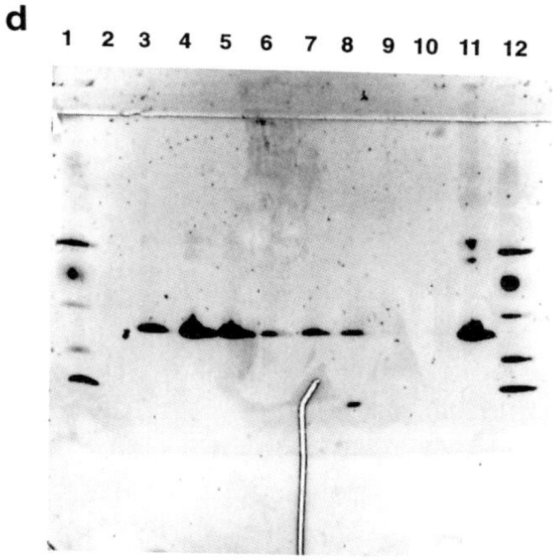

is shown with the percent acetonitrile given on the right-hand axis. Absorbance units (214 nm wavelength) are shown on the left axis. (b) The acetonitrile concentration line has been shifted to demonstrate elution concentration, rather than pump concentration, of acetonitrile by placing the salt peak at zero to compensate for the void volume of the column. (c) Fractions taken across the apoA-Ib peak were subjected to SDS–polyacrylamide gel electrophoresis using a PHAST system and 8–25% gels (Pharmacia). The lanes shown are low molecular weight standard (lane 1) and peak fractions 4–20 (every other fraction; lanes 3–11). Peak fractions 4–20 were collected during the period shown in (a) by the solid line below the apoA-Ib peak. The gel was stained with silver, and the pure fractions were pooled and lyophilized. Multiple bands in lanes 4–8 of the gel with higher molecular weights than apoA-I have been shown by immunoblotting to be aggregates of apoA-I, apparently caused by gel overloading. (d) Fractions taken across the apoA-Ib peak of a separate preparative isolation in which the gel was not overloaded show the high degree of purity which can be achieved. Conditions are similar to those in (c). Lanes 1 and 12 are low molecular weight standards, lane 11 represents an isolated apoA-I preparation which was overloaded and which demonstrates aggregated apoA-I, whereas lanes 3–8 were taken from the apoA-Ib peak of a preparative isolation. Note that the aggregates in lane 11 are similar but not identical in size to the albumin standard, and could lead to the erroneous conclusion that the sample is contaminated with albumin.

syringe and short-bevel needle. The HDL is separated using a modification of the density gradient ultracentrifugation method.[27] The total lipoprotein fraction from the previous centrifugation is adjusted to a density of 1.35 g/ml with solid KBr. In a Beckman Quik-Seal ultracentrifuge tube, a three-layer density gradient is prepared: the bottom layer is the density-adjusted lipoprotein fraction (10 ml); the middle layer is phosphate-buffered saline (PBS: 10 m*M* Na_2PO_4, 100 m*M* NaCl, pH 7.4) containing PSEEG (10 ml), adjusted to a density of 1.24 g/ml; the top layer is PBS containing PSEEG at a density of 1.006 g/ml (~18.5 ml). Tubes are sealed and placed in a 70Ti rotor for ultracentrifugation to 8×10^{11} rad^2/sec at 70,000 rpm and a temperature of 15°. Tubes are emptied by being placed in a Beckman fractionator; chloroform is pumped into the bottom of the tube, and effluent from the top of the tube flows through a photometric detector (280 mm) into a fraction collector. The HDL fractions are dialyzed against distilled water containing PSEEG, and the protein concentration is determined using the Lowry method.[28] Dialysis against water rather than PBS reduces the concentration of salts during the lyophilization procedure. The HDL fraction is then lyophilized. Lyophilization prior to delipidation improves protein recovery dramatically.

Delipidation

Lipids are extracted from the lyophilized HDL fraction using 30 ml of 1:1 (v/v) ice-cold chloroform–methanol. The tube containing 200–400 mg dry weight lyophilized HDL is placed on ice for 30 min, then centrifuged 20 min at 2000 rpm and 4° to pellet the protein. The organic solvents are decanted or aspirated, and the pellet is then washed with a second 30 ml of 2:1 chloroform–methanol, and recentrifuged. Residual organic solvents are immediately removed by lyophilization or drying under nitrogen. Use of diethyl ether in the delipidation process results in considerable oxidation of the apolipoproteins, and is therefore avoided.

Preparative High-Performance Liquid Chromatography

The protein pellet is weighed, then dissolved in 6 *M* guanidine hydrochloride. A small aliquot is taken for analytical reversed-phase HPLC as described later. The remainder is used for preparative reversed-phase HPLC using a reversed-phase C_{18} preparative column (22 × 250 mm,

[27] B. H. Chung, J. P. Segrest, M. J. Ray, J. D. Brunzell, J. E. Hokanson, R. M. Krauss, K. Beaudrie, and J. T. Cone, this series, Vol. 128, p. 181.

[28] O. H. Lowry, N. J. Rosebrough, A. I. Farr, and J. R. Randall, *J. Biol. Chem.* **193,** 265 (1951).

Vydac) with a C_{18} guard column. The gradient HPLC uses two running solvents: solvent A is HPLC-grade acetonitrile plus 0.1% trifluoroacetic acid, and solvent B is HPLC-grade water plus 0.1% trifluoroacetic acid. The column is equilibrated with 25% solvent A at a flow rate of 4.8 ml/min, and up to 20 mg protein in 6 *M* guanidine hydrochloride is injected per separation. In our system, a salt peak appears at approximately 11 min; this is assumed to be the void volume delay. The acetonitrile gradient is increased at 0.5% A/min to 44% A, then the gradient increase is reduced to 0.125% A/min to 47.5% A. The gradient increase then returns to 0.5% A/min to 66% A. The acetonitrile concentration is then increased to 90% in order to wash the column, after which it is reequilibrated at 25% A. A summary of the gradient is presented in Table II. This gradient produces baseline separation of oxidized (A-Ia) and reduced (A-Ib) apoA-I. In addition, the apoA-Ib peak is considerably expanded. Fractions are collected across the apoA-Ib peak at 0.5 min per fraction and are assessed for purity.

A sample separation using this procedure is shown in Fig. 1a. The large volume of the column creates a disparity between the acetonitrile percentage shown at the pump and that actually eluting at the end of the column. The 11-min delay of the salt peak should correspond with this disparity; thus, the eluting acetonitrile concentration at any given time should equal the acetonitrile concentration at the pump 11 min earlier. Figure 1b demonstrates the actual elution acetonitrile percentage by shifting the gradient to the left so that the salt peak is shown as being eluted at zero time. This shows that the apoA-Ib peak is in the center of the shallow gradient area, as is desirable.

TABLE III

GRADIENT FOR SEPARATION AND QUANTITATION OF APOLIPOPROTEIN A-I BY C_{18} REVERSED-PHASE HPLC[a]

Step	Time (min)	Final %A	Final %B	Gradient rate (%A/min)	Cumulative time (min)
0	1	25	75	0	1
1	33	58	42	1	34
2	2	90	10	16	36
3	2	90	10	0	38
4	2	25	75	−32.5	40
5	5	25	75	0	45

[a] Flow rate is 1.8 ml/min. Solvents A is acetonitrile plus 0.1% trifluoroacetic acid; solvent B is HPLC-grade water +0.1% trifluoroacetic acid. The column used is a Vydac C_{18} analytical column (0.46 × 25 cm).

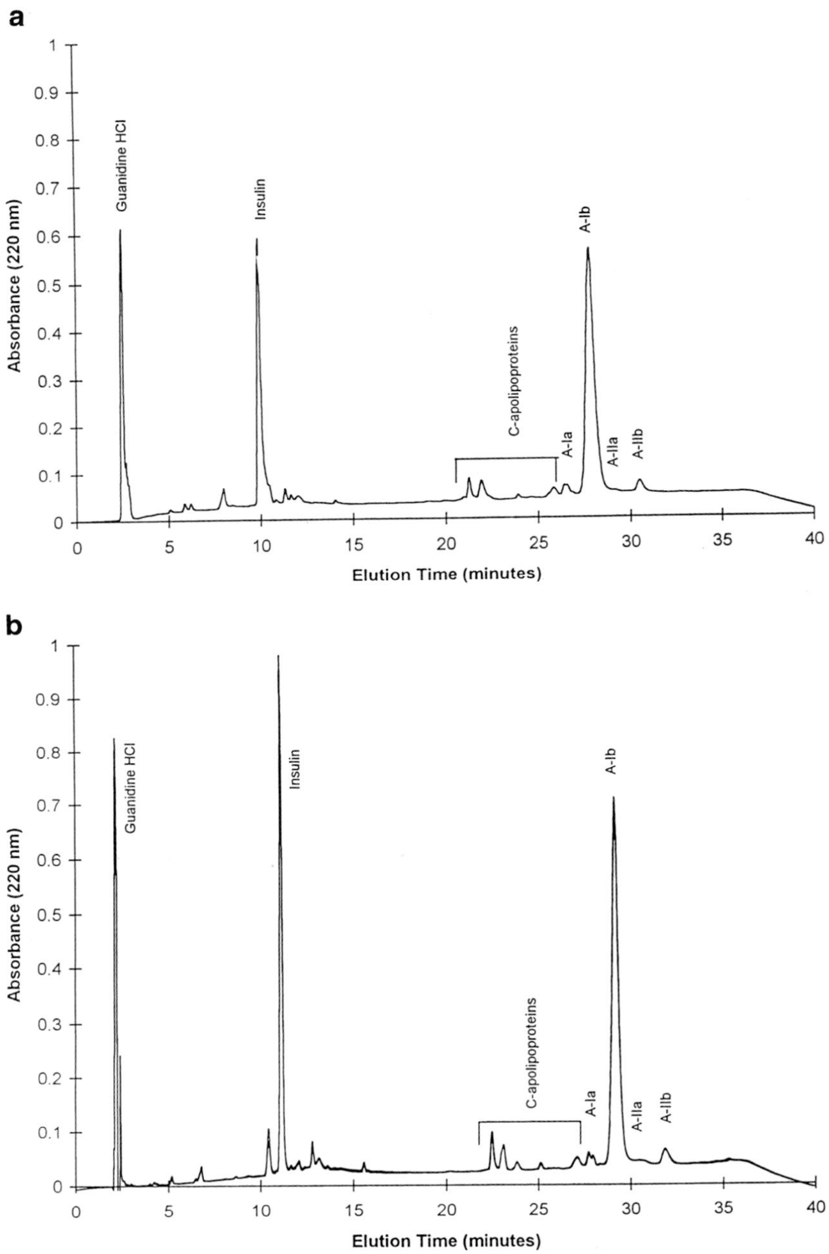

FIG. 2. (a) Quantitation of apoA-I present in HDL using a C_4 analytical column (Vydac). Table III gives details of the HPLC conditions. Insulin is the internal standard that is used to quantitate the amount of apoA-I. (b) Two consecutive runs on a C_{18} column using the conditions described in Table III. Note that the two chromatograms are superimposable and

Assessment of Purity

The purity of apoA-I is assessed by SDS–polyacrylamide gel electrophoresis and silver staining (PHAST system, Pharmacia, Piscataway, NJ). Previous preparations made using a linear acetonitrile gradient were commonly contaminated with an unknown low molecular weight material or with apoA-II. Using the discontinuous acetonitrile gradient described above, the low molecular weight contaminant is confined to the ascending limb of the apoA-I peak, whereas apoA-II contamination is confined to the descending limb (Fig. 1c). Thus, pure apoA-Ib can be collected from fractions at the top of the expanded apoA-Ib peak. Absence of oxidized apoA-Ia confirmed by analytical C_{18} reversed-phase HPLC. Appropriate fractions are pooled, lyophilized, weighed, and stored in the absence of moisture at $-70°$ until use.

Quantitation of Apolipoprotein A-I

Isolation of High-Density Lipoprotein

For reproducibility in HPLC analysis of apoA-I, it is important to fractionate lipoproteins and delipidate (or lyophilize) HDL within 3 days of the drawing of blood. As described by Hughes *et al.*[11] and Miltrop *et al.*,[20] storage of plasma even in the frozen state produces multiple peaks of apolipoproteins present in HDL for reasons described elsewhere.[19] Once delipidation is carried out, however, the lyophilized mixture of apolipoproteins as a dry powder is stable and has produced reproducible chromatograms. Although multiple peaks produced by different proteins have been characterized earlier,[19] it has also been noted that there will be considerable overlapping of the peaks arising from different isoforms of apolipoproteins; thus, integrations of these peaks will not be accurate. This has been demonstrated earlier by delipidation of freshly drawn plasma fractionated within a day and the same plasma frozen before fractionation and analysis of the delipidated HDL by HPLC.[11] It has also been demonstrated that the application of increasing amounts of the same sample produces peaks whose integration is linear. The range over which the proteins can be quantitated

look like a single chromatogram, indicating the high level of reproducibility of this procedure in the quantitation of apoA-I. Details of the data analysis are shown in Table IV. The results are highly reproducible not only within two consecutive C_{18} runs, but also compared to C_4 column runs, indicating that either a C_4 or C_{18} column can be used for the quantitation of apoA-I. However, better separation was observed with the C_{18} column compared to the C_4 column in the C-apolipoprotein regions.

is 0.1 to 50 μg protein per peak per injection. Because apoA-I is the major protein present in HDL, quantitation of this protein has therefore been consistent.

Centrifugation conditions for fractionation of lipoproteins were described earlier. A known amount of insulin is added to HDL, with the insulin concentration being adjusted so that the peak size is neither too large nor too small compared to the apoA-I peak. This is best done by quantitating the total protein concentration of HDL using Lowry protein analysis[28] and adding insulin, about one-fourth of the total protein concentration. Insulin is chosen because its elution time is much earlier than any apolipoprotein, and because it is readily available in pure form. Any other small protein that does not coelute with any of the apolipoproteins can be used for this purpose. Typically, about 250 μg insulin is added per milligram HDL. The peak area for insulin can also be used to estimate protein loss during the delipidation and analytical procedure.

Gradient conditions for HPLC quantitation of apoA-I are given in Table III. The solvents are identical to those used in preparative HPLC as described above. The flow rate is lower (1.2 ml/min) and a continuous gradient is used, rather than the discontinuous gradient used in preparative HPLC. The column is a 4.6 × 250 mm Vydac C_{18} column with a C_{18} guard column. Samples containing 0.1 to 50 μg protein in 20 μl of 6 *M* guanidine hydrochloride are injected. Although Hughes *et al.*[11] reported that the HPLC analysis should be carried out at 38°, in our experience, analysis at ambient temperature produces a satisfactory HPLC profile, although column backpressure is high. Eluted protein peaks are detected using a spectrophotometric HPLC detector at 214 nm. This wavelength is more suitable than 280 nm for the detection of the C apolipoproteins. The signal from the detector is digitized and integrated, and the areas of the protein peaks are determined. Protein concentration in each peak is determined by comparison with the area of the insulin peak, after compensation for variable detector response as described by Hughes *et al.*[11]

The HPLC profiles of the same sample produced using two types of reversed-phase columns, C_4 and C_{18}, are given in Fig. 2 (see also Table IV). The volume of sample injected onto the columns was kept constant by injecting an excess of material (35 μl) into a 20-μl sample loop, ensuring that the loop was completely filled. Figure 2b (C_{18} column) shows two column runs superimposed and demonstrates the striking reproducibility of the analysis. As expected, the retention times for insulin and other peaks are lower using the C_4 column than those obtained using C_{18} column, owing to the lesser hydrophobic interaction of protein molecules with the C_4 matrix than with the C_{18} matrix. Although both types of columns give similar results, resolution of peaks is somewhat better with the C_{18} column.

This is noticeable in the separations of the C apolipoproteins, in which a shoulder from the C_4 column is resolved as a separate peak in the C_{18} column. Although not readily apparent from Fig. 2, apoA-Ib and apoA-IIa are also better resolved using the C_{18} column. It is apparent from Fig. 2 that the total area under all peaks, as well as the areas under each peak, are somewhat lower in the C_4 columns than in the C_{18} columns. As identical volumes of sample are injected, this would suggest that the C_4 column may be retaining material, although the ratio of the areas of the apoA-Ib and insulin peaks was consistent between the two columns. On the basis of these results, for the accurate estimation of apoA-I, the C_{18} analytical column should be used.

It is also important to note that each individual column gives different retention times depending on the extent of use. The retention time decreases with increased numbers of injections, probably due to decreased

TABLE IV

HPLC ANALYSIS OF APO-HIGH-DENSITY LIPOPROTEIN AND ESTIMATION OF APOLIPOPROTEIN A-I USING C_4 AND C_{18} REVERSED-PHASE COLUMNS

Absorbance areas	C_4	$C_{18}(1)$[a]	$C_{18}(2)$[a]
Total	602	722	719
Salt peak	111	156	157
Insulin	139	157	154
A-Ia	7.05	8.65	9.25
A-Ib	245	276	277
A-IIb	9.83	12.2	11.6
Insulin			
Total area	23.0%	21.8%	21.4%
Total − salt	28.3%	27.7%	27.4%
A-Ia			
Total	1.17%	1.20%	1.29%
Total − salt	1.44%	1.53%	1.65%
Insulin	5.09%	5.51%	6.01%
A-Ib			
Total	40.7%	38.2%	38.5%
Total − salt	50.0%	48.6%	49.3%
Insulin	177%	176%	180%
A-IIb			
Total	1.63%	1.69%	1.61%
Total − salt	2.01%	2.16%	2.06%
Insulin	7.09%	7.79%	7.53%

[a] $C_{18}(1)$ and $C_{18}(2)$ represent results obtained from two consecutive injections of 20 μl of the same sample of apoHDL to demonstrate reproducibility.

efficiency of the column. Furthermore, even if columns are purchased from the same manufacturer, column-to-column variations will occur. As long as the internal standard gives consistent values, this variation will not interfere with the quantitation of apoA-I.

[18] Purification, Isoform Characterization, and Quantitation of Human Apolipoprotein A-IV

By RICHARD B. WEINBERG, RACHEL A. HOPKINS, and JENNIFER B. JONES

Introduction

Apolipoprotein A-IV (apoA-IV) is a 46,000-dalton plasma protein that has several distinctive biological characteristics. Apolipoprotein A-IV is synthesized in enterocytes of the small intestine and is initially incorporated onto the surface of nascent chylomicrons.[1–3] The subsequent intravascular metabolism of apoA-IV is unusual in that it rapidly dissociates from the surface of chylomicrons following their entry into the bloodstream and thereafter circulates primarily unassociated with serum lipoproteins,[3–5] with a plasma residence time of less than 24 hr.[6] The liver and kidney are the major sites of catabolism.[7]

Although the specific function of apoA-IV in human lipid metabolism has not been determined, a growing body of evidence suggests that apoA-IV may play a role in the metabolism of high-density lipoproteins (HDL). Apolipoprotein A-IV binds with high affinity to cell surfaces[8–10] and facili-

[1] K. H. Weisgraber, T. P. Bersot, and R. W. Mahley, *Biochem. Biophys. Res. Commun.* **85,** 287 (1978).

[2] U. Beisiegel and G. Utermann, *Eur. J. Biochem.* **93,** 601 (1979).

[3] P. H. Green, R. M. Glickman, C. D. Saudek, C. B. Blum, and A. R. Tall, *J. Clin. Invest.* **64,** 233 (1979).

[4] G. Utermann and U. Beisiegel, *Eur. J. Biochem.* **99,** 333 (1979).

[5] P. H. Green, R. M. Glickman, J. W. Riley, and E. Quinet, *J. Clin. Invest.* **65,** 911 (1980).

[6] G. Ghiselli, S. Krishnan, Y. Beigel, and A. M. Gotto, *J. Lipid Res.* **27,** 813 (1986).

[7] G. M. Dallinga-Thie, F. M. Van't Hooft, and A. Van Tol, *Arteriosclerosis (Dallas)* **6,** 277 (1986).

[8] E. Dvorin, N. L. Gorder, D. M. Benson, and A. M. Gotto, *J. Biol. Chem.* **261,** 15714 (1986).

[9] N. Savion and A. Gamliel, *Arteriosclerosis (Dallas)* **8,** 178 (1988).

[10] R. B. Weinberg and C. Patton, *Biochim. Biophys. Acta* **1044,** 255 (1990).

tates phospholipid and cholesterol efflux from cells.[11,12] A distinctive discoidal HDL containing free cholesterol and apoA-IV appears in high concentrations the peripheral nodal lymph of cholesterol-fed dogs,[13] and similar apoA-IV/lipid complexes have been observed in humans.[14] Apolipoprotein A-IV activates lecithin–cholesterol acyltransferase (LCAT; phosphatidylcholine–sterol *O*-acyltransferase),[15] a key enzyme in the metabolism of HDL and accelerates the rate of HDL speciation catalyzed by cholesteryl ester transfer protein (CETP).[16,17] These observations, taken together, suggest that apoA-IV may play a role in the earliest stages of peripheral HDL assembly, and may participate in the process of reverse cholesterol transport.

The signature biophysical property of human apoA-IV is its labile interaction with lipid surfaces. Apolipoprotein A-IV is an extremely hydrophilic apolipoprotein,[18] and its affinity for lipid surfaces is considerably weaker than other apolipoproteins[19]; consequently, it is easily displaced from lipoproteins by other apolipoproteins[20] and by physical separation techniques. A variety of approaches have been used to quantitate the relative amounts of free and lipoprotein-bound apoA-IV in circulation: ultracentrifugation,[4,5] crossed immunoelectrophoresis,[2] nondenaturing polyacrylamide gradient electrophoresis with immunoblotting,[21] gel-filtration chromatography,[14] and high-performance liquid chromatography (HPLC).[22] Nonetheless, the distribution of apoA-IV determined by these techniques must be interpreted with caution, for the binding of apoA-IV to lipoproteins is a tenuous interaction, easily perturbed by physical forces. Apolipoprotein A-IV also displays an unusually high affinity for self-association, and it readily forms dimers in the concentration ranges found in biological fluids.[23]

[11] A. Steinmetz, R. Barbaras, N. Ghalim, V. Clavey, J. C. Fruchart, and G. Ailhaud, *J. Biol. Chem.* **265,** 7859 (1990).

[12] J. K. Bielicki, W. J. Johnson, R. B. Weinberg, J. M. Glick, and G. H. Rothblat, *J. Lipid Res.* **33,** 1699 (1992).

[13] C. H. Sloop, L. Dory, R. Hamilton, B. Krause, and P. S. Roheim, *J. Lipid Res.* **24,** 1429 (1983).

[14] C. L. Bisgaier, O. P. Sachdev, L. Megna, and R. M. Glickman, *J. Lipid Res.* **26,** 11 (1985).

[15] A. Steinmetz and G. Utermann, *J. Biol. Chem.* **260,** 2258 (1985).

[16] C. L. Bisgaier, M. V. Siebenkas, C. B. Hesler, T. L. Swenson, C. B. Blum, Y. L. Marcel, R. W. Milne, R. M. Glickman, and A. R. Tall, *J. Lipid Res.* **30,** 1025 (1989).

[17] K. A. Rye, K. H. Garrety, and P. J. Barter, *J. Lipid Res.* **33,** 215 (1992).

[18] R. B. Weinberg, *Biochim. Biophys. Acta* **918,** 299 (1987).

[19] R. B. Weinberg, J. A. Ibdah, and M. C. Phillips, *J. Biol. Chem.* **267,** 8977 (1992).

[20] R. B. Weinberg and M. S. Spector, *J. Lipid Res.* **26,** 26 (1985).

[21] R. B. Weinberg and M. S. Spector, *Biochem. Biophys. Res. Commun.* **135,** 756 (1986).

[22] L. Lagrost, P. Gambert, M. Boquillon, and C. Lallemant, *J. Lipid Res.* **30,** 1525 (1989).

[23] R. B. Weinberg and M. S. Spector, *J. Biol. Chem.* **260,** 14279 (1985).

Several human genetic polymorphisms of apoA-IV have been described. The most common variant is a glutamine to histidine substitution at residue 360[24] that generates a basic isoform, apoA-IV-2. Genetic studies have established that the apoA-IV-2 allele has a population gene frequency of 0.07–0.09,[25-28] and some studies have found that subjects heterozygous for the apoA-IV-2 allele have higher HDL cholesterol levels[26,28] and lower triglyceride levels[26-28] than the population average. These observations suggest that the apoA-IV-2 allele may have a significant impact on human lipoprotein metabolism. We have shown that this substitution changes the structure, lipid affinity, and LCAT catalysis of apoA-IV.[29] Therefore, it is imperative that apoA-IV used in biophysical and biochemical studies be isomorphically pure.

A variety of antibody-based techniques have been used to quantitate human apoA-IV concentrations in biological fluids, including electroimmunoassay,[5,30] radioimmunoassay,[14] and sandwich[31] and competitive[32,33] enzyme-linked immunosorbent assays. With the exception of radioimmunoassay, the average serum concentration measured by these techniques has been in the range of 12.7–16.4 mg/dl. In humans, the gene for apoA-IV is expressed only in the enterocytes of the small intestine,[34] and the synthesis of apoA-IV is specifically stimulated by fat absorption.[5,35,36] Consequently, plasma apoA-IV levels increase in the postpartum period,[37] transiently after a fatty meal,[5] and chronically in response to high-fat diets.[38] Con-

[24] P. Lohse, M. R. Kindt, D. J. Rader, and H. B. Brewer, *J. Biol. Chem.* **265,** 10061 (1990).
[25] M. I. Kamboh and R. E. Ferrell, *Am. J. Hum. Genet.* **41,** 119 (1987).
[26] H. J. Menzel, E. Boerwinkle, S. Schrangl-Will, and G. Utermann, *Hum. Genet.* **79,** 368 (1988).
[27] J. E. Eichner, L. H. Kuller, R. E. Ferrell, and M. I. Kamboh, *Genet. Epidemiol.* **6,** 493 (1989).
[28] H. J. Menzel, G. Sigurdsson, E. Boerwinkle, S. Schrangl-Will, H. Dieplinger, and G. Utermann, *Hum. Genet.* **84,** 344 (1990).
[29] R. B. Weinberg, M. Jordan, and A. Steinmetz, *J. Biol. Chem.* **265,** 18372 (1990).
[30] R. B. Weinberg and A. M. Scanu, *J. Lipid Res.* **24,** 52 (1983).
[31] M. Rosseneu, G. Michiels, W. DeKeersgieter, J. Bury, J. P. DeSlypere, H. Dieplinger, and G. Utermann, *Clin. Chem.* **34,** 739 (1988).
[32] L. Lagrost, P. Gambert, S. Meunier, J. Morgado, J. Desgres, P. D'Athis, and C. Lallemant, *J. Lipid Res.* **30,** 701 (1989).
[33] K. Kondo, C. Allan, and N. Fidge, *J. Lipid Res.* **30,** 939 (1989).
[34] N. A. Elshourbagy, D. W. Walker, M. S. Bogusky, J. I. Gordon, and J. M. Taylor, *J. Biol. Chem.* **261,** 1988 (1986).
[35] T. F. Apfelbaum, N. O. Davidson, and R. M. Glickman, *Am. J. Physiol.* **252,** G662 (1987).
[36] M. F. Go, G. Schonfeld, B. Pfleger, T. G. Cole, N. L. Sussman, and D. H. Alpers, *J. Clin. Invest.* **81,** 1615 (1988).
[37] A. Steinmetz, P. Czekelius, E. Thiemann, S. Motzny, and H. Kaffarnick, *Atherosclerosis* **69,** 21 (1988).
[38] R. B. Weinberg, C. Dantzker, and C. Patton, *Gastroenterology* **98,** 17 (1990).

versely, plasma levels decrease during fasting[39] and in patients with malabsorption[40] and short bowel syndrome.[39] Plasma levels are elevated up to 4-fold in renal failure[41] and in diseases characterized by the accumulation of abnormal cholesterol-enriched lipoproteins.[42] Apolipoprotein A-IV levels are correlated with serum triglycerides,[32] and weakly with total serum cholesterol.[39]

Apolipoprotein A-IV has been purified from thoracic duct chylomicrons,[1] chylous effusions,[4] hypertriglyceridemic serum,[2] and the urine of patients with chyluria.[5] We have found that adsorption of apoA-IV from lipoprotein-depleted plasma to a triglyceride–phospholipid emulsion affords a convenient and rapid method for the initial purification of this apolipoprotein.[30] Below we describe effective methods for the preparation, isoform characterization, and quantitation of human apoA-IV.

Purification of Human Apolipoprotein A-IV from Human Serum

Adsorption of Apolipoprotein A-IV to Triglyceride–Phospholipid Emulsion Particles

Equipment

Low-speed preparative centrifuge (e.g., Sorvall, Wilmington, DE, RC-5B)
Refrigerated ultracentrifuge capable of 50,000 rpm
Preparative fixed-angle and swinging-bucket rotors (Beckman, Fullerton, CA, 50.2Ti and SW-28, or equivalent)

Reagents

5% (w/v) Sodium ethylenediaminetetraacetic acid (EDTA), pH 7.4
10% (w/v) Sodium azide
Chloramphenicol solution (200 μg/ml)
Gentamicin sulfate solution (40 mg/ml)
0.3 *M* NaCl, pH 7.4
1.0 *M* Benzamidine
0.2 *M* Phenylmethylsulfonyl fluoride (PMSF) in anhydrous ethanol
Intralipid 10% (Baxter Healthcare, Deerfield, IL)
Phosphate-buffered saline [PBS: 50 m*M* sodium phosphate, 0.9% (w/v) NaCl, 0.05% EDTA, pH 7.5]

[39] J. R. Sherman and R. B. Weinberg, *Gastroenterology* **95,** 394 (1988).
[40] S. Koga, Y. Miyata, A. Funakoshi, and H. Ibayashi, *Digestion* **32,** 19 (1985).
[41] P. J. Nestel, N. H. Fidge, and M. H. Tan, *N. Engl. J. Med.* **307,** 329 (1982).
[42] G. Ghiselli, E. J. Schaefer, P. Gascon, and H. B. Brewer, *Science* **214,** 1239 (1981).

Blood Collection. Five hundred milliliters of blood is collected directly into 250-ml polyallomer centrifuge bottles with sealing caps to which the following preservatives have been added:

Solution	Volume/250 ml	Final concentration
5% EDTA, pH 7.4	6 ml	1.2 g/liter (3 m*M*)
10% Sodium azide	250 μl	0.1 g/liter (1.5 m*M*)
Chloramphenicol (200 μg/ml)	100 μl	80 μg/liter
Gentamicin sulfate (40 mg/ml)	0.5 ml	80 μg/ml
0.3 *M* NaCl, pH 7.4	7 ml	150 m*M*

The bottles are maintained on ice during the collection and are periodically agitated to ensure complete mixing. The bottles are spun at 5000 rpm at 4° for 20 min, and the plasma is removed by pipette, taking care not to disturb the red cells. The plasma is respun in clean bottles, and the following preservatives are then added:

Solution	Volume/250 ml	Final concentration
1 *M* Benzamidine	250 μl	1 m*M*
0.2 *M* PMSF	1.25 ml	1 m*M*

Plasma from different donors is kept separate until the apoA-IV phenotypes have been determined.

Preparation of Lipoprotein-Depleted Plasma. Plasma density is raised to 1.25 g/ml by the addition of 9.8 g anhydrous NaCl and 25.6 g anhydrous NaBr per 100 ml plasma. Plasma is centrifuged in polycarbonate centrifuge tubes at 45,000 rpm for 60 hr at 4° in a 50.2Ti rotor. The floating orange lipoprotein layer at the top of each tube is carefully removed by pipette and set aside for other uses. The bottom fractions are pooled and diluted 1 : 1 with PBS. Two donors will yield approximately 500–600 ml of diluted lipoprotein-depleted plasma (LPDP).

Preparation of Triglyceride-Rich Particles. Three hundred milliliters of 10% Intralipid are distributed into polyallomer centrifuge tubes in 35-ml aliquots. The tubes are spun at 27,000 rpm for 35 min at 4° in a SW-28 rotor, and the thick cream layer of triglyceride-rich particles (TRP) at the top of each tube is pooled and resuspended in 300 ml PBS. The opalescent bottom fractions are discarded.

Adsorption of Apolipoprotein A-IV to Triglyceride-Rich Particles. The TRP and LPDP are combined in a ratio of 100 ml TRP to 500 ml diluted

LPDP. Then NaCl is added at a rate of 22.5 g for every 100 ml TRP and 10.75 g for every 100 ml diluted LPDP. The solution is stirred gently for 1 hr at 4°; the pH is adjusted to 7.5 if necessary with NaOH, and the mixture is degassed. Polyallomer tubes are filled with 30 ml LPDP/TRP mixture, and 5 ml cold PBS is floated on top by carefully pipetting down the side of each tube. The tubes are spun at 27,000 rpm for 35 min at 4° in a SW-28 rotor. The white TRP cake at the top of each tube is removed with a spatula, taking care not to disrupt the LPDP below; the infranatants are pooled. A second adsorption–flotation cycle may be carried out by adding an additional 100 TRP to every 500 ml pooled LPDP infranatant. Additional NaCl is added at a rate of 22.5 g for every 100 ml TRP and 10.75 g for every 100 ml PBS layered on top of the tubes in the first flotation (i.e., 5 ml × number of tubes). The TRP collected from the first and second floats are maintained separately, as the first spin will give purer and higher yields.

Delipidation of Triglyceride-Rich Particle/Apolipoprotein A-IV Complex. For every 100 ml TRP used, a 2-liter round-bottom flask, fitted with a ground glass stopper and a stirring bar, is filled with 3:1 (v/v) anhydrous diethyl ether–anhydrous ethanol and cooled to 4°. Recovered TRP are suspended in sufficient PBS to allow transfer with a glass pipette and then added dropwise to the flasks with rapid magnetic stirring. The flasks are stoppered and stirred overnight at 4°. A fine white protein precipitate will form as the TRP are delipidated. The following day, the stirring is stopped, and the precipitates are allowed to settle to the bottom of the flasks. The flasks are carefully moved to a fume hood, and as much solvent as possible is removed without disturbing the precipitate using a water aspirator connected to a side-arm flask. The remaining precipitate and solvent are then transferred to 50-ml thick-walled Pyrex centrifuge tubes with Teflon cap inserts and spun at 2000 rpm for 15 min at 0°. The solvent is aspirated from the tubes, and the precipitates are collected together by rinsing the tubes with ice-cold diethyl ether. The collected precipitate is washed a final time with 50 ml of cold 3:1 (v/v) diethyl ether–ethanol. Final traces of solvent are removed under vacuum.

Comments. With this method, crude apoA-IV can be obtained in 6 days: if blood collection is performed on a Friday, the ultracentrifugation to prepare the LPDP may proceed over the weekend; the TRP adsorption and delipidation, the settling and collection of the precipitates, and the preparation for HPLC separation each take 1 day. Under the described conditions, only apoA-IV, apoA-I, and several higher molecular weight proteins are adsorbed to the Intralipid particles. Pure apoA-IV may be separated from the crude protein mixture by preparative electrophoresis,[30] gel-filtration chromatography,[14] anion-exchange chromatography,[43] or reversed-phase HPLC.[44] Below, a rapid separation using anion-exchange HPLC is described. Because of the tendency of apoA-IV to self-associate and to bind phospholipid, this separation is performed under denaturing conditions.

Purification of Apolipoprotein A-IV by High-Performance Liquid Chromatography

Equipment

HPLC pump with computer-controlled gradient mixer
Multiwavelength UV detector
Mono Q HR 10/10 anion-exchange column (Pharmacia, Piscataway, NJ)
Chart recorder

Reagents

Solvent A: 7 *M* deionized urea, 10 m*M* Tris, pH 7.4
Solvent B: 7 *M* deionized urea, 10 m*M* Tris, pH 7.4, 1 *M* NaCl
1 m*M* Ammonium bicarbonate, pH 8.2

Sample Preparation. Sufficient 50 m*M* Tris, pH 7.4 (usually 6–8 ml), is added to the delipidated TRP residue to dissolve the protein. The solution is dialyzed against several changes of solvent A. The dialyzed protein solution is centrifuged in 13 × 64 mm Ultraclear (Beckman) tubes at 45,000 rpm for 24 hr at 4° to remove insoluble aggregates that could ruin the HPLC column. The clear supernatants are carefully removed, without disturbing the layer of sludge at the bottom of the tubes, and are filtered through a 0.45-μm Millex HV 25-mm Acrodisk (Millipore, Bedford, MA) using a 10-ml disposable plastic Luer-lock syringe.

Chromatography. A Mono Q HR 10/10 anion-exchange column is connected to the HPLC pump, and the following linear gradient program is entered into the controller:

Time (min)	Flow (ml/min)	Solvent A (%)	Solvent B (%)
Initial	2.00	95	5
40	2.00	85	15
45	2.00	0	100
50	2.00	0	100
55	2.00	95	5

[43] R. B. Weinberg and M. S. Spector, *J. Biol. Chem.* **260,** 4914 (1985).
[44] R. B. Weinberg, C. Patton, and B. DaGue, *J. Lipid Res.* **29,** 819 (1988).

The column is primed by running the gradient program without a sample injection. A test injection of 200 μl crude apoA-IV is made, and absorbance is monitored at 210 nm at a sensitivity of 0.2 and a chart speed of 0.5 cm/min. If cleanly separated peaks are obtained, 2-ml injections are made with appropriate reduction of detector sensitivity. Even larger amounts can be injected with special sample loops.

Apolipoprotein A-I and minor isoforms elute with a retention time of 23–27 min, and apoA-IV elutes later between 35 and 38 min; a washout peak of higher molecular weight proteins appears at the end of the gradient program (Fig. 1). Peaks are collected in flasks set on ice. Pooled apolipoprotein fractions are dialyzed against 1 m*M* ammonium bicarbonate, pH 8.2, concentrated by lyophilization or ultrafiltration, and stored at 4° or frozen at −70°. Average yields are 2–10 mg per donor. At the end of the run, the column is flushed extensively with deionized water.

Comments. Apolipoprotein A-IV thus produced is detergent-free and isomorphically pure, and it is suitable for biophysical and biochemical studies or for antibody production. The major factor affecting yields is the care with which the crude apoA-IV precipitate is collected. Apolipoprotein A-IV readily adsorbs to glass, plastic, and dialysis tubing, so it is advisable to minimize the surface area of these materials that comes in contact with apoA-IV solutions. Even with extensive delipidation, considerable phospholipid remains in the crude apoA-IV solution, and it will be necessary to clean the column after several runs to remove lipid residues and aggregated proteins. Apolipoprotein A-IV is an excellent antigen, and high-titer polyclonal antisera can easily be raised in rabbits or goats using standard immunization protocols.

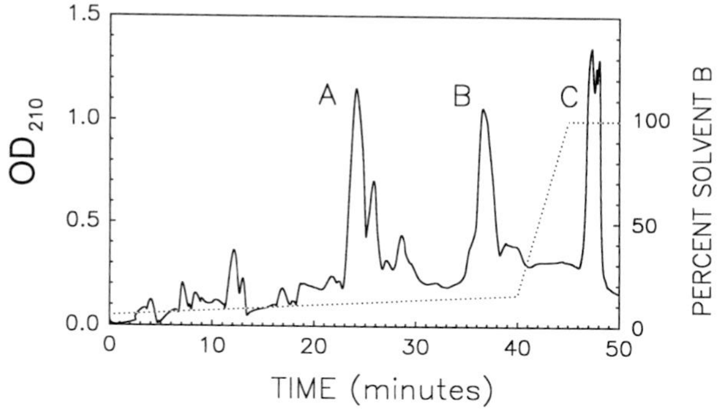

FIG. 1. Purification of crude human apoA-IV by high-performance liquid chromatography on a Mono Q HR 10/10 anion-exchange column. The solid line is the optical density at 210 nm. The dotted line is percent solvent B. Peak A, apoA-I (smaller shoulder peaks are apoA-I isoforms); peak B, apoA-IV; peak C, washout.

Apolipoprotein A-IV Phenotyping by Isoelectric Focusing and Immunoblotting

Isoelectric Focusing

Equipment

Vertical gel electrophoresis apparatus (e.g., Hoefer, San Francisco, CA, SE600)
16 × 18 cm Vertical gel cassettes, with 1.5-mm spacers
Refrigerated water circulator
Electrophoresis power source

Reagents

9 *M* Urea solution
TMD-8 mixed bed resin (Sigma, St. Louis, MO)
50% (w/v) Acrylamide solution
pH 4.5–5.4 ampholytes (Pharmalyte, Pharmacia LKB, Piscataway, NJ)
TEMED (*N,N,N',N'*-tetramethylethylenediamine)
2% (w/v) Ammonium persulfate solution (freshly prepared)
Sucrose crystals
0.5% (w/v) Evan's Blue dye
10 *N* NaOH
Concentrated phosphoric acid (85% w/w)

Gel Preparation. The urea is deionized by adding 3 g of TMD-8 mixed bed resin to 50 ml of 9 *M* urea. The slurry is gently swirled for 5 min, and the resin is removed by filtration. A 16 × 18 cm cassette with 1.5-mm spacers and a 10-well comb requires 35 ml gel solution: 27.1 ml of deionized 9 *M* urea, 5.25 ml of 50% acrylamide, and 1.75 ml of 4.5–5.4 ampholytes are combined and thoroughly degassed using a water aspirator. Then 60 μl TEMED and 0.9 ml of 2% freshly prepared ammonium persulfate are added, and the monomer solution is immediately poured into the cassette and allowed to polymerize for 1 hr. Carefully layering water over the exposed monomer solution between the teeth of the well comb will aid polymerization. The final concentrations in the gel are 7.5% (w/v) acrylamide, 7 *M* urea, 5% (v/v) 4.5–5.4 ampholytes, 0.17% (v/v) TEMED, and 0.05% (w/v) ammonium persulfate.

When the gel has polymerized, the well comb is removed, and the cassette is inserted into the lower buffer chamber of the electrophoresis apparatus, equipped with a heat exchanger connected to a refrigerated water circulator. The lower chamber is filled with 0.01 *N* H_3PO_4 (2.7 ml of 85% phosphoric acid diluted to 4 liters with deionized water) as an anode buffer; the upper chamber is filled with 0.02 *N* NaOH (2 ml of 10 *N* NaOH diluted to 1 liter with deionized water) as a cathode buffer. The buffers are cooled to 10°, and the gel is prefocused for 15 min at 100 V, 15 min at 200 V, 30 min at 300 V, and 60 min at 400 V.

Sample Preparation. Serum samples are prepared by combining 10 μl serum with 40 μl of deionized 9 *M* urea, 2 μl of 0.5% Evan's Blue dye, and several crystals of sucrose. After the gel has been prefocused, the samples are carefully loaded into the wells using a syringe and thin needle or an adjustable pipette and specially designed disposable applicator tips. Finally, 25 μl of an overlay solution (0.7 ml of 9 *M* urea diluted to 1 ml with deionized water and combined with 10 μl of 4.5–5.4 ampholytes) is layered over each sample, taking care to avoid mixing.

Electrophoresis. Isoelectric focusing is conducted at 400 V (constant voltage) for 20 hr at 10°. When focusing is complete, the tracking dye will be concentrated in sharp bands at the bottom of the gel. The cassette is removed from the buffer tank, and the gel is gently freed from the glass plates in preparation for immunoblotting.

Comments. Thorough and even cooling of the cassette is essential to prevent uneven focusing ("smile" artifacts). This may be achieved by stirring the lower buffer chamber with a magnetic stirrer and by assuring that the level of anode buffer covers the entire cassette. Isoelectric focusing is conveniently performed overnight, so that electroblotting may be performed the following morning.

Immunoblotting

Equipment

Electroblotting cell with high-capacity heat exchanger (e.g., Hoefer TE 42)
Refrigerated water circulator
High-intensity electroblotting power source
Rectangular plastic trays
Rocker platform

Reagents

Blotting buffer (20× Tris–borate–EDTA, TBE): 109.0 g Tris base, 31.0 g boric acid, and 5.8 g sodium ethylenediaminetetraacetic acid, to 2 liters with deionized water

Rinsing buffer (5× Tris-buffered saline, TBS): 24.2 g Tris base, 292.3 g NaCl, to 2 liters with deionized water; adjust to pH 7.5 with concentrated HCl

Blocking solution: 1% (w/v) bovine serum albumin in 1× TBS
Washing solution: 0.05% Tween 20 in 1× TBS
Nitrocellulose membrane (Schleicher & Schuell, Keene, NH)
Blotting paper (Schleicher & Schuell)
Polyclonal rabbit anti-human apoA-IV antiserum
Horseradish peroxidase-conjugated goat anti-rabbit antiserum (Bio-Rad, Richmond, CA)
Horseradish peroxidase (HRP) color development reagent (Bio-Rad)
30% (w/v) Hydrogen peroxide

Electrophoretic Transfer. The blotting cassette is assembled in a rectangular plastic tray filled with 1× blotting buffer in the following sequence: (1) plastic cassette screen; (2) plastic Scotch Brite sponge; (3) one piece of blotting paper cut to the size of the gel; (4) isoelectric focusing gel; (5) prewetted nitrocellulose membrane; (6) one piece blotting paper; (7) plastic Scotch Brite sponge; and (8) top plastic screen. The tank of the Hoefer TE42 blotting cell is filled with 5 liters blotting buffer made by diluting the stock 1:20 with deionized water. The blotting assembly is placed in the tank with the nitrocellulose membrane closest to the anode, and electrophoretic transfer is performed at 100 V (constant voltage) for 2 hr at 0° (the current will rise up to 760 mA). The transfer generates considerable heat, and it may be necessary to prechill the tank buffer and set the circulator thermostat to 0° to establish and maintain adequate cooling.

Visualization of Isoform Bands. The nitrocellulose membrane is removed from the cassette, placed gel-contact side up in a plastic tray containing 75 ml blocking solution, and gently agitated on a rocker platform for 30 min before proceeding; if convenient, it may be left in blocking solution overnight. The blocking solution is then poured off and rabbit anti-human apoA-IV antiserum is added, diluted 1:1000 in 75 ml of blocking solution. The membrane is agitated for 1 hr, then washed three times, 15 min each time, with 75 ml washing solution and once with 75 ml of 1× TBS. Horseradish peroxidase-conjugated goat anti-rabbit second antibody is added, diluted 1:2000 in 75 ml blocking solution. The membrane is agitated for 90 min, then washed two times for 15 min with 75 ml washing solution, two times for 15 min with 75 ml of 1× TBS, and once for 5 min with deionized water. The bands are visualized by peroxide oxidation of a color reagent. Sixty milligrams HRP color development reagent is mixed with 10 ml cold methanol; separately, 60 μl of 30% H_2O_2 is added to 100 ml of 1× TBS. The two solutions are combined and immediately poured over the membrane. The membrane is agitated for 5–15 min until the bands develop to the desired darkness.

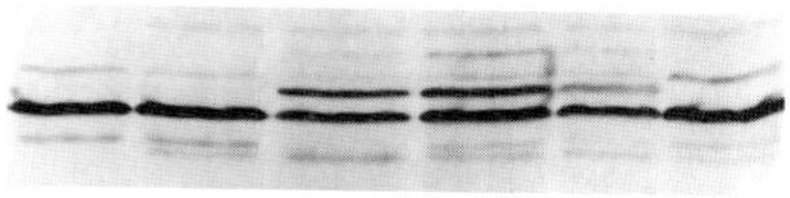

FIG. 2. Immunoblot of human sera subjected to isoelectric focusing in the pH range 4.5–5.4. Phenotypes are indicated below each lane. The faint upper band in lane 6 is a glycosylation isoform.

The reagent is then decanted, and the blot is rinsed twice with water and air dried. Developed blots must be stored in the dark to prevent fading.

Interpretation of Band Patterns. In homozygous subjects with the 1/1 phenotype, a single major band with a p*I* of 4.97 will be present near the middle of the membrane (Fig. 2); glycosylation heterogeneity may produce several fainter bands above and below the major band. In subjects who are 1/2 heterozygotes, a second dark band will be visible one charge unit above (i.e., in the basic, cathodal direction) the main isoform band at a p*I* of 5.02. The basic isoform is generated by addition of a single positive charge by substitution of histidine for glutamine at position 360.[24] Several other rare patterns may also be found. Apolipoprotein A-IV-3 is a basic isoform charge-shifted 2 units toward the cathode as a result of the substitution of lysine for glutamic acid at position 230.[45] A more acidic (anodal) band is noted in individuals carrying the apoA-IV-0 allele, which is caused by the addition of an extra Glu-Gln-Gln-Gln sequence at the carboxyl terminus[45]; this allele is more common in individuals of African ancestry.[46]

Comments. Proper technique in assembling the transfer cassette is critical for the success of this technique. The nitrocellulose membrane must be handled with gloves and must be prewetted by careful immersion in buffer before blotting. Care must be taken that no air bubbles are trapped in the gel sandwich before fastening the cassette together, or the transfer will be uneven or incomplete. Cutting a notch in one corner of both the gel and membrane will assure proper identification of samples. The dilution of the first and second antibodies may need to be adjusted depending on the titer of the anti-human apoA-IV antiserum. Other detection systems may easily be used if desired.

[45] P. Lohse, M. R. Kindt, D. J. Rader, and H. B. Brewer, *J. Biol. Chem.* **265,** 12734 (1990).

[46] M. I. Kamboh, E. R. Williams, J. C. Law, C. E. Aston, C. H. Bunker, R. E. Ferrell, and W. S. Pollitzer, *Genet. Epidemiol.* **9,** 379 (1992).

Quantitation of Human Apolipoprotein A-IV by Electroimmunoassay

Equipment

Horizontal electrophoresis apparatus (e.g., LKB, Multiphor II, Piscataway, NJ)
Refrigerated water circulator
Electrophoresis power source
Flat level metal pouring table (LKB)
Hydrophilic coated plastic film (e.g., Gelbond film, FMC, Rockland, ME)
Rocker table
Gel hole-punch template and punches (e.g., LKB immuno-electrophoresis kit)
Dacron electrophoresis wicks

Reagents

Low electroendosmosis agarose (FMC)
25 m*M* Tris–Tricine buffer, pH 8.6 (19.37 g Tris base, 8.60 g Tricine, 1.23 g calcium lactate, to 2 liters with water)
Triton X-100
Rabbit anti-human apoA-IV antiserum
Rinsing solution: 150 m*M* NaCl
Staining solution: 2.5 g Coomassie Brilliant Blue R-250, 225 ml ethanol, 50 ml acetic acid, to 500 ml with water
Destaining solution: 45% (v/v) ethanol, 10% (v/v) acetic acid in water

Preparation of Gels. For each 85 × 100 mm precut plastic film, 125 mg agarose and 12.5 ml Tris–Tricine buffer are added to a Pyrex test tube and heated in a beaker of simmering water. When the agarose has melted, the beaker is removed from the heat. The films are laid hydrophilic side up on the metal pouring platform, which has been heated to 60°. One by one, the test tubes containing the molten agarose are removed from the water bath and stirred with a short thermometer. When the agarose has cooled to 55°, 150 μl antibody is pipetted into the tube, the contents are vigorously stirred, and the agarose–antibody solution is poured onto the center of a film. The edge of the tube is used to distribute the solution to all corners of the film if it does not flow spontaneously. The films are allowed to cool to room temperature; when gelled, the agarose will turn opalescent. When the gels are cool, a row of 5-μl wells is evenly punched 1.5 cm from one edge using a template and punch connected to a water aspirator. The gels may then be stored at 4° until use in sealable plastic containers containing a piece of moistened filter paper.

Preparation of Samples and Standards. A 1 mg/ml stock solution of purified human apoA-IV is diluted with 25 m*M* Tris–Tricine buffer con-

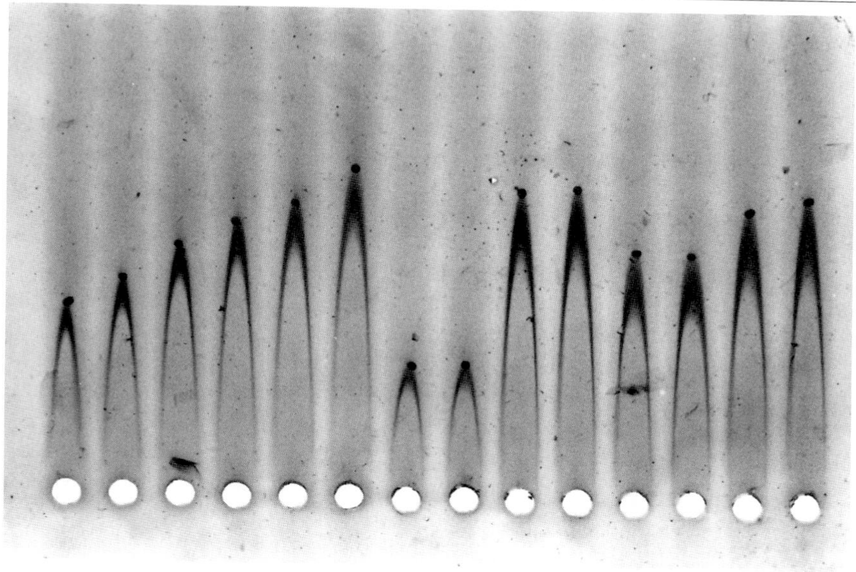

FIG. 3. Electroimmunoassay of apoA-IV. The first six lanes on the left contain apoA-IV standards, at 45, 60, 75, 90, 105, and 120 μg/ml. The next eight lanes contain duplicate plasma samples from four subjects. The dots at the tips of the "rockets" were added after staining to facilitate measurement.

taining 2% Triton X-100 to create a panel of standards with concentrations of 10 to 120 μg/ml. If only relative apoA-IV concentrations are to be determined, then standards may be prepared by dilution of a pooled reference serum (defined as having an apoA-IV concentration of 100 arbitrary units) in the range of 20–80% serum (2% final Triton X-100 concentration). Serum or plasma samples are diluted 1:2 with Tris–Tricine buffer containing 4% Triton X-100. Other body fluids such as urine, bile, cerebrospinal fluid, or lymph must first be concentrated by ultrafiltration or precipitation.[47]

Electrophoresis. The gels are laid on the cooling plate of the flat bed electrophoresis apparatus. The plate is cooled to 10°. The buffer tanks are filled with Tris–Tricine buffer, and fabric wicks soaked in buffer are placed on the top and bottom of the gels and gently pressed into place to make firm contact. The trailing ends are immersed in the buffer tanks. Using a microsyringe, the first six wells of each gel are filled with 5 μl of the apoA-IV standards. The remaining wells are filled with the samples in duplicate.

[47] G. M. Dallinga-Thie, P. H. Groot, and A. Van Tol, *J. Lipid Res.* **26,** 889 (1985).

The cover of the unit is closed, and the gels are run for 4 hr at constant voltage at 10 V/cm.

Development. The gels are removed from the electrophoresis apparatus and placed on a flat solid surface such as a sheet of thick glass. The surfaces of the gels are moistened with water, and two sheets of wet filter paper are carefully laid over the gels. A 1-inch-thick layer of dry paper towels is laid on top and covered with a second sheet of glass and small weight to press out excess buffer and unbound antibody. After 20 min, the papers are removed, and the gels are placed into plastic trays, covered with 100 ml rinsing solution, and gently agitated overnight. The following morning, the gels are pressed again for 20 min, and then are dried with a stream of hot air from a hair dryer. The gels are immersed in staining solution for 20 min and destained with several rinses of destaining solution until background staining is cleared and the immunoprecipitates are clearly visible (Fig. 3).

Quantitation. The concentration of apoA-IV in each sample is proportional to the height (or area) of the rocket-shaped immunoprecipitate. The height is measured from the upper lip of the sample well to the tip of the precipitate; the area can be calculated as height times width at half-height in millimeters. A standard curve is constructed from the heights or areas of the rockets formed by the pure apoA-IV standards, and the concentration in the biological samples is calculated from the standard curve. With this technique coefficients of variation of 2.7–3.5% for intraassay and 7.6–9.4% interassay variation are obtained.[30,39] The lower limit of sensitivity is 2.5 μg/ml.

Comments. Because of local variations in the electrical field, it is advisable to run apoA-IV standards on each gel. The amount of antiserum added to the gels, the concentration of the standards, and the dilution ratio of the samples will depend on the titer of the antisera; exact parameters must be optimized experimentally. If the standard curve flattens out at the higher standard concentrations, then the electrophoresis run time should be lengthened. Using a single electrophoresis unit, 12 samples may be run in duplicate in a single day. The chief advantages of electroimmunoassay are the easy availability and low cost of the necessary equipment and reagents. Moreover, it is not necessary to use affinity-purified antisera. The range of sensitivity of this technique is particularly suitable for the analysis of serum or plasma, and its accuracy and reproducibility compare well with radioimmunoassay or enzyme-linked immunosorbent assay. The disadvantages of electroimmunoassay are that the technique cannot be automated, and relatively large amounts of antibody are required.

[19] Immunochemical Methods for Quantification of Apolipoprotein A-IV

By NOEL H. FIDGE

Introduction

Apolipoprotein (apo) A-IV, unlike most of the other apolipoproteins, continues to elude definition of its physiological role in lipoprotein transport. The most recent position about apoA-IV is that it probably shares functions with other apolipoproteins; like apoA-I it may be involved at some level in reverse cholesterol transport, and in support of this suggestion apoA-IV was found to promote cholesterol efflux from cells[1,2] and to bind to putative high-density lipoprotein (HDL) receptors on hepatocytes[3] and adipose cells.[1] Apolipoprotein A-IV may be an activator of lecithin–cholesterol acyltransferase (LCAT; phosphatidylcholine–sterol *O*-acyltransferase),[4] although the strength of activation is considered controversial,[5] and it stimulates cholesteryl ester transfer protein (CEPT)-mediated interconversion of particle sizes within the human HDL population.[6]

With respect to lipid transport, apoA-IV may play dual roles, as there is evidence that its plasma levels and distribution among lipoproteins are associated with both cholesterol and triglyceride metabolism. Delamatre and Roheim[7] found that apoA-IV metabolism was profoundly affected by feeding rats cholesterol. Earlier studies showed that plasma levels and intestinal production of apoA-IV were elevated following fat feeding,[8] thus connecting apoA-IV metabolism with triglyceride transport. Studies[9] have confirmed that apoA-IV synthesis in rat intestine is regulated by dietary triglyceride. Because apoB has been assigned the dominant role in assembly and secretion of triglyceride-rich lipoproteins from the intes-

[1] A. Steinmetz, R. Barbaras, N. Ghalim, V. Clavey, J. C. Fruchart, and G. Ailhaud, *J. Biol. Chem.* **265,** 7859 (1990).
[2] H. Hara, H. Hara, A. Komaba, and S. Yokoyama, *Lipids* **27,** 302 (1992).
[3] E. Dvorin, N. L. Gorder, D. M. Benson, and A. M. Gotto, Jr., *J. Biol. Chem.* **261,** 15714 (1986).
[4] A. Steinmetz and G. Utermann, *J. Biol. Chem.* **260,** 2258 (1985).
[5] H. Tenkanen, M. Lukka, M. Jauhiainen, J. Metso, M. Bauman, L. Peltonen, and C. Ehnholm, *Arterioscler. Thromb.* **11,** 851 (1991).
[6] P. I. Barter, O. V. Rajaram, L. B. F. Chang, K. A. Rye, P. Gambert, L. Lagrost, C. Ehnholm, and N. H. Fidge, *Biochem. J.* **254,** 179 (1988).
[7] J. G. Delamatre and P. S. Roheim, *Biochim. Biophys. Acta* **751,** 210 (1983).
[8] C. L. Bisgaier, O. P. Sachdev, L. Megna, and R. M. Glickman, *J. Lipid Res.* **26,** 11 (1985).
[9] T. F. Appelbaum, N. O. Davidson, and R. M. Glickman, *Am. J. Physiol.* **252,** G662 (1987).

tine and liver, the additional contribution of apoA-IV to this process has been interpreted cautiously. However, evidence for the existence of a function in triglyceride metabolism is strengthening; intestinal apoA-IV mRNA is expressed maximally in the newborn rat, coinciding with the onset of suckling and triglyceride synthesis in the gut[10]; hepatic apoB mRNA editing increases in the second postnatal week. Inui *et al.*[11] suggest that there is evidence for a strong association between apoA-IV gene expression and triglyceride synthesis and secretion, and that apoA-IV may assume a functional role in triglyceride export from cells. Further studies to explore coordinated regulation of hepatic apoA-IV gene expression and apoB mRNA editing under varying conditions of altered hepatic triglyceride content will be needed to clarify this role. One unique function of apoA-IV, not shared by apoA-I, is its anorectic effect. Tso *et al.*[12] propose that apoA-IV is a circulating signal for satiation and is released in response to fat feeding.

Thus, it is conceivable that the frustration associated with attempts to assign a precise physiological role for apoA-IV is due to the fact that this apoprotein is multifunctional and serves many biological masters/mistresses. It is quite clear, however, that further explorations into apoA-IV function will depend, in part, on measurement of concentrations of this apoprotein at various biological levels, from that present in plasma to that secreted by cells, or the amount synthesized following modulation of mRNA levels by various stimuli.

Immunoassay provides a convenient method for this purpose, and several groups have addressed this subject. Production of antibodies to apoA-IV is quite easy; the antigen is not difficult to purify and can be obtained in sufficient quantities from the plasma of humans or species used as animal models for investigating cardiovascular disease. Several isoforms of apoA-IV exist,[13,14] but their distribution appears similar whether the source is lymph chylomicrons, plasma HDL, or the lipoprotein-free fraction of human plasma; it is unlikely that any antibodies, except the most residue-specific monoclonal antibody, will fail to recognize most forms of apoA-IV. The following section gives a brief review of methods used to isolate the antigen (apoA-IV).

[10] S. C. Frost, W. A. Clark, and M. A. Webb, *J. Lipid Res.* **24,** 899 (1983).

[11] Y. Inui, A. M. L. Hausman, N. Nanthakamar, S. J. Henning, and N. O. Davidson, *J. Lipid Res.* **33,** 1843 (1992).

[12] P. Tso, J. A. Cardelli, and K. Fujimoto, *FASEB J.* **6,** A1205 (1992).

[13] T. Ohta, N. H. Fidge, and P. J. Nestel, *J. Biol. Chem.* **259,** 14888 (1984).

[14] H. Tenkanen and C. Ehnholm, *Curr. Opin. Lipidol.* **4,** 95 (1993).

Purification of A-IV Apoprotein from Chylomicrons

Lymph duct chylomicrons are the best source of apoA-IV because the apoprotein is rapidly removed from chylomicrons once they enter the plasma compartment. Although such material is reasonably easy to obtain from animals following lymph duct cannulation, human material is unfortunately not so easily acquired. Occasionally, lymph can be obtained from a patient undergoing chylous pleural effusion which, if the fluid obtained is milky, provides a good source of apoA-IV-rich chylomicrons.

The chylomicrons are floated by ultracentrifugation at a density of 1.006 g/ml, using conditions similar to those used for very-low-density lipoprotein (VLDL) fractionation; all lymph triglyceride-rich lipoproteins (all of which contain apoA-IV) are thus harvested (increasing the yield of this apolipoprotein) and washed several times through buffered saline to remove albumin or other contaminating proteins. A typical apolipoprotein profile, obtained after sodium dodecyl sulfate–polyacrylamide gel electrophoresis (SDS–PAGE) of chylomicron apoproteins, and exhibiting the presence of apoA-IV, is shown in Fig. 1. Several chromatographic methods can then be used to isolate apoA-IV; most of these techniques do not result in complete purification but are useful steps for enrichment of apoA-IV. After delipidation,[13] the apolipoprotein mixture is applied to gel-filtration columns; the column medium can be Sephacryl S-300 or Sepharose CL-6B or equivalents, and it should be equilibrated with a dissociation agent such as 6 *M* urea or 4 *M* guanidine hydrochloride in 50 m*M* Tris-HCl, pH 8.0. None of these size-exclusion methods completely resolve apoproteins in

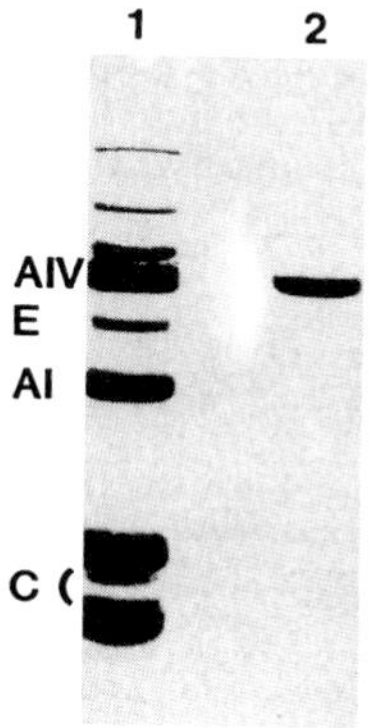

FIG. 1. Separation of apolipoproteins on SDS–polyacrylamide (15%) gels of (lane 1) 65 μg chylomicron apolipoproteins and (lane 2) 5 μg apoA-IV purified as described in the text. Gels were stained with Coomassie blue.

the 27- to 46-kDa range. Other methods can be used such as ion-exchange chromatography; we have found that QAE (quaternary aminoethyl)-Sephadex A-50 (Pharmacia, Piscataway, NJ) is useful for this purpose[13] (Fig. 2). Unless the researcher is prepared to carry out repetitive chromatographic

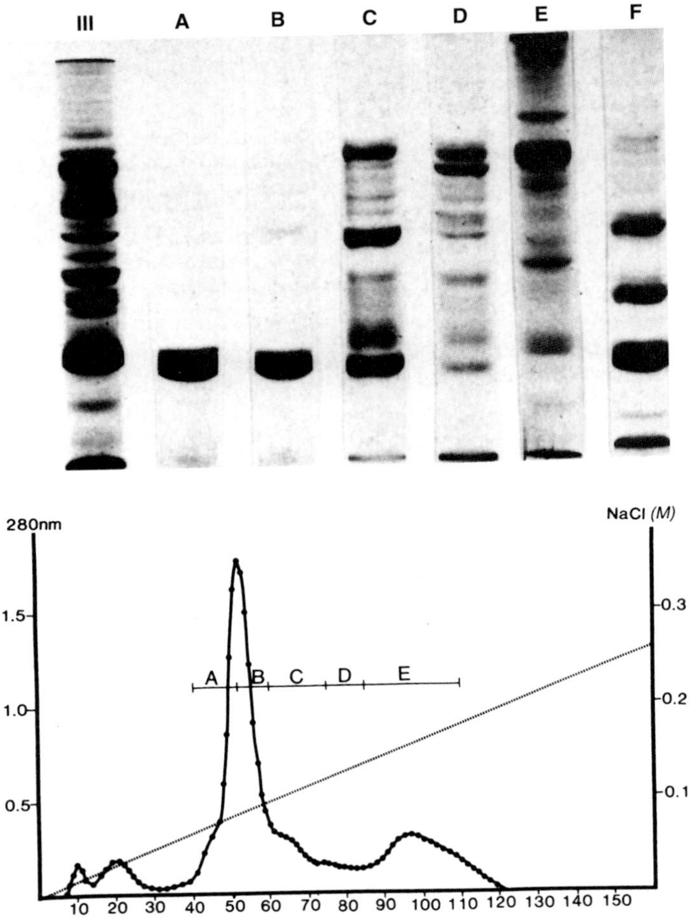

FIG. 2. Column chromatography on QAE-Sephadex A-50 of delipidated Neutralipid-bound proteins (*lower*). Forty milligrams protein was applied to the column. Analysis by SDS–polyacrylamide gel electrophoresis of each fraction (lanes A, B, C, D, and E) (*upper*), delipidated Neutralipid-bound proteins (lane III), and rat HDL (lane F) was performed. Rat HDL was used to show the positions of apolipoprotein A-IV, apolipoprotein E, and apolipoprotein A-I (from top to bottom).

fractionation, final purification requires preparative gel electrophoresis, followed by elution of the protein from gel slices.[13]

Smaller quantities of apoA-IV can, however, be isolated by high-performance liquid chromatography (HPLC). We have reported HPLC methods for the rapid separation of medium molecular weight apolipoproteins from human or rat serum following removal of low molecular weight C and A-II apoproteins by size-exclusion chromatography.[15] Human apoA-IV can be isolated in pure form in one step by this method; a typical separation is shown in Fig. 3.

Lipoprotein-Deficient Plasma

Another valuable source of apoA-IV is lipoprotein-deficient plasma (LDP).[13,16] Existing in an apparent "free form," the apoprotein can be isolated by exploiting its lipid binding properties. After removing all lipoproteins of density below 1.25 g/ml by ultracentrifugation, the infranatant is dialyzed against 0.15 *M* NaCl, 50 m*M* Tris-HCl, pH 7.4, and mixed with lipids. A liposome or micelle form is used which can be isolated (floated) by ultracentrifugation. In our experiments LDP is incubated with 0.3 volumes artificial fat emulsion (Neutralipid, Pharmacia) for 2 hr at 37°. The fat is recovered by centrifugation at 100,000 *g* in the Beckman ultracentrifuge (45Ti rotor) for 30 min at 4°, dispersed in buffered saline to the original volume, and centifuged twice further to remove unbound plasma protein. The lipid fraction is then delipidated and subjected to chromatographic procedures as described above.

Preparation of Antiserum

Details of antisera preparation are described in other issues of this series.[17] In general terms, antisera with relative high titers are prepared by injecting rabbits, sheep, or other animals used for antibody production with 50–100 μg apoA-IV emulsified with Freund's adjuvant following traditional immunization procedures. Our experience has shown[13] that three to four immunizations are necessary to produce high titers and a sufficient quantity of apoA-IV immunoglobulin G (IgG) that can be harvested for use in general immunochemical procedures such as immunoaffinity column chromatography, immunoprecipitation, and related techniques. It is necessary to establish the monospecificity of the antibody before its use in a quantitative immunoassay. Immunoblotting (Western blotting) is the most sensitive

[15] T. Tetaz, E. Kecorius, B. Grego, and N. Fidge, *J. Chromatogr.* **511,** 147 (1990).
[16] R. B. Weinberg and A. M. Scanu, *J. Lipid Res.* **24,** 52 (1983).
[17] P. H. Maurer and H. J. Callahan, this series, Vol. 70, p. 49.

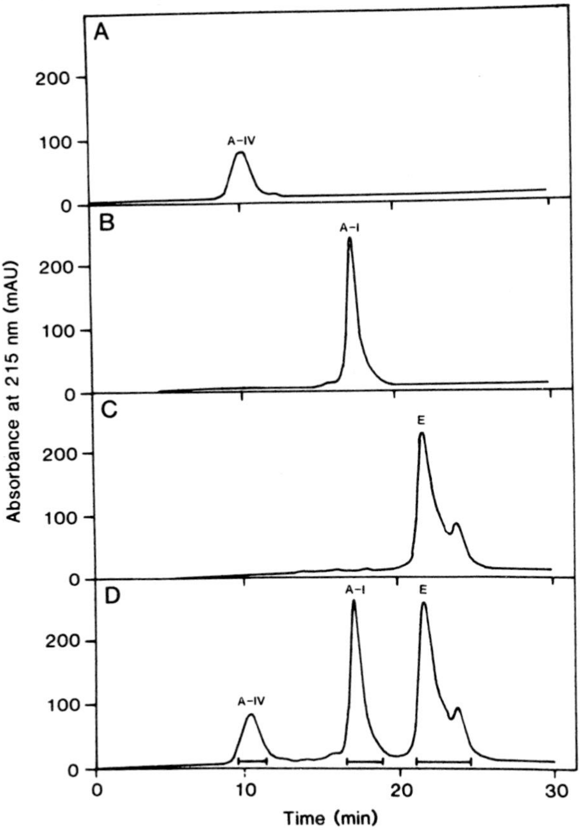

FIG. 3. Reversed-phase HPLC chromatograms of apolipoprotein separations on a TSK Phenyl-5PW column. Gradient conditions were as described in the text. Apolipoproteins A-IV (A, 30 μg), A-I (B, 50 μg), and E (C, 50 μg) were applied to the column separately (A–C) and as a mixture (D). Fractions were collected as indicated by bars in (D).

technique for this purpose and has been described in detail elsewhere.[18] For checking apoA-IV antiserum, several antigen mixtures should be tested, including the original apoA-IV source material, plasma, and HDL which contain other antigens that may be potential immunogenic contaminants.

Immunoassay

Most of the standard methods used for quantitative immunoassay have been adapted for apoA-IV measurement (Table I). The essential principles

[18] V. C. W. Tsang, J. M. Peralta, and A. R. Simons, this series, Vol. 92, p. 377.

TABLE I
IMMUNOASSAY OF A-IV APOLIPOPROTEINS

Immunoassay procedure	Plasma level[a] (mg/dl)	Assay conditions	Ref.
Electroimmunoassay	14.2 ± 3.7	Samples treated with 2% Triton X-100	b
	13.1 ± 1.8	Samples treated with Triton X-100	16
	15.7 ± 0.9	Plasma delipidated with ethanol–ether (3:2, v/v)	c
Radioimmunoassay	37.4 ± 4.0	Plasma incubated in 1% SDS in phosphate-buffered saline containing 1% bovine serum albumin	8
	17.2 ± 3.2	Final concentration of 0.35% Tween 20 in samples	d
ELISA (competitive)	16.4 ± 5.4	No treatment necessary	e
	13.0 ± 2.6	No treatment necessary	f
	143 ± 52	No treatment necessary	g

[a] Data represent normal levels ± S.D.
[b] G. Utermann and U. Beisiegel, *Eur. J. Biochem.* **99,** 333 (1979).
[c] P. H. R. Green, R. M. Glickman, J. W. Riley, and E. Quinet, *J. Clin. Invest.* **65,** 911 (1980).
[d] G. Ghiselli, S. Krishnan, Y. Beigel, and A. M. Gotto, Jr., *J. Lipid Res.* **27,** 813 (1986).
[e] K. Kondo, C. Allan, and N. Fidge, *J. Lipid Res.* **30,** 939 (1989).
[f] L. Lagrost, P. Sambert, S. Meumier, P. Morgado, J. Desgnes, P. d'Athis, and C. Lallemont, *J. Lipid Res.* **30,** 701 (1989).
[g] M. Rosseneu, G. Michiels, W. De Keersgieter, J. Bury, J. P. De Slypere, H. Dieplinger, and G. Utermann, *Clin. Chem.* **34,** 739 (1988).

of these procedures are described in immunology texts,[19] but as is usual for different antigens, some modifications are required to optimize assays. For these reasons, details of procedures for estimating apoA-IV are described, but not rigid protocols that would, at any rate, need to be altered in each laboratory according to the question asked and the characteristics of the antibody used.

Electroimmunoassay

Also known as "rocket" immunoelectrophoresis, the electroimmunoassay method of Laurell[20] was one of the first used to provide values for apoA-IV levels in plasma. In this method, antiserum is incorporated into molten 1% (w/v) agarose of medium electroendosmosis grade in 20 m*M* barbital buffer, pH 8.6 (Bio-Rad, Richmond, CA, barbital buffer II), which has been cooled to 45° and the mixture poured onto glass plates, the size

[19] H. V. Vunakis and J. J. Langone (eds.), this series, Vol. 70.
[20] C. B. Laurell, *Anal. Biochem.* **15,** 45 (1966).

of which depends on the number of samples to be assayed. After the agarose sets, wells are punched and 10 μl sample (generally 1 : 10 dilution of plasma) loaded into each well. Pure apoA-IV, in the range 50–1000 ng, is used as standard and the electrophoresis performed at 110 V for 3 hr. Varying concentrations of apoA-IV antiserum in 1% (w/v) agarose have been used, ranging from 1.7 to 5% (w/v) (Table I).

Some laboratories have observed that delipidation procedures using either detergents[21] or organic solvents[22] are necessary to unmask antigenic sites hidden in lipid environments of lipoproteins, thus providing complete values for plasma levels of apoA-IV. These variations are listed in Table I; our experience has shown that a requirement for pretreatment of plasma or lipoproteins is dependent on the comprehensive nature of the antisera because some batches of polyclonal antisera apparently recognize epitopes of apoA-IV exposed in all lipoproteins.

Radioimmunoassay

Radioimmunoassays have been used to quantitate apoA-IV in plasma and in column fractions following separation of the lipoprotein components of plasma (Table I). One protocol based on the method of Ghiselli *et al.*[23] is described as follows.

Standards and test samples are diluted in borate buffer [0.125 *M* sodium borate, 0.1% (w/v) bovine serum albumin (BSA), 0.35% (v/v) Tween 20, pH 8.0] and preincubated overnight. Diluted standards and unknowns are brought to 1 ml with the same buffer, and 0.1 ml ^{125}I-labeled apoA-IV [20,000 counts/min (cpm); 1–2 ng] and 0.1 ml anti-apoA-IV antibody (appropriately diluted following preliminary testing to determine the concentration that precipitates 50% of labeled antigen) are added. Labeled apoA-IV is prepared by radioiodination using standard methods. To obtain a preparation of high specific radioactivity (*SR*), the chloramine-T method of Hunter and Greenwood[24] using iodobeads (Pierce) is appropriate. Preparations with a *SR* of 10,000–20,000 cpm per nanogram protein can be achieved. After removal of free iodine by gel filtration (Sephadex G-50), it is advisable to dialyze the sample and then check for the presence of residual unbound iodine by precipitating the labeled apoA-IV (with added carrier) with 10% trichloroacetic acid (TCA) and counting both the supernatant and precipitate. To minimize self-irradiation, which is a penalty of obtaining high *SR*, it is recommended that a suitable carrier be added, such

[21] G. Utermann and U. Beisegel, *Eur. J. Biochem.* **99,** 333 (1979).
[22] P. H. R. Green, R. M. Glickman, J. W. Riley, and E. Quinet, *J. Clin. Invest.* **65,** 911 (1980).
[23] G. Ghiselli, S. Krishnan, Y. Beigel, and A. M. Gotto, Jr., *J. Lipid Res.* **27,** 813 (1986).
[24] W. M. Hunter and F. C. Greenwood, *Nature* (*London*) **194,** 495 (1962).

as bovine serum albumin. However, such preparations must be free of apoA-IV and the absence of this peptide established by Western blotting.

Labeled and unlabeled apoA-IV are incubated with the antibody for 2 days at 4°. Then 0.1 ml normal rabbit serum (diluted 1 : 100) and 0.1 ml goat or sheep anti-rabbit IgG (diluted 1 : 15) are added and the mixture incubated for another 2 days at 4°. The immunoprecipitate is pelleted by centrifugation, the supernatant removed, and the radioactivity in the pellet counted in a gamma counter. The intraassay coefficient of variation should be determined for normolipidemic and a hypertriglyceridemic control serum, and the interassay coefficient of variation is determined as the average of the mean value of the control sera concentrations for all of the assays performed. This assay should enable measurement of apoA-IV in the range 5–50 ng. It is important to compare values of apoA-IV present in both the lipoprotein and lipoprotein-free form to ensure that full antigenicity has been exposed by the detergent. With some antisera preparations, Tween 20 may not be required.

Enzyme-Linked Immunosorbent Assay

Competitive ELISA

At least two variations of competitive enzyme-linked immunosorbent assay (ELISA) for apoA-IV (human) have been reported.[25,26] In this section, a competitive ELISA for apoA-IV in human plasma is described that is easy, rapid, and inexpensive to perform.[26] One advantage is the convenience of preparing minimal dilutions of plasma (1 : 8 to 1 : 32), but the assay remains sensitive enough to quantitate apoA-IV levels in lipoproteins following fractionation by column chromatographic procedures of only 0.3–0.5 ml plasma. The method, using precoated plates, provides answers within a few hours compared with 3–4 days required for the radioimmunoassay. No radiolabel is used, the assay is sensitive, and it requires smaller quantities of antibody than electroimmunoassays or immunoturbidimetric methods.

As stated above, most conditions for immunoassays need to be reestablished in each laboratory, especially when polyclonal antisera are used. The antibody should not discriminate in recognition of apoA-IV whether present in whole plasma or in the lipoprotein-rich or -poor fractions of plasma; there should be no differences in displacement curves between

[25] L. Lagrost, P. Sambert, S. Meumier, P. Morgado, J. Desgnes, P. d'Athis, and C. Lallemont, *J. Lipid Res.* **30,** 701 (1989).

[26] K. Kondo, C. Allan, and N. Fidge, *J. Lipid Res.* **30,** 939 (1989).

these two fractions (which comprise the major pools of circulating apoA-IV). Variations in titer of the antisera produced will necessitate changes in antiserum dilutions and probably also in the amount of antigen required to coat the ELISA plates. Protocols for optimization of these conditions are described in a previous volume of this series.[27] In the author's laboratory we found that treatments designed to expose potentially masked antigenic sites of apoA-IV such as delipidation with ethanol–ether, dissociating agents, detergents, or heating did not significantly expose further antigenicity of plasma apoA-IV. This is a desirable situation since no pretreatments of plasma are required. Details of the methods are as follows.

Coating of Plates. Apolipoprotein A-IV is dissolved in coating buffer (35 m*M* $NaHCO_3$, 14 m*M* Na_2CO_3, pH 9.6) at a concentration of 50 μg/dl and 100 μl added to wells of 96-well microtiter plates (Immulon 2, Dynatech, Chantilly, VA), which are incubated overnight at 37°. The coating capacity of apoA-IV was observed to vary somewhat for different plates and plastic surfaces and should be assessed in each laboratory. After removing the unbound apoA-IV, wells are blocked with buffer [10 m*M* phosphate-buffered saline (PBS) containing 0.5% (w/v) BSA and 0.05% (v/v) Tween 20 (PBS–BSA–Tween 20)] for 1 hr at room temperature, then washed three times with PBS–Tween 20.

Plasma samples are diluted 1:16 in PBS and 50 μl added to the plates. Standards of apoA-IV (for purification, see above) ranging from 1 to 20 μg/ml in PBS are included in each plate. The concentration of apoA-IV present in stock solutions was determined accurately from the amino acyl mass following acid hydrolysis and amino acid composition analysis on a Beckman (Fullerton, CA) 6300 analyzer. Fifty microliters anti-apoA-IV antiserum diluted 1:50,000 in PBS–BSA–Tween 20 is then added and the samples incubated for 2 hr at 37°. After washing three times with PBS–Tween 20, 100 μl horseradish peroxidase-conjugated goat anti-rabbit immunoglobulin (Bio-Rad), diluted (1:2000) in PBS–BSA–Tween 20, is added to each well. Plates are incubated at room temperature for 1 hr and washed as above to remove excess enzyme-labeled antibody. Bound peroxidase is quantitated using H_2O_2 as substrate and *o*-phenylenediamine (OPD) as hydrogen donor. Ten microliters of 30% (v/v) H_2O_2 and 1 ml of 1% OPD in methanol (freshly prepared before each assay) is mixed with 49 ml distilled water, and 150 μl solution is pipetted into each well. Color development continues for 30 min at room temperature in the dark and is stopped by the addition of 50 μl of 8 *M* H_2SO_4 to each well. The plates are read in an ELISA plate reader at 492 nm. Results of standard assays are generally plotted as absorbance of B/B_0 against log apoA-IV concentration, and

[27] E. Engvall, this series, Vol. 70, p. 419.

values of test samples are read either manually or by computer derivations of the linear portion of the standard curve (Fig. 4).

Plasma apoA-IV concentrations are determined on fresh samples, but we have observed identical values in the same samples after storage at −20° for 2 months. To demonstrate that apoA-IV in the lipoprotein-free form shows the same immunoreactivity as apoA-IV in plasma lipoproteins, particularly HDL which transports most of the lipoprotein-bound apoA-IV, 0.5 ml plasma was subjected to fractionation by gel-permeation chromatography (as described below). The fractions can be concentrated by ultrafiltration if apoA-IV levels are low, so that values fit within the linear range of the apoA-IV standard curve.

Column Fractionation of Plasma Lipoproteins. To determine apoA-IV distribution in plasma, samples are fractionated on Superose 12 to separate

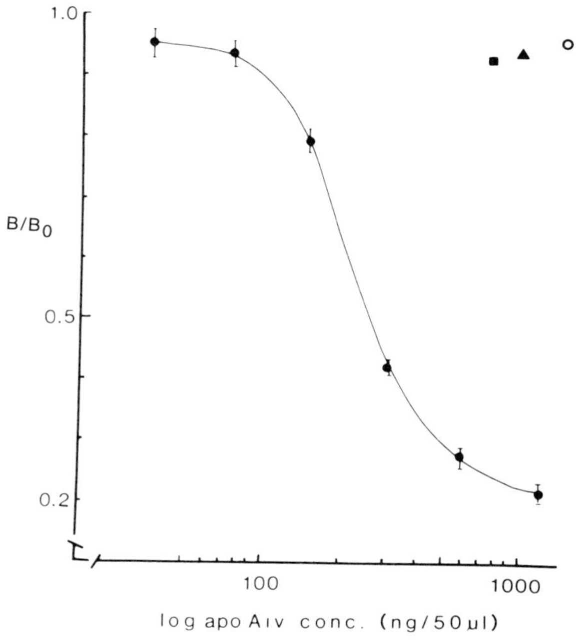

FIG. 4. Standard curve obtained with human apoA-IV in the competitive ELISA. Results are mean ± S.D. of five different displacement assays performed on different microwell plates; B, absorbance, at 492 nm, in the presence of apoA-IV standard; and B_0, absorbance in the absence of standard. Lack of competition of other apolipoproteins is demonstrated by the high B/B_0 values after addition of high concentrations of apoA-I (▲), apoE (■), and apoA-II (○).

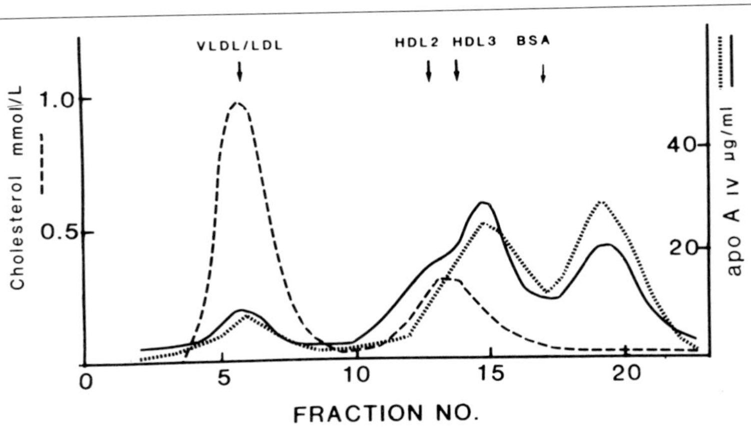

FIG. 5. Determination of apoA-IV in two samples of fresh unincubated human plasma following size-exclusion chromatography on Superose 12 (Pharmacia/FPLC). Plasma samples (0.5 ml) were loaded, and after elution the fractions were assayed for apoA-IV (dotted and solid lines represent subjects 1 and 2, respectively) or for cholesterol (dashed line, subject 1 only).

individual lipoproteins from plasma proteins (Fig. 5). Plasma (0.5 ml) is loaded onto a column (0.5 × 20 cm) attached to the Pharmacia FPLC (fast protein liquid chromatography) system. The fractions are eluted with 0.15 *M* NaCl in 50 m*M* Tris-HCl, pH 7.4, and monitored at 280 nm. Columns are calibrated with VLDL, low-density lipoprotein (LDL), HDL_2, HDL_3, and albumin to determine elution volumes of lipoprotein fractions.[26]

Sandwich ELISA

Another type of ELISA that has been developed for measuring apoA-IV levels in plasma[28] is the sandwich ELISA.[27] In this method, antibody is first purified by passing antiserum (anti-apoA-IV) through an apoA-IV Sepharose (affinity) column. It is then coated onto the plate, using 100 μl of 15 ng/ml purified antibody in coating buffer (see above). Plasma samples are diluted 8000-fold and incubated with a 5000-fold dilution of conjugated anti-apoA-IV antibody. The conjugate is prepared by covalent coupling of horseradish peroxidase (EC 1.11.1.7) to immunoaffinity-purified anti-apoA-IV IgG obtained as described above. The conjugation is performed by the periodate coupling procedure,[29] and the conjugate is stable for 6 months when stored in 50% glycerol at −70°.

[28] M. Rosseneu, G. Michiels, W. De Keersgieter, J. Bury, J. P. De Slypere, H. Dieplinger, and G. Utermann, *Clin. Chem.* **34,** 739 (1988).

[29] J. Bury and M. Rosseneu, *Clin. Chem.* **31,** 247 (1985).

This method is very sensitive and can detect as low as 0.2 ng apoA-IV. The working range of the method described[29] is 0.5–7 ng. However, to achieve levels within that range, plasma needs to be diluted 8000- to 10,000-fold.

Summary

Several methods are available for the immunoassay of apoA-IV levels in plasma, or lipoproteins. The method of choice depends on the question being asked. If sensitivity is not a major determinant, simple immunoelectrophoresis is probably sufficient. To determine apoA-IV levels in plasma or lipoprotein fractions, either radioimmunoassay or a competitive ELISA is indicated. The competitive ELISA described above, however, offers sensitivity as well as rapidity and ease of performance. When very low levels of apoA-IV are present (such as those produced by cultured cells), the higher sensitivity of the sandwich ELISA may be required.

[20] Quantitation of Plasma Apolipoprotein J

By SARAH H. JENKINS, WILLIAM D. STUART, L. A. BOTTOMS, and JUDITH A. K. HARMONY

Introduction

Apolipoprotein J (apoJ),* a fascinating secretory protein present in most, if not all, physiological fluids, including plasma, semen, cerebrospinal fluid, aqueous humor, breast milk, synovial fluid, and urine, has been linked with chronic degenerative diseases in humans.[1,2] Notable among these are atherosclerosis, glomerulonephritis and other renal pathologies, Alzheimer's disease and other neurological pathologies, and immune complex-related skin disorders.[3] The apoJ protein accumulates in the pathological lesions, often in association with components of the terminal complement complex. In human tissues and in tissues obtained from animal models of

* Apolipoprotein J is also known as clusterin, SGP-2, SP-40,40, complement lysis inhibitor, TRPM-2, GP-III, Gp80, and T64.

[1] T. C. Jordan-Starck, D. P. Witte, B. J. Aronow, and J. A. K. Harmony, *Curr. Opin. Lipidol.* **3,** 75 (1992).

[2] D. E. Jenne and J. Tschopp, *Trends Biochem. Sci.* **17,** 154 (1992).

[3] J. A. K. Harmony (ed.), "Clusterin: Role in Vertebrate Development, Function and Adaptation." R. G. Landes, Austin, Texas, 1995.

these diseases, the apoJ gene is dramatically upregulated in the affected organ, and message is strikingly abundant in cell types which border the lesions. The apoJ gene is also induced in hormone-responsive tissues undergoing involution, for example, in ductal epithelium of the regressing mammary gland following weaning and in luminal epithelial cells of the involuting prostrate gland following castration. The inducibility of the apoJ gene in response to or concomitant with tissue damage, the predominant constitutive pattern of gene expression at epithelial barriers between tissues and harsh biologically active fluids,[4] and the biochemical properties of the apoJ protein taken together suggest that apoJ functions to "police" and protect tissue boundaries from injury.[1] As its elevation may therefore be indicative of tissue injury, it will be important to determine the relationship between apoJ concentration in specific fluids and evidence of morphological and/or functional tissue pathologies.

In blood there are two sources of apoJ: high-density lipoproteins (HDL)[5] and platelets.[6,7] Of the circulating apoJ, 30–40% appears to be sequestered within platelets.[7] High-level apoJ gene expression occurs in megakaryocytes, the precursors of blood platelets,[7] and platelet apoJ is stored in alpha granules and released by platelet activation.[6,7] The positive correlation between plasma apoJ levels[8] and total cholesterol, triglyceride, low-density lipoprotein (LDL) cholesterol, apoB, and apoE may reflect the extent of platelet activation at sites of vascular injury. The HDL particles containing apoJ comprise unique and previously undescribed subclasses. The apoJ-HDL particles have $\alpha 2$ electrophoretic mobility,[9] in contrast to bulk plasma HDL with $\alpha 1$ mobility and to a minor subclass of HDL with pre-β mobility. Within these $\alpha 2$ lipoproteins, apoJ-HDL are heterogenous in size and composition. Nondenaturing gradient gel electrophoresis shows that apoJ-HDL consist of roughly equal amounts of larger particles with apparent molecular masses over 240 kDa and smaller particles of 70–90 kDa.[9] These results are consistent with the 3- to 16-nm diameter range of apoJ-HDL particles, estimated by negative staining electron microscopy, and with the biphasic distribution of apoJ into the HDL_2 and HDL_3/very-high-density

[4] B. J. Aronow, S. D. Lund, T. L. Brown, J. A. K. Harmony, and D. P. Witte, *Proc. Natl. Acad. Sci. U.S.A.* **90,** 725 (1993).

[5] H. V. de Silva, W. D. Stuart, C. R. Duvic, J. R. Wetterau, M. J. Ray, D. G. Ferguson, H. W. Albers, W. R. Smith, and J. A. K. Harmony, *J. Biol. Chem.* **265,** 13240 (1990).

[6] J. Tschopp, D. E. Jenne, S. Hertig, K. T. Preissner, H. Morgenstern, A.-P. Sappino, and L. E. French, *Blood* **82,** 118 (1993).

[7] D. P. Witte, B. J. Aronow, M. L. Stauderman, W. D. Stuart, M. A. Clay, R. A. Gruppo, S. H. Jenkins, and J. A. K. Harmony, *Am. J. Pathol.* **143,** 763 (1993).

[8] S. H. Jenkins, W. D. Stuart, J. A. K. Harmony, and L. A. Kaplan, *Clin. Chem.* **36,** 963 (1990).

[9] W. D. Stuart, B. Krol, S. H. Jenkins, and J. A. K. Harmony, *Biochemistry* **31,** 8552 (1992).

lipoprotein (VHDL) density ranges by rapid vertical gradient density centrifugation.[5] The apoJ-HDL as a whole are poor in lipid compared to bulk HDL, although individual particles may be lipid-rich (e.g., those within the HDL_2 density class). Apolipoprotein A-I is associated with larger (350–400 kDa) apoJ-HDL,[9] whereas the smaller particles appear to lack this apolipoprotein. Cholesteryl ester transfer protein (CETP) is also associated with apoJ-HDL.[5] Because apoJ has little or no effect on CETP activity, the physiological relevance of this interaction is uncertain. Paraoxonase, the only known function of which is to metabolize the organophosphate insecticide paraoxon, has also been identified in apoJ-HDL,[10,11] particularly in association with the smaller particles.[11] Paraoxonase[11] and apoA-I[12] have been reported to bind to apoJ directly. In addition, apoJ can interact with the complement proteins C7, C8, and C9. *In vitro* this interaction inhibits the assembly of C5b-9 complexes at the cell membrane, thus inhibiting complement-mediated cytolysis.[2]

Human plasma apoJ is a 70- to 80-kDa disulfide-linked heterodimer,[13] comprising two nonidentical subunits translated from a single transcript (pro-apoJ) and cleaved prior to secretion.[14] The subunits are 34–36 and (apoJα) and 36–39 kDa (apoJβ), both of which exhibit charge heterogeneity and have pI values of 4.9–5.4.[13] Each subunit is about 30% by weight carbohydrate. Interestingly, there appears to be tissue-specific glycosylation, which, for example, allows platelet apoJ to be distinguished from plasma apoJ.[5] Such differences suggest the possibility of developing tissue-specific assays of apoJ to determine the contribution from a particular organ to the total fluid concentration. Genetic polymorphism exists at the apoJ structural locus.[15] By isoelectric focusing, two common apoJ alleles, APOJ*1 and APOJ*2, have been identified. This two-allele polymorphism appears to be restricted to populations of African ancestry, where allele frequencies of APOJ*1 and APOJ*2 are about 0.75 and 0.25, respectively. Among American blacks a single rare allele, APOJ*3, has also been reported.[15] No impact of the common polymorphisms was encountered with

[10] M.-C. Blatter, R. W. James, S. Messmer, F. Barja, and D. Pometta, *Eur. J. Biochem.* **211,** 871 (1993).

[11] G. J. Kelso, W. D. Stuart, R. J. Richter, C. E. Furlong, T. C. Jordan-Starck, and J. A. K. Harmony, *Biochemistry* **33,** 832 (1994).

[12] D. E. Jenne, B. Lowin, M. C. Peitsch, A. Bottcher, G. Schmitz, and J. Tschopp, *J. Biol. Chem.* **266,** 11030 (1991).

[13] H. V. de Silva, W. D. Stuart, Y. B. Park, S. J. T. Mao, C. M. Gil, J. R. Wetterau, S. J. Busch, and J. A. K. Harmony, *J. Biol. Chem.* **265,** 14292 (1990).

[14] H. V. de Silva, J. A. K. Harmony, W. D. Stuart, C. M. Gil, and J. Robbins, *Biochemistry* **29,** 5380 (1990).

[15] M. I. Kamboh, J. A. K. Harmony, M. Nwankwo, and R. E. Ferrell, *Am. J. Hum. Genet.* **49,** 1167 (1991).

total cholesterol, LDL-cholesterol, HDL-cholesterol (either HDL_2 or HDL_3), very-low-density lipoprotein (VLDL)-cholesterol, or triglycerides. Among Nigerian males, APOJ*2 did show a marginal (18% level) effect on HDL_2-cholesterol.

Although it has not been unequivocally established that apoJ is an apolipoprotein, the available evidence favors this classification. On the basis of gene structure and the sequences of potential amphipathic helices, apoJ is clearly not a member of the apolipoprotein gene family[16] including the apoA's, apoC's, and apoE. However, where evaluated, for example, in plasma[5] and cerebrospinal fluid (B. Murphy, unpublished results, 1994), apoJ is associated with lipid. Moreover, the protein is secreted by cultured hepatocytes as a lipoprotein, in fact considerably enriched in triglyceride compared to the plasma form.[17] The most compelling evidence is the finding by Argraves and co-workers[18] that apoJ is recognized as a ligand by gp330, a member of the LDL receptor gene family. Internalization and degradation of apoJ by cultured F9 cells[18] is mediated by gp330.

Purification of Apolipoprotein J Standard

We purify standard apoJ from 3- to 5-day-old plasma obtained from the Hoxworth Blood Center, Cincinnati, OH; all plasmas have tested negative for HBsAg (hepatitis B antigens), HIV (human immunodeficiency virus), and human cytomegalovirus (HCV). Plasma, diluted 1 : 3 with plasma density buffer (20 m*M* Tris-HCl, 150 m*M* NaCl, 0.02% w/v NaN_3, pH 7.4) containing 1 m*M* EDTA and protease inhibitors [1 m*M* benzamindine, 1 m*M* phenylmethylsulfonyl fluoride (PMSF)], is stirred overnight at 4° and filtered (Whatman, Clifton, NJ, No. 1 paper); 400 ml is loaded onto a 40-ml MAb11–Affi-Gel 10 immunoaffinity column[5] and eluted with 300 ml of 1 *M* acetic acid, pH 2.8. The eluate, termed partially purified apoJ, is dialyzed against 4 liters of 100 m*M* NH_4HCO_3 containing 1 m*M* EDTA and 0.02% (w/v) NaN_3, then dialyzed against two changes of liters of 10 m*M* NH_4HCO_3 with EDTA and NaN_3, concentrated with an Amicon (Danvers, MA) YM10 membrane at 4°, and filtered with a 0.45-μm syringe filter. The apoJ is purified from the MAb11 eluate by reversed-phase high-performance liquid chromatography (HPLC), using a C_4 or C_{18} column and eluting with a gradient of 0.1% (v/v) trifluoroacetic acid (TFA) in H_2O (solvent A) and 0.1% TFA in CH_3CN (solvent B). The peak corresponding

[16] T. C. Jordan-Starck, S. D. Lund, B. J. Aronow, C. A. Ley, D. P. Witte, S. F. Sells, and J. A. K. Harmony, *J. Lipid Res.* **35,** 194 (1994).

[17] B. F. Burkey, W. D. Stuart, and J. A. K. Harmony, *J. Lipid Res.* **33,** 1517 (1992).

[18] M. Z. Kounnas, E. B. Loukinova, S. Stefansson, J. A. K. Harmony, B. H. Brewer, D. K. Strickland, and W. S. Argraves, *J. Biol. Chem.* **270,** 13070 (1995).

to apoJ, which elutes at approximately 40% CH_3CN, is neutralized with 1 *M* NH_4HCO_3, dialyzed against five changes of 4 liters water and lyophilized. The purity of the standard apoJ is determined by gel electrophoresis, and the concentration is determined by amino acid analysis.

Procedure

Apolipoprotein J Assay

Apolipoprotein J is quantitated by using a competitive enzyme-linked immunosorbent assay (ELISA). Partially purified apoJ, diluted in saline to a concentration of approximately 5 μg/ml, is used as the coating solution. The coating solution (125 μl) is incubated in covered polystyrene microtiter plates (Dynatech Immulon 2, Fisher Scientific, Springfield, NJ), overnight at room temperature. All wells except for those to be used as blanks are coated with apoJ in this manner. After the overnight incubation, the coating solution is removed, and 200 μl blocking buffer (pH 7.4) is added to all wells. The blocking buffer contains the following:

2.4 g/liter Tris(hydroxymethyl)aminomethane (THAM)
8.0 g/liter NaCl
0.2 g/liter KCl
0.5% (v/v) Tween 20
1% (w/v) Bovine seruim albumin (BSA)
0.2 g/liter NaN_3

After a 4-hr incubation at room temperature, the blocking solution is removed, and plates are sealed in plastic bags and frozen at −70° until needed. The plates are allowed to warm to room temperature before removal from the plastic bags.

Standards and samples are analyzed in triplicate. Standard apoJ solution is diluted with blocking buffer to final concentrations of 1, 0.5, 0.25, 0.125, 0.063, 0.031, and 0.0156 mg/dl. Plasma samples are diluted 1 : 80 in blocking buffer. Blocking buffer (50 μl) is added to the blank wells and zero standard wells. Diluted samples and standards (50 μl) are added to the remaining wells. Monoclonal antibody (MAb11)[5] is diluted 1 : 1000 in blocking buffer, and 50 μl of this solution is added to each well. The plates are incubated 2 hr at room temperature with gentle shaking. Solutions are aspirated and the plates are washed three times with blocking buffer before the addition of conjugate solution. Anti-mouse alkaline phosphatase immunoglobulin G (IgG; Sigma, St. Louis, MO) is diluted 1 : 500 in blocking buffer, and 100 μl is added to each well. The plates are incubated for 1 hr at room temperature with gentle shaking.

Substrate, *p*-nitrophenyl phosphate (Sigma), is dissolved in DEA buffer to a final concentration of 1 mg/dl. The DEA buffer contains 97 ml diethanolamine, 0.1 g $MgCl_2$, and 0.2 g NaN_3 per liter distilled, deionized water. The substrate solution (100 μl) is added to each well and incubated for 15 min or until the optical density of the zero standard is 1.5. Optical densities are determined at 410 nm by using a microtiter plate reader (Dynatech MR 7000). The standard curve is determined with a four-parameter logistics curve-fitting model.

Assay Performance

A typical standard curve for the apoJ ELISA is shown in Fig. 1. The curve is linear from 1.0 to 0.015 mg/dl. Intraassay precision ranges from 2.9 to 3.5%, and interassay precision ranges from 4.7 to 10%. Recovery of purified apoJ spiked into plasma is shown in Table I. The normal range (mean ± 2 S.D.) for females is 6.2–15.4 mg/dl; for males, it is 4.7–15.1 mg/dl.

Specimen

Samples treated with the anticoagulant EDTA exhibit marked increases in apoJ levels that become significant after 48 hr (40%). The increase can be as high as 5–10 times the baseline levels when samples are stored for 2 weeks at 4°. The rate of increase varies considerably between individuals but is reproducible for each individual. The rate of increase in immunoreac-

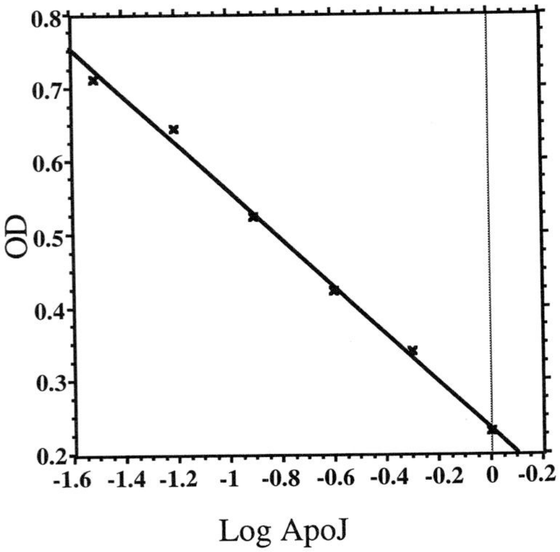

FIG. 1. Standard curve for apoJ ELISA.

TABLE I
RECOVERY OF APOLIPOPROTEIN J SPIKED INTO PLASMA SAMPLES

Calculated apoJ	Measured apoJ	Recovery (%)
6.5	6.13	94
15.3	15.0	98
26.8	27.3	102
32.6	36.5	112

tivity is slowed slightly if samples are stored at −20° or −70°. Example profiles of this "aging effect" are shown in Fig. 2 for samples from the same individual stored at 4° in different anticoagulants. The reason for the increase in immunoreactivity is not known, but it correlates with the aggregation of apoJ in plasma, as determined by gradient gel electrophoresis.[9] A similar increase is detected by polyclonal anti-apoJ antibodies. Because there is no change in the amount of plasma apoJ immunoreactivity over time as judged by electroimmunoblotting, we suspect that aggregation alters the affinity of apoJ for the antibody. Some contribution to the "aging effect" may also be attributable to apoJ in the residual platelets.

The sample of choice for analysis of apoJ is sodium citrate-treated plasma. Sample preparation includes a two-step centrifugation. Citrated blood samples should be centrifuged at 2500 rpm for 15 min in a refrigerated centrifuge. Plasma should be separated from cells and centrifuged a second time under identical conditions. Samples prepared in this manner show no significant change for up to 3 months when stored at −70°. Studies to date

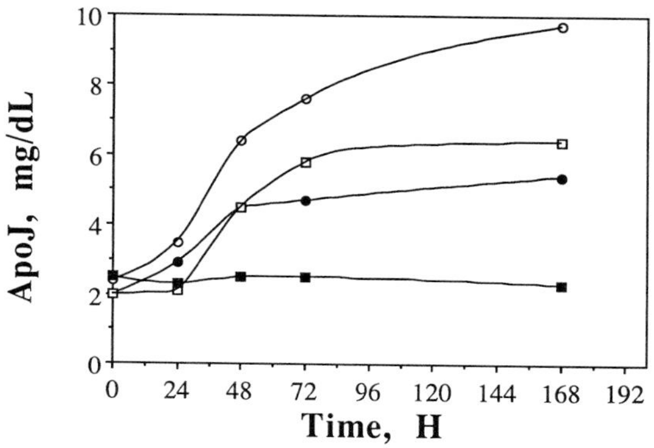

FIG. 2. Aging profile of samples collected with different anticoagulants and stored at 4°: EDTA, ○; serum, ●; heparin, □; citrate, ■.

have not extended beyond 3 months. In addition, citrated plasma samples (twice centrifuged) may be stored for at least 1 week at 4°.

Acknowledgment

Support for this project was provided by the Department of Pathology and Laboratory Medicine, College of Medicine, University of Cincinnati.

Section III

Lipases and Cholesteryl Ester Transfer Protein

[21] Human Lipoprotein Lipase: Production *in Vitro*, Purification, and Generation of Polyclonal Antibody

By RICA POTENZ, JING-YI LO, EVA ZSIGMOND, LOUIS C. SMITH, and LAWRENCE CHAN

Introduction

Lipoprotein lipase (LPL) is a hydrolytic enzyme produced by adipose tissue, heart, skeletal muscle, as well as in small amounts by many other tissues.[1] LPL is a homodimeric protein located on the luminal surface of the vascular endothelium, being anchored to the latter by interaction with cell surface glycosaminoglycans.[2] Lipoprotein lipase deficiency is an autosomal-recessive disorder associated with severe hypertriacylglycerolemia and decreased plasma high-density lipoproteins (HDL).[3] Heterozygous LPL deficiency may contribute to lipid abnormalities present in some cases of familial combined hyperlipidemia.[4] Defects in plasma lipoprotein metabolism are also characteristic of a number of metabolic disorders, such as diabetes mellitus, hypothyroidism, obesity, and renal disease. Changes in LPL activity are thought to be involved in the malfunctions of lipid metabolism associated with these diseases.[5,6]

Lipoprotein lipase is the rate-limiting enzyme for the metabolism of the triacylglycerol-rich lipoproteins, chylomicrons, and very-low-density lipoproteins (VLDL).[3] Triacylglycerols and monoglycerols are preferred substrates for LPL, whereas phosphatidylcholine is hydrolyzed at a slower rate. The action of LPL on lipoproteins releases free fatty acids that become the major source of energy for extrahepatic tissues. The metabolic role of LPL is not limited to triacylglycerol catabolism, because the major cholesterol-carrying lipoproteins, low-density lipoproteins, intermediate-density lipoproteins, and HDL are either the end products of the lipase action or

[1] C. F. Semenkovich, S.-H. Chen, M. Wims, C.-C. Luo, W.-H. Li, and L. Chan, *J. Lipid Res.* **30,** 423 (1989).

[2] T. Olivecrona and G. Bengtsson-Olivecrona, *in* "Lipoprotein Lipase" (J. Borensztaja, ed.), p. 15. Evener, Chicago, 1987.

[3] J. D. Brunzell, *in* "The Metabolic Basis of Inherited Disease" (C. R. Scriver, A. L. Beaudet, W. S. Sly, and D. Valle, eds.), 6th Ed., p. 1165. McGraw-Hill, New York, 1989.

[4] S. P. Babirak, P.-H. Iverius, W. Y. Fujimoto, and J. D. Brunzell, *Arteriosclerosis (Dallas)* **9,** 326 (1989).

[5] R. H. Eckel, *N. Engl. J. Med.* **370,** 1060 (1989).

[6] M. R. Taskinan, *in* "Lipoprotein Lipase" (J. Borensztajn, ed.), p. 201. Evener, Chicago, 1987.

require components released by LPL action on VLDL or chylomicrons for their formation.[3,7,8]

The activity of LPL is markedly stimulated in the presence of apolipoprotein C-II. Lipoprotein lipase gene expression is complex and involves control at both transcriptional and posttranscriptional levels.[9,10] The expression of LPL is constitutive, but its activity is regulated in a tissue-specific manner by different hormones such as insulin, catecholamines, corticosteroids, sex steroids, and thyroid hormone.[11] The LPL protein is produced by many different tissues and cell types,[1,12,13] as well as by a variety of primary cell cultures.[14–17] The expression and secretion of active LPL have also been demonstrated in various cell lines,[18–20] including Chinese hamster ovary (CHO) cells.[21]

Lipoprotein lipase has been purified to apparent homogeneity from bovine milk, human milk, as well as post-heparin plasma.[22,23] Many of the biochemical and enzymatic properties of LPL have been determined from studies on LPL isolated from these sources. However, LPLs prepared from these sources are often relatively unstable due to the copurification of contaminating proteases. Our experience with bovine milk LPL is summarized as follows. To investigate the nature of protease contamination, we incubated highly purified (>90% pure by 15% polyacrylamide reducing gels stained with Coomassie blue) bovine LPL at 4° in the presence of various protease inhibitors. The LPL activity was measured[24] at days 1, 4, and 7, and the integrity of the LPL was analyzed by visual inspection of gels run

[7] I. J. Goldberg, N. A. Le, H. N. Ginsberg, R. M. Krauss, and F. T. Lindgren, *J. Clin. Invest.* **81,** 561 (1988).
[8] A. M. Gotto, Jr., H. J. Pownall, and R. J. Havel, this series, Vol. 128, p. 3.
[9] C. F. Semenkovich, M. Wims, L. Noe, J. Etienne, and L. Chan, *J. Biol. Chem.* **264,** 9030 (1989).
[10] I. Gouni, K. Oka, J. Etienne, and L. Chan, *J. Lipid Res.* **34,** 139 (1993).
[11] A. Cryer, *Int. J. Biochem.* **13,** 525 (1981).
[12] A. Cryer, *in* "Lipoprotein Lipase" (J. A. Borensztajn, ed.), p. 277. Evener, Chicago, 1987.
[13] J. H. Auwerx, Z. Deeb, J. D. Brunzell, R. Peng, and A. Chait, *Biochemistry* **27,** 2651 (1988).
[14] P. A. Kern, S. Marshall, and R. H. Eckel, *J. Clin. Invest.* **75,** 199 (1985).
[15] M. Cupp, A. Bensadoun, and K. Melford, *J. Biol. Chem.* **262,** 8390 (1987).
[16] H. Semb and T. Olivecrona, *J. Biol. Chem.* **264,** 4195 (1989).
[17] R. B. Simsolo, J. M. Ong, and P. A. Kern, *J. Lipid Res.* **33,** 1777 (1992).
[18] L. S. Wise and M. Greene, *Cell (Cambridge, Mass.)* **13,** 233 (1978).
[19] O. Stein, E. Leiterdord, T. Chajek-Shaul, G. Friedman, and Y. Stern, *Monogr. Atheroscler.* **14,** 124 (1986).
[20] C. Vannier and G. Ailhaud, *J. Biol. Chem.* **264,** 13206 (1989).
[21] C. Rojas, S. Enerback, and G. Bengtsson-Olivecrona, *Biochem. J.* **271,** 11 (1990).
[22] R. L. Jackson, *in* "The Enzymes" (P. D. Boyer, ed.), Vol. 16, p. 141. Academic Press, New York, 1983.
[23] A. S. Garfinkel and M. C. Schotz, *in* "Lipoproteins" (A. M. Gotto, Jr., ed.), p. 335. Elsevier, Amsterdam, 1987.
[24] P. H. Iverius and A. M. Östlund-Lindqvist, this series, Vol. 129, p. 691.

on the same samples. With 40 and 80 μM aprotinin, 1 and 5 mM methyl benzimidate, and 0.5 and 1 mM phenylmethylsulfonyl fluoride (PMSF), there was substantial proteolysis, even after just 24 hr. At 7 days, with 1 mM PMSF, by sodium dodecyl sulfate (SDS)–polyacrylamide gel electrophoresis, about half of the original material remained at 55 kDa, compared to only barely detectable amounts with the other two protease inhibitors. Neither aprotinin nor methyl benzimidate inhibited LPL activity, whereas 1 mM PMSF completely inhibited triacylglycerol hydrolysis. After 7 days, about 25% of the original LPL activity remained in the sample incubated with buffer alone. Therefore, the contaminating protease is PMSF-inhibitable and likely has chymotrypsin-like activity.

Because of the difficulty of obtaining a homogeneous LPL preparation from natural sources that is stable and free from protease contamination, we have turned to using purified recombinant LPL produced *in vitro*. We have expressed human LPL (hLPL) in yeast and in baculovirus and found that the immunoreactive LPL produced by such systems is enzymatically inactive. The reason for the inactivity was not investigated, but it could result from aberrant glycosylation as proper N-linked glycosylation of Asn-43 appears to be essential for LPL activity.[25]

Enzymatically active recombinant LPL can be produced in mammalian cells (such as COS-1 cells, ATCC, Rockville, MD) by transient transfection.[25] The advantages of using COS cells are as follows: (1) they do not produce any detectable LPL activity or protein prior to transfection, (2) the amount of LPL produced is generally sufficient for biochemical analysis because it can be readily detected by activity assays and immunoassays, and (3) experiments can be completed within a few days. Such systems are suited for structure–function analyses of large numbers of mutant LPL molecules.[25–29] However, the absolute amount of LPL produced is small so that purification to homogeneity is impractical. If pure LPL is needed for structural studies or for production of polyclonal antibody, a stable cell line transfected with a human LPL expression vector is the method of choice. Here we describe production of hLPL by CHO-K1 cells transfected with hLPL cDNA in the expression vector pEE14,[30] and use of heparin-Sepharose affinity chromatography to purify the enzyme secreted by CHO-

[25] C. F. Semenkovich, C.-C. Luo, M. K. Nakanishi, S.-H. Chen, L. C. Smith, and L. Chan, *J. Biol. Chem.* **265,** 5429 (1990).

[26] F. Faustinella, L. C. Smith, C. F. Semenkovich, and L. Chan, *J. Biol. Chem.* **266,** 9481 (1991).

[27] F. Faustinella, A. Chang, J. P. Van Biervliet, M. Rosseneu, N. Vinaimont, L. C. Smith, S.-H. Chen, and L. Chan, *J. Biol. Chem.* **266,** 14418 (1991).

[28] F. Faustinella, L. C. Smith, and L. Chan, *Biochemistry* **31,** 7219 (1992).

[29] J.-Y. Lo, L. C. Smith, and L. Chan, *Biochem. Biophys. Res. Commun.* **206,** 266 (1995).

[30] E. Zsigmond, E. Scheffler, T. M. Forte, R. Potenz, W. Wu, and L. Chan, *J. Biol. Chem.* **269,** 18752 (1994).

K1 cells into the medium. A drawback of the use of CHO cells is that they produce small amounts of LPL before transfection. However, following the production of a hLPL-producing stable cell line, the amount of LPL produced is over 1000-fold more than that in the basal state. Therefore, the theoretical contamination of the hLPL by the CHO LPL has not created any difficulty in our experiments.

The availability of purified LPL has permitted the production of monoclonal and polyclonal antibodies against LPL.[4,31–35] The method for the production of monoclonal antibodies to hLPL is not different from that used for other antigens.[36,37] Here we describe our experience with the production of a polyclonal antibody to hLPL.

Lipoprotein lipase is a highly conserved molecule.[38] When human (or bovine) LPL is injected into rabbits, the antibody produced is generally of low titer and affinity because of the high sequence similarity among LPLs of different mammals. Of all the mammalian LPL sequences known, the guinea pig LPL sequence is the most different from those of other species (including human, cow, mouse, and rat). Therefore, guinea pig would be a suitable animal for production of antibodies against LPLs isolated from these other mammals. However, we found that the chicken is an even better animal for this purpose. The avian lineage separated from mammals about 300 million years ago. A comparison of chicken LPL to various mammalian LPLs indicates that the chicken LPL has the highest amount of sequence divergence.[38] Mammalian LPLs are highly immunogenic in the chicken.[31] Furthermore, immunoglobulins are concentrated in the egg yolk and can be readily purified by a precipitation method.[39] In this chapter, we also present a detailed procedure for raising of antiserum against hLPL in laying hens and rapid purification of the anti-hLPL antibody (immunoglobulin Y, IgY) from egg yolk using a simple extraction protocol.

[31] T. Olivecrona and G. Bengtsson-Olivecrona, *Biochim. Biophys. Acta* **752,** 38 (1983).

[32] J. C. Voyta, D. P. Via, P. K. J. Kinunen, J. T. Sparrow, A. M. Gotto, Jr., and L. C. Smith, *J. Biol. Chem.* **260,** 893 (1985).

[33] I. Goldberg, J. R. Paterniti, D. S. France, G. Martinelli, and J. A. Cornicelli, *Biochim. Biophys. Acta* **878,** 168 (1986).

[34] J. F. Goers, M. E. Petersen, P. A. Kern, J. Ong, and M. C. Schotz, *Anal. Biochem.* **166,** 27 (1987).

[35] C. Vannier, S. Deslex, A. Pradines-Figueres, and G. Ailhaud, *J. Biol. Chem.* **264,** 13199 (1989).

[36] G. Köhler and C. Milstein, *Nature* (*London*) **256,** 495 (1975).

[37] E. Harlow and D. Lane, "Antibodies: A Laboratory Manual." Cold Spring Harbor Laboratory, Cold Spring Harbor, New York, 1988.

[38] W. A. Hide, L. Chan, and W.-H. Li, *J. Lipid Res.* **33,** 167 (1992).

[39] J. C. Jensenius, I. Andersen, J. Hau, M. Crone, and C. Koch, *J. Immunol. Methods* **46,** 63 (1981).

Production of Human Lipoprotein Lipase in Stable CHO Cell Line

Reagents

25 μM Methionine sulfoximine (MSX): A stock solution of 100 mM methionine sulfoximine is prepared by dissolving 18 mg in 1 ml PBS. The solution is sterile filtered and diluted to 25 μM prior to use. The MSX should be stored at −20°.

Phosphate-buffered saline (PBS): prepared by combining sodium chloride (8 g), potassium chloride (0.2 g), disodium phosphate (1.15 g), and monopotassium phosphate (0.2 g) in 1 liter water at pH 7.3; the buffer is autoclaved and stored at room temperature

No-salt buffer: 5 mM sodium barbital, pH 7.4, 25% glycerol, and 1 mM Na_2EDTA

Low-salt buffer: 5 mM sodium barbital, pH 7.4, 25% glycerol, 1 mM Na_2EDTA, and 0.5 M NaCl

High-salt buffer: 5 mM sodium barbital, pH 7.4, 25% glycerol, 1 mM Na_2EDTA, and 1.5 M NaCl

Dialyzed fetal calf serum (dFCS): FCS is heat-inactivated at 56° for 30 min and dialyzed against 0.1 M sodium bicarbonate and 0.5 M sodium chloride, pH 8.3; dialyzed FCS is divided into 1-ml aliquots in tubes and stored at −20°

Expression Vector Preparation and Transfection of CHO-K1 Cells

The human LPL cDNA construct contains the entire coding sequence for human LPL, including the 27-amino acid signal peptide, 42 bp of the 3′ untranslated sequence, and 320 bp of the 5′ untranslated region. This DNA fragment is gel purified and subcloned into the linearized 9.4-kb pEE14 vector (Celltech Ltd., Berkshire, U.K., glutamine synthetase gene amplification system). This expression vector contains the human cytomegalovirus promoter/enhancer and the SV40 (simian virus 40) polyadenylation signal for transcription termination. To allow for the selection and amplification of transfectants, the plasmid contains the glutamine synthetase (glutamate–ammonia ligase) minigene which is transcribed from the SV40 late promoter.[40,41]

The expression construct (10 μg) is transfected into 10^6 exponentially growing CHO-K1 (ATCC, Rockville, MD) cells by calcium phosphate coprecipitation (GIBCO–BRL, Gaithersburg, MD, calcium phosphate transfection kit), followed by glycerol shock in 10-cm^2 dishes. A mock

[40] P. E. Stephens and M. I. Cockett, *Nucleic Acids Res.* **17,** 7110 (1989).
[41] M. I. Cockett, C. R. Bebbington, and G. T. Yarranton, *Nucleic Acids Res.* **19,** 319 (1990).

transfection with no vector, and with the pEE14 expression vector alone, is simultaneously performed. Cells are incubated in glutamine-free GMEM-S medium (GIBCO–BRL), containing 10% dFCS (GIBCO–BRL) and 5000 U/ml penicillin/streptomycin (Pharmacia-LKB, Piscataway, NJ) at 37° with 5% (v/v) CO_2. Twenty-four hours posttransfection, fresh medium containing 25 μM methionine sufoximine (MSX), a specific inhibitor of glutamine synthetase that is lethal in concentrations greater than 3 μM to non-transfected cells, is added to the flasks. The medium is changed 1–2 times per week until stable colonies appear in 2–3 weeks. Stable colonies are picked by using cloning cylinders to 24-well plates and grown until confluent.

Testing of Stable Transfectants for Production of Human Lipoprotein Lipase

Transfectants grown in GMEM-S with 10% dFCS are plated in 24-, 12-, or 6-well plates for 5–7 days until confluent. Fresh medium containing 40 μg/ml heparin is added and left on the cells 24–48 hr. The medium is assayed for LPL activity using a [^{3}H]triacylglycerol emulsion,[24] and transfectants with the highest LPL activity are further tested. The vector is amplified by selecting cells grown in increasing concentrations of MSX up to 1000 μM. The highest producers are checked by limiting dilution cloning to ensure their authenticity, and production conditions are then optimized for the highest producing cell lines.

Purification of Human Lipoprotein Lipase by Heparin-Sepharose Affinity Chromatography

Growth of Cells. Culture flasks (T-175 cm^2 flasks) are inoculated with 0.5–1 $\times$ 10^6 cells of the highest producing cell line in GMEM-S containing 5% dFCS. After 24 hr, MSX is added to a final concentration of 25 μM. Cells are grown at 37° with 5% CO_2 to 80–100% confluency (~1 $\times$ 10^7 cells/flask). Fresh GMEM-S (18 ml/flask), containing 2% dFCS, 5 mg/ml heparin, 25 μM MSX, and 5000 U/ml penicillin/streptomycin is added to each flask and incubated for 5 or 19 hr. The medium is removed from the flasks, and remaining cells are pelleted by centrifugation (1500 *g*) at 4° for 10 min. Medium containing LPL is placed into fresh bottles, and glycerol (100%) is added to a final concentration of 25%. An aliquot is removed for determining LPL activity, and the rest of the sample is frozen on dry ice and stored at −20° until the purification is performed.

Primary Purification. Thawed medium (3000–3500 ml) containing LPL is mixed with 5 *M* NaCl to a final concentration of 0.3 *M* NaCl and divided into aliquots in 500-ml spinner flasks (Bellco, Vineland, NJ). Heparin-

Sepharose affinity chromatography columns are prepared by covalently coupling heparin to Sepharose 4B (Pharmacia) using the cyanogen bromide activation method.[42] Following the addition of 2–3% (v/v) heparin-Sepharose 4B to each spinner flask, the samples are placed on magnetic stir plates and mixed at approximately 50 rpm at 4° for 2 hr. Flasks are stored overnight at 4°, and heparin-Sepharose-4B is then collected on 350-ml coarse sintered glass filter funnels and resuspended in 50 ml low-salt buffer. Collected medium is placed back into spinner flasks, and the adsorption procedure is repeated with fresh heparin-Sepharose 4B. The heparin-Sepharose 4B is collected on the sintered glass filter funnel and combined with heparin-Sepharose 4B from the first adsorption step. The LPL bound to the heparin-Sepharose 4B is washed with 10 volumes of ice-cold low-salt buffer, resuspended in the same buffer, and packed into columns (2.5 × 150 cm) at a flow rate of approximately 80 ml/hr. The column is washed overnight (10 ml/hr) in the cold room with the low-salt buffer. Samples are eluted with high-salt buffer (80 ml/hr), and 10 ml fractions are collected. Fractions are assayed for protein content[43] (Bio-Rad, Richmond, CA) and for enzyme activity.[24]

Secondary Purification. Fractions with the most bioactive LPL from six separate primary purifications are combined, then diluted with no-salt buffer to a final concentration of 0.5 *M* NaCl. Pooled fractions are loaded onto a 50-ml heparin-Sepharose 4B column equilibrated with the low-salt buffer (50–70 ml/hr). The column is washed with 6 volumes of low-salt buffer, and samples are eluted with a 0.5–1.5 *M* salt gradient buffer at a flow rate of 50 ml/hr. The protein content[43] (Bio-Rad) and LPL activity[24] of the collected fractions are measured. Average specific activity (milliunits/microgram) is calculated for each fraction. On SDS–polyacrylamide gels, the purified LPL protein migrates as a single band of the expected molecular weight, and the protein exhibits apolipoprotein C-II-dependent, NaCl-sensitive lipase activity.

Production of Polyclonal Chicken Anti-human Lipoprotein Lipase

Injection of Chickens. Purified hLPL is diluted with an equal volume of Freund's complete adjuvant (Sigma, St. Louis, MO) and emulsified. White Leghorn laying hens are injected subcutaneously with 70–80 μg purified hLPL. Booster injections of hLPL (40–80 μg) emulsified with Freund's incomplete adjuvant (Sigma) are given to the hens 14, 45, and 100 days after the first injection.

[42] S. C. March, I. Parikh, and P. Cuatrecasas, *Anal. Biochem.* **60,** 149 (1974).
[43] M. M. Bradford, *Anal. Biochem.* **72,** 248 (1976).

Preparation of Immunoglobulin Y from Egg Yolks. Eggs are collected 13–20 days after the third booster injection. Egg yolks are carefully separated and transferred to 50-ml Sorvall tubes containing 10 ml PBS. Tubes are covered and gently shaken until homogeneous yellow solutions are obtained. The tubes are incubated for 10 min at room temperature, and the shaking procedure is repeated three more times. Tubes are centrifuged at 12,000 *g* for 20 min; the supernatant is transferred to fresh tubes and combined with 2 volumes chloroform. Each tube is covered with a rubber stopper, vigorously shaken, and then allowed to settle at room temperature for 30 min. The shaking of the supernatant is repeated an additional three times. Following centrifugation (5000 *g*) for 10 min, the upper aqueous phase is carefully removed and may be stored at 4° with 0.01% sodium azide.

To purify IgY further, 25 ml washed Avid AL (Unisyn Technologies Corp., San Diego, CA) is packed into 1.5-cm diameter columns and equilibrated with PBS. Fifty milliliters of the sample is applied to the column at a flow rate of 0.5 ml/min. The column is washed with 5 volumes PBS and eluted with Avid AL Neutral Elution Buffer (Unisyn Technologies Corp.). The IgY is concentrated by mixing it with an equal volume of 100% ammonium sulfate, pH 7.0, at 4°, overnight. The suspension is then centrifuged (12,000 *g*), and the pellet is resuspended in PBS with 0.01% (w/v) sodium azide.

A final purification procedure is advisable in order to remove any antibodies that may bind to albumin. The contaminating binding activities can be removed by passage through a fetal calf serum column. Twenty milliliters dFCS is coupled to 80 ml cyanogen bromide-activated Sepharose 4B.[42] The concentrated IgY fraction (20–25 ml) is mixed with the dFCS-Sepharose 4B at 4° overnight, and the suspension is poured into glass columns with sintered glass disks at the bottom (Bio-Rad, 2 × 15 cm). The columns are washed with PBS (~120 ml final volume), and the recovered solution contains the purified IgY (~1–2 mg/ml) free of contaminating activity.

Acknowledgments

We thank Bill Farley for assistance with the LPL purification and antibody production, Dr. Michael S. Kilberg of the University of Florida, Gainesville, for providing the initial protocol for purification of chicken antibody, and Sally Tobola for expert secretarial assistance. The protocols described in this chapter were developed with support from the National Institutes of Health (Grants HL16512 and HL27341 for a Specialized Center of Research in Arteriosclerosis at Baylor College of Medicine).

[22] Immunochemical Quantitation of Lipoprotein Lipase

By EVA ZSIGMOND, JING-YI LO, LOUIS C. SMITH, and LAWRENCE CHAN

Introduction

Lipoprotein lipase (LPL) has an important function in the regulation of plasma lipoprotein metabolism and is also involved in adipose tissue differentiation and in the prediction of adipocyte cell size.[1] The enzyme hydrolyzes the triacylglycerol components of very-low-density lipoproteins (VLDL) and chylomicrons to glycerol and free fatty acids, and hence it regulates the supply of the latter to various tissues.

In view of the critical role that LPL plays in normal metabolism, as well as in pathophysiological states, assays have been developed that measure LPL enzyme activity and immunoreactive mass. In a previous volume of this series, Iverius and Östlund-Lindqvist[2] presented a detailed protocol for an LPL activity assay. Here we concentrate on the assessment of LPL immunoreactive material by immunological methods. Various immunoassays have previously been described in the literature.[3–8] This chapter presents two immunoassay protocols designed to quantitate human LPL (hLPL), making use of a monoclonal antibody (M40) raised against bovine milk LPL that cross-reacts with hLPL[9] and a polyclonal anti-hLPL antibody (immunoglobulin Y, IgY) prepared in chickens.[10,11] The antibody capture enzyme-linked immunosorbent assay (ELISA) is designed to detect LPL

[1] J. D. Brunzell and M. R. C. Greenwood, *in* "Biochemical Pharmacology of Metabolic Disease" (P. B. Curtis-Prior, ed.), Vol. 1, p. 175. Elsevier Amsterdam, 1983.
[2] P. H. Iverius and A. M. Östlund-Lindqvist, this series, Vol. 129, p. 691.
[3] T. Olivecrona and G. Bengtsson, *Biochim. Biophys. Acta* **752,** 38 (1983).
[4] J. W. F. Goers, M. E. Pedersen, P. A. Kern, J. Ong, and M. C. Schotz, *Anal. Biochem.* **166,** 27 (1987).
[5] P. A. Kern, J. M. Ong, J. W. F. Goers, and M. E. Pedersen, *J. Clin. Invest.* **81,** 398 (1988).
[6] E. Vilella, J. Joven, M. Fernández, S. Vilaró, J. D. Brunzell, T. Olivecrona, and G. Bengtsson-Olivecrona, *J. Lipid Res.* **34,** 1555 (1993).
[7] I. J. Goldberg, J. R. Paterniti, Jr., D. S. France, J. A. Martinelli, and J. D. Cornicelli, *Biochim. Biophys. Acta* **878,** 168 (1986).
[8] S. P. Babirak, P.-H. Iverius, W. Y. Fukimoto, and J. D. Brunzell, *Arteriosclerosis* (*Dallas*) **9,** 326 (1989).
[9] J. C. Voyta, D. P. Via, P. K. J. Kinunen, J. T. Sparrow, A. M. Gotto, Jr., and L. C. Smith, *J. Biol. Chem.* **260,** 893 (1985).
[10] E. Zsigmond, E. Scheffler, T. M. Forte, R. Potenz, W. Wu, and L. Chan, *J. Biol. Chem.* **269,** 18757 (1994).
[11] R. Potenz, J.-Y. Lo, E. Zsigmond, L. C. Smith, and L. Chan, this series, in press.

in samples with low-level contamination by other proteins, such as culture medium, whereas the two-antibody sandwich ELISA is more sensitive and is best suited for detecting LPL in more complex samples, such as cellular extracts or post-heparin plasma. We also present a procedure for preparing LPL from post-heparin plasma and detecting the protein by immunoblot analysis. Immunoblotting provides a semiquantitative measure of LPL immunoreactive mass as well as the apparent molecular weight of the protein.

Antibody Capture ELISA

The simple antibody capture ELISA to detect immunoreactive mass may be used with monoclonal or polyclonal antibodies.[12] The assay involves binding of the antigen to a solid phase, such as polyvinyl chloride (PVC) microtiter plates, and then adding an antibody that recognizes the antigen. After thorough washing to remove unbound antibodies, the bound antibodies are quantitated with a labeled second antibody. Advantages of this assay are its simplicity and the possibility of determining either antigen or antibody levels. Problems associated with the antibody capture assay are high background and recognition of unrelated proteins by polyclonal antibodies that are not highly purified.

In our laboratory we have used an antibody capture ELISA to quantitate LPL secreted into the culture medium by transfected COS.M6 cells (ATCC, Rockville, MD).[13] The assay works well whether we use a monoclonal or a polyclonal antibody against LPL.

Reagents

Phosphate-buffered saline (PBS) with 0.25% bovine serum albumin (PBSA): NaCl (8 g), KCl (0.2 g), Na_2HPO_4 (1.44 g), and KH_2PO_4 (0.24 g), pH 7.2, are dissolved in 1 liter water and autoclaved; bovine serum albumin (BSA; Sigma, St. Louis, MO) with a low free fatty acid content is dissolved in PBS to make a final concentration of 0.25% (w/v)

Blocking buffer: 3% BSA in PBS and 0.02% sodium azide

2,2′-Azinobis(3-ethylbenzthiazoline sulfonate) (ABTS) substrate: The ABTS substrate buffer is prepared fresh immediately prior to use; to 15 ml autoclaved citrate buffer (0.1 *M*, pH 4.0) are added 3 mg ABTS (Sigma) and 9.9 μl hydrogen peroxide

Sodium fluoride buffer: NaF (4.2 g) is dissolved in water (100 ml)

[12] E. Harlow and D. Lane, "Antibodies: A Laboratory Manual." Cold Spring Harbor Laboratory, Cold Spring Harbor, New York, 1988.

[13] F. Faustinella, L. C. Smith, C. F. Semenkovich, and L. Chan, *J. Biol. Chem.* **266,** 9481 (1991).

by heating at 50°; the solution is filtered prior to use to remove undissolved powder

Microtiter plates: Microtiter plates from different manufacturers differ markedly, and many give unacceptably high background binding; we have found that Immulon 4 (Fisher, Houston, TX) gives the best results, but each investigator should perform comparison testing to select for the best microtiter plates

Sodium azide should not be used as a bacteriostatic agent in the conjugate dilution buffer or the substrate buffer when horseradish peroxidase (HRP) is used for detection

Procedure. To prepare the samples, the medium is collected from COS.M6 cells cultured in Dulbecco's modified Eagle's medium (DMEM) with 4% fetal bovine serum (FBS) and 40 μg/ml heparin.[13] The conditioned medium containing the antigen is then added to the wells of an Immulon 4 (Fisher) microtiter plate (50 μl/well) and serially diluted with PBS. To allow for quantitation of antigen levels, serially diluted purified LPL of known concentrations added to the wells of another plate will be used as standards. For even binding, plates are shaken on a microtiter plate shaker (Fisher, Hyperion micromix) for 10 min prior to each incubation. The antigen-coated plate is incubated at 4°, overnight. After wells are washed four times with PBSA (200 μl/well), the unoccupied binding sites are blocked with 3% BSA–PBS (200 μl/well) by incubating plates at room temperature for 4 hr, at the end of which the blocking buffer is removed and the plate washed twice with PBSA. Then the monoclonal or polyclonal antibody (15 μg/ml in 1% BSA–PBS; 50 μl/well) is added, and the plate is incubated at 4° overnight. Following four washes with PBSA, the labeled antibody is added and the plates are incubated at 37° for 2 hr. If the assay is performed with the M40 monoclonal antibody, the labeled second antibody used is goat anti-mouse IgG–HRP conjugate (Bio-Rad, Richmond, CA), diluted 1:1000 in 1% BSA–PBS. For experiments with the chicken anti-hLPL polyclonal antibody, rabbit anti-chicken IgG–HRP conjugates (Cappel, Durham, NC) are used at a 1:1000 to 1:5000 dilution in 1% BSA–PBS. After four washes with PBSA, the ABTS substrate is added. When the color development is complete (1–5 min), the reaction is stopped with 1 *M* NaF. Absorbance is read at 410 nm with an ELISA plate reader (Titertek, Multiscan MCC, ICN, Irvine, CA).

Calculations. To calculate immunoreactive mass, a parallel run of standard samples of known LPL protein concentration (1 ng/μl) in PBS is conducted for each experiment. The standard is serially diluted along with the samples, and at the end of the assay a standard curve of LPL concentrations versus optical density is constructed. The LPL mass in the samples is determined from the standard curve.

Two-Antibody Sandwich ELISA

The two-antibody sandwich ELISA is a sensitive and quick procedure to quantitate antigens in unknown samples.[12] The principle of the assay is to bind the first antibody to a solid support (e.g., PVC microtiter plate), allow the antigen to interact with the first antibody, and then add a second antibody that recognizes a different epitope on the antigen. Finally, the amount of antigen is quantitated by the amount of second antibody bound as indicated with a labeled antibody that binds to the second antibody. The antibodies used may be monoclonal or affinity-purified polyclonal antibodies. The advantage of the sandwich ELISA over the antibody capture ELISA is that it has greater sensitivity of detection and lower background. The limitation is that two antibodies that interact with independent epitopes on the antigen are needed.

The following procedure describes a sandwich ELISA that uses a monoclonal anti-LPL antibody to coat the microtiter plate, prior to the addition of the sample, and a second antibody consisting of a polyclonal antibody that recognizes hLPL. As an example, we use this assay to quantitate the amount of hLPL in cellular extracts of COS.M6 cells transfected with a hLPL expression vector.[13]

Reagents. The reagents used for the two-antibody sandwich ELISA are the same as described for the antibody capture assay. For the two-antibody sandwich ELISA, we found that the microtiter plates from Corning (Corning, NY) work better than other brands.

Procedure. To prepare the samples, COS.M6 cells are cultured as reported by Faustinella *et al.*[13] Prior to harvesting the transfected COS.M6 cells, each flask is washed with 10 ml PBS. Cells are harvested in 2 ml ice-cold 25 m*M* ammonia buffer, pH 8.2, 5 m*M* EDTA, containing 16 mg Triton X-100, 80 μg heparin, 20 μg leupeptin, and 2 μg pepstatin. The sides of the flask are scraped, and the cells are sonicated for 30 sec. Cellular debris is removed by a brief centrifugation before the assay.

For even binding, plates are shaken on a microtiter plate shaker (Fisher, Hyperion micromix) for 10 min prior to each incubation. The M40 monoclonal antibody (10 μg/ml, 50 μl/well) is coated onto wells of the microtiter plate (Corning) at 4°, overnight. The plate is washed four times with PBSA (200 μl/well), blocked with 3% BSA–PBS (200 μl/well) for 4 hr at room temperature, and then washed again twice with PBSA. Cell extracts are serially diluted with 1% BSA–PBS with 0.02% sodium azide in another plate and then transferred to the blocked plate to incubate at 4°, overnight.

Following four washes with PBSA, the polyclonal antibody is added (7.5 μg/ml in 1% BSA–PBS; 50 μl/well) and incubated at room temperature for 2 hr and then at 37° for 1 hr. At the end of the incubation, the plate is washed 4 times with PBSA, and the HRP conjugated rabbit anti-chicken IgG antibody (Cappel; 1:1000 to 1:5000 dilution in 1% BSA–PBS) is added. The plate is incubated at 37° for 2 hr and then washed four times with PBSA. The ABTS substrate is added (100 μl/well), and color is developed for 1–5 min. The reaction is stopped with 1 *M* NaF (50 μl/well). Optical density is measured at 410 nm (Titertek, Multiscan MCC).

Calculations. Each plate contains a standard prepared with known LPL protein concentration (1 ng/μl) in the same medium that the samples are in (e.g., control cell extract) and serially diluted along with the samples. At the end of the assay, a standard curve is constructed. The LPL mass in the samples is determined from the standard curve.

Immunoblot Analysis of Lipoprotein Lipase from Post-heparin Plasma

Because immunoblot analysis combines gel electrophoresis with immunochemical detection, it is a valuable technique for the determination of molecular weights of proteins, as well as the specificity of the antigen for polyclonal antibodies. Antigens are separated on a sodium dodecyl sulfate (SDS)–polyacrylamide gel, transferred to a support membrane, washed, and incubated with antibodies in the presence of blocking reagents that reduce nonspecific binding. One limitation of the technique is that many monoclonal antibodies do not recognize antigens that have been denatured prior to the electrophoresis. Because LPL concentrations in unknown samples are usually quite low, polyclonal sera often react with extraneous, unrelated antigens present in high concentrations. To minimize this problem, partial purification of the sample and antiserum may be necessary.

In the following description of the procedure, we use post-heparin plasma as an example. In this case, contamination by other unrelated proteins, especially albumin (which has a molecular weight very similar to that of LPL), can be reduced by partial purification of the LPL from plasma by adsorbing it to heparin-Sepharose. This simple maneuver greatly reduces the background and improves the sensitivity of the immunoblot detection method.

Reagents

Low-salt buffer: 25% glycerol, 5 m*M* sodium barbital, pH 7.4, 1 m*M* EDTA, and 0.5 *M* NaCl

High-salt buffer: 25% glycerol, 5 mM sodium barbital, pH 7.4, 1 mM EDTA, and 1.5 M NaCl

Heparin-Sepharose: The resin is resuspended and stored in 10 mM Tris-HCl, pH 7.4, 1 mM EDTA, 0.15 M NaCl, and 0.01% sodium azide; storage buffer is removed by centrifuging a 100-μl aliquot of the heparin-Sepharose at 1000 g for 3 min, and the pellet is washed three times with low-salt buffer

Equilibration buffer: 20 mM Tris, pH 8.0, 200 mM glycine, 0.1% (w/v) SDS, and 20% (v/v) methanol

Transfer buffer: 20 mM Tris, pH 8.0, 200 mM glycine, and 20% (v/v) methanol

Blocking buffer: 50 mM Tris, pH 8.0, 150 mM NaCl, 1 mM EDTA, 5% (w/v) BSA, and 0.1% (v/v) Tween 20

Washing buffer (TBS-T): 50 mM Tris, pH 8.0, 150 mM NaCl, 1 mM EDTA, and 0.1% (v/v) Tween 20.

Sample Preparation. Post-heparin plasma is collected on EDTA, or in Li-heparinized Vacutainer tubes.[2] The tube of blood is immediately immersed in ice; plasma is separated by centrifugation (10,000 g; 1 min) and either used fresh or snap-frozen in liquid nitrogen. Sixty microliters plasma is diluted to 1 ml with low-salt buffer and incubated with 0.1 ml heparin-Sepharose at 4° for 30 min, with intermittent mixing. Following a 5-min centrifugation at 1000 g, the supernatant is removed and the pellet is washed three times with 1 ml low-salt buffer. The LPL is eluted from the heparin-Sepharose by mixing it with 40 μl high-salt buffer for 5 min. The sample is centrifuged (1000 g) for 5 min, and the supernatant is removed. To improve recovery, the elution is repeated one more time.

Procedure. The eluted sample (14 μl) is heated at 100° for 3 min in SDS–polyacrylamide sample buffer and loaded onto a 4–20% gradient SDS–polyacrylamide gel. Electrophoresis is at 200 V for 1 hr. The gel is equilibrated for 1 hr in the equilibration buffer and transferred at 30 V to a polyvinyl difluoride membrane (PVDF; Millipore, Bedford, MA) in the transfer buffer at 4°, overnight. After the transfer, the PVDF membrane is air dried, wetted in methanol and then in water, and incubated at 37° for 1 hr in the blocking buffer. The membrane is washed twice with TBS-T for 5 min and exposed to the chicken anti-hLPL antibody (2 μg/ml) for 1 hr, at room temperature. After five washes with 0.25% BSA–TBS-T, anti-chicken IgG–HRP (1:10,000 dilution) is added and incubated for 1 hr, at room temperature. The PVDF membrane is rinsed briefly with TBS-T and then washed twice for 15 min and four times for 5 min at room temperature with TBS-T. To visualize the protein, a second gel with the same samples is stained with Coomassie blue dye.

Immunodetection is performed using the DuPont Western blot chemiluminescence protocol (DuPont–NEN, Boston, MA). Equal volumes of the enhanced luminol reagent and oxidizing reagent are mixed and added to the PVDF membrane for 1 min, at room temperature. The membrane is then blotted dry, placed between the covers of a polypropylene sheet protector (Boise Cascade, Houston, TX) and exposed to film for 1–5 min.

Acknowledgments

We thank Sally Tobola for expert secretarial assistance. The work described in this chapter was supported by National Institute of Health Grants HL16512 and HL27341 for a Specialized Center of Research in Arteriosclerosis at Baylor College of Medicine.

[23] Sandwich Immunoassay for Measurement of Human Hepatic Lipase

By ANDRÉ BENSADOUN

Introduction

Hepatic triglyceride lipase (HL; triacylglycerol kinase, EC 3.1.1.3) is an essential enzyme for the metabolism of intermediate-density lipoproteins (IDL) and high-density lipoproteins (HDL) in plasma.[1,2] Overexpression of human HL in transgenic rabbits decreases plasma total cholesterol by approximately 50%.[2] In humans, HL deficiency results in elevated plasma cholesterol and triglyceride, with increased concentrations of IDL and large HDL.[3] Liver perfusion studies suggest that HL enhances the initial uptake of chylomicrons by rat liver.[4] These findings demonstrate that HL plays a major role in the catabolism and uptake of plasma lipoproteins.

Quantitative measurements of HL catalytic activity in human post-heparin plasma are complicated by the presence of lipoprotein lipase (LPL), which is also released by heparin injections. Another confounding factor is the presence of variable amounts of plasma triglyceride, which can com-

[1] B. P. Daggy and A. Bensadoun, *Biochim. Biophys. Acta* **877,** 252 (1986).

[2] J. Fan, J. Wang, A. Bensadoun, S. J. Lauer, Q. Dang, R. W. Mahley, and J. M. Taylor, *Proc. Natl. Acad. Sci. U.S.A.* **91,** 8724 (1994).

[3] R. A. Hegele, J. A. Little, C. Vezina, and P. W. Connelly, *Arterioscler. Thromb.* **134,** 720 (1993).

[4] S. Shafi, S. E. Brady, A. Bensadoun, and R. J. Havel, *J. Lipid Res.* **35,** 709 (1994).

pete with labeled substrates used in triglyceride lipase determinations. Methods for the measurement of HL and LPL activities are based on the partial purification and separation of these enzymes on small heparin-Sepharose columns[5] or on the use of specific antibodies inhibiting HL catalytic activity.[6] Enzyme-linked immunosorbent assays (ELISAs) of HL in human or rat plasma have been described and are much more sensitive[7,8] than measurements based on catalytic activity. In this chapter a sandwich ELISA for human HL is outlined. It is based on the use of two monoclonal antibodies against human HL. It is sensitive and repeatable, and can detect HL even in pre-heparin plasma.

Purification of Human Hepatic Lipase

A detailed purification of human HL from post-heparin plasma has been published.[9] The purification procedure involves chromatography steps on heparin-Sepharose 4B and DEAE-Sephacel, and a gel-permeation step on AcA-34. All three resins are obtained from Pharmacia (Piscataway, NJ). The highly purified enzyme is homogeneous by sodium dodecyl sulfate–polyacrylamide gel electrophoresis (SDS–PAGE) and is free of antithrombin, a common contaminant of lipases purified from post-heparin plasma. Antithrombin is not detectable by Western blotting, activity assay, or probing with concanavalin A.[9] Others have also devised procedures for purifying HL from post-heparin plasma.[10,11]

Purification of Monoclonal Antibodies

The ELISA described utilizes two monoclonal antibodies, XHL-3-6 and XHL-1-6. These mouse monoclonal antibodies react with HL but not with LPL by Western blotting.[9] The monoclonal antibodies are purified from ascites fluid on protein G-Sepharose.

[5] J. Boberg, J. Augustin, M. L. Baginsky, P. Tejada, and W. V. Brown, *J. Lipid Res.* **18,** 544 (1977).

[6] J. K. Huttunen, C. Ehnholm, P. K. Kinnunen, and E. A. Nikkila, *Clin. Chim. Acta* **63,** 335 (1975).

[7] Y. A. Ikeda, A. Takagi, Y. Ohkaru, K. Nogi, T. Iwanaga, S. Kurooka, and A. Yamamoto, *J. Lipid Res.* **31,** 1911 (1990).

[8] L. A. Cisar and A. Bensadoun, *J. Lipid Res.* **26,** 380 (1985).

[9] C.-F. Cheng, A. Bensadoun, T. Bersot, J. S. T. Hsu, and K. H. Melford, *J. Biol. Chem.* **260,** 10720 (1985).

[10] C. Ehnholm and T. Kuusi, this series, Vol. 129, p. 716.

[11] R. L. Jackson and L. R. McLean, this series, Vol. 197, p. 339.

Reagents

Wash buffer: 20 m*M* sodium phosphate, pH 7.0
Elution buffer: 0.1 *M* glycine, pH 2.7, adjusted to pH 2.7 with concentrated HCl
Collection buffer: 1 *M* Tris-HCl, pH 9.0

All three buffers are filter-sterilized.

Procedure. One milliliter ascites fluid is centrifuged at 12,000 *g* in a microcentrifuge tube for 15 min, filter sterilized, and loaded on a 2-ml protein G-Sepharose column equilibrated with wash buffer. The column is washed with 50 ml wash buffer, and the antibodies are eluted with elution buffer. The immunoglobulins are collected in 3-ml fractions in tubes containing 0.2 ml collection buffer to neutralize the elution buffer. The fractions containing immunoglobulins are identified by absorption at 280 nm. The fractions containing immunoglobulins are pooled and dialyzed against phosphate-buffered saline supplemented with 0.2% (w/v) sodium azide. The purified immunoglobulins are concentrated by ultrafiltration to 1 to 10 mg/ml and stored at −20° in small aliquots. For long-term storage, the samples are first diluted with an equal volume of glycerol and then stored at −20°.

Preparation of Conjugate

The purified XHL-1-6 is conjugated to horseradish peroxidase by the method of Avrameas and Ternynck.[12]

Day 1. Two to five milligrams XHL-1-6 in 1 ml is dialyzed overnight against 2 liters of 0.15 *M* NaCl, 10 m*M* phosphate, pH 7.4 (phosphate-buffered saline, PBS), at 4°. Ten milligrams horseradish peroxidase is dissolved in 200 μl PBS, 1.25% (v/v) glutaraldehyde and incubated overnight at room temperature.

Day 2. Apply the horseradish peroxidase solution to a Sephadex G-25 column (PD-10 column, Pharmacia) equilibrated in PBS. Elute the peroxidase with the same buffer. The peroxidase is monitored by its color and is collected in 1 ml. If the color band elutes in less than 1 ml, the volume is adjusted to 1 ml with PBS. The XHL-1-6 (in 1 ml) is added to the peroxidase, 100 μl of 1 *M* carbonate–bicarbonate buffer, pH 9.5, is added, and the mixture is incubated at 4° for 24 hr.

Day 3. Add 0.1 ml of 0.2 *M* lysine to the monoclonal antibody–horseradish peroxidase mixture. Incubate at room temperature for 2 hr. Dialyze overnight against PBS.

[12] S. Avrameas and T. Ternynck, *Immunochemistry* **8,** 1175 (1971).

Day 4. Precipitate the conjugated monoclonal antibody by adding an equal volume of saturated ammonium sulfate, pH 7.0. Incubate at 4° for 6 hr. Centrifuge at 3000 *g* for 30 min. Discard the supernatant and dissolve the pellet in 1 ml PBS. Dialyze overnight against 4 liters of PBS at 4°.

Day 5. Transfer the sample to a microcentrifuge tube and centrifuge at 10,000 *g* for 10 min. Remove the supernatant and adjust to 1% BSA (ELISA grade) and 0.02% thimerosal. The conjugate is then filter-sterilized (0.45-μm filter), divided in small aliquots, and stored at 4°.

ELISA Procedure for Human Hepatic Lipase

Coating Microtiter Plates with XHL-3-6. Wells are coated overnight at 4° with 200 μl XHL-3-6 that has been diluted to 2.5 μg purified immunoglobulin/ml with 0.02% NaN_3, 0.1 *M* carbonate–bicarbonate, pH 9.6. The wells are aspirated and washed six times with 0.05% (w/v) Tween 20 in PBS, pH 7.4.

Blocking Microtiter Plates. The wells are blocked with 300 μl of 1% bovine serum albumin (BSA; ELISA grade from Sigma, St. Louis, MO), 0.05% Tween 20 in PBS, pH 7.4, for 2 hr at 37° or overnight at 4°. The wells are aspirated and washed six times as described in the previous paragraph.

Sample Preparation. To express antigenic determinants fully, the standards and samples are preincubated at 37° for 1 hr in the presence of 10 m*M* sodium dodecyl sulfate (SDS), 1% BSA, 0.05% Tween 20, 10 m*M* phosphate, pH 7.4, and then diluted to 1 m*M* SDS, 1% BSA, 1 *M* NaCl, 0.05% Tween 20, 10 m*M* phosphate, pH 7.4, before addition to wells. This is accomplished conveniently by preparing the following buffers:

- Buffer 1: 1% BSA, 0.05% Tween 20, 10 m*M* phosphate, pH 7.4 (prepare also a 10× buffer 1)
- Buffer 2: 1% BSA, 2 *M* NaCl, 0.05% Tween 20, 10 m*M* phosphate, pH 7.4
- Buffer 3: 1% BSA, 1 *M* NaCl, 0.05% Tween 20, 10 m*M* phosphate, pH 7.4
- Buffer 4: 1% BSA, 1 *M* NaCl, 1 m*M* SDS, 0.05% Tween 20, 10 m*M* phosphate, pH 7.4

For the processing of plasma samples, mix 42.5 μl plasma or plasma diluted in PBS, 2.5 μl of 200 m*M* SDS, and 5.0 μl of 10× buffer 1. Incubate at 37° for 1 hr. Dilute by adding 50 μl buffer 2 and 400 μl buffer 3. Similarly, the standard HL is preincubated in buffer 1 supplemented with 10 m*M* SDS for 1 hr at 37°, adjusted to 1 *M* NaCl and 1 m*M* SDS with buffer 2 and buffer 3, respectively. The standard solution is then diluted with buffer 4 to generate solutions containing 0 to 1 ng HL per 200 μl.

Assay. Two hundred microliters processed plasma sample and standards are then added per well to microtiter plates that have been blocked as described above and incubated overnight at 4°. Following the overnight incubation the plates are aspirated and washed six times with PBS, 0.05% Tween 20. Monoclonal antibody XHL-1-6, conjugated to horseradish peroxidase, is then diluted (1:2000 or 1:3000) in 1% BSA, PBS, 0.05% Tween 20, added to wells, and incubated overnight at 4°. After removal of solutions from the plates, they are first washed six times with PBS, 0.05% Tween 20 and then six times with 0.75 *M* NaCl, 0.05% Tween 20, 10 m*M* phosphate, pH 7.4. Then, 200 μl substrate solution is added to each well with a multichannel pipetting device. The horseradish peroxidase substrate contains 0.4 mg/ml *o*-phenylenediamine in 0.01% H_2O_2, 0.1 *M* citrate–phosphate buffer, pH 5.0. The microtiter plate is incubated at room temperature 10 to 30 min. The reaction is stopped by the addition of 50 μl of 2.5 *M* H_2SO_4 to each well, and the plate is read at 490 nm on an ELISA reader. A typical standard curve is presented in Fig. 1.

Comments

The precision of the method is highly dependent on the proper washing of plates. Washing of plates six times reduces the value of the chemical blanks containing all constituents except plasma or standard HL. Inclusion

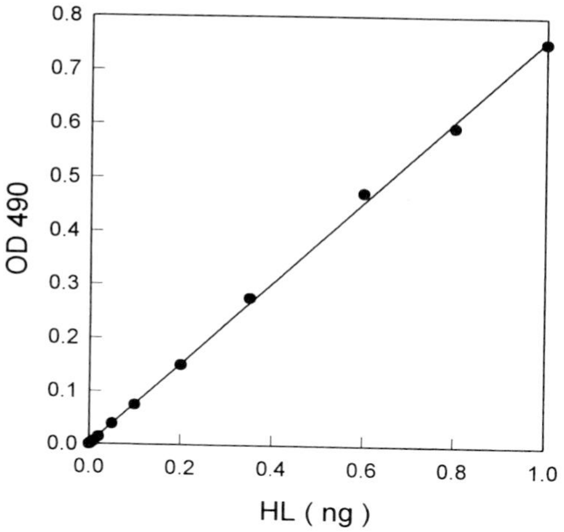

FIG. 1. Standard curve for highly purified human HL assayed as described in the text. The conjugate was diluted 1:4000 and the plate developed for 18 min. Each point represents the average of two observations which differed on average by less than 2.7%.

of 0.75 *M* NaCl in the wash solution following conjugate removal also contributes to reduction of blank values. The average chemical blank is 0.079 ± 0.008 (mean ± S.D.; $n = 7$). The wells are washed with either an automated plate washer or a manifold connected to a vacuum trap. In either case it is important to achieve a complete emptying of wells. It is also essential to never allow the plates to dry and to standardize the length of time the wells are left empty. When the above precautions are followed, all wells in the plate can be employed and no edge effect is observed.

The samples are incubated initially in buffer 4, a buffer containing 1% BSA, 1 *M* NaCl, 1 m*M* SDS, 0.05% Tween 20, 10 m*M* phosphate, pH 7.4. The presence of BSA, high salt molarity, and ionic and nonionic detergents decreases nonspecific interactions.

The standard curves representing OD_{490} values versus amount of HL per well are adequately fitted by a linear function. The mean r^2 for five standard curves was 0.998 ± 0.001 with a slope of 1.10 ± 0.05 ng HL/1 OD_{490} unit. For five normolipemic subjects, the pre- and post-heparin plasma values were 7.5 ± 5.1 and 189 ± 70 ng/ml, respectively. In the same subjects, time courses of HL plasma values following intravenous injections of 100 U heparin/kg body weight indicated that the peak values were between 15 and 20 min. The presence of heparin in plasma or standards, between 0 and 10 U/ml, does not affect the readings observed. Varying the SDS concentration in the initial incubation of samples in wells, between 0 and 10 m*M*, shows that a maximum in HL expression is achieved at 1 m*M* SDS for both purified HL and post-heparin plasma samples.

The ELISA described above is very sensitive and repeatable. Duplicates of 0.05 ng HL analyzed on seven plates differed by 1.1 ± 0.8%. For duplicate wells containing 1 ng HL the difference between the readings was 2.0 ± 2.5%. The plate to plate variation is quite minimal. For five replicates of 0.2 ng/well analyzed the same day on five different plates, the OD_{490} was 0.235 with a coefficient of variation of 7%; at 0.8 ng/well, the OD_{490} was 0.876 and the coefficient of variation 5%. It is advisable to have a standard curve on each plate when high precision is required. For most samples, HL can be detected reliably in 17 μl of pre-heparin plasma and 1 μl of post-heparin plasma per well.

[24] Quantification of Cholesteryl Ester Transfer Protein: Activity and Immunochemical Assay

By KEVIN C. GLENN and MICHELE A. MELTON

Introduction

Plasma cholesteryl ester transfer protein (CETP) plays an important role in establishing the lipid composition and levels for plasma lipoproteins [high-density lipoproteins (HDL), low-density lipoproteins (LDL), and very-low-density lipoproteins (VLDL)].[1] The 70- to 74-kD acidic plasma glycoprotein CETP mediates balanced exchange of cholesteryl esters (CE) and triglycerides (TG) between lipoproteins. A number of experimental lines of evidence have established CETP as a critical factor governing plasma lipoprotein cholesterol levels including transgenic animal studies,[2,3] work with antibodies that block CETP activity,[4,5] studies of CETP involvement in diet-induced coronary artery disease,[6] and species variation in plasma CETP.[7]

Perhaps the most compelling data, however, come from the identification of several genetic deficiencies in the human CETP gene (splicing and missense defects).[8,9] Individuals that are homozygous for CETP gene defects have been shown to have HDL cholesterol levels up to 3- to 4-fold higher than normal (100–250 mg/dl) and LDL cholesterol levels that are

[1] A. R. Tall, *J. Lipid Res.* **34,** 1255 (1993).

[2] K. R. Marotti, C. K. Castle, R. W. Murray, E. F. Rehberg, H. G. Polites, and G. W. Melchior, *Arterioscler. Thromb.* **12,** 736 (1992).

[3] T. Hayek, N. Azrolan, R. B. Verdery, A. Welsh, T. Chajek-Shaul, L. B. Agellon, A. R. Tall, and J. L. Breslow, *J. Clin. Invest.* **92,** 1143 (1993).

[4] M. E. Whitlock, T. L. Swenson, R. Ramakrishnan, M. T. Leonard, Y. L. Marcel, R. W. Milne, and A. R. Tall, *J. Clin. Invest.* **84,** 129 (1989).

[5] T. Swenson, C. B. Hesler, M. L. Brown, E. Quinet, P. P. Trotta, M. F. Haslanger, F. C. Gaeta, Y. L. Marcel, R. W. Milne, and A. R. Tall, *J. Biol. Chem.* **264,** 14318 (1989).

[6] E. Quinet, A. R. Tall, R. Ramakrishnan, and L. Rudel, *J. Clin. Invest.* **87,** 1559 (1991).

[7] Y. C. Ha and P. J. Barter, *Comp. Biochem. Physiol.* **71B,** 265 (1982).

[8] M. L. Brown, A. Inazu, C. B. Hesler, L. B. Agellon, C. Mann, M. E. Whitlock, Y. L. Marcel, R. W. Milne, J. Koizumi, H. Mabuchi, R. Takeda, and A. R. Tall, *Nature (London)* **342,** 448 (1989).

[9] K. Takahashi, X.-C. Jiang, N. Sakai, S. Yamashita, K. Hirano, H. Bujo, H. Yamazaki, J. Kusunoki, T. Miura, P. Kussie, Y. Matsuzawa, Y. Saito, and A. Tall, *J. Clin. Invest.* **92,** 2060 (1993).

METHODS IN ENZYMOLOGY, VOL. 263

up to 2-fold lower than normal (35–150 mg/dl).[10] Individuals homozygous for the CETP defect lack detectable levels of CETP protein or activity. Homozygotes have enlarged apolipoprotein (apoE)-rich HDL_2 that appears to be able to bind the apoB/E or LDL receptor.[11] The CETP gene defects appear in 0.81% of the Japanese population (48% of these people have HDL of >120 mg/dl). Heterozygotes show intermediately elevated HDL_2 and decreased LDL levels. Both heterozygotes and homozygotes appear to be otherwise healthy individuals.[10]

In addition, numerous demographic and physiological studies have linked regulation of CETP levels to changes in plasma HDL cholesterol. For example, it is well established that aerobic exercise and alcohol consumption are ways to raise plasma HDL cholesterol. A decrease in plasma CETP activity is one of the strongest changes shown to accompany the rise in HDL cholesterol associated with increased exercise.[12] By comparison, changes in other lipid- or lipoprotein-metabolizing enzymes such as lipoprotein lipase (LPL), hepatic lipase (HL), and lecithin–cholesterol acyltransferase (LCAT; phosphatidylcholine–sterol *O*-acyltransferase) correlated poorly, if at all, with exercise-induced changes in HDL. Likewise, a 28% increase in HDL due to alcohol consumption correlated with a 27% decrease in CETP concentration and 22% decrease in CETP activity.[13] Probucol,[14] insulin-dependent diabetes mellitus (IDDM),[15] and obesity[16] have been shown to correlate with elevated CETP protein levels, activity, or tissue mRNA and reduced HDL. Nephrotic syndrome (NS) appears to lead to elevated CETP levels that correlate well with high VLDL, LDL, and apoB.[17] Corticosteroid treatment of NS patients normalized their CETP activity levels and, concomitantly, normalized their VLDL plus LDL cholesterol ester and apoB levels.

[10] A. Inazu, M. L. Brown, C. B. Hesler, L. B. Agellon, J. Koizumi, K. Takata, Y. Maruhama, H. Mabuchi, and A. R. Tall, *N. Engl. J. Med.* **323,** 1234 (1990).

[11] S. Yamashita, D. L. Sprecher, N. Sakai, Y. Matsuzawa, S. Tarui, and D. Y. Hui, *J. Clin. Invest.* **86,** 688 (1990).

[12] R. L. Seip, P. Moulin, T. Cocke, A. Tall, W. M. Kohrt, R. Ostlund, and G. Schonfeld, *Arterioscler. Thromb.* **13,** 1359 (1993).

[13] M. Hannuksela, Y. L. Marcel, Y. A. Kesäniemi, and M. J. Savolainen, *J. Lipid Res.* **33,** 737 (1992).

[14] G. Franceschini, M. Sirtori, V. Vaccarino, G. Gianfranceschi, L. Rezzonico, G. Chlesa, and C. R. Sirtori, *Arteriosclerosis* (*Dallas*) **9,** 462 (1989).

[15] J. Kahri, G. C. Viberti, P.-H. Groop, M.-R. Taskinen, and T. Elliott, *Diabetes Care* **17,** 412 (1994).

[16] T. Arai, S. Yamashita, K.-I. Hirano, N. Sakai, K. Kotani, S. Fujioka, S. Nozaki, Y. Keno, M. Yamane, E. Shinohara, A. H. M. Waliul Islam, M. Ishigami, T. Nakamura, K. Kameda-Takemura, K. Tokunaga, and Y. Matsuzawa, *Arterioscler. Thromb.* **14,** 1129 (1994).

[17] P. Moulin, G. B. Appel, H. N. Ginsberg, and A. R. Tall, *J. Lipid Res.* **33,** 1817 (1992).

Given the close relationship between CETP and lipoprotein cholesterol levels in animals and humans, and because VLDL and LDL are considered atherogenic while HDL is associated with decreased atherogenic risk, accurate measurement of CETP activity and protein levels is becoming increasingly important.

Cholesteryl Ester Transfer Protein Activity Assay

Overview

The assay for CETP activity common to most publications consists of incubating [^{3}H]HDL (1–10 μg cholesterol) and unlabeled LDL (1–10 μg cholesterol) at various donor-to-acceptor ratios from 1:1 to 1:10 with a source of CETP activity for 2 to 24 hr.[18] Aliquots are removed from the reaction mixture at timed intervals or at the end of the incubation period, and the LDL fraction is precipitated by the addition of heparin/$MnCl_2$. The amount of CE transferred from HDL to LDL is measured as the difference between the amount of [^{3}H]CE radioactivity remaining in the supernatant HDL fraction and the amount of [^{3}H]CE HDL in a control reaction lacking CETP.

We have established a rapid, high sample capacity CETP activity assay utilizing the Packard (Downer Grove, IL) TopCount 96-well plate liquid scintillation counter and Millipore (Bedford, MA) MultiScreen filtration plates that permit direct measurement of CETP-mediated transfer of CE into LDL by counting the radioactivity in the precipitated LDL. In this filter plate-based CETP activity assay, higher or lower levels of CETP activity directly correspond to higher or lower amounts of labeled CE incorporated into the precipitated LDL.

Materials

- Tris–saline–EDTA buffer (TSE): 50 m*M* Tris, pH 7.4, 150 m*M* NaCl, 2 m*M* ethylenediaminetetraacetic acid (EDTA)
- [^{3}H]Cholesteryl ester-labeled high-density lipoprotein ([^{3}H]CE-HDL): EDTA is added to 150 ml fresh plasma for a final concentration of 2 m*M*. The density of the EDTA-treated plasma is adjusted to 1.11 g/ml by the addition of solid NaBr prior to subjecting the plasma to centrifugation at 45,000 rpm at 10°, in a Beckman (Fullerton, CA) 50.2Ti rotor for 24 hr. The $d > 1.11$ g/ml fraction (HDL plus plasma proteins) is recovered from the bottom one-third of each tube, pooled, and dialyzed against 12 liters (4 liters, three changes) of TSE.

[18] R. E. Morton and J. V. Steinbrunner, *J. Lipid Res.* **31,** 1559 (1990).

The dialyzed plasma fraction is transferred to a vessel containing a small stir bar. Using a tuberculin syringe and a fine gauge needle (26 gauge or smaller), approximately 1 ml ethanol containing 1 mCi [1,2,6,7-^{3}H(N)] cholesterol (New England Nuclear, Boston, MA; 70–100 Ci/mmol) is injected into the plasma with stirring. 2-Mercaptoethanol (2-ME) is added to the plasma fraction at a rate of 26.7 μl 2-ME/100 ml plasma fraction. Likewise, penicillin and streptomycin are added to the plasma fraction for final concentrations of 50 units/ml and 50 μg/ml, respectively. The container is flushed with argon gas, sealed tightly, and incubated at 37° in a shaking water bath for 18 hr to allow esterification of the radiolabeled cholesterol by LCAT endogenous in the $d > 1.11$ g/ml plasma fraction.[19] The [^{3}H]CE-HDL is isolated by adjusting the density of the incubated plasma fraction to 1.21 g/ml with solid NaBr prior to subjecting it to ultracentrifugation at 45,000 rpm for 40 hr, at 10° in Beckman 50.2Ti rotor. The top one-third of each tube is removed, pooled, and dialyzed against 12 liters of TSE containing 0.02% (w/v) sodium azide (4 liters, three changes). Routinely, greater than 90% of the radiolabeled material in the purified [^{3}H]CE-HDL preparation is esterified cholesterol. The specific activity of the [^{3}H]CE-HDL is typically about 35,000 disintegrations per minute (dpm)/μg total cholesterol (total cholesterol levels in the preparation are determined enzymatically)

Low-density lipoprotein (LDL): Lipoproteins in the density range 1.019 to 1.063 g/ml are isolated from human plasma by sequential ultracentrifugation.[20] The isolated lipoprotein fraction is extensively dialyzed against 0.9% NaCl, 0.02% (w/v) EDTA, 0.02% (w/v) sodium azide, pH 8.5. The total cholesterol level in the preparation is determined enzymatically

Assay buffer: TSE containing 1% (w/v) bovine serum albumin (BSA; Sigma, St. Louis, MO)

Precipitation reagent: The working reagent is prepared by mixing equal volumes of dextran sulfate stock and $MgCl_2$ stock described below in a container suitable for storage at 4°. The working reagent is stable for at least 4 months[21]

Dextran sulfate stock: Dissolve 2.0 g dextran sulfate (Dextralip 50, Sigma) in 80 ml distilled, deionized water and adjust to pH 7.0 with HCl. [Note: The dry dextran sulfate powder is stored at 4° in a

[19] A. Tall, D. Sammett, and E. Granot, *J. Clin. Invest.* **77,** 1163 (1986).

[20] F. T. Hatch and R. S. Leese, *Adv. Lipid Res.* **6,** 1 (1968).

[21] P. S. Bachorik and J. J. Albers, this series, Vol. 129, p. 78.

desiccator.] Add 1.0 ml preservative solution [0.05% (w/v) gentamicin sulfate, 0.1% (w/v) chloramphenicol, and 5% NaN_3 (w/v) in distilled, deionized water] to the dextran sulfate solution and adjust the final volume to 100 ml with distilled, deionized water. The dextran sulfate stock solution can be stored at 4° for up to 6 months

$MgCl_2$ stock: A 1.0 *M* solution is made by dissolving 20.3 g $MgCl_2 \cdot 6H_2O$ (this hygroscopic reagent should be stored in a tightly closed container) in 80 ml distilled, deionized water and adjusting the pH to 7.0 with NaOH. The final volume is adjusted to 100 ml with distilled, deionized water for storage at 4° for up to 6 months

CETP standard: Chinese hamster ovary (CHO) cells stably transfected with human CETP cDNA were obtained from Dr. Alan Tall (Columbia University, New York, NY).[22] The "working" standards for the CETP activity assay are CETP preparations such as either unfractionated CHO-conditioned medium or recombinant CETP partially purified by butyl-Toyopearl column chromatography.[22]

Cholesteryl Ester Transfer Protein Activity Assay Procedure

To each well of a 96-well microtiter plate is added 150 μl assay buffer containing 66.7 μg/ml LDL cholesterol and 8.33 μg/ml [^{3}H]CE-HDL cholesterol. As such, LDL (acceptor lipoprotein) and [^{3}H]CE-HDL (donor lipoprotein) are delivered at an 8:1 ratio in each well. Dilutions of CETP standard in assay buffer, dilutions of test sample in assay buffer, and/or assay buffer alone (blank) are made such that 50 μl can be added per well, in triplicate. The final volume for each well is 200 μl (adjusted with assay buffer if necessary). The 96-well plate is subjected to gentle mixing on a multitube vortex mixer (Baxter Scientific Products, McGaw Park, IL) at level 1 to 2 for 15 to 20 sec prior to incubation at 37° for 2 hr. After incubation, 20 μl precipitating reagent is added per well. The plate is subjected to gentle vortex mixing as described above and incubated at room temperature for 10 min. From each well, a 180-μl aliquot is transferred to a MultiScreen 96-well filter plate (0.45-μm hydrophilic Durapore membrane, opaque plate (Millipore), and the filter plate is placed on a manifold (Millipore) for vacuum filtration as described by the manufacturer. Following filtration, 100 μl scintillation cocktail (Microscint 20, Packard) is added per well. The plate is sealed with Top Seal A (Packard) prior to being subjected to mixing on a Titer Plate Shaker (Lab-Line Instruments, Melrose Park, IL) at the highest speed for 30 min at room temperature. Radioactivity

[22] A. Inazu, E. M. Quinet, S. Wang, M. L. Brown, S. Stevenson, M. L. Barr, P. Moulin, and A. R. Tall, *Biochemistry* **31,** 2352 (1992).

within each well is counted directly with a TopCount scintillation counter (Packard). Transfer activity is expressed as the percentage [^{3}H]CE transferred from the supernatant donor HDL to the precipitated acceptor LDL using the following equation:

$$\% \text{ Transfer} = \frac{(\text{dpm}_{\text{Test}} - \text{dpm}_{\text{Blank}})}{(\text{Total dpm}_{\text{added/well}})} \times 100$$

Assay Performance

Approximately 5–10% of the labeled material in the [^{3}H]CE-HDL preparations typically transfers to LDL over the 2- to 4-hr assay incubation period at 37° in the absence of CETP and represents "background" or non-CETP-mediated lipid exchange. "Background" disintegrations per minute are routinely subtracted from the disintegrations per minute measured with test CETP samples prior to calculation of CETP activity in the test samples as given in the above equation. As shown in Fig. 1, the amount of [^{3}H]CE transferred from HDL to LDL is directly dependent on the concentration of CETP added to the incubation mixture. The amount of transferred [^{3}H]CE is linear with respect to CETP concentration over a range of 5–30% CE transferred in a 2-hr assay. The intraassay coefficient of variation (*CV*) for the activity assay was 9.91% for replicate samples within a plate. The interassay *CV* for a CHO-conditioned medium sample was 13.9%. The higher interassay *CV* for the activity assay highlights some of the variability seen with the MultiScreen 96-well plates that has forced a need to include controls on each individual plate.

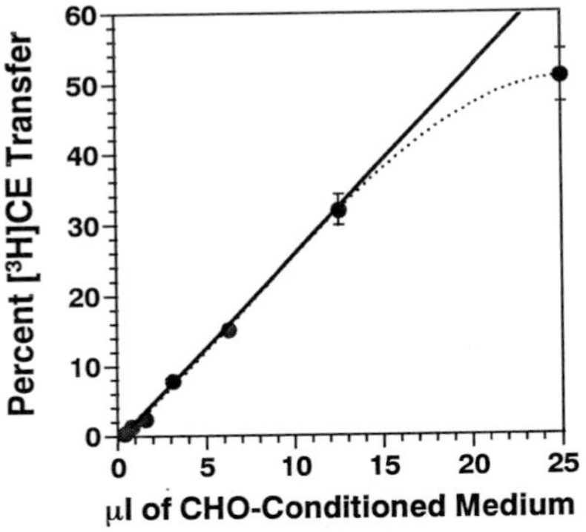

FIG. 1. Effect of CETP concentration in CHO-conditioned medium on transfer of [^{3}H]CE from HDL to LDL in the CETP activity assay. Means and standard deviations of triplicates are shown for each data point. The CETP activity assay is linear with respect to CETP concentration up to 30% CE transferred. $f(x) = 2.63x \pm 0.906$; $r^2 = 0.998$.

With respect to incubation time for the assay, it was found that CETP-mediated CE transfer was essentially linear for up to 24 hr at 37°. Typically, the amount of [^{3}H]CE transferred from HDL to LDL by CETP within 2 to 4 hr is sufficient for accurate measurement in the TopCount instrument.

Previous studies have demonstrated that CETP-mediated CE transfer activity is greatly influenced by the ratio of donor to acceptor lipoproteins.[23] In a series of experiments we investigated the effect of various donor : acceptor ratios on CETP activity. The apparent activity in our assay procedure increased with increasing donor : acceptor ratios from 1 : 1 to 1 : 16 (HDL : LDL). From these experiments we concluded that a donor : acceptor ratio of 1 : 8 (HDL : LDL) gave the best signal-to-noise ratio, while conserving lipoproteins. The results suggested that final assay concentrations of 6.25 μg/ml HDL cholesterol and 50 μg/ml LDL cholesterol were optimal. As configured above, the present CETP activity assay produces results that agree well with what has been reported previously.[18]

This assay can be used to measure CETP activity in a variety of serum or plasma samples as well as conditioned medium samples and samples from various purification schemes. We have used this assay to measure CETP activity in human, hamster, and rabbit serum samples. When measuring CETP activity in serum samples, up to 10 μl/well is assayed for 18–24 hr at 37°. The use of no more than 10 μl/well of plasma minimizes the amount of serum-derived lipoproteins in the assay. Testing only small amounts of whole plasma in the assay allows for minimal shifts in the critical donor : acceptor ratio, permitting transfer activity to remain in the linear range of the assay. However, this assay configuration of testing a small volume of plasma in the presence of calibrated levels of exogenous lipoproteins may not accurately reflect the level of CETP transfer activity *in vivo* since the endogenous lipoprotein ratios and amounts are not preserved.[24]

Discussion

The present CETP activity assay has proved to be reliable provided the following details are noted. (1) The best results are obtained when the initial incubation step is conducted in a separate 96-well microtiter plate and the reaction mixture is transferred to the Millipore MultiScreen only immediately prior to the final stages of the assay as described above. The assay cannot be incubated in the 0.45-μm Millipore MultiScreen plates. The filter membrane used in the 0.45-μm plate is not designed to retain

[23] J. Ihm, D. M. Quinn, S. J. Busch, B. Chataing, and J. A. K. Harmony, *J. Lipid Res.* **23,** 1328 (1982).

[24] M. Guerin, P. J. Dolphin, and M. J. Chapman, *Arterioscler. Thromb.* **14,** 199 (1994).

aqueous liquid in the wells for prolonged time periods required for the assay incubations. (2) The small inner diameter and volumes intrinsic to 96-well plates necessitates thorough mixing at each stage of the assay, especially during the initial setup and reaction incubation and again after addition of precipitation reagent for LDL/HDL separation as described above. In addition, the Millipore MultiScreen plates must be vigorously mixed after the addition of LS cocktail prior to counting for the most accurate measurements. With regard to the final mixing of the LDL precipitate with the LS cocktail, incubation of the plate overnight at room temperature is a suitable alternative to the vigorous mixing step discussed above. The advantage of vigorous mixing is that it permits counting of the plates 1 day earlier than would be possible by the alternative overnight method. (3) The presence of BSA in the assay buffer is critical. The most consistently reliable BSA preparation in this assay configuration has been fraction V from Sigma. (4) Significant lot-to-lot variability has been observed with the Millipore MultiScreen filter plates. Care should be taken, therefore, to ensure that new lots of plates are tested for complete trapping of the LDL precipitate. One method to assess retention of the LDL precipitate by the Millipore MultiScreen plates is to capture the supernatant flowthrough in a 96-well plate placed in the vacuum manifold at the time of filtration. Poorly performing Millipore MultiScreen filter plates will result in a visible level of precipitate being collected in the lower 96-well plate. (5) The highest level of accuracy for measuring radioactivity in the Millipore Multiscreen plates from the present ^{3}H-based CETP activity assay on the Packard TopCount necessitates generating a quench curve in order to calculate the disintegrations per minute per well. Weekly calibration of the Packard TopCount is also highly recommended by the manufacturer for optimal assay accuracy.

Cholesteryl Ester Transfer Protein ELISA

Overview

A widely used assay for measuring levels of CETP protein in samples is a radioimmunoassay (RIA) that utilizes the TP-2 monoclonal antibody to CETP.[25] This quantitative radioimmunoassay for CETP has been extensively used to measure CETP amounts in plasma samples from normolipemic subjects and numerous clinical situations.[1]

We have modified the TP-2-based CETP RIA to eliminate the need

[25] Y. L. Marcel, R. McPherson, M. Hogue, H. Czamecka, Z. Zawadzki, P. Weech, M. Whitlock, A. R. Tall, and R. W. Milne, *J. Clin. Invest.* **85,** 10 (1990).

for radiolabeled TP-2 while retaining comparable speed (results within 6 hr), assay sensitivity, and intra- and interassay coefficients of variance (*CV*). The present assay is an enzyme-linked immunosorbent assay (ELISA) involving competitive immunoreaction for unlabeled TP-2 monoclonal antibody between CETP absorbed to 96-well polystyrene plates and soluble samples of CETP. The amount of TP-2 bound to immobilized CETP is detected with an immunoconjugate of affinity-purified goat anti-mouse immunoglobulin G (IgG) linked to alkaline phosphatase. The measured level of TP-2 bound to immobilized CETP is indirectly proportional to the amount of soluble CETP competing for immunoreaction with a finite amount of TP-2.

Reagents

Coating buffer (CB): 15 m*M* Na_2CO_3, 35 m*M* $NaHCO_3$, 0.02% (w/v) NaN_3, pH 9.8

Washing buffer (WB): 138 m*M* NaCl, 8.1 m*M* Na_2HPO_4, 1.2 m*M* KH_2PO_4, 2.7 m*M* KCl, pH 7.2

Saturation buffer (SB): WB containing 1% (w/v) BSA, 1 m*M* EDTA, 0.02% (w/v) NaN_3

Triton buffer (TB): SB containing 0.2% (w/v) Triton X-100

Plates: Costar (Cambridge, MA) Stripette or Nunc (Naperville, IL) Maxisorp brand 96-well plates are recommended

CETP standard: Homogeneously pure recombinant CETP [i.e., >95% pure as judged by silver-stained sodium dodecyl sulfate (SDS)–polyacrylamide gel] was used to calibrate less purified recombinant CETP preparations that serve as reagents in the assay. Amino acid analysis of the pure recombinant CETP was performed to determine the protein concentration. Either homogeneously pure recombinant CETP or CETP partially purified by butyl-Toyopearl column chromatography[22] were used for absorption to 96-well plates and as soluble CETP standards

TP-2: Mouse hybridoma cells producing the TP-2 monoclonal antibody were obtained from Dr. Alan Tall.[26] TP-2 was purified from ascites fluid by protein A-Sepharose affinity chromatography[27]

Affinity-purified goat anti-mouse alkaline phosphatase conjugate (Promega, Madison, WI)

Phosphatase substrate [*p*-nitrophenyl phosphate from Kirkegaard & Perry Laboratories, Inc. (KPL), Gaithersburg, MD]

[26] C. B. Hesler, R. W. Milne, T. L. Swenson, P. K. Wech, Y. L. Marcel, and A. R. Tall, *J. Biol. Chem.* **263,** 5020 (1988).

[27] G. G. Krivi and E. Rowold, *Hybridoma* **3,** 151 (1984).

Procedure

Plate Coating. Partially purified CETP is diluted in CB to 0.4 μg/ml and dispensed at 100 μl/well into all but the top left three wells of 96-well plates. The same volume of CB, but lacking CETP, is added to the three upper left wells of the plate for later use as spectrophotometer blank. Following an overnight incubation at room temperature, the plates are rinsed twice with WB and 250 μl/well SB is added per well prior to incubation at 37° for 2 hr, room temperature for 4 hr, or 4° overnight. Plates are stored without further handling at −20° or are used immediately. No significant differences in assay performance were noted when purified CETP was used for plate coatings in place of partially purified CETP.

ELISA Procedure. The SB-treated plates are rinsed four times with WB. Dilutions of competing CETP standard or test sample are prepared in TB such that 100 μl is added per well, in triplicate. For each 96-well plate, the three uncoated wells and three CETP-coated wells each receive 100 μl TB to serve as spectrophotometer blanks and "0" CETP blanks, respectively. Six serial dilutions of frozen aliquots of working CETP standard are dispensed as 100-μl aliquots into triplicate wells [standards (ng/well) of 10, 20, 40, 80, 160, and 320]. The working CETP standards have previously been calibrated against highly purified CETP preparations that have been quantified by amino acid analysis. For every test sample, at least two dilutions are prepared and dispensed into triplicate wells per dilution. A solution of TP-2, 10 ml at a concentration of 100 ng/ml in TB, is prepared from an aliquot of stock TP-2 (typically 2–4 mg/ml). After dispensing 100 μl/well of the TP-2 solution into all 96 wells of each plate, the plates are incubated at 37° for 4 hr.

Each plate is rinsed four times with WB prior to addition of 100 μl/well goat anti-mouse alkaline phosphatase conjugate (a 1 : 7000 dilution of the Promega stock prepared using SB). The plate is incubated at 37° for 1 hr and rinsed four times with WB. To each well, 100 μl alkaline phosphatase substrate is added, and the plates are incubated at 37° for 30 min. The substrate is prepared fresh prior to use by dissolving 1 tablet per 1 ml of 5× alkaline phosphatase buffer concentrate and 4 ml Milli-Q-purified water. Typically, a 30-min incubation at 37° of plates containing the alkaline phosphatase substrate should produce an optical density (OD) of at least 0.300 in the "0" CETP blank wells when read at 405 nm with a reference wavelength of 630 nm. The non-CETP coated wells are used to blank the 96-well plate reader. The CETP standard curve results are subjected to linear regression analysis by a Microsoft Excel spreadsheet template. The equation produced by performing linear regression analysis of the log of CETP concentration relative to the mean OD from the ELISA is then used to calculate CETP levels in test samples from the mean of the triplicate ELISA OD values for each sample. For each sample, the final value for CETP protein mass is obtained by averaging the CETP concentrations measured for each sample dilution that generated OD values within the linear range of the CETP standard curve.

Assay Performance

As shown in Fig. 2, the linear range for the CETP ELISA is between typically 10 and 300 ng/well, or 0.1 to 3 μg/ml for a test sample. Partially purified CETP is used as the routine "working" standard for the ELISA. Highly purified CETP, quantified by amino acid analysis, has been used to calibrate the CETP preparations used as ELISA standards. The intraassay *CV* for the ELISA for more than one assay operator and over a 6-month period (more than a dozen assays) is 10.5% for replicate determinations within a single assay. The interassay *CV* for a CHO-conditioned medium sample tested by more than one assay operator and over a 6-month period (more than a dozen assays) was 11.5%.

The present ELISA method was used to measure the amount of CETP in the plasma of 10 normolipemic individuals. The mean ($\pm$ standard devia-

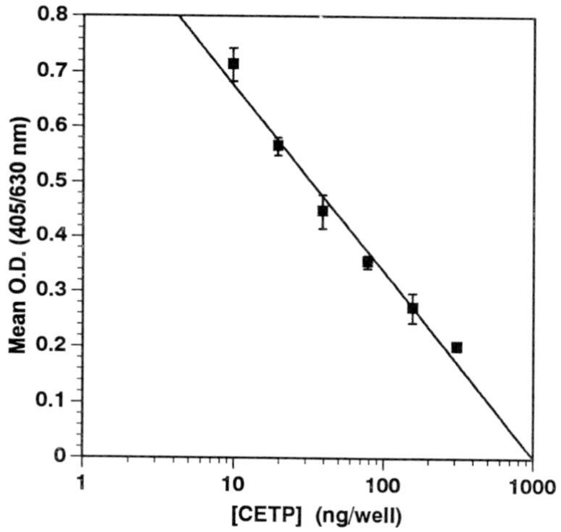

FIG. 2. Effect of CETP concentration on measured optical density in the CETP ELISA using the TP-2 monoclonal antibody. Mean and standard deviations of triplicates are shown for each data point. The linear regression equation derived from the log of the CETP concentration and ELISA OD is $f(x) = 0.146 \ln(x) + 1.01$; $r^2 = 0.981$. Substitution of test sample OD values into the equation permits calculation of CETP protein mass for the given dilution of a test sample.

tion) was 0.93 ± 0.44 μg/ml. The absolute value for the CETP concentration in human plasma appears lower in the present ELISA compared to the previously reported TP-2-based RIA,[25] possibly owing to differences in calibration of the working CETP standards used in the ELISA relative to the RIA. However, the mean value is in close agreement with results from a two-site enzyme immunoassay of CETP concentration in human plasma.[28]

Discussion

The present CETP ELISA has proved to be reliable and durable provided the following details are noted. (1) The brand and style of the 96-well immunoassay plate are critical for adequate and consistent absorption of CETP to the solid phase during plate preparation. Deviation from the recommended plate style is advisable only following side-by-side comparison of ELISA results between the recommended plates and any new plate style. (2) Each batch of CETP-coated plates and CETP standards requires titration of the CETP preparation being used relative to the previous batch of plates or standards to ensure consistency over time. (3) Sufficient quantities of CETP-coated 96-well plates and CETP standards should be prepared per batch to last 4 to 6 months when stored frozen at or below −20°. Preparation of plates and standards on a more frequent basis makes it more difficult to maintain consistency in results over time. Conversely, storage of batches for more than 6 months has been found to be problematic owing to degradation of coated plates or standards with prolonged storage. (4) The presence of Triton X-100 in the TB buffer used during the initial incubation is essential for accurate quantitation of CETP. Triton X-100 and BSA prevent nonspecific absorption of soluble CETP to the 96-well plates and prevent apoA-I interference in the measurement of CETP levels in serum sample as well.[25] (5) The length of time for the incubation of the alkaline phosphatase substrate at the final stage of the ELISA is not critical provided that the OD for the "0" CETP blank does not exceed the upper absorbance limit of the ELISA reader or that the OD for the highest concentration of the CETP standard is above the lower limits for accurate measurement by the plate reader. If time limitations prevent reading of the plate at the conclusion of the ELISA procedure, plates containing alkaline phosphatase substrate can be quick frozen and read at a future date following thawing. (6) Samples containing chaotropic or denaturing materials can significantly interfere with the ELISA by either impairing TP-2 reaction with CETP or dissolving absorbed CETP from the solid phase of the 96-well plate.

[28] H. Mezdour, I. Kora, H. J. Parra, A. Tartar, Y. L. Marcel and J.-C. Fruchart, *Clin. Chem.* **40,** 593 (1994).

Conclusions

The present CETP activity assay and ELISA are largely based on previously developed activity and immunoassay procedures.[18,25] However, in their present configurations, the CETP activity assay and ELISA embody significant improvements. The present activity assay significantly increases sample capacity compared to previous methods. The ELISA improves on the RIA by eliminating the need for radioactively labeled TP-2. Both assays are rapid, accurate analytical tools for measuring levels of CETP in a variety of serum or plasma samples as well as conditioned medium samples and samples from various purification schemes. We have used both assays to measure CETP activity and mass levels in human, hamster, and rabbit serum samples.

Acknowledgments

We are grateful to Dr. Elaine Krul for guidance and advice during preparation of the manuscript. We thank Dr. Daniel Connolly, Scott Vogt, Debbie Heuvelman, Roger Monsell, Charles Lewis, and Linh Vu for providing various reagents including conditioned medium containing recombinant CETP protein and TP-2, or preparations of partially purified or highly purified CETP preparations as CETP standards for both assays. We also wish to thank Dr. Alan Tall and Dr. Suke Wang for providing the CETP-transfected CHO cell line and TP-2 hybridoma cell line as well as technical advice for using these reagents in published RIA and activity assays.

[25] Determination of Apolipoprotein mRNA Levels by Ribonuclease Protection Assay

By ALANA MITCHELL and NOEL FIDGE

Introduction

The recognition that apolipoproteins (apo) are the major determinants of lipoprotein structure and function, together with emerging evidence of roles for some apolipoproteins, such as apoA-I and apoE, in cellular processes not directly related to their functions in the transport of cholesterol, ensure a continued research interest in the regulation of genes encoding these important proteins. The term "gene regulation" encompasses a number of different biochemical levels at which studies may be undertaken. For example, detailed studies of the specific sequences which define the functional promoter of genes are yielding information on which of the known transcription factors and hormone receptors bind the DNA of regulatory regions to activate gene transcription. The role of putative regulatory elements identified in upstream regions can now be tested using *in vitro* systems in which the promoter under investigation is linked to a reporter gene, the product of which provides a sensitive assay for determining promoter responsiveness to hormone or drug treatments.

For the protein constituents of lipoproteins, and for the plasma factors known to influence the metabolism of circulating lipoproteins, such as lecithin–cholesterol acyltransferase (LCAT, phosphatidylcholine–sterol *O*-acyltransferase), lipoprotein lipase, hepatic lipase, and cholesteryl ester transfer protein (CETP), considerable information on gene structure, and particularly those regions of the gene which regulate tissue-specific transcription, has been reported.[1–3] With the exception of experiments involving transgenic animals that overexpress genes in the appropriate tissues, much of the work designed to define promoters and regulatory regions adjacent to genes has, of necessity, been performed in artificial or reconstituted systems. The need remains to assess the effectiveness of possible regulatory regions by testing them *in vitro* in tissue culture models and *in vivo* in whole animals.

Another level of gene expression relates to the synthesis of the end product, that is, the protein encoded by a particular gene. This is important in the lipoprotein field, where investigators frequently study the effect of environmental factors such as dietary components, drugs, and hormones on the synthesis and secretion of apoproteins. In combination with protein chemistry and immunology, molecular biology provides information about any changes in content of the functional end product of gene expression and about the mechanism by which such changes may arise.

A basic component in the understanding of regulatory mechanisms is the measurement of an early product of gene transcription, namely, cytoplasmic mRNA. In cases where transcriptional regulation is thought to dominate the control process, the level of mRNA tends to parallel that of the mature protein end product. Where posttranscriptional mechanisms are operative, this simple relationship is lost, and alternative explanations for observed changes must be sought.

This chapter considers a basic method for measuring specific mRNA sequences in samples of total RNA extracted from animal tissue or cells in culture. It describes a modification to the ribonuclease protection assay

[1] W. S. Simonet, N. Bucay, S. J. Lauer, and J. M. Taylor, *J. Biol. Chem.* **268,** 8221 (1993).

[2] A. Walsh, N. Azrolan, K. Wang, A. Marcigliano, A. O'Connell, and J. L. Breslow, *J. Lipid Res.* **34,** 617 (1993).

[3] L. B. Agellon, A. Walsh, T. Hayek, P. Moulin, X. C. Jiang, S. A. Shelanski, J. L. Breslow, and A. R. Tall, *J. Biol. Chem.* **266,** 10796 (1991).

(or RPA) that simplifies the performance of multiple RPAs in a single sample of RNA.

A number of methods has been developed for the estimation of specific mRNA levels, including Northern blot analysis,[4] dot-blot hybridization,[5] polymerase chain reaction (PCR)[1]-based techniques,[6,7] and solution hybridization methods such as ribonuclease or S1 nuclease protection assays.[8–10] Northern blot analysis is relatively insensitive, a disadvantage that often necessitates the extra step of selecting mRNA from total cytoplasmic RNA (>90% of which is ribosomal RNA) before the quantitation of mRNA with a low level of expression. Dot-blot hybridization may be similarly insensitive. Competitive reverse transcriptase (RT)-PCR offers the greatest sensitivity for quantitating mRNA concentrations[11]; however, technical difficulties may arise related to the availability of a suitable fragment of DNA to provide the competing internal control.[11]

Ribonuclease protection assay (RPA) has the advantages of being sensitive and not requiring specialized instrumentation. This chapter describes a procedure for applying RPA to the semiquantitative determination of mRNA in a sample of total RNA isolated from cells in tissue culture or from animal tissue.

Principle

The assay of mRNA by RPA is achievable because double-stranded RNA is not susceptible to degradation by ribonuclease (RNase). The hybrid fragment formed when RNA is incubated with a small fragment of its antisense sequence is not digested by RNase, whereas all other RNA in the sample is degraded. A preliminary requirement in the setup of the RPA is a suitable cDNA clone for the preparation of a region of antisense mRNA for the protein of interest. In the presence of RNA polymerase, the cDNA, and the four ribonucleoside triphosphates (rNTPs), one of which is labeled in the α position with ^{32}P, a short antisense fragment (~150–400 bases) is transcribed. Incubation of the target RNA with a substantial molar

[4] P. S. Thomas, *Proc. Natl. Acad. Sci. U.S.A.* **77,** 5201 (1980).

[5] B. A. White and F. C. Bancroft, *J. Biol. Chem.* **257,** 8569 (1982).

[6] M. Becker-Andre and K. Hahlbrock, *Nucleic Acids Res.* **17,** 9437 (1989).

[7] G. Gilliand, S. Perrin, K. Blanchard, and H. F. Bunn, *Proc. Natl. Acad. Sci. U.S.A.* **87,** 2729 (1990).

[8] D. A. Melton, P. A. Krieg, M. R. Rebagliati, T. Maniatis, K. Zinn, and M. R. Green, *Nucleic Acids Res.* **12,** 7035 (1984).

[9] D. B. Thompson and J. Sommercorn, *J. Biol. Chem.* **267,** 5921 (1992).

[10] M. C. O'Donovan, P. R. Buckland, and P. McGuffin, *Nucleic Acids Res.* **19,** 3466 (1991).

[11] P. D. Siebert and J. W. Larrick, *Nature (London)* **359,** 557 (1992).

excess of the fragment drives the formation of hybrid regions of double-stranded RNA, the approximate length of the added probe. The addition of RNase to the hybridization mixture causes digestion of single-stranded RNA, originating from either the target or the probe, to leave a protected fragment that is isolated by polyacrylamide gel electrophoresis.

Ribonuclease Protection Assay

Materials

cDNA in a vector with RNA polymerase promoter sequence(s), prepared as described below
^{32}P-Labeled antisense RNA, prepared as described below
Target RNA
Formamide hybridization buffer: 80% (v/v) formamide, 40 m*M* piperazine-*N*,*N*′-bis(2-ethanesulfonic acid) (PIPES), pH 6.5, 400 m*M* NaCl, 1 m*M* EDTA
Yeast tRNA, 10 mg/ml in water
RNase digestion mix: 300 m*M* NaCl, 5 m*M* EDTA, 10 m*M* Tris–HCl, pH 7.5, containing 100 μg/ml RNase A (50 units/mg) and approximately 4.0 μg/ml RNase T1 (>350,000 units/mg)
Proteinase K, 10 mg/ml in water
10% (w/v) SDS
Ficoll loading dye: 0.05% (w/v) bromphenol blue, 0.05% (w/v) xylene cyanol in 3% (w/v) Ficoll
5% (w/v) Polyacrylamide gel: 15.5 × 13.5 × 0.1 cm, in TBE buffer (90 m*M* Tris–borate, 2 m*M* EDTA, pH 8.0)

Preparation of Vector. The cDNA, either full-length or partial, coding for the protein of interest is first cloned into a plasmid vector that contains one or two promoters for the bacteriophage RNA polymerases SP6, T7, or T3. A number of such vectors is commercially available, for example, the pGEM series from Promega (Madison, WI) which contain both SP6 and T7 promoters, one on either side of the multiple cloning site. The various "SP6/T7" vectors may possess additional features for use in applications (e.g., blue/white color screening or the ability to direct single-stranded DNA synthesis) other than those described here.

To prevent the transcription of vector sequences contiguous with the inserted cDNA sequences, it is first necessary to linearize the plasmid at a point distal to the RNA polymerase promoter to be used for transcription. The choice of promoter and therefore RNA polymerase depends on the orientation of the inserted cDNA fragment to allow synthesis of antisense RNA in the *in vitro* transcription reaction. Another consideration is the length of the RNA to be synthesized. Although the promoters are highly specific for the appropriate bacteriophage RNA polymerase, eukaryotic DNA may contain sequences that cause failure in elongation. In our experience, there is no way of determining beforehand which cloned sequences will fail to give rise to a full-length transcript. Each potential RNA probe should be tested empirically for the presence of termination regions. The choice of cDNA of less than 500 bp is a precaution against the occurrence of premature termination.

In preparation for transcription, the template is linearized using a restriction enzyme with a digestion site present in the multiple cloning region on the opposite side to the promoter of choice, or within the cDNA insert if the clone is over 500 bp. The restriction enzyme chosen should yield either a 5′ overhang or a blunt end on digestion at its recognition site. The creation of a 3′ overhang should be avoided as it is likely to cause problems by promoting the occurrence of artifactual protected fragments following RPA, which are independent of the target mRNA. Complete linearization of the template is essential to avoid long transcripts that extend into vector sequences and may deplete the reaction mixture of substrate. It is worthwhile to run a small aliquot of the reaction mixture on an agarose gel to check for complete digestion before proceeding with the transcription reaction.

Although only 0.5 to 1.0 μg linearized template is used in the transcription reaction, it is efficient to cut enough plasmid DNA for a number of probe syntheses. The linearized template is extracted with phenol–chloroform, the aqueous phase is precipitated with ethanol, and the DNA is resuspended at a concentration of 0.25 μg/μl. It is advisable to check the recovery of plasmid DNA by agarose gel electrophoresis before proceeding.

Preparation of ^{32}P-Labeled Antisense RNA Probe. The conditions for the preparation of ^{32}P-labeled RNA are adapted from those of Melton *et al.*[8] The reaction mixture contains, in a volume of 20 μl, 0.5 μg template DNA, 40 m*M* Tris-HCl, pH 7.5, 6 m*M* $MgCl_2$, 2 m*M* spermidine, 10 m*M* NaCl, 10 m*M* dithiothreitol (DTT), 20 units RNasin ribonuclease inhibitor (Promega), 0.5 m*M* ATP, GTP, and CTP, 15 μM UTP, 30–60 μCi [α-^{32}P]UTP (3000 mCi/mol), and 20 units RNA polymerase. To avoid the possibility of DNA precipitating in the presence of 2 m*M* spermidine at 4°, the reaction mixture is maintained at room temperature during the addition of the various components. The reaction continues for 30 to 45 min at 37°. To remove the template, DNase 1 (Worthington, Freehold, NJ) is added to a concentration of 50 μg/ml and incubation continued for 10 min at 37°. The reaction mixture is diluted to 100 μl with 10 m*M* Tris-HCl, 1 m*M* EDTA, pH 8.0 (TE) and two 1-μl samples are applied to small squares of DEAE paper to allow assessment of incorporation of the radioactive precursor. One square is set aside as a measure of total radioactivity. The second is washed three times for 5 min in 0.5 *M* phosphate buffer, pH 7, to remove unincorporated [α-^{32}P]UTP and so give a measure of incorporated label. Carrier yeast tRNA (1 μl of 10 mg/ml) is added to the remainder of the reaction mixture, which is then extracted once with phenol–chloroform and the RNA in the aqueous phase precipitated by the addition of 0.5 volumes of 8 *M* ammonium acetate and 3 volumes ethanol. After 30–60 min at −20°, RNA is collected by centrifugation; the pellet is air-dried and resuspended in 100 μl hybridization solution (see below).

From the percent incorporation of precursor UTP and using the assumptions that the average molecular weight of a nucleotide residue in RNA is 330 and that UTP incorporation represents one-quarter of the total nucleotide in the newly synthesized RNA, it is possible to calculate the amount of labeled RNA produced and its specific radioactivity. Under the conditions described, we routinely attain specific activities of 1 to 4 $\times$ 10^8 counts/min (cpm)/μg RNA. Higher values may be obtained by including additional [^{32}P]UTP or by lowering the concentration of unlabeled UTP, although this has limitations as the conditions for transcription are severely compromised by the low availability of UTP. We have made RNA probes of higher specific radioactivity by using UTP at 7.5 μM and found most of the product to be full-length; however, these conditions should be tested for individual probes. The yield of ^{32}P-labeled RNA, as measured by the percent incorporation of [^{32}P]UTP, tends to vary appreciably among probe preparations, even when the same batch of linearized template is used. For T7 RNA polymerase, we have found that the range is from 30 to 80% incorporation, whereas with SP6 RNA polymerase, incorporation is routinely less than 30%. Usually, the yield of antisense RNA probe greatly exceeds the amount required, so a lower level of incorporation does not present a problem.

RNA Extraction Procedures. To extract cytoplasmic RNA from rat tissue, we use the method of Chomcynski and Sacchi.[12] This method, which employs acid phenol extraction for the purification of RNA from DNA and protein, is not difficult to adapt to a relatively large number of samples, and we find that 0.5 g liver tissue (which is first frozen in liquid nitrogen and then stored at −70°) is a convenient amount to process. The method may be scaled up or down as required. Total RNA from cultured cells is prepared by the method of Gough[13] starting with cells grown to 70–80% confluence in dishes of 6 cm diameter, in Dulbecco's modified Eagle's medium (DMEM) supplemented with 10% (v/v) fetal bovine serum.

Assay Procedure. The conditions for RPA are based on those of Melton *et al.*[8] An aliquot of total RNA containing from 1 to 20 μg is ethanol-

[12] P. Chomczynski and N. Sacchi, *Anal. Biochem.* **162,** 156 (1987).

[13] N. M. Gough, *Anal. Biochem.* **173,** 93 (1988).

precipitated and redissolved in 20 μl of 80% formamide hybridization solution which already contains the ^{32}P-labeled antisense RNA. A control hybridization containing the same amount of yeast tRNA as sample RNA is also performed; it serves as a confirmation that no bands are detected where no target sequence is present, and may provide sound diagnostic information if artifactual bands appear. Various published methods for RPA recommend that the probe and the target RNA be coprecipitated at this first stage. However, we have found no difference in the results from either method, and suggest that there is considerable advantage in minimizing exposure to radioactivity by adding the labeled probe at the time of resuspension of the sample RNA. This procedure also ensures uniform hybridization conditions in all samples. The amount of probe added is probably best determined empirically, since the major consideration of including a molar excess of probe in the hybridization must be weighed against minimizing the possibility of high background on gels and unnecessary radiation hazards. For reasonably abundant mRNA species, such as apoA-I in liver extracts, we use, per hybridization, approximately 0.15% of the total probe synthesized in a standard reaction.

Samples are heated to 85° for 5 min, then quickly transferred to an incubator at 45° to allow hybridization to proceed for approximately 16 hr. The removal of single-stranded RNA is effected by the addition of 200 μl RNase digestion buffer and incubation for 30 min at 37°. The RNase is inactivated by the addition of 13 μl of 10% sodium dodecyl sulfate (SDS) and 5 μl of 10 mg/ml proteinase K and a further 10-min incubation at 37°. At this stage, an alteration to the established methods effects an appreciable saving in time and handling. The phenol–chloroform extraction step following proteinase K digestion is omitted, and, instead, RNase-digested samples are diluted with 250 μl TE. The protected RNA fragment(s) are precipitated in the presence of 10 μg carrier yeast tRNA by the addition of 1 ml ethanol and storage at −20° for 20 min. This time is critical in order to avoid the coprecipitation of NaCl and/or SDS, which interferes with electrophoretic migration of the samples.

The RNA is pelleted by centrifugation at 14,000 *g* for 10 min at room temperature. The supernatant is carefully and quantitatively removed from the RNA pellet and from the walls of the microcentrifuge tube by aspiration with a finely drawn-out Pasteur pipette. The air-dried pellets are resuspended in 10 μl Ficoll loading dye for polyacrylamide gel electrophoresis on a 5%, nondenaturing gel. In the event of large pellets arising after the centrifugation step, extra Ficoll loading dye should be used to lower the salt concentration of the sample (e.g., 20–25 μl). We have found that despite some distortion of the electrophoretic front by excess salt, the final separation is usually only minimally affected.

Quantitation of protected fragments on the dried gels is by autoradiography to locate the regions of gel corresponding to protected RNA species and excision of the bands for liquid scintillation counting or by phosphorimaging analysis on a Bioimager BAS-1000 (Fuji, Japan).

Adaptation of Basic Ribonuclease Protection Assay Method to Allow Multiple Assays in Single Sample. Depending on the research intentions, it may be advantageous to assay a number of specific mRNA species in a limited amount of test material, in which case a desirable feature of the RPA is that it can be adapted for multiple assays in a single sample. However, we found, when first attempting multiple assays, that problems arise when the mRNAs of interest differ over a wide range of expression levels. In this situation, the specific radioactivities of the antisense RNA probes will undoubtedly require adjustment to decrease interference by background radioactivity from the more highly expressed RNA. An example of this approach is illustrated below in the specific example provided (Figs. 1A and 2A). Another requirement for the performance of multiple assays in a single hybridization mixture is that the sizes of the protected fragments differ sufficiently to allow their separation by nondenaturing polyacrylamide gel electrophoresis. Consequently, we developed an empirical approach to determining the appropriate dilution of specific radioactivity.

For each cDNA, antisense RNA is prepared in two parallel reactions which differ only in that one of them contains no labeled UTP. The yield of unlabeled RNA is assumed to be equal to that for the reaction containing labeled UTP. For each different probe/target RNA system of study, RPAs are initially performed with the probes present individually. Quantitation of each protected fragment then indicates the relative representation of the corresponding mRNA in the target RNA sample. This information is used to calculate the appropriate dilution of the radioactive RNA preparation with the unlabeled RNA.

Alternative methods for the assay of multiple mRNA species based on solution hybridization have been described elsewhere.[9,10] These methods employed 5′-end-labeled oligonucleotides as probes. However, where cDNA is available, the use of RNA probes has significant advantages over the use of oligonucleotide probes. The latter requires the synthesis of oligonucleotides of increasing length as the number of mRNAs increases, and it necessitates higher resolution polyacrylamide gel electrophoresis. Additionally, the greater stability of RNA–RNA over DNA–RNA hybrids reduces the background as a consequence of the use of high stringency hybridization conditions, and the scope exists for preparing RNA probes of very high specific radioactivity. With either method, a single preparation of radioactively labeled probe is sufficient for the hybridization of hundreds of samples.

Inclusion of Internal Standard. Although the RPA method described minimizes the number of steps at which RNA may be lost during the assay procedure, two steps remain in which RNA is collected by precipitation and resuspended that are potential contributors to loss of material. These occur at the first step in setting up the hybridization reaction and during the final pelleting of the protected fragment(s) before electrophoresis. Such losses can be determined by including an internal standard to measure recovery. Actin mRNA is frequently the internal standard of choice since it is assumed to be constitutively synthesized and therefore unlikely to change during treatments. We have found that this assumption is not necessarily correct. Certain hormonal treatments do alter the mRNA levels of actin in rat liver. On the other hand, for situations where actin mRNA levels are shown not to change as a result of treatment, the inclusion of an internal standard tends to improve the quality of data. Where actin mRNA levels are subject to change between test groups and the inclusion of an internal standard is desirable, the suitability of glyceraldehyde-3-phosphate dehydrogenase mRNA, which is also thought to be constitutively synthesized, should be explored. With careful application of the RPA method, the reproducibility should obviate the need to correct for losses of sample or final protected product.

Specific Application of Ribonuclease Protection Method to Measure Apolipoproteins A-I and B and Hepatic Lipase

We have used both the basic RPA method and its adaptation in the assay of multiple samples for the determination of mRNA levels of apoA-I, apoB, and hepatic lipase in extracts of rat and human hepatoma cell lines and rat liver. The fragments (and vectors) for the various templates were as follows: rat apoA-I, a 326-bp *Eco*RI–*Hin*dII fragment in pGEM3Z; rat apoB, a fragment of approximately 200 bp in pGEM4Z; rat hepatic lipase, a 236-bp *Eco*RI–*Pst*I fragment in pGEM4Z; human apoA-I, a 254-bp *Sac*I–*Xho*I fragment in pGEM4Z; human hepatic lipase, a 453-bp *Sac*I–*Eco*RV fragment in pGEM4Z; and rat actin, a 389-bp *Eco*RI fragment in pGEM3Z.

Figures 1 and 2 illustrate application of RPA to the measurement of mRNA for apoA-I and hepatic lipase in cytoplasmic RNA from Hep G2 (human hepatocellular carcinoma) cells or total RNA from rat liver, using a single probe or multiple probes with adjusted specific radioactivities, respectively. Note that the presence of multiple bands in Fig. 1A, lane 5, and Fig. 2A, lane 4, is due to incomplete homology of the rat actin probe with

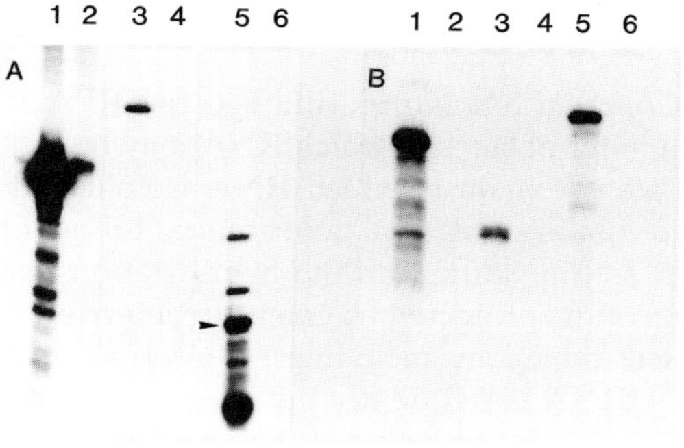

FIG. 1. Ribonuclease protection assay of apoA-I and hepatic lipase. Ten micrograms total RNA or, for controls, yeast tRNA was used for each assay. (A) Hep G2 cell RNA. For lanes 1, 3, and 5, RNA was hybridized with probes to human apoA-I, human hepatic lipase, and rat actin, respectively. The arrow in lane 5 indicates the protected actin band selected for quantitation. Lanes 2, 4, and 6 are the corresponding tRNA controls. (B) Rat liver RNA. For lanes 1, 3, and 5, RNA was hybridized with probes to rat apoA-I, rat hepatic lipase, and rat actin, respectively. Lanes 2, 4, and 6 are the corresponding tRNA controls.

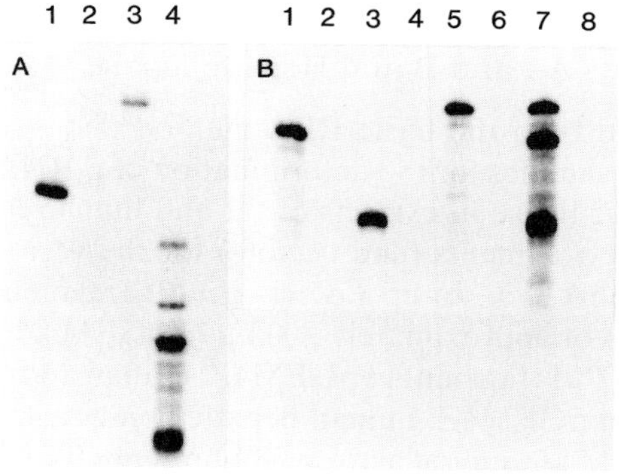

FIG. 2. Ribonuclease protection assay of apoA-I and hepatic lipase mRNA after adjusting the specific radioactivity of probes. (A) Hep G2 cell RNA. For lanes 1, 3, and 4, RNA was hybridized with probes to human apoA-I, human hepatic lipase, and rat actin, respectively, after dilution of the specific radioactivities as described in the text. Lane 2 shows a tRNA control to which the three probes were added. (B) Rat liver RNA. For lanes 1, 3, and 5, RNA was hybridized with probes to rat apoA-I, rat hepatic lipase, and rat actin, respectively. For lane 7, RNA was hybridized with all three probes. Lanes 2, 4, 6, and 8 are the corresponding tRNA controls.

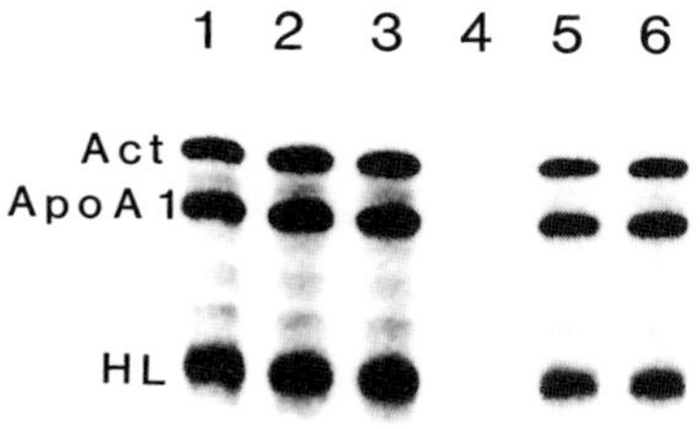

FIG. 3. Replicate RPAs on rat liver RNA. Twenty micrograms total RNA was hybridized simultaneously with probes to rat apoA-I, rat hepatic lipase, and actin. Lanes 1–3 are replicate assays from one sample of rat liver RNA. Lanes 5 and 6 are assays of a sample of liver RNA extracted from a second rat. Lane 4 is the tRNA control.

the human actin mRNA present in the target sample. The nonhomologous regions are digested during treatment of the hybridization mixtures with RNase, whereas stretches of homology appear as a reproducible pattern of protected fragments. As shown in Fig. 1A, the level of apoA-I mRNA in Hep G2 cells is very much higher than that of hepatic lipase (or actin). Quantitation of this difference showed that the ratio of counts per minute in apoA-I versus hepatic lipase was 50:1. Figure 2A shows the same data after adjusting the specific radioactivities by dilution with the corresponding unlabeled antisense RNA. Similar data are presented for rat liver in Figs. 1B and 2B. For rat liver, the disparity between levels of individual mRNAs

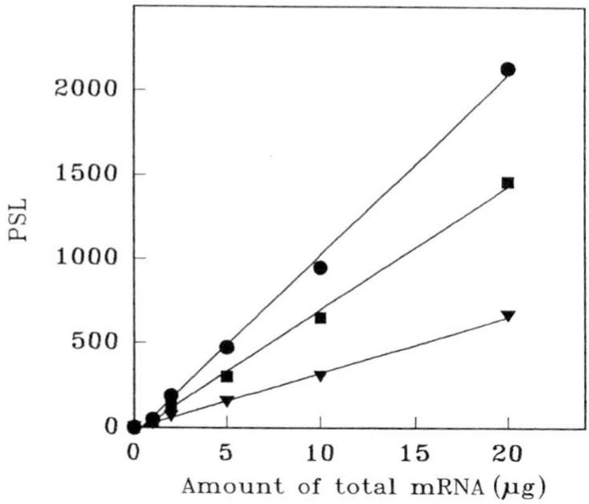

FIG. 4. Sensitivity of the multiple RPA. Amounts of total rat liver RNA in the range 1 to 20 μg were adjusted by the addition of yeast tRNA so that all assays contained 20 μg RNA and were hybridized simultaneously with probes to rat apoA-I (▼), rat hepatic lipase (■), and actin (●). The final dried gels were analyzed by phosphorimaging.

within the sample is much less pronounced. Note that the exposure of gels to autoradiographic film was eight times longer for the assay of Hep G2 cell RNA than that for assay of rat liver RNA.

The reproducibility and sensitivity of the method are demonstrated by Figs. 3 and 4, respectively. Figure 3 shows replicate data from total RNA prepared from two rats, and Fig. 4 demonstrates that over the range of 1–20 μg total rat liver RNA, the assay is linear. Finally, Fig. 5, showing the effect of treatment of rats with simvastatin and cholestyramine on the level of hepatic apoB mRNA, demonstrates the relative "robustness" of the assay, which provided usable data where the quality of the RNA preparations did not allow measurement of the very large apoB mRNA by Northern analysis.

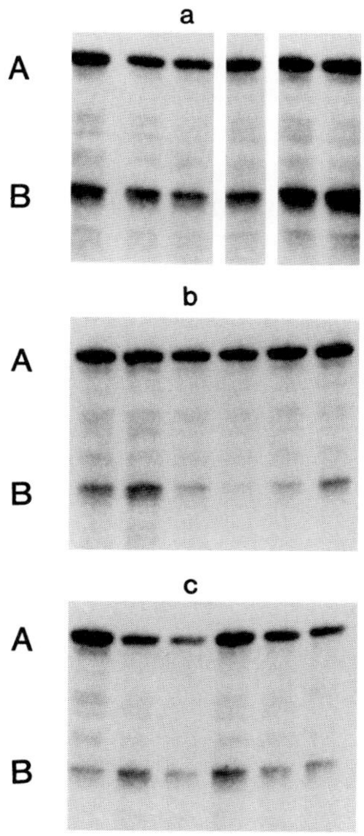

FIG. 5. Ribonuclease protection assay of apoB mRNA in samples of rat liver total RNA: (a) control, (b) simvastatin treatment, and (c) cholestyramine plus simvastatin treatment. (A) Fragment protected by actin probe. (B) Fragment protected by apoB probe. [Reprinted with permission from A. Mitchell, N. Fidge, and P. Griffiths, *Biochim. Biophys. Acta* **1167,** 9 (1993).]

Summary

The ribonuclease protection assay procedure described enables the relative quantitation of either single mRNAs or multiple mRNA species simultaneously in a sample of total RNA and demonstrates its applicability to two systems of relevance to the study of apolipoproteins, namely, liver tissue and liver-derived cell lines in culture. The main requirements of the method are the availability of cDNA cloned into a vector that directs the transcription of antisense RNA for the preparation of radioactive probes, and choice of suitable restriction endonuclease sites for linearizing the cDNA so that the final protected products of the various mRNA species are sufficiently different in size to allow their separation. For moderately abundant apolipoprotein mRNAs in rat liver, the method is sensitive down to 1 μg total RNA. Other experimental sources of RNA or the assay of less abundant mRNA species may require a larger amount of starting material. These aspects of the assay need to be determined for each probe/tissue system to be studied. The need for adjustments to the specific radioactivity of the probes will depend on the relative abundance of the target mRNA molecules and can be readily determined empirically. The rationale for varying the specific radioactivities to facilitate multiple assays as suggested here is both simple and effective. A preliminary assay provides information on the relative levels of the mRNA species of interest, and the step of preparing, in parallel, a sample of unlabeled antisense RNA provides the means for quantitative dilution of specific radioactivity of the ^{32}P-labeled RNA probe where required.

Acknowledgments

This work was supported by funding from the National Health and Medical Research Council of Australia. We thank Dr. Domenic Autelitano for helpful discussion of the method used for the ribonuclease protection assay.

Author Index

Numbers in parentheses are footnote reference numbers and indicate that an author's work is referred to although the name is not cited in the text.

A

Aalto-Setala, K., 178, 209
Aalto-Setälä, K., 23, 25–26
Adams, P. C., 231, 232(60), 233(60), 237(60)
Aden, D. P., 177
Adolphonson, J. L., 65, 77(5)
Agard, D. A., 174, 182(33)
Agellon, L. B., 339–340, 352
Aggerbeck, L. P., 10
Agnani, G., 51, 59(156)
Ailhaud, G., 15–16, 47, 51(139), 59(148), 283, 297, 320, 322
Akanuma, Y., 179
Aladjem, F., 38
Alaupovic, P., 22, 32, 36, 38–40, 40(65), 41, 41(34), 42(74, 75), 47–48, 48(75), 49(62, 74, 75, 141), 50, 50(75), 51, 51(138, 150), 52–53, 53(75, 140), 54–56, 56(140), 57, 57(140, 157, 160), 58(172, 173), 59, 59(75, 140, 150, 157, 168, 178, 191), 67, 75, 163, 167, 168(3), 169(3), 211, 218(13)
Albee, D., 263
Albers, H. W., 43, 310, 311(5), 312(5), 313(5)
Albers, J. J., 11, 14–15, 41, 51, 53, 55–56, 56(183), 57(183), 58, 59(167), 65–66, 77(5), 78, 144, 221, 225(15), 231(15), 232(15), 233(15), 234(15), 235(15), 237(15), 253, 260, 269, 342
Alberti, K. G. M. M., 231, 232(60), 233(60), 237(60)
Alihaud, G., 49
Alix, J. F., 42
Allan, C., 284, 303, 305, 308(26)
Allan, C. M., 262
Alpers, D. H., 8, 16, 284
Alsayed, N., 197, 211, 217(20)
Altkemper, I., 92
Altrocchi, P. H., 39
Amouyel, P., 53, 59(165)
Anantharamaiah, G. M., 11, 261, 267, 270–271, 272(19), 279(19)
Anchors, J. M., 148
Andersen, I., 322
Anderson, D. W., 255, 262
Anderson, G. E., 120
Anderson, R. A., 53, 57(163)
Andersson, S., 139, 140(39)
Andres, D. W., 130
Andrews, E. P., 235
Antonarakis, S. E., 19
Aoki, T., 154, 159(59), 160(59)
Aoyama, T., 32
Apfelbaum, M., 26
Apfelbaum, T. F., 284
Appel, G. B., 340
Appelbaum, T. F., 297
Aragon, I., 262
Arai, H., 23, 173
Arai, T., 340
Arbeiter, K., 17
Argaves, W. S., 312
Argimon, J. M., 182
Armstrong, V. W., 29–30, 59, 180, 221, 222(16), 234, 234(16), 240
Arnold, K. S., 19, 65, 172, 177
Aronow, B. J., 28, 309–310, 310(1), 312
Arveiler, D., 53, 59(165)
Ashworth, J. N., 34
Assmann, G., 12–13, 32, 44, 48(105), 204, 253, 261, 271, 279(20)
Aston, C. E., 293
Asztalos, B. F., 252, 255(7), 257(7), 262

Atkinson, D., 18, 150
Atland, K., 12
Atmeh, R. F., 54
Attar, M., 43
Attie, A. D., 13
Auget, J. L., 53, 59(166)
Augustin, J., 334
Austin, M. A., 19–20, 58
Auwerx, J. H., 320
Avigan, J., 38, 40(55)
Avioli, L. V., 39
Avogaro, P., 154, 180, 181(63)
Avrameas, S., 247, 335
Aweryckx, J. P., 42
Ayrault-Jarrier, M., 42, 50, 51(150), 59(150), 262
Azen, S. P., 54, 58(172)
Azrolan, N., 22, 28, 30, 209, 339, 352

B

Babirak, S. P., 20, 319, 322(4), 327
Bachorik, P. S., 55, 56(182), 57(182), 77, 342
Backus, J. W., 8
Badimon, J. J., 31, 267
Badimon, L., 31, 267
Baer, B., 4
Baginsky, M. L., 334
Bailey, M. C., 11
Baker, H. N., 11, 21(68)
Baker, N., 146, 150(5)
Balasubramaniam, S., 154
Ball, M. J., 23
Bancroft, F. C., 353
Banka, C. L., 11
Baralle, F. E., 6–7
Barbaras, R., 15–16, 47, 51(139), 283, 297
Bard, J. M., 48, 51, 53(75, 140), 54–55, 56(140), 57(140), 59, 59(140), 59(156), 219
Bardon, S., 26
Barenghi, L., 6
Barja, F., 26, 311
Barker, W. C., 3
Barkia, A., 47, 51(139), 203, 210, 213(9), 268
Barlett, A., 197
Barlett, G. R., 133
Barnett, J., 12
Barnhart, R. L., 14
Baron, D., 154
Barr, D. P., 37, 53(47)
Barr, M. L., 343
Barr, S. I., 199
Barrett, P. H. R., 126, 132(25), 138(25)
Barter, P. I., 297
Barter, P. J., 47, 283, 339
Bartlett, A., 183
Baum, C. L., 148, 149(25)
Baumann, M., 17, 297
Bausserman, L. L., 203
Bayliss, J. D., 20
Bayse, D. D., 68
Beaty, T. H., 19
Beaubatie, L., 235
Beaudrie, K., 243, 276
Bebbington, C. R., 323
Becker-Andre, M., 353
Beg, Z. H., 9, 261
Beigel, Y., 282, 303–304
Beisaier, C. L., 42
Beisiegel, U., 42, 47, 150, 172, 226, 282, 283(2, 4), 285(2, 4), 303–304
Bekaert, E. D., 50, 51(150), 59, 59(150), 262
Bell, G. T., 105
Belpaire, F., 211, 218(21)
Bendayan, M., 147, 153
Benditt, E. P., 26
Bengtsson, G., 327
Bengtsson-Olivecrona, G., 14, 319–320, 322, 327
Benninghoven, A., 12, 204, 261, 271, 279(20)
Bensadoun, A., 15, 69–70, 73(34), 320, 333–334
Bensen, M. D., 12
Benson, D. M., 16, 282, 297
Berg, K., 29, 43
Bergeron, N., 82–83, 86(7), 87(7), 89(7), 90(7), 91(7), 92(7), 93, 94(7)
Bernini, F., 262(34), 263
Bernstein, L., 154
Bernstein, S., 155
Bersot, T., 12, 14, 150, 175, 178(34), 262, 282, 285(1), 334
Bertics, S. J., 120, 122(2), 123(2), 129, 129(1), 144(1, 2)
Bertiere, M. C., 26
Bertrand, M., 14, 15(105), 53, 59(164), 253
Betsholta, C., 148
Bhatnager, P. K., 211
Bhown, A. S., 270

Bidwel, D., 183
Bidwell, D. E., 197
Bieber, L. L., 67, 70, 111, 112(12)
Bielicki, J. K., 283
Bier, D. M., 47
Bierman, E. L., 15, 19, 20(149)
Bihari-Varga, M., 30
Bilheimer, D. W., 150
Billardson, C., 221, 222(16), 234(16)
Billington, T., 150, 154(31)
Bingham, A., 53, 59(165)
Bisgaier, C. L., 12, 17, 51, 57(158), 59(157), 283, 284(16), 287(14), 297
Bishop, C., 203, 268, 270(9), 272
Bishop, C. A., 270
Bishop, J. M., 6
Bittolo Bon, G., 180, 181(63)
Black, A. S., 11
Black, D. D., 92
Blackett, P., 50, 51(150), 59(150)
Blackhart, B. D., 4, 10, 18, 64, 121, 125(11), 139, 139(11), 140(11), 142(11), 143(11), 144(11), 150, 167
Blades, B. L., 261
Blanchard, K., 353
Blanche, P. J., 130, 262
Blanchette, L., 167, 168(3), 169(3)
Blanco-Vaca, F., 46, 50(118)
Blankenhorn, D. H., 53–54, 58(172, 173), 59(168, 178)
Blatter, M.-C., 26, 311
Blix, G., 34
Blomback, M., 32
Blouquit, Y., 225
Blum, C., 44, 282–283
Boberg, J., 334
Bocan, T. M. A., 30, 225, 226(40), 227(40), 229(40)
Bode, C., 260
Boersma, W. J. A., 176
Boerwinkle, E., 17, 24, 28–29, 31, 56, 132, 133(32), 138(32), 225, 231, 238, 284
Boettcher, A., 11
Boguski, M. S., 3, 23, 45, 46(117), 284
Bohlen, P., 80, 81(55)
Bojanovski, D., 147, 149(12), 155(12)
Bolzano, K., 29
Bon, G. B., 154
Bond, M. G., 227
Bondjers, G., 57
Bonnet, D. J., 11
Boomsma, D., 29
Boquillon, M., 283
Borén, J., 57
Borensztain, J., 146, 147(2)
Borth, W., 223, 234(23)
Boström, K., 8
Bottcher, A., 311
Bottoms, L. A., 309
Boulet, L., 55
Bouma, M.-E., 10
Boyles, J. K., 136, 137(36), 138(36), 171, 173(9), 177
Bradford, M. M., 80, 325
Bradley, W. A., 18, 162, 171, 177
Brady, D. W., 80, 81(58), 94, 209
Brady, S. E., 333
Bragdon, J. H., 36, 37(35), 84, 252, 265
Brandstatter, E., 20
Branson, J., 154
Brasaemle, D. L., 15
Brasitus, T. A., 52
Brasseur, R., 45, 48(110)
Breckenridge, W. C., 22, 188, 207, 235
Brennhausen, B., 13
Breslow, J. L., 4–5, 7, 12, 19, 22–25, 28, 31, 44–45, 47(108), 121, 170, 176–178, 209, 260, 339, 352
Brewer, H. B., Jr., 9, 12–14, 17–18, 24–26, 28–29, 44–47, 51, 51(138), 59(151), 105, 147–149, 149(12), 150, 153, 155(12), 156(56), 178, 203, 252–253, 261, 268, 270(9), 272, 284–285, 293, 293(24)
Briner, W. W., 38
Brinton, E. A., 7, 19
Britten, M. L., 28, 30(216), 232
Brooks, A. R., 5
Brouillette, C. G., 11, 51, 57(158), 59(157), 270
Brousseau, M. E., 9
Brown, A. J., 26
Brown, B. G., 53, 59(167), 253
Brown, M. L., 339–340, 343
Brown, M. S., 21–22, 29, 44, 65, 177
Brown, R. K., 37, 38(41, 42)
Brown, S. A., 16, 24, 29, 31, 56
Brown, T. L., 310
Brown, V. W., 75
Brown, W. V., 22, 39, 40(70), 47, 80, 81(57), 85, 87(15), 144, 177, 212, 334
Bruckert, E., 54

Bruening, T., 11
Brun, J.-M., 180, 181(65)
Brunck, T., 239, 240(11)
Bruns, G. A. P., 170
Brunzell, J., 219, 220(3)
Brunzell, J. D., 19–20, 55, 56(183), 57(183), 243, 276, 319–320, 320(3), 322(4), 327
Bucay, N., 352, 353(1)
Bucholtz, C., 81, 190
Buchwald, W. F., 157, 164(76)
Buckland, P. R., 353, 358(10)
Bujo, H., 339
Bukberg, P. R., 75
Bunker, C. H., 293
Bunn, H. F., 353
Burant, C. F., 8
Burdick, B. J., 30, 225, 226(40), 227(40), 229(40)
Burkey, B. F., 12, 312
Burstein, M., 42
Bury, J., 176, 197, 199(15), 211, 217(17), 218(17, 21), 284, 303, 308, 309(29)
Busch, S. J., 311, 345
Butler, S., 21, 179
Byers, M., 148
Byrne, R. E., 162

C

Cachera, C., 203
Cahoon, B. E., 52, 57(161)
Cai, S. J., 8, 18(35), 45, 95, 105, 148
Caiati, L., 4, 64, 121
Cain, W., 29
Caines, P. S., 261
Callahan, H. J., 301
Camare, R., 54
Cambien, F., 53, 59(165)
Cambou, J. P., 53, 59(165)
Cameron, A. H., 39
Campbell, M., 209
Candelier, L., 51, 59(156)
Capurso, A., 12
Cardelli, J. A., 17, 298
Cardin, A. D., 84, 123, 127(18), 162
Cardot, P., 6
Carew, T. E., 21
Carlos, R. C., 148
Carlson, K., 73
Carlson, L. A., 13, 73, 83, 95–96, 96(4), 99(10)
Carlsson, J., 180
Carrasquillo, B., 16
Carson, S. D., 32
Carter, R. J., 68
Casal, D. C., 155
Casari, G., 7
Caslake, M. H., 20
Cassidy, D. K., 12
Castellino, F. J., 248
Castiñeiras, M. J., 182
Castle, C. K., 263, 339
Castro, G. R., 4, 15, 51, 59(151), 253, 263
Catapano, A. L., 254–255, 259(24), 262
Catty, D., 192
Cazzolato, G., 154, 180, 181(63)
Cerrone, A., 262
Chait, A., 320
Chajek-Shaul, T., 23, 209, 320, 339
Champagne, J., 11
Chan, L., 4, 8, 18, 18(35), 22(7), 45, 63, 95, 105, 148–151, 151(36), 152, 166, 319–320, 320(1), 321–322, 327–328, 329(13), 330(13)
Chang, A., 321
Chang, J. Y., 199
Chang, L. B. F., 297
Chang, V. T., 223, 234(23)
Chantler, S. M., 192
Chao, J., 84, 123, 127(18), 162
Chao, Y., 208
Chapman, J., 197, 211, 217(20)
Chapman, M. J., 221, 222(16), 225, 234(16), 345
Chappell, D. A., 24, 122, 128(17), 151, 173
Chataing, B., 345
Chatterton, J. E., 105, 121
Chen, C., 140, 141(41)
Chen, C. Y., 209
Chen, E., 4
Chen, E. Y., 28, 31(211), 220, 239
Chen, G., 263, 264(46, 47)
Chen, G. C., 86
Chen, P.-F., 150, 151(36), 152
Chen, S. H., 8, 18, 18(35), 45, 60, 63, 95, 105, 148–151, 151(36), 152, 319, 320(1), 321
Chen, X., 23, 209
Cheng, C.-F., 334
Cheng, J., 6, 12
Chen-Liu, L. W., 121
Chesebro, J. H., 31

Chesterman, C. N., 27
Cheung, M. C., 15, 41, 51, 53, 57(158), 59(157, 167), 253, 260
Chick, H., 33
Chiesa, G., 28, 30, 223, 231, 232(21), 238
Child, J. S., 21
Chin, H. P., 54, 58(172)
Chin-On, S., 10, 25(62)
Chivot, L., 53, 59(166)
Chlesa, G., 340
Chomcynski, P., 356
Chow, A., 22, 188
Christenson, R. H., 180
Christophe, J., 42, 211, 218(21)
Chu, J. W., 261
Chu, M. L., 4, 64, 121
Chuck, S. L., 10
Chung, B. H., 36, 204, 243, 268, 270, 272(11), 276, 279(11, 19), 280(11)
Chung, W., 23
Cianflone, K., 77
Cisar, L. A., 334
Civeira, F., 12
Claassen, E., 176
Cladaras, C., 5–6, 11, 18, 25, 121, 150
Claeys, H., 239
Claiborne, F., 122, 128(17), 151
Clark, W. A., 298
Clarke, H. R. G., 19
Clarke, J. G., 30
Clarkson, T. B., 30
Clavey, V., 16, 51, 59(156), 210–211, 213(9), 217(16), 219, 262, 268, 283, 297
Clay, M. A., 310
Clift, S. M., 13, 268
Cobbaert, C., 56
Cochran, S., 30
Cocke, T., 340
Cockett, M. I., 323
Cohen, J. C., 231, 238
Cohn, E. J., 34
Cohn, J. S., 95, 155, 167, 169(7), 219
Cohn, S. D., 95, 167, 169(7)
Colburn, K. A., 153
Cole, T., 8, 132–133, 170, 173, 284
Collet, X., 11, 261, 262(9)
Collins, D. R., 121
Comstock, L., 4
Cone, E. J., 157, 164(76)
Cone, J. T., 243, 276
Connelly, P. W., 22, 188, 204, 207, 207(25), 208, 333
Cook, K., 132, 133(32), 138(32)
Cook, M. E., 13
Cooper, A. D., 150
Cooper, G. R., 68, 78(33), 161, 264
Cooper, J. A., 31
Cooper, W. H., 161
Corder, C., 55
Corder, C. N., 164
Coresh, J., 19
Cornely-Moss, K., 15
Cornicelli, J. A., 322
Cornicelli, J. D., 327
Cornwell, D. G., 36–38
Coste-Burel, M., 53, 59(166)
Courtneidge, S. A., 6
Cox, D. W., 22
Craddock, N., 174
Craig, W. Y., 229, 235(56)
Cremer, P., 30
Cresswell, S. R., 10, 25(62)
Creutzfeldt, C., 29
Creutzfeldt, W., 29
Cristol, L. S., 237
Crone, M., 322
Cruickshank, J. K., 31
Cryer, A., 320
Csaszar, A., 29, 225
Cuatrecasas, P., 194, 325, 326(42)
Cullen, P., 20
Culver, H., 4
Cummings, A. M., 24
Cupp, M., 320
Curry, M. D., 211, 218(13)
Curtis, L. K., 65
Curtiss, L. K., 11, 19, 120, 122(2), 123(2), 129, 129(1), 132, 136, 137(36), 138(36), 144(1, 2), 155, 173, 179, 262, 264
Cushing, G. L., 30, 225, 226(35, 40), 227, 227(40), 229(40), 237(35)
Czamecka, H., 346, 350(25), 351(25)
Czazar, A., 24
Czekelius, P., 284

D

Dac, N. V., 262
Dachet, C., 54
Dack, N. V., 261, 262(9)

Daggy, B. P., 333
DaGue, B., 288
Dahlbäck, H., 139, 140(39)
Dahlén, G. H., 28, 43, 220, 225(9), 235(9)
Dairman, W., 80, 81(55)
Dalal, K. B., 151
Dallinga-Thie, G. M., 282, 295
Dallongeville, J., 26
Dammerman, M., 23
Dang, Q., 333
Dangerfield, W. G., 35
Dantzker, C., 284
D'Apice, A.J.F., 27, 43
Darnell, J. E., Jr., 6
Darwin, W. D., 157, 164(76)
Das, H. K., 170
Dashti, N., 51–52, 57, 57(157, 160), 59(157), 147, 156(9), 157(9), 162, 162(9)
D'Athis, P., 284, 285(32), 303, 305
Datta, S., 45
Daugherty, A., 18, 132–133, 177
Davidson, N. O., 8, 92, 148–149, 149(25), 154, 284, 297–298
Davignon, J., 26, 31, 55, 154, 167, 168(3), 169(3, 4), 170, 175(4), 223, 237(27)
Davis, C. E., 56
Davis, D. C., 36
Davis, D. L., 139, 140(39)
Davis, R. A., 148
Day, A. J., 46
Dayhoff, M. O., 3
Dean, R. T., 125, 127(23), 132(23)
de Backer, G., 260
DeBakey, M. E., 30, 225, 226(40), 227(40), 229(40)
Deckelbaum, R. J., 21, 121, 153
Deeb, S., 14, 19, 148
Deeb, Z., 320
Deegan, T., 261
De Farie, U., 32
DeGennes, J. L., 54
Deines, M., 148
De Keersgieter, W., 211, 218(21), 284, 303, 308
De Knijff, P., 29, 176
de Knijff, P., 176
DeLalla, L., 37, 38(42)
DeLalla, O., 35, 36(25, 26), 53(25)
DeLalla, O. F., 35, 37(25, 26)
Delamatre, J. G., 297
Delattre, S., 51, 59(156)
Delattre-Lestavel, S., 219
Delcroix, C., 46, 51(123)
DeLoof, H., 18
DeLy, C., 26
Demarquilly, C., 59, 253
Demosky, S. J., 9
Demosky, S. J., Jr., 9, 147, 149(12), 153, 155(12)
Demossky, J. S. J., 261
Deng, W. P., 139
Denke, M. A., 66
DePauw, M., 45, 48(110)
Dergunov, A. D., 178
Derr, J. E., 67
Derudas, B., 219
Desgnes, J., 303, 305
Desgres, J., 284, 285(32)
deSilva, H. V., 43, 46, 51(124), 209, 310–311, 311(5), 312(5), 313(5)
Deslex, S., 322
De Slypere, J. P., 8, 18(35), 45, 95, 105, 148, 260, 284, 303, 308
Deufel, T., 261
Deutsch, D. G., 241, 248(15)
Dieplinger, H., 17, 29, 220, 225, 232(37), 284, 303, 308
Dietel, M., 226
Dieterich, J.-H., 92
Dinh, D. M., 11
Disteche, C., 148
Dixon, J. L., 10
Dolphin, P. J., 47, 345
Donahue, R. P., 56
Donaldson, V. H., 84, 123, 127(18), 162
Dong, L.-M., 174
Donnerhak, B., 179
Dontula, K., 26
Doolittle, M. H., 116
Doolittle, R. F., 152, 153(46)
Dory, L., 283
Douglas, T. C., 17
Doumas, B. T., 68
Doumas, T. B., 68
Douste-Blazy, P., 53–54, 59(165), 176
Dowdee, M., 36
Downs, D., 36, 47, 51(138), 57, 59(191), 135, 158
Drabkin, H. A., 4, 45
Draisey, T. F., 261

Drauss, X. H., 10, 25(62)
Drayna, D. T., 4, 45
Drelich, M., 12
Drewek, M. J., 154
Driscoll, D. M., 171
Driscoll, R., 145
Drittenden, L. B., 6
Drouin, P., 54
Duba, H. C., 229, 230(53), 234(53), 235(53), 238
Dubois, B. W., 120, 129, 129(1), 136, 137(36), 138(36), 144(2)
Ducimetière, P., 53, 59(165)
Dudman, N. P. B., 261
Dufour, R., 31
Dujovne, C. A., 181
Dullaart, R. P. F., 18, 147
Dunn, D. S., 145
Dunn, J. K., 10, 25(63)
Duriez, P., 211, 217(16)
Duvic, C. R., 43, 220, 228(11), 310, 311(5), 312(5), 313(5)
Dvorin, E., 16, 282, 297
Dwulet, F. E., 12
Dwyer, T., 154
Dyer, C. A., 174, 179

E

East, C., 150
Eaton, D. L., 28–29, 31(211), 220, 221(8), 235(8), 236(8), 239
Eberle, E., 30
Eckel, R. H., 319–320
Eddy, R., 148
Edelstein, C. J., 9, 188, 260, 270
Eder, H. A., 36–37, 37(35), 42, 53(47), 84, 252, 265
Edge, S. B., 9, 105
Edgington, T. S., 132, 262
Edsall, J. T., 34
Edwards, P. A., 21
Edwards, T. J., 45
Edwards, Y. H., 8, 18(34), 95, 105, 148
Egusa, G., 80, 81(58), 94
Ehnholm, C., 17, 24, 26, 30, 173, 176, 297–298, 334
Eichberg, J., 67
Eichner, J. E., 284
Eisenbart, J. E., 223, 225(24), 234(24), 239, 240(11)
Eisenberg, C., 10
Eisenberg, S., 21, 24, 252
Eisenhauer, T., 59
Elam, R. L., 261
Ellefson, R. D., 224, 235(32)
Elliott, H., 35, 36(24), 58(44)
Elliott, T., 340
Elo, O., 180
Elovson, J., 105, 116, 121, 146, 148, 150(5), 154
Elshourbagy, N. A., 3, 284
Enerback, S., 320
Engchild, J., 32
Engvall, E., 306, 308(27)
Erickson, E. E., 227
Eriksen, N., 26
Erkelens, D. W., 18, 147
Erlich, H. A., 122, 128(16)
Esser, V., 177
Esterbauer, H., 223, 225(26)
Estlack, L. E., 30
Etienne, J., 320
Evans, A. E., 53, 59(165)
Evans, A. J., 207–208
Everson, B., 26
Ewens, S. L., 41
Ewing, A. M., 36

F

Fadley, A. M., 6
Fager, G., 14
Fairbanks, G., 156
Fairclough, G. F., 158, 162(79), 163(79)
Fairwell, T., 9, 153, 203, 252, 261
Falko, J. M., 185
Fan, J., 333
Fang, D., 211, 217(18)
Fantappie, S., 262
Farese, R. V., Jr., 120, 121(4), 122(4), 123, 123(4), 124, 126(4, 22), 127(4), 128(22), 129(22), 141, 143(7), 144(7)
Farmer, J. A., 43
Farnier, M., 180, 181(65)
Farr, A. I., 276, 280(28)
Farr, A. L., 66, 85, 90(17), 96, 161, 196, 200(13)
Farral, M., 20
Farren, B., 20
Farrer, M., 231, 232(60), 233(60), 237(60)
Faustinella, F., 321, 328, 329(13), 330(13)
Felber, B. K., 25

Felgenhauer, K., 180
Fellowes, A. P., 123
Felts, J. M., 263
Ferguson, D. G., 43, 310, 311(5), 312(5), 313(5)
Fernández, M., 327
Ferrell, R. E., 229, 232(57), 233(57), 234(57), 284, 293, 311
Ferris, E., 12
Ferrone, S., 183
Fesmire, J. D., 48, 53(140), 56(140), 57(140), 59, 59(140), 211, 218(13)
Feussner, G., 179, 219, 220(3)
Fidge, N. H., 150, 154(31), 262, 262(35), 263, 284–285, 297–298, 299(13), 300(13), 301, 301(13), 303, 305, 308(26), 351, 362
Fielding, C. J., 4, 11, 15, 23, 47–48, 49(143), 51, 178, 251–256, 259(24), 263, 272
Fielding, P. E., 11, 47–48, 49(143), 51, 178, 251–256, 259(24)
Fievet, C., 211, 217(16), 224, 225(33)
Fievet, P., 14, 15(105), 53, 59, 59(164), 253
Fiol, C., 182
Fisher, E. A., 23, 209
Fisher, M., 21, 179, 261
Fisher, W., 80, 81, 153, 162
Fiske, C. H., 37
Fitch, W. M., 3
Fitzpatrick, T., 16
Fless, G. M., 28, 31, 31(211), 220–221, 223–224, 225(24), 234(24), 238–240, 240(11), 244(1), 246, 248, 249(1)
Flynn, F. V., 27
Flynn, L. M., 141
Fogelman, A. M., 21
Föger, B., 16, 20
Fojo, S. S., 47, 203
Folsom, A. R., 28
Fong, L. G., 12, 163
Ford, A. L., 130
Foreback, C. C., 261
Forsbrooke, A. S., 39
Forster, N. R., 229, 235(56)
Forte, T. M., 52, 57(161), 75, 255, 262, 321, 327
Fortier, C., 4, 64
Fourrier, J. L., 14, 15(105), 53, 59(164), 253
France, D. S., 12, 322, 327
Franceschini, G., 12, 59, 262, 340
Francone, O. L., 23, 51, 252, 263
Frangione, B., 174
Frankenthal, K., 33
Franklin, F. A., 77
Frants, R. R., 29, 176, 179
Franz, S., 9
Fredrickson, D. S., 22, 38–39, 40(70), 41, 57(76), 75, 85, 87(15), 101, 212
Freeman, D. A., 223, 237(29, 30)
Freeman, N. K., 35–36, 36(25), 37(25), 53(25), 255
French, L. E., 310
Frick, M. H., 180
Friedl, C., 210
Friedman, G., 320
Frolich, J., 256, 272
Frost, P. H., 181
Frost, S. C., 298
Fruchart, J.-C., 14–15, 15(105), 16, 47, 49, 51, 51(139), 53–55, 59, 59(148, 151, 156, 164, 165), 125, 127(23), 132(23), 203, 210, 211, 213(9), 217(16), 219, 221, 224, 225(17, 33), 235(17), 253, 261–262, 262(9), 268, 283, 297, 350
Fryer, P. J., 81
Fu, M., 211, 217(18)
Fujimoto, K., 17, 298
Fujimoto, W. Y., 20, 319, 322(4)
Fujioka, S., 340
Fukasawa, M., 23, 173
Fukazawa, C., 179
Fukimoto, W. Y., 327
Fukuzaki, H., 153
Fuller, M., 18
Fumeron, F., 26
Funahashi, T., 172
Funakoshi, A., 285
Funke, H., 12–13, 32, 44, 48(105), 261
Furlong, C. E., 311
Furman, R. H., 36, 39, 40(65), 41, 67
Fuster, V., 31, 267

G

Gaeta, F. C., 339
Gainsborough, H., 33
Galloner, W. R., 220, 228(11)
Gambert, P., 23, 180, 181(65), 283–284, 285(32), 297
Game, F. L., 231, 232(60), 233(60), 237(60)
Gamliel, A., 282

Garber, D. W., 267, 273
Garcia, Z., 8, 150
Gardner, J. A., 33
Garfinkel, A. S., 320
Garg, A., 124, 126(22), 128(22), 129(22)
Garner, C. W., 11, 21(68)
Garrety, K. H., 283
Gartside, P. S., 209, 210(6), 211(6), 212(6)
Gascon, P., 178, 285
Gaubatz, J. W., 28, 30, 218, 220, 222, 225, 225(9), 226(35, 40), 227, 227(40), 229, 229(40), 230(54), 231(54), 232(54), 234(54), 235(50, 54), 237(9, 35)
Gaur, V. P., 221, 225(15), 231(15), 232(15), 233(15), 234(15), 235(15), 237(15)
Gavish, D., 19, 28, 30–31
Gawish, A., 261, 271, 272(19)
Gay, G., 10
Gedde-Dahl, T., Jr., 170
Geer, J. C., 36
Geigelson, H. S., 267
Gelfand, D. H., 122, 128(16)
Gelman, B. B., 180
Genest, J. J., Jr., 59
George, P. K., 211
George, P. M., 123
Gerritse, K., 176
Gesquiere, J. C., 203
Gesteland, R. F., 145
Getz, G. S., 146, 147(2), 171
Geuze, H. J., 18
Ghalim, N., 16, 47, 51(139), 59, 253, 283, 297
Ghanem, K. I., 28, 229, 230(54), 231(54), 232(54), 234(54), 235(54)
Ghiselli, G., 9, 26, 178, 261, 271, 282, 285, 303–304
Gianfranceschi, G., 262, 340
Gianturco, S. H., 18, 162, 171, 177
Gibson, J. C., 22, 47, 75, 177
Gijbels, M. J., 179
Gil, C. M., 46, 51(124), 311
Gil, G. M., 311
Gilbert, T., 11
Gilliand, G., 353
Gilligan, M., 23
Ginsberg, H. N., 10, 22–23, 47, 80, 81(57), 177, 209, 320, 340
Ginsburg, G. S., 6
Girardot, G., 180, 181(65)
Gitlin, D., 37–38
Giudici, G. A., 6
Glazier, F., 35, 36(25), 37(25), 53(25)
Glenn, K. C., 339
Glick, J. M., 283
Glickman, J. N., 147
Glickman, R. M., 42, 52, 147, 154, 282–283, 283(5), 284, 284(5, 16), 285(5), 287(14), 297, 303–304
Gliemaun, J., 150
Glover, J. S., 213
Glueck, C. J., 120, 209
Go, H.-C., 221
Go, M. F., 8, 284
Goate, A., 174
Goers, J. F., 322, 327
Gofman, J. W., 35, 36(24–26), 37, 37(25, 26), 38, 53(25, 45), 54(45), 58(44, 45)
Gohde, M. F., 9
Goldberg, A. C., 132, 133(32), 138(32)
Goldberg, I. A., 22
Goldberg, I. J., 17, 47, 320, 322, 327
Goldgaber, D., 174
Goldstein, J. L., 19, 20(149), 21–22, 29, 44, 65, 177
Gong, E. L., 15, 75, 262, 268
Goodman, D. S., 39
Gorder, N. L., 7, 16, 282, 297
Gordon, J., 3, 9, 16, 113, 154, 190, 255, 258(22), 260, 265, 284
Gordon, R. S., 188
Gorecki, M., 179
Gospodarowicz, D., 12
Gotham, S. M., 81
Goto, Y., 211, 218(22)
Gotoda, T., 25, 178–179
Gotohda, T., 179
Gotto, A. M., Jr., 3, 7–11, 14–18, 18(35), 21, 21(68), 22, 25(63), 28–30, 43, 45, 56, 63, 95, 105, 148, 150–151, 151(36), 152, 162, 171, 177, 211, 220, 222, 225(9), 227, 235(9, 50), 261, 262(34), 263, 270, 271, 282, 297, 303–304, 320, 322, 327
Gough, N. M., 356
Gould, B. J., 83, 95, 155, 167
Goulinet, S., 221, 222(16), 234(16)
Gouni, I., 320
Gracia, V., 182
Graham, D. L., 121
Graham, D. M., 37, 53(45), 54(45), 58(45)
Graiser, S. R., 264

Granot, E., 21, 342
Grant, S. G., 16
Grant, S. M., 18, 150
Gratzl, R., 17
Gray, M. E., 147
Green, M. R., 353, 355(8), 356(8)
Green, P. H., 282, 283(5), 284(5), 285(5), 303–304
Greene, M., 320
Greenman, B., 179
Greenwood, F. C., 213, 304
Greenwood, M. R. C., 327
Greeve, J., 8, 92
Gregg, R. E., 9–10, 24, 26, 47, 51(138), 147, 149(12), 153, 154(56), 155(12), 170, 175(4)
Grego, B., 301
Greguski, R. A., 181
Greten, H., 92
Gries, A., 220, 224, 225(10, 33)
Griffin, B. A., 20
Griffiths, P., 362
Grinstead, G. F., 224, 235(32)
Grob, E., 225
Grob, W., 225
Groenewegen, W. A., 128
Groot, P. H., 10, 25(62), 295, 340
Gross, W., 170, 179
Grothe, A. M., 55
Gruber, E., 30
Grundy, S. M., 19, 52, 58, 63, 65–67, 70(20), 76–77, 80, 81(58, 60), 94, 123, 144, 223, 237, 237(29, 30), 252
Gruppo, R. A., 310
Gylling, H., 252
Gu, Z.-W., 8, 18, 18(35), 45, 63, 95, 105, 148, 151
Gualandiri, V., 262
Gudnason, V., 6
Guerin, M., 345
Guevara, J., Jr., 220, 222, 229, 230(54), 231(54), 232(54), 234(54), 235(54), 239
Gueze, H. J., 147
Guo, H. C., 221, 222(16), 225, 234(16)
Guo, S., 19
Gurakar, A., 23, 28, 51, 263
Gurd, F. R. N., 34
Gurin, S., 153
Gustafson, A., 36, 39, 40(65), 67
Guyton, J. R., 30, 43, 225, 226(40), 227(40), 229(40)

H

Ha, Y. C., 339
Haapa, K., 180
Haas, J., 12
Haas, S. H., 70
Haas, S. M., 67, 111, 112(12)
Habib, G., 8, 18(35), 45, 95, 105, 148
Hachey, D. L., 7
Haddad, I. A., 12, 16
Hadzopoulou-Cladaras, M., 6, 9, 18, 25, 121, 150
Hagiwara, N., 262
Hajjar, K. A., 31
Halaas, J. L., 23
Halbrock, K., 353
Hallman, D. M., 24, 29, 225
Hamada, N., 260
Hamilton, R. L., 11, 47, 57, 67, 72, 73(40), 94, 151, 263, 264(44, 45), 283
Hammer, R. E., 30, 223, 232(21)
Hammer, R. L., 29
Hamsten, A., 32, 57, 83, 85(6), 92(6), 94(6), 95–96, 96(4), 100(9), 101(9), 103, 103(9), 104(9)
Hancock, W. S., 270
Handelmann, G. E., 171, 173(9)
Hanis, C. L., 17
Hanks, J., 22
Hannon, W. H., 58, 68, 78, 78(33), 161, 264
Hannuksela, M., 340
Hara, H., 297
Hara, I., 262, 270
Harada, K., 25, 178–179
Harano, Y., 154, 159(59), 160(59)
Harders-Spengel, K., 30, 232
Harding, D. R., 270
Hardman, A., 151
Hardman, D. A., 101, 105, 118(5), 122, 124(15), 146–147, 150(7), 151, 151(3), 263, 264(47)
Hardy, W. B., 33
Harlow, E., 322, 328
Harmony, J. A. K., 28, 43, 46, 51(124), 253, 309–311, 311(5, 9), 312, 312(5), 313(5), 315(9), 345
Harpel, P. C., 223, 234(23)
Harrach, B., 29, 240
Harry, P. J., 23
Haschemeyer, R. H., 158, 162(79), 163(79)
Hashimoto, S., 211, 218(14)

Hashimoto, Y., 23
Haslam, H. C., 33
Haslanger, M. F., 339
Haslauer, F., 29
Hata, Y., 211, 218(22)
Hatch, F. T., 36, 342
Hattori, S., 59
Hau, J., 322
Haubold, K. W., 167
Haupt, H., 42
Hausman, A.M.L., 298
Havekes, L. M., 10, 18, 25(62), 29, 147, 176, 179
Havel, R. J., 22, 36, 37(35), 44, 47–48, 57, 67, 72, 73(40), 82–86, 86(7), 87(7), 89(5, 7), 90(7), 91(7), 92(5, 7), 93–94, 94(5, 7), 95, 147, 150(7), 167, 169(6), 171, 178, 188–189, 208–209, 252, 265, 267, 320, 333
Hayden, M., 12
Hayek, T., 23, 25, 178, 209, 339, 352
Hazzard, W. R., 19, 20(149), 44
Hegele, R. A., 22, 207–208, 333
Heid, K., 42
Heideman, C., 28, 220, 225(9), 235(9)
Heideman, C. L., 227, 235(50)
Heimberg, M., 36
Heinonen, O. P., 180
Heinrich, J., 32
Heinsalmi, P., 180
Heiss, G., 24, 28
Helenius, A., 73
Helmhold, M., 29, 240
Helo, P., 180
Hemphill, L. C., 54, 58(173)
Henderson, L. O., 58, 68, 78, 78(33), 161
Henke, A., 12
Hennessy, L. K., 9
Henning, S. J., 298
Henry, M. L., 26
Hensel, W., 4
Hensley, W. J., 219
Herbert, P., 14, 22, 75, 101, 203
Hercaud, E., 262
Heremans, F. J., 67
Hermier, M., 10
Herscovitz, H., 121
Hertig, S., 310
Hervaud, E., 262
Herz, J., 22, 172, 177
Hesler, C. B., 283, 339–340, 347
Hess, H. H., 67
Hewett-Emmet, D., 17
Hidaka, H., 79, 83, 154, 159(59), 160(59)
Hide, W. A., 322
Higashi, A., 59
Higuchi, K., 18, 45, 105, 147–148, 149(12), 150, 155(12)
Higuchi, R., 122, 128(16)
Hilderman, H. L., 36
Hinman, J., 223, 225(24), 234(24)
Hirano, K., 339, 340
Hirata, M., 211, 218(14)
Hisatomi, A., 181
Hitchens, J., 226
Hitchens, O., 173
Hixson, J. E., 24, 28, 30, 30(216), 232
Hjerten, S., 157
Hnatiuk, O., 105
Hobbs, H. H., 28, 223, 231, 232(21), 238
Hodges, P. E., 8
Hodis, H. N., 54, 58(173), 59(178)
Hoeg, J. M., 9, 28, 47, 105, 147, 149, 149(12), 153, 155(12), 156(56), 203, 261
Hoff, H. F., 220, 223–224, 226(31), 227, 235(50), 237(25)
Hoffman, T., 204, 207(25)
Hofker, M. H., 179
Hofmann, S. A., 29
Hofmann, T., 22
Hofmann-Radvani, H., 123
HogenEsch, H., 179
Hogle, D. M., 65
Hogue, M., 346, 350(25), 351(25)
Hokanson, J. E., 243, 276
Hokland, B., 11
Hokom, M., 21
Holdsworth, G., 188, 191(5)
Holly, R., 11
Holme, I., 56
Holmes, D. R., 11
Holmquist, L., 73, 211, 217(15)
Holz, H., 204, 261, 271, 279(20)
Holzl, B., 20
Honda, S., 239
Hoover-Plow, J. L., 239, 240(11)
Hopferwieser, T., 10, 25(63), 28
Hopkins, P. N., 181
Hopkins, R. A., 282
Hoppe, C. A., 81
Hoppichler, F., 29, 31, 220, 232
Horiuchi, S., 16

Horn, G. T., 122, 128(16)
Hospattankar, A., 18, 45, 105, 148, 150
Howard, B. V., 80, 81(58), 94
Hoye, E., 225, 232(37)
Hsu, J. S. T., 334
Hsu, M. J., 32
Hu, A., 11
Huang, D. Y., 174
Huang, G., 153, 162(48)
Huang, L.-S., 121
Huang, S. S., 157, 162(75), 163(75)
Huang, Y., 253
Huang, Y. O., 146, 150(5)
Huber, C., 150
Hubl, S. T., 122–123, 126(19), 128(17), 132(19), 136(19), 141, 143(19), 151
Huff, M. W., 207–208
Hughes, T. A., 204, 261, 268, 270–271, 272(11, 19), 279(11, 19), 280(11)
Hughes, T. E., 12, 57
Hughes, W. L., Jr., 34, 37
Hui, D. Y., 167, 169(5), 172, 175, 177, 178(34), 340
Hui, Y., 171, 182(10)
Hulley, S. B., 56
Hulshof, M. M., 18, 147
Humphries, S., 6, 121, 132, 133(32), 138(32)
Hunter, W. M., 213, 304
Hurn, B. A. L., 192
Hurtado, I., 182
Husby, G., 42
Hussain, M. M., 172, 179
Hussarn, M. M., 150
Hutchinson, R., 31, 56
Huttunen, J. K., 180, 334
Hwang, S.-L. C., 171, 177
Hyman, R., 14
Hynd, B. A., 209, 210(6), 211(6), 212(6)

I

Ibayashi, H., 285
Ibdah, J. A., 252, 283
Ierssel, G. v., 176
Ihm, J., 345
Ihrke, G., 172
Ikeda, Y., 16
Ikeda, Y. A., 334
Ikewaki, K., 252
Illingworth, D. R., 181
Inazu, A., 339–340, 343
Innerarity, T. L., 8, 18–19, 65, 127, 148, 150, 167, 169(5), 171–172, 175, 177, 178(34), 182(10)
Inoue, I., 25, 178
Inoue, K., 23, 173, 181
Inui, Y., 298
Iqbal, M., 261, 271, 272(19)
Ishibashi, S., 179
Ishida, B. Y., 13, 256, 263, 268, 272
Ishigami, M., 340
Ishii, K., 21
Ishikawa, Y., 153, 178
Itakura, K., 211, 218(14)
Ito, K., 211, 218(14)
Ito, Y., 22, 209
Iverius, P. H., 20, 319–320, 322(4), 324(24), 325(24), 327, 332(2)
Iwanaga, T., 334

J

Jäckle, S., 150
Jackson, E., 239, 240(11)
Jackson, R. L., 3, 11, 14, 21(68), 22, 45, 84, 123, 127(18), 188, 191(5), 227, 235, 320, 334
Jacobs, D. R., Jr., 56
Jacobson, S. F., 148
Jacotot, B., 54
Jahn, C. E., 14
James, R. W., 15, 26, 311
Jan, A. Y., 220
Janas, A., 11
Janeway, C. H., 37
Jansen, H., 10, 25(62)
Jaritt, N. B., 57
Jauhiainen, M., 17, 26, 30, 297
Jeenah, M., 6
Jenkins, S. H., 253, 309–310, 311(9), 315(9)
Jenne, D. E., 309–310, 310(1), 311, 311(2)
Jensen, L. C., 36
Jensenius, J. C., 322
Jerstedt, S., 157
Jessup, W. K., 125, 127(23), 132(23)
Jewer, D., 261, 262(9), 262(36), 263
Ji, Z.-S., 150
Jialal, I., 223, 237, 237(29)
Jiang, X. C., 339, 352
John, K. M., 13, 44

Johnson, C. J., 11
Johnson, D. F., 8, 18, 116, 148, 167
Johnson, E. J., 167, 169(7)
Johnson, J. P., 20
Johnson, P., 261
Johnson, W. J., 15, 283
Jones, A. L., 67, 85
Jones, H. B., 37, 53(45), 54(45), 58(44, 45)
Jones, J. B., 282
Jones, J. L., 270
Jones, T. C., 121
Jordan, M. K., 17, 284
Jordan-Starck, T. C., 28, 309, 310(1), 311–312
Jörnvall, H., 139, 140(39)
Joven, J., 327
Joyner, J., 148
Jungner, I., 56

K

Kaffarnik, H., 173, 284
Kageshita, T., 183
Kahl, G., 11
Kahn, R., 83
Kahri, J., 59, 340
Kaitaniemi, P., 180
Kajiyama, G., 14
Kamanna, V. S., 153, 162(49)
Kamboh, M. I., 229, 232(57), 233(57), 234(57), 284, 293, 311
Kameda-Takemura, K., 164, 340
Kan, W. Y., 148
Kandoussi, A., 14, 15(105), 53, 59, 59(164), 253
Kane, J. P., 63, 67, 72, 73(39, 40), 84, 85(12), 86, 90(12), 94, 101, 105, 118(5), 122–123, 124(15), 146–147, 150(7), 151, 151(3), 162, 171, 189, 209, 252, 263, 264(37, 44–47)
Kanellos, J., 262
Kannan, R., 146, 150(5)
Kanno, H., 211, 218(14)
Kantor, M. A., 203
Kantz, J. A., 43
Kaplan, L. A., 310
Kaptein, A., 29
Kaptein, J. S., 148
Karathanasis, S., 4, 6, 6(8), 12, 16, 170
Kardassis, D., 6
Karlin, J. B., 171, 177
Karpe, F., 83, 85(6), 92(6), 94(6), 95–96, 96(4), 100(9), 101(9), 103, 103(9), 104(9)
Kashyap, M. L., 208–209, 210(6), 211, 211(6), 212(6)
Kassart, E., 261, 262(9)
Katsuki, M., 179
Katz, A. J., 260
Kauffman, D. L., 37, 38(41)
Kauppinen-Mäkelin, R., 17, 26
Kawabata, T., 154, 159(59), 160(59)
Kawai, C., 21, 32
Kawai, Y., 178
Kawakami, M., 179
Kawamura, M., 25, 178
Kawano, M., 254–256, 259(24)
Kawarabayasi, Y., 172
Kayden, H., 121
Kecorius, E., 301
Keller, U., 54
Kelly, K., 120
Kelso, G. J., 311
Kemmler, H. G., 229, 230(53), 234(53), 235(53), 238
Kennedy, H., 66
Keno, Y., 340
Kern, P. A., 320, 322, 327
Kesäniemi, Y. A., 21, 176, 340
Kessling, A., 6
Kesteloot, H., 56
Khan, M., 226
Khoo, J. C., 21
Khoury, P., 120
Kilgore, L. L., 80
Kilsdonk, E. P. C., 15
Kim, T. W., 63
Kindt, M. R., 284, 293, 293(24)
King, M. C., 20, 58
Kinniburgh, A. J., 16
Kinnunen, P. K., 22, 322, 327, 334
Kinoshita, M., 18, 23, 125, 127(24), 132, 133(24, 32), 138(32), 173
Kirk, L., 67
Kirszbaum, I., 27
Kirszbaum, L., 43
Kita, T., 21
Kitchens, R., 132–133
Kjeldsen, K., 54
Klasen, E., 176
Klein, R. L., 80, 81(59)
Kleinman, Y., 125, 127(24)

Kloer, H. U., 22, 47, 51
Knapp, E., 10, 25(63)
Knapp, R. D., 220, 222, 239
Knerer, B., 17
Knight, B. L., 30, 232
Knight-Gibson, C., 47, 50, 51(138, 150), 52, 54, 57(160), 59, 59(150, 178)
Knopp, R. H., 55, 56(183), 57(183), 181
Knott, J., 18
Knott, T. J., 4, 8, 18(34), 45, 64, 95, 105, 121, 148, 150
Knowles, B. B., 177
Knyrim, K., 260
Koch, C., 322
Kodama, M., 59
Koffigan, M., 210, 213(9), 268
Koga, S., 285
Kohler, E., 32
Köhler, G., 322
Kohler, M., 8
Kohno, M., 172
Kohr, W., 4
Kohr, W. J., 239
Kohrt, W. M., 340
Kohzaki, K., 25, 178
Koivisto, V., 59
Koizumi, J., 339–340
Kojima, H., 79, 154, 159(59), 160(59)
Kolar, J. B., 185
Kolar, W., 56
Kollmer, M. E., 154
Komaba, A., 297
Komanduri, P., 155
Konda, K., 284, 303, 305, 308(26)
Kontula, K., 17
Koo, C., 150
Kora, I., 210–211, 213(9), 217(16), 268, 350
Koren, E., 14, 15(105), 41, 47–48, 51, 51(138), 52–53, 57(157, 160), 59, 59(157, 164), 147, 156(8), 157(8), 158(8), 162, 162(8), 253
Korman, S. H., 121
Korn, E. D., 188
Koschinsky, M. L., 220, 221(8), 222(8), 223, 223(8), 232(21), 235(8), 236(8)
Koskinen, P., 17, 180
Kostner, G., 28–29, 30, 40–41, 42, 46, 52(93), 180, 181(63), 220, 223–225, 225(10, 26, 33)
Kotani, K., 340
Kotite, L., 82–83, 86, 86(7), 87(7), 89(5, 7), 90(7), 91(7), 92(5, 7), 93, 94(5, 7), 95, 167, 169(6), 189
Kottke, B. A., 11, 56, 199, 229, 232(57), 233(57), 234(57)
Kounnas, M. Z., 312
Kouvatsi, A., 9
Kowal, R. C., 22, 177
Kraft, H. G., 28, 225, 229, 230(53), 231–232, 232(37), 234(53), 235(53), 238
Kramsch, D. M., 54, 59(178)
Krasinski, S. D., 155
Kratky, O., 243
Krause, B., 283
Krauss, R. M., 13, 19–20, 54, 58, 58(170), 65, 130, 243, 268, 276, 320
Krebber, H., 226
Krempler, F., 29
Krezer, H., 30
Krieg, P. A., 353, 355(8), 356(8)
Krieger, M., 9
Krilis, S. A., 27
Krishman, S., 261
Krishnaiah, K. V., 146, 147(2), 162
Krishnan, S., 9, 271, 282, 303–304
Krivi, G. G., 347
Krol, B., 253, 310, 311(9), 315(9)
Krul, E. S., 120, 125–126, 127(24), 128, 132, 132(25), 133(24, 32), 138(25, 32), 170, 176, 177(39), 178
Kuang, W.-J., 28, 31(211), 220, 239
Kuberger, M. B., 147, 149(12), 155(12)
Kuchmak, M., 264
Kuhlenschmidt, T. B., 81
Kulkarni, K. R., 273
Kuller, L. H., 284
Kume, N., 21
Kung, S., 263, 264(47)
Kunisaki, M., 181
Kunitake, S. T., 252, 260, 263–264, 264(37, 44–47)
Kunkel, H. G., 35, 37, 38(40)
Kuo, C. F., 6
Kuo, W. H., 157, 158(73), 162(73), 163(73), 164(73)
Kuo, W. U., 273
Kurokawa, K., 23, 173
Kurooka, S., 334
Kussie, P., 339
Kusunoki, J., 339
Kuusi, T., 26, 334

Kwiterovich, P. O., 19, 20(148)
Kwiterovich, P. O., Jr., 19, 55, 56(182), 57(182), 66, 77
Kyte, J., 152, 153(46)

L

Labeur, C., 55, 220, 226
Lackner, C., 28–29, 220, 225, 231, 238
Lackner, K., 18, 150
Lackner, K. J., 9, 148
Ladias, J. A. A., 6
Laemmli, U. K., 88, 109, 112(11), 156, 160(68), 265
Lagrost, L., 23, 283–284, 285(32), 297, 303, 305
Lai, E., 6
Laker, M. F., 231, 232(60), 233(60), 237(60)
Lallemant, C., 180, 181(65), 283–284, 285(32), 303, 305
Lam, C. W. K., 219
Lamphugh, S. M., 270
Landry, D., 241
Lane, D., 322, 328
Lanford, R. E., 30, 223, 232(22)
Langone, J. J., 303
LaPiana, M. J., 75
Laplaud, P. M., 235
Laprade, M., 262
Lapsley, M., 27
LaRosa, J. C., 22, 25
Larrick, J. W., 353
La Sala, K. J., 263, 264(37, 46)
Laskarzewski, P. M., 120
Laskarzwski, P. M., 267
Lau, P. P., 149
Lauer, S. J., 209, 333, 352, 353(1)
Laughton, C., 14
Laurell, C. B., 226, 303
Laux, M. J., 59
Law, A., 18, 125, 127(23, 24), 132(23)
Law, J. C., 293
Law, S. W., 18, 45, 105, 148, 150
Lawn, R. M., 4, 28, 30, 31(211), 45, 219–220, 221(8), 223, 232(21), 235(8), 236(8), 239
Lawrie, G. M., 30, 225, 226(40), 227(40), 229(40)
Le, N. A., 22, 47, 80, 81(57), 320
Leblond, L., 173
Lebo, R. V., 148
LeBoeuf, R. C., 14, 148
Ledue, T. B., 229, 235(56)
Lee, B. R., 8, 18(35), 45, 95, 105, 148
Lee, D. M., 36, 41(34), 51, 75, 135, 146–147, 153, 156(8, 9, 10), 157, 157(8, 9, 10), 158, 158(8, 10, 73), 162, 162(8, 9, 10, 47–49, 73, 75), 163, 163(73, 75), 164, 273
Lee, F. S., 18
Lee, J. A., 263
Lee, L. T., 220, 228(11)
Lee, N., 18, 150
Lee, N. S., 46
Lee, S. H., 177
Lees, M. B., 67, 70
Leese, R. S., 342
Leete, T. H., 16
Lefevre, M., 263
Leff, T., 4–5
Leffert, C. C., 28, 231, 238
Legmann, P., 42
Legna, L., 283
Leighton, J. K., 148
Leiter, L., 208
Leiterdord, E., 320
Lendon, C., 174
Lentz, S. R., 260
Leopold, H., 243
Lepage, G., 153
LeQuire, V. S., 147
Levin, E. G., 31
Levine, L., 37, 38(41, 42)
Levinson, S. S., 57
Levy, E., 147, 153
Levy, R. I., 22, 39, 40(70), 75, 85, 87(15), 101, 212
Levy-Wilson, B., 4–5, 18, 64, 121, 148, 150
Lewis, B., 53, 54(169)
Lewis, E., 67
Lewis, L. A., 37, 38(43)
Lewis, S. B., 42, 50(95)
Ley, C. A., 312
Li, W.-H., 4, 18, 22(7), 45, 105, 148, 150–151, 151(36), 152, 319, 320(1), 322
Lieberman, M., 38
Liepnieks, J., 12
Lin, A. H. Y., 171, 177
Lin, J. P., 28, 238
Lindahl, G. E., 30
Lindgren, F. T., 35–36, 36(24), 37(25, 28), 53(25, 45), 54(45), 58(44, 45), 255, 262, 320

Lingappa, V. R., 10
Lin-Lee, Y.-C., 7–8, 10, 17
Lins, L., 45, 48(110)
Linton, M. F., 120–121, 121(4), 122(4), 123, 123(4), 124, 125(11), 126(4), 127, 127(4), 139(11, 26), 140, 140(11), 141, 142(11), 143, 143(7, 11), 144(7, 11, 26), 145(26), 151
Lipton, B. A., 179
Lissens, W., 56
Little, J. A., 22, 188, 204, 207, 207(25), 208, 333
Liu, A. C., 30
Liu, B., 211, 217(18)
Liu, S. W., 63, 151
Lloyd, J. K., 39, 121, 220
Lo, J.-Y., 319, 321, 327
Lohse, P., 17, 25, 284, 293, 293(24)
Londono, I., 147
Lontie, J.-F., 46, 51(123)
Loof, H. D., 262
Loscalzo, J., 43
Lou, H. C., 120
Lowe, W. F., 130
Lowin, B., 311
Lown, J. S., 153
Lowry, O. H., 66, 85, 90(17), 161, 196, 200(13), 276, 280(28)
Lowry, T. G., 96
Lozier, J., 46
Luc, G., 53–54, 59(165), 221, 222(16), 225(17), 234(16), 235, 235(17)
Ludwig, E., 24, 173
Ludwig, E. H., 4, 19, 64, 121
Ludwig, E. M., 4
Lukaszewicz, A. M., 148
Lukka, M., 17, 297
Lum, K. D., 51, 57(158), 59(157), 253
Lund, B. J., 310
Lund, S. D., 312
Lund-Katz, S., 19, 252
Luo, C.-C., 4, 22(7), 45, 319, 320(1), 321
Lusis, A. J., 8, 14, 16, 18, 116, 148, 150
Lussier-Cacan, S., 26, 55
Lutalo-Bosa, A. J., 65, 77(5)
Lutmer, R. F., 209
Lux, S. E., 13, 22, 44
Luyeye, I., 221, 225(17), 235(17)
Lynch, K. M., 203
Lyon, T. P., 37, 53(45), 54(45), 58(44, 45)

M

Ma, X., 12
Maartmann-Moe, K., 29
Mabuchi, H., 339–340
Macheboeuf, M., 33, 34(10, 11)
Maciejko, J. J., 11
Mack, W. J., 54, 58(173), 59(178)
Madec, Y., 53, 59(166)
Maeda, H., 157, 158(73), 162(73), 163(73), 164(73), 273
Maeda, N., 13, 25, 178
Mäenpää, H., 180
Magnani, H. N., 226
Magnusson, S., 239
Maguire, G. F., 22, 188, 204, 207, 207(25), 208
Magun, A. M., 52
Mahdavi, V., 6
Mahley, R. W., 4, 12, 18–19, 22–24, 24(185), 25(185), 32, 44, 64–65, 86, 121, 127, 148, 150, 167, 169(5), 171–173, 173(9), 174–175, 177, 178(34), 179, 182(10, 33), 209, 262, 282, 285(1), 333
Mahrabian, M., 148
Maika, S. D., 223, 232(21)
Mainard, F., 53, 59(166)
Makino, K., 223, 225(24), 234(24)
Makrides, S. C., 12
Mälkönen, M., 180
Malloy, M. J., 123, 151
Malmendier, C. L., 42, 46, 51(123)
Maniatis, T., 353, 355(8), 356(8)
Manis, G. S., 28, 30(216), 229, 231(55), 232, 234(54)
Manke, A., 30
Mann, C., 339
Manninen, V., 17, 180
Mänttäri, M., 17, 180
Mao, J. T., 199
Mao, S. J. T., 11, 211, 311
Marcel, Y., 18, 23, 121, 125, 127, 127(23), 132(23), 147, 153, 154–155, 167, 168(3, 11), 169(3–5, 7), 173, 176, 260, 261, 262(4), 262(9, 36), 263, 271, 283, 339–340, 346–347, 350, 350(25), 351(25)
March, S. C., 194, 325, 326(42)
Marchesi, V. T., 235
Marcigliano, A., 352
Marcovina, S., 58, 66, 68, 78, 78(33), 161, 221, 224, 225(15, 33), 231(15), 232(15),

233(15), 234(15), 235(15), 237(15), 254–255, 259(24), 262, 269, 273
Margolius, H. S., 84, 123, 127(18), 162
Markwell, M. K., 67, 70, 111, 112(12), 255, 258(23)
Marlow, C. T., 27
Marotti, K. R., 263, 339
Marpole, D., 19, 20(148), 66
Marsh, J. B., 105, 110(1), 116(10), 118(10), 146, 150(4)
Marshall, S., 320
Martin, C., 203
Martin, J. C., 31
Martinelli, G., 322
Martinelli, J. A., 327
Maruhama, Y., 340
Marz, W., 225
März, W., 170, 179
Más-Oliva, J., 150
Massey, J. B., 46, 50(118)
Mather, I. H., 67
Matsuda, I., 16, 59
Matsudate, T., 211, 218(14)
Matsuzawa, Y., 172, 339–340
Mattila, K. J., 26
Maurer, P. H., 301
Maxfield, F. R., 150
Mayes, P. A., 158
McCall, M. R., 268
McCarthy, B. J., 4, 10, 18–19, 64, 120–121, 125(11), 139, 139(11), 140(11), 142(11), 143(11), 144(11), 150, 167
McCarthy, S., 31
McConathy, W. J., 22, 29, 39, 41, 47–48, 49(141), 53(75, 140), 56(140), 57, 57(140), 59(140, 191), 211, 218(13), 219–220, 260, 262(4)
McCormick, S., 123, 137–138
McFarlane, A. S., 35
McGinley, J. P., 37
McGuffin, P., 353, 358(10)
McKernan, P. A., 11
McKnight, G. L., 11
McLachlan, A. D., 3
McLean, J. W., 4, 28, 31(211), 45, 220, 239
McLean, L. R., 334
McMaster, D., 53, 59(165)
McNamara, J. R., 95
McNeil, H. P., 27
McPherson, J., 170
McPherson, R., 346, 350(25), 351(25)
McVicar, J. P., 263, 264(44, 45)
Meade, T. W., 31
Medley, M. F., 204, 261, 268, 271, 272(11, 19), 279(11, 19), 280(11)
Meer, K., 30
Meglin, N., 45, 105, 148
Megna, L., 42, 283, 284(16), 287(14), 297
Mehrabian, M., 14
Melchior, G. W., 263, 339
Melford, K. H., 320, 334
Melian, A., 4–5
Melin, M., 34
Melish, J. S., 80, 81(57)
Mellies, M. J., 120
Melo, C., 7
Melton, D. A., 353, 355(8), 356(8)
Melton, M. A., 339
Mendel, C., 263, 264(46)
Mendez, A. J., 15
Meng, M., 9, 25, 147, 149(12), 153, 154(56), 155(12), 203, 261, 268, 270(9), 272
Menotti, A., 262
Menzel, H. J., 17, 24, 28, 220, 225, 229, 230(53), 231–232, 234(53), 235(53), 238, 284
Mertz, E. T., 241, 248(15)
Messmer, S., 26, 311
Metso, J., 17, 30, 297
Meumier, S., 303, 305
Meunier, S., 284, 285(32)
Meyers, W., 22
Mezdour, H., 59, 210, 213(9), 221, 224, 225(17, 33), 235(17), 268, 350
Michel, J.-B., 225
Michels, V. M., 56
Michiels, G., 197, 199(15), 284, 303, 308
Miesenbock, G., 10, 20, 25(63)
Miettinen, T. A., 176
Mietus-Snyder, M., 6
Miida, T., 255–256
Milane, R. W., 271
Miles, L. A., 31, 239, 240(11)
Millar, J. S., 167, 169(7)
Miller, C. G., 239
Miller, G. J., 31
Miller, J., 211
Miller, N., 6
Miller, N. E., 121
Miller, N. H., 80
Milne, R. W., 18, 23, 121, 125, 127, 127(23),

132(23), 147, 153–155, 166–167, 168(3, 11), 169(3–5, 7), 176, 260, 262(4), 283, 339, 346–347, 350(25), 351(25)
Milstein, C., 322
Milthorp, P., 155, 262(36), 263, 271
Mimura, K., 181
Mindham, M. A., 158
Minnich, A., 11
Mital, P., 218
Mitchel, Y. B., 181
Mitchell, A., 351, 362
Mitchell, D., 14
Mitchell, W. M., 157
Mitropoulos, K. A., 31
Miura, T., 339
Miyabo, S., 172
Miyata, Y., 285
Miyazaki, A., 16
Mock, P., 154
Moestrup, S., 150
Moger, W. H., 174
Mok, H. Y. I., 144
Mok, T., 147, 156(8, 9), 157(8, 9), 158(8), 162(8, 9)
Mokrach, L. C., 67
Mokuno, H., 179
Moll, V. M., 56
Molloy, M. J., 150
Monaci, P., 5
Moore, M. A., 204, 268, 272(11), 279(11, 19), 280(11)
Moore, M. N., 4, 22(7)
Moorehouse, A., 57
Morgado, J., 284, 285(32)
Morgado, P., 303, 305
Morgan, J., 154
Morgan, L. M., 155
Morgenstern, H., 310
Morishita, H., 32
Morris, J. C., 174
Morris, M. D., 263
Morrisett, J. D., 3, 28–31, 45, 56, 218, 220, 222, 225, 225(9), 226(35, 40), 227, 227(40), 229, 229(40), 230(54), 231(54), 232(54), 234(54), 235(9, 54), 237(35), 239
Morrison, J. A., 120
Morrison, J. R., 262
Morrow, A., 14
Morton, R. E., 341, 345(18), 351(18)
Motulski, A. G., 19, 20(149), 148
Motzny, S., 284
Moulin, P., 340, 343, 352
Mowri, H. O., 14, 16
Muehlberger, V., 10, 25(63)
Mueller, G., 59
Muirhead, R. A., 261
Mulford, D. J., 34
Muller, D., 121, 220
Mullis, K. B., 122, 128(16)
Munck, A., 10
Murase, T., 172, 179
Murphy, B., 27, 43, 312
Murray, M. D., 220, 228(11)
Murray, R. W., 339
Musliner, T. A., 54, 58(170), 130

N

Nabulsi, A. A., 28
Nachman, R. L., 30–31
Nadal-Ginard, B., 6
Nagano, Y., 21
Naggert, J. K., 20
Nagy, B. P., 5
Nakai, H., 148
Nakai, T., 172
Nakajima, Y., 79, 154, 159(59), 160(59)
Nakamura, K., 211, 218(14)
Nakamura, R., 16, 59
Nakamura, T., 154, 159(59), 160(59), 340
Nakane, P., 265
Nakanishi, M. K., 321
Nakano, T., 154, 159(59), 160(59)
Nakasoto, D., 37
Nakaya, Y., 47
Nanthakamar, N., 298
Naruszewicz, M., 31, 223, 237(27)
Nathans, A., 36
National Committee for Clinical Laboratory Standards, 203
National Institute of Standards and Technology Certificate of Standard Reference Material, 68
Natvig, J. B., 42
Nava, M. L., 28, 30, 225, 226(40), 227(40), 229, 229(40), 230(54), 231(54), 232(54), 234(54), 235(54)
Navaratnam, N., 8
Nawata, H., 181
Naya-Vigne, J., 260, 263
Naylor, S. L., 148

Neame, P. J., 204, 261, 268, 271, 272(11, 19), 279(11, 19), 280(11)
Nelson, J. W., 232
Nemes, P., 190
Nerking, J., 33
Nermes, P., 81
Nestel, P. J., 48, 150, 154(31), 285, 298, 299(13), 300(13), 301(13)
Nestruck, A. C., 55
Neveux, L. M., 229, 235(56)
Newman, B., 20
Nguyen, T.-D., 260, 262(4)
Nichols, A. V., 15, 35–36, 36(25), 37(25, 28, 33), 53(25, 45), 54(45), 75, 130, 252, 255, 262, 268
Nichols, W. C., 12
Nickoloff, J. A., 139
Nicolosi, R. J., 9, 203
Nicosia, A., 5
Nicosia, M., 174
Nieminen, M. S., 26
Niendorf, A., 226
Nikkilä, E. A., 180, 334
Nimpf, J., 220, 224, 225(10), 225(33)
Nimpf, M., 220, 225(10)
Nishina, P. M., 20
Nishiyama, S., 59
Noe, L., 320
Nogi, K., 334
Nolte, R. T., 18, 150
Noma, A., 211, 218(22)
Norden, A. G., 27
Nordestgaard, B. G., 53–54, 54(169)
Norola, S., 180
North, J. D., 23
Northey, S. T., 120
Northup, S. R., 239
Norum, R., 22, 47, 59
Nowicka, G., 11
Nozaki, S., 340
Nugent, M. J. T., 262(35), 263
Nussbaum, A. L., 12
Nusser, E., 92
Nustede, R., 221, 222(16), 234(16)
Nwankwo, M., 311

O

Obasohan, A. O., 56
O'Connell, A., 22, 209, 352
O'Connor, P., 260, 263
O'Donovan, M. C., 353, 358(10)
Ogami, K., 5
O'Hara, P. J., 11
Ohkaru, Y., 334
Ohta, T., 16, 59, 298, 299(13), 300(13), 301(13)
Oida, K., 125, 127(24), 172
Oikawa, S., 15
Oka, K., 320
Okada, H., 183
Okayama, H., 140, 141(41)
Okazaki, M., 262, 270
Okubo, M., 172
Olaisen, B., 170
Olivecrona, Y., 83, 95, 96(4), 319–320, 322, 327
Olofsson, S. O., 14, 41, 57
Olomu, A., 56
Olson, L. M., 174
Onasch, M. A., 4, 64, 121
Oncley, J. L., 3, 34, 36–37
O'Neil, J. A., 220, 223–224, 226(31), 237(25)
Ong, J., 146, 320, 322, 327
Ono, R., 183
Ooshima, A., 21
Oram, J. F., 11, 15
Ord, V. A., 179
Ordovas, J. M., 9, 12, 16, 57, 59, 95
Orekhov, A. N., 222
Orsini, G. B., 262
Osada, J., 9
Osborn, M., 101, 156
Osborne, J. C., Jr., 14, 46
Oschry, Y., 21
Ostlund, R., 340
Östlund-Lindqvist, A. M., 320, 324(24), 325(24), 327, 332(2)
Oviasu, V. O., 56

P

Packard, C. J., 20, 48, 54, 59
Paetzold, R., 12
Pagani, F., 6
Page, I. H., 37, 38(43)
Paigen, B., 14, 263
Palinski, W., 21, 179
Paolucci, F., 262
Papadopoulos, N. M., 28
Papazafiri, P., 5

Pape, M. E., 263
Parhofer, K. G., 18, 126, 132(25), 138(25)
Parikh, I., 194, 325, 326(42)
Park, Y. B., 311
Parra, H. J., 53–54, 59, 59(165), 219, 221, 225(17), 235(17), 350
Parsy, D., 211, 217(16)
Parthasarathy, S., 12, 21, 163, 223, 237(28)
Pasternack, A., 180
Pastier, D., 262
Paterniti, J. R., 12, 322, 327
Paterson, W. R., 81
Patsch, J. R., 10, 14–16, 20–21, 25(63), 29
Patsch, W., 3, 7–10, 14–17, 21, 24, 25(63), 28–29, 31–32, 56, 229, 230(54), 231(54), 232(54), 234(54), 235(54)
Patterson, A. P., 149
Patterson, B. W., 80
Patton, C., 282, 284, 288
Pau, B., 262
Paulus, H. E., 101, 105, 118(5), 122, 124(15), 146, 151(3)
Paulweber, B., 5, 20
Pavlakis, G., 25
Paxam, S., 67
Payne, G. S., 6
Pazman, S., 70
PDAY Research Group, 24
Pease, R., 4, 64, 121, 125, 127(24)
Pease, R. J., 8, 18, 18(34), 45, 64, 95, 105, 121, 125, 127(23), 132(23), 148, 150
Peavy, D. L., 162
Pedersen, K. O., 35
Pedersen, M. E., 327
Peel, A. S., 83, 95, 155, 167
Peitsch, M. C., 23, 45, 46(117), 311
Pelachyk, J. M., 59
Peltonen, L., 17, 297
Peng, R., 14, 320
Pepe, M. G., 173
Pepin, J. M., 224, 226(31)
Peralta, F. P., 136, 137(36), 138(36)
Peralta, J. M., 302
Pericak-Vance, M., 32, 174
Perisutti, G., 209, 210(6), 211(6), 212(6)
Perosa, F., 183
Perret, B., 261, 262(9)
Perrin, S., 353
Peschke, B., 170
Peters, T., Jr., 68
Petersen, M. E., 322
Petersen, T. E., 239
Peterson, G. L., 67, 70, 73(35)
Petit, E., 262
Pfaffinger, D., 223, 225(24), 234(24), 239, 240(11)
Pfleger, B., 8, 125, 127(24), 173, 284
Phillips, G. B., 35
Phillips, M. C., 15, 19, 252, 283
Phillips, N., 83, 86, 267
Piedrahita, J. A., 25, 178
Pierotti, V., 127, 139(26), 143, 144(26), 145(26)
Pierotti, V. R., 4, 64, 121, 124, 126(22), 128(22), 129(22)
Pikkarainen, J., 180
Pintó, X., 182
Pio, F., 262
Pitas, R. E., 150, 171–172, 173(9), 177
Plow, E. F., 31, 239, 240(11)
Plump, A. S., 25, 178
Poapst, M., 22, 79, 83, 96, 101, 103(7), 104(7), 159, 161(80), 188
Poduslo, J. F., 235
Poensgen, J., 22
Poernama, F., 13
Poetro, T., 261
Pogoda, J. M., 54, 58(173)
Poksay, K. S., 8, 127, 167
Polacek, D., 248
Polites, H. G., 339
Pollack, R. L., 70
Pollitzer, W. S., 293
Polonovski, J., 42, 262
Polz, E., 42, 46, 52(93)
Pometta, D., 15, 26, 311
Pont, P., 219
Poonia, N. S., 154
Potenz, R., 319, 321, 327
Poulin, S. E., 229, 235(56)
Powell, L., 8, 18, 18(34), 45, 64, 95, 105, 121, 148, 150
Powell, M. K., 68, 78(33), 161
Pownall, H. J., 11, 45–46, 50(118), 63, 151, 270, 320
Pownell, L., 4
Prack, M. M., 174
Pradines-Figueres, A., 322

Prasad, S., 14, 21, 171, 177
Priessner, K. T., 310
Priestly, L. M., 148
Princen, H. M. G., 29
Proceedings of the Workshop on Apolipoprotein, 55
Proudfoot, A., 15
Proudfoot, N. J., 6
Provost, P., 125, 127(23), 132(23), 261, 262(9)
Puchois, P., 14–15, 15(105), 47, 51(139), 53, 59, 59(164), 203, 253
Pullinger, C. R., 121, 123, 151
Puppione, D. L., 105
Putnam, F. W., 46

Q

Qiao, J.-H., 14
Quarford, S. H., 36, 48, 49(142)
Quarfordt, S. H., 22, 41
Quinet, E., 282, 283(5), 284(5), 285(5), 303–304, 339, 343
Quinn, D. M., 345

R

Rader, D. J., 10, 29, 51, 59(151), 252–253, 284, 293, 293(24)
Radosavljevic, M., 7, 17
Raffai, E., 261, 262(9)
Rahmig, T., 28, 231, 238
Rainwater, D. L., 28, 30, 30(216), 130, 223, 229, 231(55), 232, 232(22), 234(54)
Rajaram, O. V., 297
Rall, L. B., 148
Rall, S. C., 12, 86, 235
Rall, S. C., Jr., 18, 23, 24(185), 25(185), 137, 148, 150, 177
Ramakrishnan, R., 23, 209, 339
Ramji, D. P., 5
Randall, J. R., 276, 280(28)
Randall, R. J., 66, 85, 90(17), 96, 161, 196, 200(13)
Rash, J. M., 105
Rassart, E., 23
Rath, M., 226
Ratnoff, O. D., 26
Ray, M. J., 43, 243, 276, 310, 311(5), 312(5), 313(5)
Raykundalia, C., 192
Reardon, M. F., 101
Reasoner, J., 11
Reaveley, D., 31
Rebagliati, M. R., 353, 355(8), 356(8)
Rebeyrotte, P., 33
Reblin, T., 226
Rebourcet, R., 197, 211, 217(20)
Reddick, R. L., 25, 178
Redfield, R., 38, 40(55)
Redgrave, T. G., 96, 99(10)
Rees, A., 23
Reese, H., 42
Reeve, J. R., Jr., 105
Reeves, B. E., 31
Reeves, R., 29
Rehberg, E. F., 339
Reiber, H., 180
Reik, K. B. M., 46
Reilly, P., 211
Reisner, A. H., 190
Reisner, P., 81
Remaley, A. T., 9
Repin, V. S., 147, 149(12), 155(12)
Report of the National Cholesterol Education Program, 57
Retegui, L., 176
Reuben, M. A., 105, 116, 148
Reue, K., 4, 14, 16
Reyland, M. E., 173
Rezzonico, L., 340
Rhee, L., 4
Richards, J. L., 53, 59(165)
Richter, R. J., 311
Richter, W. O., 160
Riddle, J. W., 38
Riesen, W. F., 211, 217(19)
Rifai, N., 180, 211, 218(22)
Rigaud, D., 26
Riley, J. W., 282, 283(5), 284(5), 285(5), 303–304
Ripps, M. E., 121
Ritchie, R. F., 229, 235(56)
Ritsch, A., 16
Roach, B. C., 80, 81(57)
Robbins, J., 46, 51(124), 311
Robenek, H., 13, 29, 240
Roberts, D. C. K., 26
Robertson, E., 148

Robertson, F. W., 24
Robertson, S., 121
Robin, H., 262
Rochette, C., 147
Rodbell, M., 38
Rogers, J. L., 80
Rogers, M., 147
Roghan, A., 11
Rohde, M. F., 261, 271
Roheim, P. S., 220, 228(11), 252, 255(7), 257(7), 262–263, 283, 297
Rojas, C., 320
Rolih, C. A., 224
Roman, R., 261
Romo, M., 180
Ronan, R., 9, 13, 44, 153, 203, 268, 270(9), 272
Roscher, A., 29
Rose, L., 21
Rosebrough, H. J., 66, 85, 90(17), 96, 161, 196, 200(13), 276, 280(28)
Rosenfeld, M. E., 21, 179
Roses, A. D., 32, 174
Rosseneu, M., 8, 18, 18(35), 45, 48(110), 55, 95, 105, 148, 176, 178, 197, 199(15), 211, 217(17), 218(17, 21), 226, 260, 262, 264, 284, 303, 308, 309(29), 321
Rothblat, G. H., 15, 105, 283
Rotheneger, M., 30
Rottman, J. N., 6
Rowold, E., 347
Roy, C. C., 147, 153
Roy, P., 154
Rubin, E. M., 13, 15, 25, 30, 178, 268
Rubin, L., 37
Rubin, R. W., 83
Rubinstein, A., 22, 47, 75, 177
Ruddle, D., 14
Rudel, L., 263, 339
Rusiñol, A., 57
Russ, E. M., 37, 53(47)
Russel, R. M., 167, 169(7)
Russell, D. W., 139, 140(39)
Russell, R. M., 155
Rutz-Opazo, N., 12
Ruysschaert, J.-M., 45, 48(110)
Ruzicka, V., 170, 179
Rye, K. A., 47, 283, 297

S

Saboureau, M., 235
Sacchi, N., 356
Sachdev, O. P., 42, 283, 287(14), 297
Safonova, I. G., 147, 149(12), 155(12)
Saha, N., 24, 225
Saiki, R. K., 122, 128(16)
Saito, Y., 339
Sakaguchi, A. Y., 148
Sakai, J., 172
Sakai, N., 172, 339–340
Sako, Y., 181
Sakurai, T., 211, 218(14)
Salomon, J., 8, 149
Salt, H. B., 39
Salvesen, G. S., 32
Sambert, P., 303, 305
Sammett, D., 342
Samuelsson, O., 59
Sanbar, S. S., 41
Sanders, S., 7, 8
Sandhofer, F., 20, 29
Sandholzer, C., 24, 29, 219, 220(3), 225
Sandkuul, L. A., 23
Sandor, G., 34
SanGiacomo, T. R., 177
Sanson, P. A., 27
Santamarina-Fojo, S., 22
Sappino, A. P., 310
Saritelli, A. L., 203
Sasak, W. V., 153
Sasaki, N., 14
Sastry, K., 4
Sata, T., 67, 72, 73(40), 85, 94
Sattler, W., 223, 225(26)
Saudek, C. D., 282
Saunders, A. M., 32, 174
Savion, N., 282
Savolainen, M. J., 340
Saxena, U., 17
Scanu, A. M., 9, 28, 31, 31(211), 37, 38(43), 162, 188, 220–221, 223–224, 225(24), 232, 234(24), 238–239, 240(11), 244(1), 246, 248, 249(1), 260, 270, 284, 285(30), 287(30), 296(30), 301
Schaaf, L., 179
Schacterle, G. R., 70
Schaefer, E. J., 9, 12–14, 24, 26, 47, 57, 59, 95, 153, 154(56), 155, 167, 169(7), 178, 285

Schaefer, J. R., 252
Schaffer, P., 53, 59(165)
Schaffer, R., 68
Scharf, S. J., 122, 128(16)
Schechter, I., Jr., 21
Scheffler, E., 321, 327
Schellekens, M. M., 176
Scheraldi, C. A., 17
Scheuermann, E., 170
Schilling, J., 263, 264(47)
Schleef, J., 29
Schleichler, E., 261
Schmechel, D., 32, 174
Schmidt, C. F., 270
Schmitz, G., 11–13, 311
Schmitz, J., 10
Schneeman, B. O., 83, 89(5), 92(5), 94(5), 95, 167, 169(6)
Schneider, P. D., 105
Schoeppe, W., 170
Schonfeld, G., 8–9, 15, 18, 44, 125–126, 127(24), 128, 132, 132(25), 133, 133(24, 32), 138(25, 32), 146, 147(2), 173, 176, 177(39), 185, 211, 284, 340
Schonfeld, R., 32
Schotz, M. C., 148, 320, 322, 327
Schrangl-Will, S., 17, 284
Schreiber, J. R., 174
Schrott, H. G., 19, 20(149)
Schuff-Werner, P., 59
Schuh, J., 158, 162(79), 163(79)
Schulman, R. S., 75
Schulte, H., 32
Schultz, E. M., 268
Schultz, J. R., 15
Schultze, H. E., 42, 67
Schulz, F. N., 33
Schulze, F., 30
Schumacher, U. K., 9
Schumaker, V. N., 81, 105, 121, 148
Schuurman, H. J., 18, 147
Schwandt, P., 160, 210
Schwartz, K., 30, 220, 221(8), 235(8), 236(8)
Scott, J., 4, 8, 18, 18(34), 20, 45, 64, 95, 105, 121, 125, 127(23, 24), 132(23), 148, 150
Seager, J., 21
Sechner, R., 209
Seeburg, P. H., 8
Seed, M., 29, 31
Seedorf, U., 4
Seelos, C., 7, 17
Segrest, J. P., 3, 11, 36, 45, 235, 243, 261, 270–271, 272(19), 273, 276
Seidel, D., 29–30, 59, 180, 221, 222(16), 234, 234(16), 240
Seidman, E., 153
Seip, R. L., 340
Seitz, C., 28, 229, 230(53), 234(53), 235(53), 238
Selinger, E., 31, 223, 237(27)
Sells, S. F., 312
Seman, L. J., 235
Semb, H., 320
Semenkovich, C. F., 319–320, 320(1), 321, 328, 329(13), 330(13)
Setzer, D., 150, 151(36), 152
Shafi, S., 333
Shames, D. S., 48
Shapiro, S., 19, 20(148), 66
Sharp, P. M., 63, 151
Sharrett, A. R., 24
Shaw, N. F., 23
Shea, T. M., 83, 86(8), 92(8), 96, 159, 160(81), 161(81)
Shears, S., 174
Shelanski, S. A., 352
Shelbourne, F. A., 22, 41, 48, 49(142)
Shepard, J., 20, 48, 54
Shepherd, J., 55, 59, 226
Sherman, J. R., 285, 296(39)
Shigeta, Y., 83, 154, 159(59), 160(59)
Shimada, M., 25, 178–179
Shimano, H., 25, 178–179
Shimdzu, I., 33
Shinohara, E., 340
Shinohara, M., 16
Shinomiya, M., 14
Shiomi, M., 25, 178
Shirai, K., 14
Shore, B., 38, 40, 40(56), 41–42, 47, 48(84), 49(84), 50(95), 171
Shore, V., 11, 40–42, 47, 48(84), 49(84), 50(95), 52, 57(161), 171
Shoulders, C. C., 23
Shows, T. B., 148
Shulman, R. S., 14, 101, 203
Shultz, J. R., 268
Sidney, S., 56

Sidoli, A., 6
Siebenkas, M. V., 283, 284(16)
Siebert, P. D., 353
Siekmeier, R., 170, 179, 225
Sigurdsson, G., 6, 29, 225, 284
Silberberg, J., 58
Silberman, S. R., 8, 18(35), 45, 95, 105, 148, 262(34), 263
Silberstein, A., 92
Silverman, L. M., 180, 211, 218(22)
Siman, B., 32
Simonet, W. S., 352, 353(1)
Simons, A. R., 302
Simons, J., 154
Simons, K., 73
Simons, L. A., 154
Simpson, R. J., 27
Sims, H. F., 9, 260
Simsolo, R. B., 320
Sing, C. F., 170, 175(4)
Singh, S., 147, 153, 156(8, 10), 157, 157(8, 10), 158(8, 10), 162, 162(8, 10, 48)
Sirtori, C. R., 12, 59, 262, 340
Sirtori, M., 340
Sisson, P., 173
Sjöberg, A., 57
Sjöblom, T., 180
Skarlatos, S. I., 25
Skinner, B., 19, 20(148), 66
Sladek, F. M., 6
Slater, R. J., 35
Slater, R. S., 226
Slavesen, G. S., 174
Sledge, W. E., 220, 228(11)
Sloop, C. H., 252, 255(7), 257(7), 262–263, 283
Small, D. M., 121
Smirnov, V. N., 147, 149(12), 155(12), 222
Smit, M., 176
Smith, C., 252
Smith, D. P., 16
Smith, E. A., 57
Smith, E. B., 30, 35, 226
Smith, G., 220, 228(11)
Smith, H. C., 8
Smith, J. D., 5, 7, 25, 178
Smith, L. C., 11, 21(68), 22, 262(34), 263, 319, 321–322, 327–328, 329(13), 330(13)
Smith, R. S., 11, 65, 122–123, 126(19), 127, 128(17), 132(19), 136(19), 141, 143(19), 151, 167, 262
Smith, S. C., 14
Smith, S. J., 68, 78(33), 161, 264
Smith, W. R., 43, 310, 311(5), 312(5), 313(5)
Sniderman, A. D., 19, 20(148), 58, 77
Snyder, M. L., 220, 238, 240, 244(1), 248, 249(1)
Snyder, S. M., 122–123, 126(19), 128(17), 132(19), 136(19), 141, 143(19), 151
Sobenin, I. A., 222
Sodetz, J. M., 248
Sokol, R. J., 121
Sommer, B., 8
Sommercorn, J., 353, 358(9)
Song, A. W., 9
Song, M., 7
Soria, L. F., 19
Sottrup-Jensen, L., 239
Soulé, P. D., 147
Sourgoutchev, A., 8
Soutar, A. K., 11, 21(68), 30, 232
Soyal, S. M., 7–8, 17
Spangler, E. A., 13, 268
Sparks, C. E., 10, 104–105, 106(10), 110(1, 10), 112, 113(13), 116(10), 118(10), 146, 150(4)
Sparks, J. D., 10, 104, 112, 113(13)
Sparrow, D. A., 18
Sparrow, J. T., 11, 18, 21(68), 22, 211, 262(34), 263, 270, 322, 327
Spector, M. S., 283, 288
Speelberg, B., 18, 147
Sperry, W. M., 37
Sprecher, D. L., 44, 172, 203, 267, 340
Sprengel, R., 8
Spring, D. J., 121
Srivastava, L. S., 209, 210(6), 211(6), 212(6)
Stabinger, H., 243
Staehelin, T., 113, 190, 255, 258(22), 265
Stanley, K. K., 172
Stark, D., 24
Stauderman, M. L., 310
Steck, T. L., 156
Steele, J.C.H., 162
Stein, E. A., 181
Stein, O., 320
Stein, S., 80, 81(55)
Steinberg, D., 21, 38, 40(55), 144, 163, 223, 237(28)
Steinberg, K. K., 264
Steinbrecher, U. P., 21, 223, 237(28), 261
Steinbrunner, J. V., 341, 345(18), 351(18)

Steiner, E., 56
Steiner, G., 22, 79, 83, 95–96, 96(4), 101, 103(7), 104(7), 159, 161(80), 188
Steinmetz, A., 11–12, 16, 16(69), 17, 32, 173, 283–284, 297
Stemerman, M. B., 12
Stender, S., 54
Stephens, P. E., 323
Stern, Y., 320
Stevenson, S., 343
Stoffel, S., 122, 128(16)
Stoffel, W., 260
Stonik, J. A., 9, 261
Stowell, C. P., 81
Straumfjord, J. V., Jr., 68
Strauss, A. W., 9, 16, 260
Strisower, B., 35, 36(25), 37, 37(25), 53(25, 45), 54(45), 58(44, 45)
Strittmatter, W. J., 32, 174
Strobl, W., 7, 8, 17
Strong, L. E., 34
Strong, W. L. P., 207
Strudemire, J. B., 48
Stuart, W. D., 43, 46, 51(124), 253, 309–311, 311(5, 9), 312, 312(5), 313(5), 315(9)
Stucchi, A. F., 9
Sturzenegger, E., 211, 217(19)
Subbarow, I., 37
Subramanian, R., 13
Sullivan, D. R., 219
Sullivan, M. L., 41
Sundquist, K. O., 11
Sussman, N. L., 8, 284
Suzuki, J., 172
Svenson, K. L., 116
Svensson, H., 34
Sviridov, D. D., 147, 149, 149(12), 155(12)
Swahn, B., 35
Swaney, J. B., 42
Swenson, T. L., 283, 339, 347
Swift, L. L., 147
Szklo, M., 28

T

Taam, L., 44, 203
Tabas, I., 150
Tae, W. K., 151
Tajima, S., 178
Tajiri, Y., 181
Takagi, A., 334
Takahashi, K., 339
Takahashi, M., 32
Takahashi, N., 46
Takahashi, S., 172
Takaku, F., 179
Takata, K., 14, 340
Takatsu, Y., 32
Takeda, R., 339
Tal, M., 92
Talbot, C., 174
Tall, A., 172, 282–283, 339–340, 342, 343, 346, 346(1), 347, 350(25), 351(25), 352
Talmud, P., 44, 48(106), 121, 132, 133(32), 138(32), 220
Tamplin, A. R., 35, 36(25), 37(25), 53(25)
Tamplin, C. B., 67
Tan, M. H., 285
Tanenbaum, S., 9, 261, 271
Taniguchi, T., 153
Tanimura, M., 18, 45, 150, 151(36), 152
Tartar, A., 203, 350
Tarui, S., 172, 340
Taskinen, M. R., 17, 26, 59, 319, 340
Tauber, J. P., 12
Tavella, M., 48, 53(75, 140), 54–55, 56(140), 57(140), 59, 59(140)
Taylor, H. L., 34
Taylor, J. M., 3, 209, 284, 333, 352, 353(1)
Taylor, M. T., 121, 125(11), 139(11), 140(11), 142(11), 143(11), 144(11)
Taylor, S. M., 167
Taylor, W. H., 18
Technical Group and Committee on Lipoproteins and Atherosclerosis, The, 57
Teisberg, P., 170
Tejada, P., 334
Temponi, M., 183
Teng, B., 8, 19, 20(148), 66, 77, 92
Teng, B.-B., 148–149, 149(25)
Tenkanen, H., 17, 297–298
Tennyson, G. E., 147, 149, 149(12), 155(12)
Teramoto, T., 23, 173
Tercé, F. R., 154–155, 167, 169(4)
Terdiman, J. F., 122–123, 126(19), 128(17), 132(19), 136(19), 141, 143(19), 151
Ternynck, T., 335
Tertov, V. V., 222
Tetaz, T., 301
Thakker, T., 20

Theiry, J., 59
i.e. Lipid Research Clinics Coronary Prevention Trial results, The, 267
Theorell, A. H. T., 33
Thibert, R. J., 261
Thiemann, E., 284
Thiery, J., 29
Thomas, P. S., 353
Thompson, D. B., 353, 358(9)
Thompson, G. R., 29, 31
Thompson, J. J., 220, 228(11)
Thompson, R. L., 31
Thompson, W. G., 58
Thorlin, T., 57
Thorpe, S. R., 177
Thrift, R. N., 52, 57(161)
Thuren, T., 173
Tijssen, P., 247
Tikkanen, M. J., 176, 177(39)
Tiselius, A., 34, 157
Tocsas, K. R., 9
Todd, K., 83, 89(5), 92(5), 94(5), 95, 167, 169(6)
Tokunaga, K., 340
Tolbert, N. E., 67, 70, 111, 112(12)
Tollefson, J. H., 41, 51, 253
Tomlinson, J. E., 28, 31(211), 220, 221(8), 222(8), 223(8), 235(8), 236(8), 239
Torpier, G., 47, 51(139)
Towbin, H., 113, 190, 255, 258(22), 265
Tozuka, M., 262(35), 263
Trent, J. M., 4, 45
Trieu, V. N., 219–220
Triplett, R. B., 162
Trojaborg, W., 120
Trotta, P. P., 339
Tsai, A. L., 32
Tsang, V. C. W., 302
Tschopp, J., 309–311, 311(2)
Tso, P., 17, 298
Tsukada, T., 25, 178
Tsukamoto, K., 23, 173
Tsunemitsu, M., 153
Tulinsky, A., 220
Tun, P., 86, 181
Turner, J. D., 144
Turner, S., 132, 133(32), 138(32)
Tuteja, R., 7
Tybjaerg-Hansen, A., 53, 54(169)

U

Uany, R., 150
Uffelman, K., 79, 83, 95–96, 101, 103(7), 104(7), 159, 161(80)
Umeda, F., 181
Underfriend, S., 80, 81(55)
Upham, F. T., 36, 37(25)
Urdea, M. S., 148
Usher, D., 29
Utermann, G., 11–12, 16(69), 17, 24, 28–29, 31, 41–44, 173, 219–220, 220(3), 225–226, 229, 230(53), 231–232, 232(37), 234(53), 235(53), 238, 282–283, 283(2, 4), 284, 285(2, 4), 297, 303–304, 308

V

Vaccarino, V., 340
Valente, A. J., 157, 158(73), 162(73), 163(73), 164(73), 273
Valle, M., 26
Van Biervliet, J. P., 321
Vance, J. E., 57
VandeBerg, J. L., 229, 231(55), 234(54)
van den Maagdenberg, A. M., 179
van der Boom, H., 179
van Eckardstein, A., 12
Van Handel, E., 37
Vanloo, B., 45, 48(110)
Vannier, C., 320, 322
van Ramshorst, E., 10, 25(62)
van Stiphuit, W. A. J. J., 10, 25(62)
Van't Hoff, F. M., 147, 150(7)
Van't Hooft, F. M., 282
Van Tol, A., 10, 15, 25(62), 282, 295
van Vlijmen, B. J., 179
Varmus, H. E., 6
Vauhkonen, M., 153
Vega, D. E., 252
Vega, G. L., 19, 58, 63, 65–67, 70(20), 76–77, 80, 81(60), 123–124, 126(22), 128(22), 129(22)
Venkatachalapathi, Y. V., 11
Verdery, R. B., 339
Vergani, C., 6
Verkade, H., 57
Verp, M., 8, 149
Verstuyft, J. G., 13, 15, 25, 30, 178, 268
Vezina, C., 155, 188, 207, 262(36), 263, 333
Via, D. P., 46, 50(118), 222, 322, 327

Viberti, G. C., 340
Vigne, J.-L., 86
Vilaró, S., 327
Vilella, E., 327
Vinaimont, N., 321
Vlaxic, N., 12
Vogel, T., 179
Vogel, W., 232
Voller, A., 183, 197
von Eckardstein, A., 13, 32, 44, 48(105), 204, 253, 261, 271, 279(20)
Voyta, J. C., 322, 327
Vranizan, K. M., 20, 58
Vu, H., 77
Vu-Dac, N., 221, 225(17), 235(17)
Vunakis, H. V., 303

W

Wade, D. P., 30, 232
Waeg, G., 223, 225(26)
Wagenknecht, L. E., 56
Wagner, L. A., 145
Wagner, R. D., 126, 132(25), 138(25)
Wagner, S. G., 57
Waite, M., 173
Waliul Islam, A.H.M., 340
Walker, D. W., 284
Walker, I. D., 27, 43
Walker, L. F., 146, 147(2)
Wallace, R. B., 53, 57(163)
Wallach, D. F. H., 156
Walldius, G., 56
Walli, A. K., 29, 234, 240
Wallis, S. C., 4, 8, 18, 18(34), 45, 64, 95, 105, 121, 148, 150
Walmsley, T. A., 123
Walraven, S. L., 41
Walsh, A., 22–23, 25, 178, 209, 352
Walsh, M. T., 121
Walter, M., 204, 261, 271, 279(20)
Walton, K. W., 36, 226
Wang, C. S., 22, 47, 51(138), 57, 59(191)
Wang, J., 261, 333
Wang, K., 352
Wang, S., 45, 343
Wang, X. L., 261
Wang-Iverson, P., 22, 47
Wardell, M. R., 174, 182(33)
Watanabe, J., 181
Watanabe, T., 23, 173
Watanabe, Y., 25, 178–179
Waters, D., 83, 267
Watson, D., 67
Webb, M., 37, 298
Weber, K., 101, 156
Weber, W., 172
Weech, P., 154–155, 167, 168(3, 11), 169(3, 4), 260–261, 262(4, 9, 36), 263, 271, 346–347, 350(25), 351(25)
Weidman, S. W., 185
Weidman, W. H., 56
Weilbaecher, D., 30, 225, 226(40), 227(40), 229(40)
Weinberg, R. B., 17, 282–285, 285(30), 287(30), 288, 296(30, 39), 301
Weinstein, D., 69–70, 73(34), 144
Weintraub, M. S., 24
Weiser, D., 57, 59(191)
Weisgraber, K. H., 12, 19, 22, 24, 44, 65, 86, 127, 137, 150, 167, 170–171, 173, 173(9), 174, 174(5), 175(5), 176(5), 177, 179, 182(33), 209, 262, 282, 285(1)
Weiss, R. B., 145
Weisweiler, P., 160, 210, 272
Welsh, A., 339
Welty, F., 138
Wen, G., 51
Weng, S.-A., 8, 18(35), 63, 105, 148
Weng, S. A., 95, 151
Wernette-Hammond, M. E., 150
West, R., 148
Wetterau, J. R., 10, 43, 310–311, 311(5), 312(5), 313(5)
Wettesten, M., 57
White, A. L., 28, 30, 223, 232(22)
White, B. A., 353
White, S. E., 23
Whitlock, M., 339, 346, 350(25), 351(25)
Wicken, D. E. L., 261
Wickham, E., 54, 58(172)
Widom, R. L., 6
Wiebe, D. A., 264
Wieding, J. U., 240
Wiegandt, H., 162
Wieland, H., 30
Wieland, O. H., 261
Wijsman, E., 19
Wiklund, O., 14, 57
Wilcken, D. E. L., 261
Wilcox, H. G., 36
Wilkes, H. C., 31

Wilkie, S. D., 241
Wilkinson, T., 36
Williams, C. M., 83, 95, 155, 167
Williams, D. L., 173–174
Williams, M. R., 293
Williams, S. C., 16
Wills, R. D., 36, 37(25)
Wilson, C., 174, 182(33)
Wilson, D., 24
Wilson, P. R., 59, 162
Wilson, R. F. W., 59
Wilson, T. M. E., 20
Wims, M., 319–320, 320(1)
Windler, E., 22, 47, 92, 208
Windler, W., 178
Windmueller, H. G., 14, 150
Wion, K. L., 4, 45
Wise, L. S., 320
Wisniewski, T., 174
Witt, K. R., 84, 123, 127(18), 162
Witte, D. P., 28, 309–310, 310(1), 312
Witztum, J. L., 21, 24, 65, 120, 122(2), 123(2), 129, 129(1), 136, 137(36), 138(36), 144(1, 2), 155, 163, 173, 179, 185, 211, 223, 237(28), 261
Wojciechowski, P., 20
Wolf, A. C., 51, 53, 59(167), 253
Wolfe, B. M., 207–208
Wolff, O. H., 39
Wolfson, C., 77
Wong, L., 57, 252, 255(7), 257(7), 262
Wu, A.-L., 150
Wu, J. T., 181
Wu, K. K., 28, 32
Wu, L. H., 181
Wu, W., 321, 327
Wurm, H., 220, 224, 225(10, 33)
Wygant, M. D., 232

X

Xiong, W., 149
Xu, Q.-T., 239

Y

Yakoub, L. K., 17
Yamada, N., 25, 48, 178–179
Yamamoto, A., 172, 178, 334
Yamamoto, T., 172
Yamane, M., 340
Yamashita, S., 172, 339–340
Yamauchi, T., 181
Yamazaki, J., 339
Yang, C.-I., 46, 50(118)
Yang, C. J., 45
Yang, C. Y., 8, 18, 18(35), 45, 63, 95, 105, 148, 150–151, 151(36), 152
Yao, Z., 10, 121, 125(11), 139, 139(11), 140(11), 142(11), 143(11), 144(11), 167
Yarranton, G. T., 323
Yashiro, A., 220, 223, 237(25)
Yazaki, Y., 25, 178–179
Yla-Herttuala, S., 21
Yokode, M., 21
Yokoyama, S., 172, 178, 297
Yoshida, H., 21
Young, N. L., 105, 106(10), 110(10)
Young, S. G., 19, 65, 120–121, 121(4), 122, 122(2, 4), 123, 123(2, 4), 124, 125(11), 126(4, 19, 22), 127, 127(4), 128(17, 22), 129, 129(1, 22), 132(19), 136, 136(19), 137, 137(36), 138(36), 139(11, 26), 140, 140(11), 141, 142(11), 143, 143(7, 11, 19), 144(1, 2, 7, 11, 26), 145(26), 151, 155, 167
Yuhasz, M. P., 148
Yui, Y., 32

Z

Zajdel, M., 239
Zamarripa, J., 148
Zampelas, A., 83, 95, 167
Zanni, E. E., 6, 9
Zannis, V. I., 5–6, 9, 11–12, 18, 25, 44, 121, 150, 176–177, 260
Zawadzki, Z., 346, 350(25), 351(25)
Zech, L. A., 24, 26, 29, 51, 59(151), 252–253
Zechner, R., 23
Zechner, T., 30
Zegers, N. D., 176
Zhang, S. H., 25, 178
Zhang, Z. H., 221, 225(15), 231(15), 232(15), 233(15), 234(15), 235(15), 237(15)
Zhong, W., 6
Zhou, X., 174
Zhu, H.-J., 149
Ziegler, O., 54

Zilversmit, D. B., 37, 80, 81(59), 83, 86(8), 92(8), 96, 154, 155(63), 159, 160(81), 161(81), 166
Zinn, K., 353, 355(8), 356(8)
Zinsmeister, A. R., 11
Zioncheck, T. F., 219–220, 221(8), 222(8), 223(8), 235(8), 236(8)
Zoppo, A., 254–255, 259(24), 262
Zorich, N., 11
Zsigmond, E., 319, 321, 327
ZumMallen, M. E., 28, 221, 246
Zupkis, R. V., 181
Zylberberg, G., 59, 253
Zysow, B. R., 30, 123

Subject Index

A

ABC nomenclature, for apolipoproteins, 40–41
Abetalipoproteinemia, 39, 56, 65, 147
Affinity chromatography
  lipoprotein lipase, 324–325
  truncated apolipoprotein B-containing lipoproteins, 138
Alzheimer's disease, apolipoprotein E-4, 32, 174
Amphipathic helix, apolipoproteins, 3–4, 45
Anion-exchange chromatography, apolipoprotein C-II separation, 190–192
Antibodies
  apolipoprotein A-IV, 298, 301–302
  apolipoprotein B, lipoprotein [a] estimation, 249–250
  apolipoprotein B-48, quantification, 155–156
  apolipoprotein C-II, 194–197
  apolipoprotein C-III, 213
  hepatic lipase, 334–335
  lipoprotein [a], 245–247
  lipoprotein lipase, 325–326
Antibody capture ELISA, lipoprotein lipase, 327–333
Antioxidants, apolipoprotein B quantitation, 163–164
Antithrombin, 334
Apolipoprotein [a]
  atherosclerosis, 30–31
  carbohydrate heterogeneity, 221–222
  classification, 43
  coronary artery disease, 31
  kringle unit heterogeneity, 220–221
  LDL, 29–30
  lipoprotein [a], 219, 238
  plasma lipid transport, 28–31
  polymorphism, 220–221, 235–237
  quantitation
    in plasma, 224–226
    in tissue, 226–228
Apolipoprotein A
  lipoproteins containing, 49–52, 53
  structure, 40
Apolipoprotein A-I
  apolipoprotein A-I/apoC-III deficiency, 56
  atherosclerosis, 11–12
  coronary artery disease, 11–13, 267
  function, 5, 32, 47, 53
  gene expression, effect of thyroid hormone, 7
  HDL, 12–13, 260–267
  heterogeneity, 260–264
  as marker, 53
  plasma lipid transport, 11–13, 252
  quantitation, 264–267
    chromatography, 267–282
  ribonuclease protection assay, 359–362
  structure, 3, 4, 251
    variability, 260–262
  synthesis, 8–9, 251
Apolipoprotein A-II
  coagulation, 32
  coronary artery disease, 13–16
  function, 5, 32, 47
  plasma lipid transport, 13–16
  structure, 3, 4, 44, 271
  synthesis, 8–9, 252
Apolipoprotein A-III, *see also* Apolipoprotein D
  discovery, 41
Apolipoprotein A-IV
  antibodies, 298, 301–302
  cholesterol, 297

discovery, 42
function, 5, 11, 32, 283, 297–298
metabolism, 282–283
plasma lipid transport, 16–18, 282–283, 297
polymorphism, 284
quantitation, 284–286
chromatography, 288–290, 300–301
electroimmunoassay, 294–296
immunochemical, 284, 297–309
structure, 4
synthesis, 282
triglycerides, 297–298
VLDL, 42
Apolipoprotein B
antibodies, in lipoprotein [a] estimation, 249–250
degradation, 10
function, 65
gene, 4, 64, 120
minigene expression, 138–139, 151
mutations, 120–122, 127–130, 151
hypobetalipoproteinemia, truncated forms of apoB, 120–145
as immunogen for antibody production, 113–116
LDL, 19–20, 75
lipoproteins containing, 49–52, 53, 64
as marker, 53
mRNA editing, 148–150, 165, 166
plasma lipid transport, 18–21
properties, 63–64
quantitation, 105–106
chemical methods, 63–82
chromatography, 83–84, 104–119, 137–138, 156–157, 159–160, 166–167
LDL and, 75
lipid peroxidation, 163–164
proteases, 162–163
SDS–PAGE, 82–94, 95–104, 156–157, 159–160
simultaneous, apoB-48 and apoB-100, 82–94
standards, 84–86, 87, 89, 98–100, 111–112
ribonuclease protection assay, 359–362
secretion
hepatic, 9–10
insulin effect, 10
structure, 40, 45, 63, 104–105
synthesis, 45
truncated species
lipoproteins with, characterization, 130–145
apolipoprotein B-31-containing, 141, 143–144
apolipoprotein B-32-containing, 137, 138
apolipoprotein B-37-containing, 137–138, 141, 143–144
apolipoprotein B-46-containing, 137, 151
apolipoprotein B-48-containing, 151
apolipoprotein B-50-containing, 151
apolipoprotein B-61-containing, 137
apolipoprotein B-67-containing, 138
apolipoprotein B-75-containing, 138
apolipoprotein B-83-containing, 137
apolipoprotein B-86-containing, 139, 142, 144–145
apolipoprotein B-89-containing, 137, 138
mutations producing, 120–121, 127–128
nomenclature, centile system, 122
quantitation in plasma, 129–130
screening in hypolipidemic subjects, 122–127
size, estimation, 127
VLDL, 72–75, 166
Apolipoprotein B-48
amino acid composition, 150–152
antibodies, 155–156
biosynthesis, 8
carbohydrate moiety, 153
chromogenicity, 91–93, 96
fasting VLDL, 153–154
function, 5, 47, 65, 166
hydropathy, 152–153
intestinal, 146–148
as intestinal marker, 154–155
metabolism, 147–150
quantitation, 79–80, 155–165
chromatography, 83–84, 104–119, 137–138, 158–160, 166–167
immunochemical, 155–156
radioiodination, 160–161
SDS–PAGE, 82–94, 95–104, 156–157
simultaneous, apolipoproteins B-100, B-48, and E, 82–94
standards, 86, 89

separation, immunoaffinity chromatography, 166–170
structure, 150–153
structure and synthesis, 18, 153, 165, 166
intestinal, 146–148
triglycerides, 65
Apolipoprotein B-100
atherosclerosis, 18–19, 30
chromogenicity, 91–93, 96
function, 5, 47, 166
gene, 147–148, 166
plasma lipid transport, 18–21
quantitation
chemical, 66–82
chromatography, 83–84, 104–119, 137–138, 159–160, 166–167
SDS–PAGE, 82–94, 95–104
simultaneous, apolipoproteins B-100, B-48, and E, 82–94
standards, 84–86, 86–87, 89, 98–100
separation, immunoaffinity chromatography, 166–170
structure and synthesis, 18, 63, 150–151, 152
Apolipoprotein C
LDL, 21
lipoproteins containing, 49–52, 64
plasma lipid transport, 21–23
structure, 39–40
VLDL, 39
Apolipoprotein C-I, function, 5, 11, 21
Apolipoprotein C-II, 192
antibodies, 194–197
deficiency, 56
function, 5, 22, 47, 188
lipoprotein lipase, 320
mutations, 204–208
quantitation
chromatography, 190–192
heterogeneity and, 203–208
immunoassay, 192–203
purification, 188–192
structure, 188
VLDL, 192
Apolipoprotein C-III
antibodies, 213
apolipoprotein A-I/apoC-III deficiency, 56
function, 5, 22–23, 47, 208
HDL, 208–209
quantitation
immunoassay, 210–218
radioimmunoassay, 211–218
structure, 3, 4, 208
VLDL, 208
Apolipoprotein D
discovery, 41
function, 5
plasma lipid transport, 23
structure, 4, 45–46
Apolipoprotein E
atherosclerosis, 24–25, 178–179
cholesterol, 170, 172–173, 175–176, 178
coronary heart disease, 179, 180
discovery, 41
function, 5, 11, 47, 171–174
gene, 170
HDL, 172–173
heterogeneity, 175–178
LDL, 23–24
plasma lipid transport, 23–26
polymorphism, 172
quantitation, 170–187
ELISA, 182–187
simultaneous, apolipoproteins B-100, B-48, and E, 82–94
standards, 86–87, 89
structure, 174
synthesis, 170
VLDL, 171–172
Apolipoprotein E-2
function, 173, 174
structure, 24, 175
Apolipoprotein E-3
function, 173–174
structure, 24
Apolipoprotein E-4
function, 32, 174
lipid metabolism, 25–26
structure, 23–24
Apolipoprotein F
discovery, 41
structure, 46
Apolipoprotein G
discovery, 42
structure, 46
Apolipoprotein H
plasma lipid transport, 26–28
structure, 46
Apolipoprotein I, 43
structure, 46

Apolipoprotein J, 43
function, 309–310
gene, 310
HDL, 310–311
plasma lipid transport, 26, 27–28
polymorphism, 311
quantification, 313–316
structure, 46, 311–312
Apolipoproteins
chromogenicity, 91–93, 96
dyslipoproteinemias, 52–60
function, 3–5, 21–23, 31–32, 46–48
genes, 45
expression, 4–10, 139–140, 351–352
isolation, 38–43
as markers for plasma lipoproteins, 48–49, 53, 55, 57–58
minor apolipoproteins
discovery, 41–43
plasma lipid transport, 26–28
mRNA levels, 351–363
nomenclature, 40–41, 44
plasma lipid transport, 11–31
polymorphism, 44, 47–48
quantitation
methods, 161–162, 269
standards, 84–87, 89, 98–100, 272–273
separation, non-immunologic, 49–52
structure, 3–4, 44–46
Apotransferrin, in apolipoprotein recovery from lipoproteins, 87–88
Arteries, coronary, *see* Coronary artery disease
Atherosclerosis
apolipoprotein [a], 30–31
apolipoprotein A-I, 11–12
apolipoprotein A-I/A-II ratio, 12–13
apolipoprotein B-100, 18–19, 30
apolipoprotein E, 24–25, 178–179
apolipoproteins and lipoproteins, 53–54, 57–59
studies, 54
Atherosclerosis susceptibility locus, LDL, 20
Ath-1 gene, function, 14

B

Bovine serum albumin, colorimetric standard, 68
Bradford procedure, protein quantitation, 80–81

C

Cardiovascular disease, lipoprotein [a], 218
Centrifugation, *see* Ultracentrifugation
Cerebrospinal fluid, apolipoprotein E, 180
CETP, *see* Cholesteryl ester transfer protein
Cholesterol
apolipoprotein A-IV, 297
apolipoprotein E, 170, 172–173, 175–176, 178
coronary artery disease, 57
Cholesterol ester transfer protein, 64
Cholesterol-lowering atherosclerosis study, 54
Cholesteryl ester transfer protein
activity assay, 341–346
ELISA, 346–350
function, 339
gene, 339–340
HDL, 340–341
Chromatography, *see also specific techniques*
apolipoprotein A-I, 267–282
apolipoprotein A-IV, 288–290, 300–301
apolipoprotein B-100 and B-48
isolation and quantitation, 104–119, 159–160
separation by immunoaffinity chromatography, 166–170
simultaneous, 83–84
apolipoprotein C-II separation, 190–192
lipoprotein [a], 239–241
lipoprotein lipase, 324–325
truncated apoB-containing lipoproteins, 137–138
Chromogenicity, apolipoproteins, 91–93, 96
Chylomicrons
apolipoproteins A-IV, 282
definition, 36
proteins, 38–39
Colorimetric assay, Lowry–Folin protein assay, 67–69
Complex lipoprotein families, definition, 49
Coronary artery disease
apolipoprotein [a], 31

apolipoprotein A-I, 11–13, 267
apolipoprotein A-I/A-II ratio, 12–13, 267
apolipoprotein A-II, 13–16
cholesterol, 57
dyslipoproteinemias, 53
lipoproteins, 58–59
postprandial lipemia, 10
Coronary heart disease
apolipoprotein E, 179, 180
hyperapo B, 66
COS-7 cells, transient transfection, 140

D

Deficiency diseases
abetalipoproteinemia, 39, 56, 65
familial apolipoprotein C deficiency, 188
hyperchylomicronemia, 188
hypobetalipoproteinemia, 19, 24, 25, 56
lipoprotein lipase, 319
Tangier disease, 12–13, 39, 56
Delipidation, lipoproteins, 67, 87–88, 157–158, 276
Density gradient ultracentrifugation
lipoprotein [a], 242–244
lipoproteins containing truncated apolipoprotein B, 134–137
Dialysis, lipoproteins, 158–159
Dye binding
apolipoprotein B, 83, 159
protein quantitation, 80–81
Dyslipoproteinemias
abetalipoproteinemia, 39, 56, 65
apolipoprotein A-I/C-III deficiency, 56
apolipoprotein C-II deficiency, 56
apolipoproteins and lipoproteins, 52–60
characterization, 52
coronary artery disease, 53
familial combined hyperlipemia, 19, 20
familial defective apoB-100, 18–19, 65
familial dysbetalipoproteinemia, 171
familial hypobetalipoproteinemia, 19
fish-eye disease, 13
hyper-apoB, 19, 20, 66
hyperlipoproteinemias, 56, 171
hypobetalipoproteinemia, 19, 24, 25, 56
truncated apoB forms in, 120–145
Tangier disease, 12–13, 39, 56

E

Electroimmunodiffusion
advantages and disadvantages, 269
apolipoprotein A-IV, 294–296, 303–304
Electrophoresis, *see also* Gel electrophoresis; Rocket immunoelectrophoresis
HDL speciation, 251–259
lipoprotein [a] quantitation, 218, 228–237
nondenaturing electrophoresis, 251–259
plasma lipoprotein classification, 34–35, 224–226, 251–259
Enzyme-linked immunosorbent assay
advantages and disadvantages, 269
apolipoprotein A-I, 264–267
apolipoprotein A-IV, 303, 305–309
apolipoprotein C-II, 197–200
apolipoprotein C-III, 211, 217–218
apolipoprotein E, 182–187
apolipoprotein J, 313–316
cholesteryl ester transfer protein, 346–350
hepatic lipase, 334, 335–338
lipoprotein [a], 244–251
lipoprotein lipase, 327–333
Expression vectors
apolipoprotein B, 139–140
lipoprotein lipase, 323–324

F

Familial apo-C deficiency, 188
Familial combined hyperlipemia, 19, 20
Familial defective apoB-100, 18–19, 65
Familial dysbetalipoproteinemia, 171
Familial hypercholesterolemia, 65
Familial hypobetalipoproteinemia, 19, 120
Fibrinolysis, apolipoproteins, 31–32
Fish-eye disease, apolipoprotein A-I, 13
Fluorescamine method, protein quantitation, 81–82

G

Gel electrophoresis, *see also* Electrophoresis; Isoelectric focusing
nondenaturing gradient polyacrylamide gel electrophoresis, 130–131
SDS-PAGE
apolipoprotein [a], 231–233

apolipoprotein A-IV, 299
apolipoproteins B-48 and B-100, 95–104
apolipoproteins B-48, B-100 and E quantitation, 82–94
apotransferrin, 87–88
chromatography columns, 112–113
rod gel technique, 101–104
slab gel technique, 97–100
Gel filtration
apolipoprotein A-I, 268
apolipoprotein B variant isolation, 104–119
lipoproteins containing truncated apolipoprotein B, 131–134
Gene expression
apolipoprotein A-I, effect of thyroid hormone, 7
apolipoprotein B expression vectors, 139–140
apolipoproteins, 4–10, 351–352
Gene regulation, mRNA assay, 351–363

H

HDL, *see* High-density lipoproteins
Heart, *see* Coronary heart disease
Heparin, binding by truncated apolipoprotein B species, 137
Hepatic lipase
activation
by apolipoprotein A-II, 14
by apolipoprotein E, 173
antibodies, 334–335
ribonuclease protection assay, 359–362
sandwich immunoassay, 333–338
Hepatoma cell line McA-RH7777, truncated apoB, 140–145
High-density lipoproteins
apolipoprotein A-I quantification and heterogeneity, 260–267
apolipoprotein A-I/A-II ratio, 12–13
apolipoprotein C-III, 208–209
apolipoprotein E, 172–173
apolipoprotein J, 310–311
cholesteryl ester transfer protein, 340–341
definition, 36
immunoaffinity chromatography, 253
immunoblotting, subfractions, 258
lipoprotein families, 49
pre-βHDL, 254–255, 256, 263–264
speciation, nondenaturing electrophoresis, 251–259
ultracentrifugation, 273, 276
High-performance liquid chromatography
apolipoprotein A-I, 267–282
apolipoprotein A-IV, 288–290, 301
apolipoprotein B, 159–160
apolipoprotein B-48 and B-100, simultaneous quantification, 83
apolipoprotein C-II, 190–191
reverse-phase, *see* Reverse-phase HPLC
Hydropathy, apolipoprotein B-48, 150–153
Hyperapobetalipoproteinemia, 19, 20, 66
Hypercholesterolemia, 53, 65–70
apolipoprotein B truncations, screening for, 122–123
Hyperchylomicronemia, 188
Hyperlipoproteinemias, 56
apolipoprotein B-48, 153–154
apolipoprotein E, 171, 175, 178
screening, 179
Hypertriacylglycerolemia, 319
Hypertriglyceridemia
apolipoprotein B, 77
apolipoprotein C-III, 209, 210
lipoproteins, 53
Hypobetalipoproteinemia, 56
apolipoprotein B, 19
screening, 122–127
apolipoprotein E, 24, 25

I

IDL, *see* Intermediate-density lipoproteins
Immunoaffinity chromatography
anti-lipoprotein C-II, 194–196
apolipoprotein B-containing lipoproteins, 137–138
apolipoprotein B-48- and B-100-containing lipoproteins, 166–170
HDL subfractions, 253
Immunoblotting
apolipoprotein A-IV, 291–293
apolipoprotein B-48, 83–84
HDL subfractions, 258
lipoprotein lipase, 331–333
Immunodiffusion, *see* Electroimmunodiffusion; Radial immunodiffusion

Immunonephlometric assay
advantages and disadvantages, 269
apolipoprotein C-III, 211, 212
Immunoquantitation
apolipoprotein A-I, 264–267
apolipoprotein A-IV, 297–309
apolipoprotein B-48, 155–156
apolipoprotein C-II, 192–203
apolipoprotein C-III, 210–218
lipoprotein lipase, 327–333
plasma apolipoproteins B and A-I, 78–79
Immunosorbent assay, enzyme-linked, *see* Enzyme-linked immunosorbent assay
Immunoturbidometric assay, advantages and disadvantages, 269
Insulin, apolipoprotein B secretion, 10
Intermediate-density lipoproteins
apolipoprotein B quantitation, 72–75
apolipoproteins in, 64
Isoelectric focusing
apolipoprotein A-IV, 290–291
apolipoprotein C-II, 189–190
apolipoprotein C-III, 214–216

K

Kallikrein, 162
Kringle unit heterogeneity, apolipoprotein [a], 220–221

L

LDL, *see* Low-density lipoproteins
Lipase, *see* Hepatic lipase; Lipoprotein lipase
Lipemia, postprandial, 10, 14
Lipid-lowering drugs, 54–55
Lipids, peroxidation, 163–164
Lipid transport
apolipoprotein A-IV, 282–283, 297
apolipoprotein functions, 11–31
dyslipoproteinemias, 52–60
Lipocalins, apolipoprotein D, 23
Lipoprotein [a]
antibodies, 245–247
cardiovascular disease, 218
heterogeneity, 219–222
oxidative and reductive modification, 222–223
quantitation
chromatography, 239–241
density-gradient centrifugation, 242–244
electrophoresis, 218–237
ELISA, 244–251
structure and function, 43, 238
Lipoprotein-deficient plasma, apolipoprotein A-IV, 301
Lipoprotein families
composition, 49–50
definition, 49
metabolic properties, 50–52
Lipoprotein lipase, 64, 319
antibodies, 325–326
apolipoprotein C-II, 188, 207
purification, 320–322
quantitation
chromatography, 324–325
immunochemical, 327–333
VLDL, 319
Lipoproteins
apolipoprotein A-containing, 49–52, 53
apolipoprotein B-containing, 49–52, 53, 64, 137–138
apolipoprotein B-31-containing, 141, 143–144
apolipoprotein B-32-containing, 137, 138
apolipoprotein B-37-containing, 137–138, 141, 143–144
apolipoprotein B-46-containing, 137
apolipoprotein B-61-containing, 137
apolipoprotein B-67-containing, 138
apolipoprotein B-75-containing, 138
apolipoprotein B-86-containing, 139, 142, 144–145
apolipoprotein B-89-containing, 137, 138
apolipoprotein C-containing, 49–52, 64
apolipoprotein E-containing, 172
apolipoprotein E metabolism, 171
apolipoprotein quantitation
chemical methods, 66–82
SDS–PAGE, 82–94
coronary artery disease, 58–59
dialysis, 158–159
dyslipoproteinemias, 52–60, 171
families, 49
function, 52
HDL, *see* High-density lipoproteins
high-density speciation, 251–259

IDL, *see* Intermediate-density lipoproteins
LDL, *see* Low-density lipoproteins
plasma
apolipoprotein B, truncated, 134–137
apolipoprotein C-II, heterogeneity, 203–208
apolipoproteins as markers, 48–49, 53
classification, 49–52
electrophoresis, 34–35
ultracentrifugation, 35–38
quantitation of total protein, 66–72
structure, 32–34
triglyceride-rich, isolation, 96–97
VLDL, *see* Very-low-density lipoproteins
Low-density lipoproteins
apolipoprotein [a], 29–30
apolipoprotein B, 19–20
quantitation, 75
apolipoprotein C, 21
apolipoprotein E, 23–24
atherosclerosis susceptibility locus, 20
definition, 36
function, 65
lipoprotein families, 49
synthesis from VLDL, 64–65
Lowry–Folin assay, 66–67, 71–72
colorimetric assay, 67–69
delipidation, 67
Markwell modification, 70–71, 75
substances interfering with, 69–70

M

Markwell modification, 70–71, 75
McA-RH7777 cell line, truncated apolipoprotein B, 140–145
Mevinolin atherosclerosis regression study, 54
Minigene expression, apolipoprotein B cDNA, 138–145
Minor apolipoproteins
discovery, 41–43
plasma lipid transport, 26–28
mRNA
apolipoprotein, assay, 351–363
apolipoprotein B, editing, modulation, 148–150, 165, 166
Multiple sclerosis, apolipoprotein E, 180

N

Nephrotic syndrome, cholesteryl ester transfer protein, 340
NIST standard, protein assay, 68
Nondenaturing electrophoresis
lipoproteins containing truncated apolipoproteins, 130–131
lipoprotein speciation, 253–259
Normolipidemia, apolipoprotein B-48, 153–154
Normotriglyceridemia, apolipoprotein B quantitation, 76

P

Plasma lipid transport
apolipoprotein A-IV, 282–283, 297
apolipoprotein functions, 11–31
Postprandial lipemia
apolipoprotein A-II, 14
coronary artery disease, 10
Pre-β HDL, 254–255, 256, 263–264
Proteases, apolipoprotein B quantitation, 162–163
Protein assay
methods, 161–162, 269
Lowry–Folin assay, 66–72, 75
standards, 84–87, 89, 98–100, 272–273
Pyridylethylcysteine, apolipoprotein C-II mutants, 204–207

R

Radial immunodiffusion, advantages and disadvantages, 269
Radioimmunoassay
advantages and disadvantages, 269
apolipoprotein A-IV, 303, 304–305
apolipoprotein C-III, 211–218
cholesteryl ester transfer protein, 346–350
Radioiodination, apolipoprotein B-48 quantitation, 160–161
Reverse-phase HPLC
apolipoprotein A-I, 268–282
apolipoprotein C-II, 191–192
Ribonuclease protection assay, apolipoprotein mRNA levels, 351–363
RNA, messenger, *see* mRNA

Rocket immunoelectrophoresis
apolipoprotein [a], 224–228
apolipoprotein A-IV, 303–304

S

Sandwich ELISA
apolipoprotein A-I, 264–267
apolipoprotein A-IV, 308–309
apolipoprotein C-II, 197–200
apolipoprotein C-III, 217–218
hepatic lipase, 334, 335–338
lipoprotein lipase, 330–331
SDS–PAGE, *see* Gel electrophoresis, SDS–PAGE
Sequential ultracentrifugation, truncated apoB-containing lipoproteins, 138
Serum amyloid protein A, discovery, 42
Simple lipoprotein families, definition, 49
Sodium dodecyl sulfate-polyacrylamide gel electrophoresis , *see* Gel electrophoresis, SDS–PAGE
Southern blotting, genomic DNA, 128

T

Tangier disease, 12–13, 39, 56
Thrombosis, apolipoproteins, 31–32
Thyroid hormone, effect on apolipoprotein A-I gene expresison, 7
Tracer kinetic studies, apolipoprotein B, 80, 81
Transfection
apolipoprotein B cDNA constructs, 138–145
lipoprotein lipase, 323–324
Triglycerides
apolipoprotein A-IV, 297–298
apolipoprotein B-48, 65

U

Ultracentrifugation
HDL, 273, 276
lipoprotein [a], 242–244
plasma lipoprotein classification, 35–38, 75
triglyceride-rich lipoproteins, 96–97
truncated apoB-containing lipoproteins, 134–137, 138
in hypolipidemics, 123–124

V

Vectors, *see* Expression vectors
Very-low-density lipoproteins
apolipoprotein A-IV, 42
apolipoprotein B, 72–75, 166
apolipoprotein B-48, 153–154
apolipoprotein C, 39
apolipoprotein C-II, 192
apolipoprotein C-III, 208
apolipoprotein E, 171–172
definition, 36
LDL synthesis, 64–65
lipoprotein families, 49
lipoprotein lipase, 319
proteins, 38–39

W

Western blotting
apolipoprotein [a], 233–235
truncated apolipoprotein B species, 125–126

X

X-protein, 35

ISBN 0-12-182164-1

90018

9 780121 821647